RESEARCH HANDBOOK ON COMMUNITY DEVELOPMENT

Research Handbook on Community Development

Edited by

Rhonda Phillips

Purdue University, West Lafayette, Indiana, USA

Eric Trevan

The Evergreen State College, Olympia, Washington, USA

Patsy Kraeger

Georgia Southern University, Statesboro, Georgia, USA

Edward Elgar
PUBLISHING

Cheltenham, UK • Northampton, MA, USA

© Rhonda Phillips, Eric Trevan and Patsy Kraeger 2020

All rights reserved. No part of this publication may be reproduced, stored in a retrieval system or transmitted in any form or by any means, electronic, mechanical or photocopying, recording, or otherwise without the prior permission of the publisher.

Published by
Edward Elgar Publishing Limited
The Lypiatts
15 Lansdown Road
Cheltenham
Glos GL50 2JA
UK

Edward Elgar Publishing, Inc.
William Pratt House
9 Dewey Court
Northampton
Massachusetts 01060
USA

A catalogue record for this book
is available from the British Library

Library of Congress Control Number: 2019956346

This book is available electronically in the Elgaronline
Social and Political Science subject collection
DOI 10.4337/9781788118477

Printed on elemental chlorine free (ECF)
recycled paper containing 30% Post-Consumer Waste

ISBN 978 1 78811 846 0 (cased)
ISBN 978 1 78811 847 7 (eBook)

Typeset by Servis Filmsetting Ltd, Stockport, Cheshire
Printed and bound in the USA

Contents

List of contributors ix

Introduction to the *Research Handbook on Community Development* 1
Rhonda Phillips, Eric Trevan and Patsy Kraeger

PART I FOUNDATIONS

1 Weaving reflection, action, and knowledge creation: lived experience as a catalyst into the cycle of praxis for community development 12
C. Bjørn Peterson, Craig A. Talmage and Richard C. Knopf

2 The study of poverty in places: scope, scale, and space 24
Elizabeth A. Dobis, Lionel J. Beaulieu and Indraneel Kumar

3 In pursuit of just communities: supporting community development for marginalized communities through regional sustainability planning 48
Jason Reece

4 Asset-Based Community Development (ABCD): core principles 67
Ivis García

5 Stepping up the ladder: reflecting on the role of nonprofit organisations in supporting community participation 76
Julia Fursova

6 Social economy, social capital, NGOs and community development: a gendered perspective 93
Dyana P. Mason

7 What can Northwest European community enterprises learn from American community-based organizations? 104
David P. Varady, Reinout Kleinhans and Nuha Al Sader

8 Community development, well-being and technology: a Kenyan village 124
Claire Wallace and Leanne Townsend

PART II RESEARCH METHODS AND FRAMEWORKS

9 Experience of group formation in Grameen Bank, Bangladesh 137
Kazi Abdur Rouf

10 How to build an "intentional community" 172
Brenda M. Elias

11	Inclusionary zoning and inclusionary housing in the United States: measuring inputs and outcomes *Katrin B. Anacker*	189
12	Enhancing evaluation capacity: lessons from faith-based community development in El Salvador *James G. Huff, Jr.*	204
13	Managing competing interests in the public participation process: lessons from an analysis of residential displacement in Buffalo, New York's transitioning neighborhoods *Robert Mark Silverman, Li Yin and Henry Louis Taylor, Jr.*	211
14	Methods and framework of participatory action research for community development in Bangladesh *M. Rezaul Islam*	224
15	Building a healthy community: the Coastal Georgia Indicators Coalition *Patsy Kraeger*	244
16	Social indicator projects for rural communities: the case of the Northwoods Quality of Life Database *Brandon Hofstedt*	273
17	An exploratory study of food deserts in Utica, Mississippi *Talya D. Thomas*	290
18	Impact of socioeconomic characteristics on neighborhood environment satisfaction in deteriorated areas *Mostafa Norouzi, Abolfazl Meshkini and Somayeh Khademi*	301
19	Downtown revitalization, livability and quality of life in Tucson, Arizona *Carlos J.L. Balsas*	319

PART III EMERGING CONSTRUCTS AND THE FUTURE OF COMMUNITY DEVELOPMENT RESEARCH

20	Theories and concepts influencing sustainable community development *Maria Spiliotopoulou and Mark Roseland*	337
21	Re-imagining community development: the Cocoa360 model *Shadrack Frimpong, Allison R. Russell and Femida Handy*	348
22	Community development and place attachment using an inductive social media approach *Justin B. Hollander and Max Page*	361
23	Re-imagining democratic research processes in community-based development: a case for photovoice *Camille Sutton-Brown*	382

24	Centering aesthetics in community development: approaches from the Banff Centre for Arts and Creativity *Jerrold McGrath*	391
25	The new role of the university in community development *Graciela Tonon*	407
26	Community innovation and small liberal arts colleges: lessons learned from local partnerships and sustainable community development *Craig A. Talmage, Robin Lewis, Kathleen Flowers and Lisa Cleckner*	416
27	Sustaining an urban education pipeline: a case study of university and community development partnership *Gloria Bonilla-Santiago*	439

Index 457

Contributors

Nuha Al Sader is a PhD candidate in the Faculty of Architecture and the Built Environment at Delft University of Technology. Her research focuses on entrepreneurial citizenship in deprived urban neighborhoods in the Netherlands. The research is based on a variety of qualitative research methods such as discourse analysis of policy documents, qualitative in-depth interviews with policy makers and entrepreneurial citizens and case studies of citizen initiatives.

Katrin B. Anacker is currently an Associate Professor in the Schar School of Policy and Government at George Mason University. She is the lead Editor of the *Routledge Handbook of Housing Policy and Planning* (Routledge, 2019) and the lead Editor of *Introduction to Housing*, second edition (University of Georgia Press, 2018). She is also the Editor of *Housing and Society*, the former North American Editor of the *International Journal of Housing Policy* and the former Co-Editor of *Housing Policy Debate*.

Carlos J.L. Balsas, PhD, AICP, is an Assistant Professor in the Department of Geography and Planning at the University at Albany. He is an urban and regional planner by training. His main research interests are comparative urban revitalization, resilience, urban governance, non-motorized transportation planning and international planning. His most recent book is *Walkable Cities Revitalization, Vibrancy, and Sustainable Consumption* (SUNY Press, 2019).

Lionel J. Beaulieu is Director of the Purdue Center for Regional Development and Professor in the Department of Agricultural Economics. He has played key roles in launching a number of innovative national research and extension programs over the course of his career, including the National e-Commerce Extension Project, the Stronger Economies Together (SET) program (in partnership with US Department of Agriculture (USDA) Rural Development) and the Food Assistance Research Program (in collaboration with the USDA Economic Research Service). He has guided the Economic Development Administration's University Center work in Indiana since 2013. Beaulieu served as president of the Rural Sociological Society (2003–04) and president of the Community Development Society (2012–13). He is the recipient of a number of major awards, including the Distinguished Rural Sociologist Award from the Rural Sociological Society, inductee in the George Washington Carver Hall of Fame for Public Engagement at Tuskegee University, Honor Award for External Partnership from the US Secretary of Agriculture, National Institute of Food and Agriculture (NIFA)/USDA National Partnership Award for Multi-State Programs for his leadership of the SET program, Distinguished Career Award from the National Association of Community Development Extension Professionals and Community Development Achievement Award from the Community Development Society. Dr Beaulieu completed his MS degree and PhD degree in Sociology from Purdue University.

Gloria Bonilla-Santiago is a Board of Governors Distinguished Service Professor, Graduate Department of Public Policy and Administration at Rutgers, The State University of New

Jersey. She also directs the Community Leadership Center and is the overseer and Board Chair of the School. Throughout her academic career, she has established a track record in coordinating large-scale programs and private and public ventures that bring together external and internal stakeholders from a range of organizations, including government, business, nonprofits and philanthropic sectors at the local, national and international levels. As a leading scholar, researcher, speaker and international cross-cultural training consultant, Dr Santiago brings over 25 years of experience in program development and innovation, social entrepreneurship, research, fundraising, strategic planning, school development and leadership training. She writes and speaks widely on the areas of community development, public policy, school leadership and education, migration, diversity management and organizational leadership. In 2016–17, Dr Santiago was the recipient of the Fulbright Specialist Award for research and professional training in Paraguay, South America. In 2017, Dr Santiago received the Cabrini Ivy Young Willis and Martha Willis Dale Award, which recognizes women who have made outstanding contributions in the field of Public Affairs and Community Development. Dr Santiago is also the recipient of the 2018 Power of Woman Award, presented by Lupe Fund. In May 2018, Dr Santiago was the Keynote Speaker at the Cabrini University Commencement for the Master's Degree Students and received an Honorary Doctor of Humane Letters.

Lisa Cleckner is Director of the Finger Lakes Institute (FLI) at Hobart and William Smith Colleges (HWS). She holds a PhD in Environmental Health Sciences from the University of Michigan. She also holds an MBA from Simon Graduate School of Business at the University of Rochester. Her research and educational efforts have collaborated with various audiences including students, professionals, scientists, community members, government agencies and businesses. She has been a pivotal player in the creation and support of the Sustainable Community Development Program at HWS. She is also a member of the faculty in the Environmental Studies Department at HWS.

Elizabeth A. Dobis is a Postdoctoral Scholar in the area of Regional Economic Growth and Development with the Northeast Regional Center for Rural Development at the Pennsylvania State University. Her research interest is spatial economic analysis, particularly in the fields of demography, poverty and health. Prior to her current position, she worked with the Purdue Center for Regional Development as a graduate research assistant and postdoctoral scholar. She holds a PhD in Agricultural Economics from Purdue University specializing in Space, Health, and Population Economics (SHaPE), which she completed in 2017. Her dissertation explores growth patterns and interactions among cities within the American system of cities. She also holds an MS in Agricultural Economics from Purdue University, which she completed in 2011. Her master's thesis explores health care utilization of vulnerable and non-vulnerable populations in the United States and was awarded an Outstanding Master's Thesis – Honorable Mention award from the Agricultural and Applied Economics Association in 2012. She was also a recipient of the Frederick N. Andrews Fellowship from Purdue University. She holds a BA in Geography, an honors BS in Economics and a minor in Mathematics from the University of Minnesota – Twin Cities, which she completed in 2008.

Brenda M. Elias was awarded a Canada Mortgage and Housing Corporation Graduate Research Fellowship upon completing her Bachelor of Applied Science at the University of Guelph. She obtained her Master of Science degree from the University of Guelph with specialty in housing and gerontology and has a Specialist Certificate in Aging, Graduate School of Social Work, University of Michigan. In 2009 she completed her PhD in Adult Education and Community Development at the Ontario Institute for Studies in Education (OISE), University of Toronto. Her thesis was on rural homelessness and subsequently she served as Project Coordinator for the Mental Health Commission of Canada on their Homelessness Research Initiative. In 2012, Brenda was a member of the Advisory Committee on the Law as it Affects Older Adults, with the Law Commission of Ontario and has been a 2018 faculty member of the Chang School of Continuing Education at Ryerson University for 5 years in the Non-profit and Voluntary Sector Management Program. For the past 15 years she has been teaching Foundations of Social Gerontology at the University of Guelph Humber degree program in Family and Community Social Sciences and is Professor and Coordinator of the Multidisciplinary Gerontology Certificate at Humber College School of Social and Community Services. Courses taught include Legislation for Social Services, Case Management, Canadian Political Process, Women and Aging at York University, Special Topics related to Indigenous Studies, Community Development and Social Policy. As an Ontario Public Servant for over 20 years Brenda has held many executive positions including 6 years as Director, Ministry of Health and Long-Term Care and Director of Economic Development and Communications for the Ontario Women's Directorate and more recently has provided consulting services on Leadership and Change and creating Age Friendly Communities to several national and provincial nonprofit boards. Brenda is interested in legal and ethical issues related to aging, social justice, community development, supportive housing (homelessness) and life span development, and has been primarily involved in policy development and community-based participatory research. She is currently Research Director for Reena, a nonprofit agency providing supportive housing to those individuals with intellectual challenges, and has served as a former Board Director with Abbeyfield of Canada, a nonprofit agency offering independent older adults a retirement housing option that balances privacy and companionship, and security and independence.

Kathleen Flowers earned her undergraduate degree in Communications from Stonehill College and her EdM in Higher Education Administration from the University at Buffalo. She joined Hobart and William Smith College in 2004 and is responsible for strategic leadership and management of the Center for Community Engagement and Service-Learning. Ms Flowers oversees the service-learning and community-based research programs, the Summer of Service internship program, and she co-facilitates the Faculty Service-Learning Advisory Council. Her purview also includes federal work-study tutoring programs, Alternative Spring Break trips and postgraduate service opportunities. She is an AmeriCorps VISTA alumna and serves on the board for five local nonprofit agencies.

Shadrack Frimpong, described by the late United Nations Secretary General, Mr Kofi Annan, as the "embodiment of youth leadership," is a proud son of a peasant farmer and charcoal seller from rural Ghana. Frimpong secured a full scholarship to attend the University of Pennsylvania, later graduating as a University Scholar and one of five students in his graduating class to be awarded the $150,000 President's Engagement Prize, Penn's highest honor. With the prize as seed funding, Mr Frimpong founded

Cocoa360 and pioneered the "farm-for-impact" model: a tuition-free girls' school and hospital sustained by proceeds from a cocoa plantation. In less than 3 years of operation, Cocoa360 has grown to 30 full-time staff members, cared for 3,000 patients and reached over 35,000 farmers in eight communities. Cocoa360 currently educates 120 young girls and is poised to scale its model into three new communities in 2019. Mr Frimpong is a recipient of many other awards, including the prestigious Samuel Huntington Public Service Award; the Queen's Young Leader Award; and the Ghana Legacy Honors, making him the youngest person ever to receive one of Ghana's highest honors. In addition to his work with Cocoa360, Mr Frimpong is currently pursuing an MS in Nonprofit Leadership at the University of Pennsylvania's School of Social Policy and Practice.

Julia Fursova is a doctoral candidate at York University, Faculty of Environmental Studies, Toronto, Canada. Her doctoral research examines community action for health justice in urban environments with the focus on the role of nonprofit organizations in advancing or impeding community action for health justice. The research integrates institutional ethnography and participatory action research. As a researcher, she is committed to anti-oppression principles and methodologies. For this reason, she chooses transparent processes and highly participatory methods of data collection and analysis, including participatory action research and arts-based methods. She has over 15 years of experience in nonprofit and public sectors in Canada and abroad. Her academic pursuits have been continuously influenced by her personal experience as an immigrant woman advocating for health equity and justice, as well as by her experience as a service provider in the nonprofit sector. As an immigrant woman, she experienced how one's social location is produced through multiple aspects of positionality that are socially and politically constructed and depend on a geographical locale and socio-political context. Critical reflection on the changing social location within systems of power continuously informs her understanding of how privilege and oppression shape individual experiences. As an academic, she is committed to examining the dialectical relationship between an individual and a system with a goal to inform progressive social change.

Ivis García, AICP, PhD, is an Assistant Professor in the City and Metropolitan Planning Department at the University of Utah. She is a fellow and board member of the Asset-Based Community Development Institute at DePaul University, Chicago. García earned her PhD in Urban Planning and Policy from the University of Illinois at Chicago. She holds dual master's degrees from the University of New Mexico in Community and Regional Planning and Latin American Studies and a bachelor's degree in Environmental Sciences from Inter-American University in Puerto Rico.

Femida Handy is Professor of Social Policy at the School of Social Policy and Practice at the University of Pennsylvania and the Director of the PhD program in Social Welfare. Her research and teaching focus on the economics of the nonprofit sector, volunteering, philanthropy, nonprofit management, entrepreneurship and microfinance. Prior to being appointed as Editor-in-Chief of *Nonprofit and Voluntary Sector Quarterly* in 2010, the premier journal in the field, she served on the editorial board of several academic journals. Professor Handy has published widely in a variety of scientific journals on a variety of nonprofit-related topics, and her work has garnered many awards. Her most recent book, *Ethics for Social Impact* (Palgrave Publishing, 2018), focuses on ethical decision-making for nonprofit leaders. Other recent books include *The Practice and Promise of Philanthropy in India* (Sage Publishing, 2016) and *The Palgrave Research Companion to Global Philanthropy* (Palgrave Publishing, 2015), which she co-edited. She has also written on environmental issues, including a children's book that introduces the concept of ecological footprint. Currently, Professor Handy's research projects include an NIH-funded project that investigates if and how autistic youth benefit by volunteering; research on the role of volunteering in the lives of immigrants; and a study of the transmission of values and behaviors across generations in three countries. Before coming to Penn, Professor Handy was Associate Professor at the Faculty of Environmental Studies at York University in Toronto, Canada.

Brandon Hofstedt is Associate Professor of Sustainable Community Development and Director of the Center for Rural Communities at Northland College since 2010. Currently, Dr Hofstedt is the Community Economic Development program manager with the Community Development Institute, Division of Extension at the University of Wisconsin, Madison. Dr Hofstedt has a PhD in Sociology from Iowa State University, and his research focuses on rural community sustainability and community resilience. He has extensive experience administering quantitative and qualitative methodologies to understand community issues and using social scientific data to inform community level decisions. He is the principal investigator and lead researcher of the Northwoods Quality of Life project.

Justin B. Hollander is a Professor of Urban and Environmental Policy and Planning and Director of the Certificate of Advanced Graduate Studies (CAGS) in Urban Justice and Sustainability at Tufts University. His research and teaching is in the areas of shrinking cities, Big Data, brownfields and the intersection between cognitive science and the design of cities. He is the author of seven books on urban planning and design, most recently *An Ordinary City: Planning for Growth and Decline in New Bedford, Massachusetts* (Palgrave, 2018) and *A Research Agenda for Shrinking Cities* (Edward Elgar, 2018). Professor Hollander has conducted extensive research exploring the relationship between urban development, planning, well-being and health, including work for the Cities of New York, Baltimore and Detroit, among others. His international research has been supported through grants from the Governments of Canada and Quebec, as well as through funding from Duy Tan University in Vietnam. He runs a podcast through iTunes, "Cognitive Urbanism," where he talks about the big ideas of urban planning and community development.

James G. Huff, Jr., PhD, serves as Associate Director and Associate Professor of Human Needs and Global Resources and Anthropology at Wheaton College (Wheaton, Illinois). His scholarly interests focus on international and rural community development, religion in Latin America, faith-rooted social movements and nongovernmental organizations (NGOs) and comprehensive community initiatives. He also has an interest in the pedagogy of study abroad programs and cross-cultural learning. His research examines the changing religious landscapes of contemporary Latin America, with a particular focus on the social and economic changes generated by the rapid growth of Pentecostal-charismatic forms of Christianity across the region. For the past decade he has regularly conducted ethnographic fieldwork in El Salvador where he has documented the involvement of Pentecostal churches in community-based development projects. As an applied anthropologist, Dr Huff has worked with various NGOs and community-based organizations to evaluate the effectiveness of their programs to alleviate poverty and to strengthen human well-being and community resiliency.

M. Rezaul Islam, PhD, is a Professor in Social Work at the Institute of Social Welfare and Research, University of Dhaka, Bangladesh and International Academic Adviser at the Department of Social Administration and Justice, University of Malaya, Malaysia. He is actively involved in social work education and the professional development of social work in Bangladesh and Malaysia. He is a community development expert and he has numerous research works and publications in the community development field. His research focuses on the migrant labor force, human rights and poverty, nongovernmental organizations (NGOs) and community development, child welfare and social development, which is embedded by a human rights approach and inspired by people who are vulnerable and affected by structural causes of poverty, unemployment and inequalities.

Somayeh Khademi is a Visiting Lecturer at the Payam-e Noor University, Shiraz branch. Her main research interest is in sustainable urban development and urban neighborhoods. She is particularly interested in issues concerning social sustainability, the urban environment and studying the historic districts. Khademi is currently working on a research project on social sustainability in the field of urban neighborhoods. She studied for her bachelor degree in Social Planning at Shiraz University in 2008 and then pursued her master's degree in Urban and Regional Planning at the Art University of Isfahan and graduated in 2011. She then joined the Department of Geography and Urban Planning at the Kharazmi University in Tehran. She has taught courses such as an Introduction to Urbanization, Fundamentals of Urbanization, Urban Infrastructure and Facilities, Fundamentals and Methods of Urban Planning, Regional Planning, Ecology, Environmental Planning and Design and Environmental Psychology since 2015 as a visiting lecturer at the Payam-e Noor University, Shiraz branch. Also, she has conducted and supervised some of the final projects of bachelor students in the same university on issues like quality of life and urban vitality. As well as this, she has published some articles in the field of community development.

Reinout Kleinhans is Associate Professor of Urban Regeneration and Neighbourhood Change at the Faculty of Architecture and the Built Environment, Delft University of Technology. His research interests and expertise include urban regeneration, social capital, citizens' self-organization, community entrepreneurship and citizen engagement. Reinout has published widely on neighborhood change, social implications of urban restructuring, community enterprise, co-production, online governance and the use of digital participatory platforms for co-production between citizens and governments.

Richard C. Knopf, PhD, is Professor of Community Resources and Development at Arizona State University (ASU) and serves as Director of ASU's Partnership for Community Development (PCD). His research has focused on grassroots community organizing, sustainable community development practices, life quality enhancement, sustainable livelihoods and sustainable economic growth. Dr Knopf is a recipient of the Community Development Achievement award from the Community Development

Society, and is known for his capacity to build innovative partnership among government organizations, NGOs, faith communities, businesses, environmental groups and other community-based organizations to help communities co-create collective vision and action to achieve that vision.

Patsy Kraeger is an Assistant Professor in the Master's Public Administration program at Georgia Southern University, in the College of Behavioral and Social Sciences, Department of Public and Nonprofit Studies. She received her PhD in Public Administration from Arizona State University (ASU), the School of Public Affairs, with a certificate in Nonprofit Leadership and Management from the School of Community Resources and Development. She also holds a master's degree in Nonprofit Studies from ASU, the School of Community Resource and Development. She has studied social enterprise, innovation and entrepreneurship at the European Summer School on Social Economy, the Department of Economics of the University of Bologna (Forlì Campus), Italy and through the EMES Research Network, University of Trento, Italy. She holds a JD from Mercer University; and an AB from Sweet Briar College with a double major in history and international relations. She is a co-editor with Dr Rhonda Phillips in the first edition of the four volume series Community Planning and Development (Critical Concepts in Built Environment) (Routledge, 2017) and the lead co-editor for *New Dimensions in Community Well-Being* (Springer, 2017). She has published in the areas of community development and quality of life, philanthropy and public policy, nonprofit education and other related areas in various journals. Her primary research interests are focused on community development, the study of philanthropy vis-à-vis democracy as well as social enterprise in a public policy context.

Indraneel Kumar is the Regional Planner: GIS (Geographic Information System) and Spatial Analysis for the Purdue Center for Regional Development (PCRD). He received undergraduate and graduate degrees in Architecture, Urban and Regional Planning and Community Planning from India and the United States. Indraneel earned a PhD in Civil Engineering with a major in transportation and infrastructure systems from Purdue University in 2014. At PCRD, he focuses on regional economic analysis, socioeconomic data snapshots, GIS databases and spatial analysis. His two decades of work experience include comprehensive planning for large metropolitan regions, districts and counties in India and the United States. He has participated in several regional economic development research grants and projects funded by the US Economic Development Administration, Rural Development (RD) and Economic Research Service (ERS) of the US Department of Agriculture (USDA) and state agencies of Indiana, such as the Office of Community and Rural Affairs (OCRA) and Indiana State Department of Agriculture (ISDA). A few noted regional economic development projects where he has participated as a team member include industry and occupation clusters in the US, motorsports clusters in Indiana, Stronger Economies Together (SET) and Resilience to the Intergenerational Transmission of Poverty.

Robin Lewis serves as an Assistant Professor in the Environmental Studies Program at Hobart and William Smith Colleges in Geneva, New York where she offers courses on biodiversity, sustainability, qualitative methods and community development. She also serves as Chair of the Colleges' Sustainable Community Development Program. Many of her courses integrate service-learning projects that focus on cultivating and sustaining reciprocal relationships between the Colleges and the broader Geneva community. In addition to her community-based work in the Finger Lakes region of upstate New York, Dr Lewis also conducts research on knowledge production and circulation in the field of bryology in collaboration with colleagues at the University of Tennessee – Knoxville. As a graduate student, she earned a Fulbright fellowship to Malaysia where she completed 13 months of ethnographic fieldwork. She completed her PhD in Geography at the University of Arizona in 2011.

Jerrold McGrath designs and delivers interventions in systems that are failing to meet the needs of those that are meant to be served by them. He works in the creative and cultural sector and his practice reintegrates fragmented disciplines and communities around issues of shared concern. He has developed partnerships, cross-sector collaborations and development programs to leverage the strengths of various sectors in addressing complex, systems-level social and cultural issues (hopelessness, economic inequality and city building). Jerrold is currently the Director of Programs for Artscape Daniels Launchpad, a 30,000-square-foot creative entrepreneurship hub in Toronto, Canada. He was previously the Director of Creative Ecology Leadership at the Banff Centre for Arts and Creativity and the Director of Innovation and Program Partnerships for leadership programming at the Banff Centre. Jerrold completed his master's in Strategic Innovation and Change at the University of Denver with a focus on strategy formulation in creative sector organizations. Jerrold writes on systems change and has been invited to speak at events such as C2-MTL; the Open Leadership Summit in Portland; Maker Faire Global in New York; the Association Forum of Chicagoland; and many more. Jerrold helped found the Toronto Arts Council Cultural Leader's Lab; ALT/Now; Hope Decoded, among 100 other programs. Jerrold is an ambassador for the STATE Festival in Berlin and is a BMW Foundation Responsible Leader.

Dyana P. Mason is an Assistant Professor at the University of Oregon's School of Planning, Public Policy and Management. Her research and teaching focus on nonprofit management and governance, including the use of diversity and inclusion practices, fundraising, community development and teaching nonprofit management. Her work has appeared in leading journals including *Nonprofit and Voluntary Sector Quarterly*, *Voluntas*, *Journal of Public Administration and Research and Theory* and the *International Journal of Cultural Policy*. She is interested in all things "nonprofit," and has also led study abroad courses to Southeast Asia and Argentina to study the role of nonprofit organizations in comparative context.

Abolfazl Meshkini is Associate Professor in Geography and Urban Planning at the Faculty of Humanities, Tarbiat Modares University, Tehran, Iran. He received his PhD from Shahid Beheshti University in Urban Geography. His main research interests are in housing and neighborhood, community development, urban development plans, Islamic cities and quality of urban environment. He has authored or co-authored several research papers in these fields for national and international conferences and peer-reviewed journals. He also has granted and conducted several projects on community and urban planning. Moreover, he has been supervisor of more than 100 academic projects, dissertations and theses. Dr Meshkini is credited with achieving awards and honors from several private and public sector organizations in Iran. In addition, he is director and reviews editor of several scientific journals as well as academic associations.

Mostafa Norouzi graduated from the Geography and Urban Planning Department at Tarbiat Modares University, Tehran, Iran. His work focuses specifically on (1) the urban quality of life (residential environment satisfaction, housing quality, livability and well-being), (2) community development (community participation, housing and neighborhood planning, capacity building and social sustainable development), and (3) urban social geography (poverty, spatial justice, geographic disparities and access to services and resources in disadvantaged communities and distressed areas). He has presented and published several research papers in the same fields for national and international conferences and peer-reviewed journals. He also has conducted several projects on community planning and quality of life.

Max Page is Professor of Architecture and Director of Historic Preservation Initiatives at the University of Massachusetts (UMass) in Amherst. Professor Page directs the Master of Design in Historic Preservation program, which trains students for careers in historic preservation and related fields. He received his education at Yale University (BA, *magna cum laude* in History, 1988) and from the University of Pennsylvania (PhD, 1995). Professor Page teaches and writes about the history of cities and architecture. He has written or edited eight books on subjects ranging from the destruction and rebuilding of New York City, in reality and in the imagination, the urbanist Jane Jacobs, the history of historic preservation, the architecture of the UMass campus and the future of higher education.

C. Bjørn Peterson, PhD, is Senior Research Associate at the Partnership for Community Development at Arizona State University and Instructor in Transformational Leadership at Seattle University. His research has focused on the relationship between identity construction among community workers and the corresponding programs, projects and interventions of these individuals and groups. Dr Peterson is an international scholar-practitioner based in Seattle, Washington.

Rhonda Phillips, PhD, FAICP, is Dean of the Purdue Honors College and Professor in the Agricultural Economics Department. Previously, she served as Associate Dean at Arizona State University and as Professor and Director, School of Community Resources and Development. She also served as Senior Sustainability Scientist in Arizona State University's Global Institute of Sustainability and as affiliate faculty in the School of Public Affairs and School of Geographical Sciences and Urban Planning. Earlier in her career, she was a member of the faculty of the University of Florida's Urban and Regional Planning Department, held visiting appointments at the University of Vermont in the Community Development and Applied Economics Department and is a three-time Fulbright award recipient. As a planning and community and economic development specialist, community quality of life, development and well-being comprise the focus of her research and outreach activities. She is author or editor of 27 academic books, including *Introduction to Community Development* (Routledge, 2015) and *The Handbook of Community Well-Being Research* (Springer, 2017). Rhonda is former editor of the journal, *Community Development*, and is founding editor of the book series, Community Development Research and Practice and Current Issues in Community Development, both in conjunction with the Community Development Society and Routledge. She is founding editor of the *International Journal of Community Well-Being*, published by Springer, and is launching a new journal, *Local Development & Society* with Taylor & Francis. Phillips' work prior to joining academe was in community and economic planning and development at the state, local and regional levels; she is a member of the College of Fellows of the American Institute of Certified Planners, nominated and inducted in 2016 for career achievements. Rhonda is past President of the International Society for Quality-of-Life Studies, www.isqols.org, and serves on the site visitor team of the national Planning Accreditation Board. She received her doctorate in city and regional planning and an MS in Economics from the Georgia Institute of Technology; and an MS in Economic Development and a BS in Geography from the University of Southern Mississippi. She was the first woman to graduate from the doctoral program in City and Regional Planning at the Georgia Institute of Technology.

Jason Reece is an Assistant Professor of City and Regional Planning in the Knowlton School at the Ohio State University. His research, teaching and professional experience focus on social equity in city planning, community development and public health. Reece has acted as an advisor and capacity builder to foundations, nonprofits, community organizations and government agencies. He has managed more than $10 million in research initiatives and contributed to more than 110 scholarly or technical publications. His work has been featured in 41 media publications and he has been an invited guest speaker at more than 300 professional engagements. Reece has worked with partner organizations in more than 30 states in the US and acted as a capacity builder for the US Department of Housing and Urban Development Sustainable Communities Initiative for 4 years. He currently is President of the Board of Directors for the Parsons Avenue Community Development Corporation. Reece was formerly the Senior Associate Director and Director of Research for the Kirwan Institute for the Study of Race and Ethnicity at the Ohio State University. He established the Opportunity Communities program at the Institute, developed the opportunity mapping methodology and established the Institute's Health Equity program.

Mark Roseland is Director and Professor, School of Community Resources and Development, Arizona State University (ASU) and Senior Sustainability Scientist, Julie Ann Wrigley Global Institute of Sustainability at ASU. Before ASU, he was at Simon Fraser University (SFU) in Vancouver, Canada, where he was professor of planning in the School of Resource and Environmental Management and Director of the Centre for Sustainable Development. He is a Registered Professional Planner and full member of the Canadian Institute of Planners, and he has worked as Chief City Planner for a municipality in the Metro Vancouver area. He has been cited by the *Vancouver Sun* as "one of Vancouver's top 50 living public intellectuals" and has received both the SFU Sustainability Network Award for Excellence in Research on Sustainability and the SFU President's Award on Leadership in Sustainability. Dr Roseland lectures internationally and advises communities and governments on sustainable development policy and planning. His best-selling book *Toward Sustainable Communities: Solutions for Citizens and Their Governments* (New Society Publishers, 2012) is in its fourth edition. He is also the founder of Pando | Sustainable Communities, a multilingual online network to promote collaboration between sustainable communities researchers and practitioners.

Kazi Abdur Rouf is a Professor at Noble International University, a Research Associate in the Center for Learning Social Economy and Workplace at the University of Toronto and York Center for Asian Research (YCAR), and International Visiting Scholar in the School of Education at Indiana University Bloomington, USA. He is researching green and social business, women's development, nongovernmental organizations (NGOs) and community economic development. Kazi has worked in Grameen Bank, Bangladesh for more than two decades. He completed his PhD from the University of Toronto and a master's in Environmental Studies, York University, Canada. Kazi is researching on renewable energy, green social enterprise development, micro-financing student higher education financial aid, social safety nets, women's human rights development, community gardening, mini cooperatives and community development in Canada and in Bangladesh.

Allison R. Russell is a postdoctoral fellow at the School of Social Policy and Practice at the University of Pennsylvania. Her research focuses on volunteering, nonprofit management, equity and ethics in the nonprofit sector and civil society development around the world. Dr Russell's work appears in edited volumes and peer-reviewed journals related to nonprofit and voluntary sector studies, nonprofit and public management and community development. She is also the co-author of the book *Ethics for Social Impact* (Palgrave Publishers, 2018), with Femida Handy. Dr Russell served as interim Managing Editor and Editorial Assistant for the academic journal *Nonprofit and Voluntary Sector Quarterly* and is a 2017 recipient of the Association for Research on Nonprofit Organizations and Voluntary Action Emerging Scholars Award.

Robert Mark Silverman is a Professor in the Department of Urban and Regional Planning at the University at Buffalo (UB). His research focuses on fair and affordable housing, community development, the nonprofit sector and education reform.

Maria Spiliotopoulou is a PhD candidate at Simon Fraser University's School of Resource and Environmental Management and an instructor in the Sustainable Development Program. Maria's doctoral research aims to advance community sustainability theory by exploring the potential of urban productivity to holistically operationalize sustainable community development. Her goal is to contribute to the global discourse on implementing and assessing local sustainability and propose new ways to support the achievement of productive – and eventually sustainable – communities. Maria also has extensive work experience as an environmental consultant in Europe. Her most recent publications, co-authored with Dr Mark Roseland, include a chapter in the *Palgrave Handbook of Sustainability* (Palgrave Publishers, 2018) and a chapter in Elsevier Publishing's *Encyclopedia of Sustainable Technologies* (2017).

Camille Sutton-Brown is an Assistant Professor of Educational Research at Kennesaw State University in Georgia. She holds a PhD in Educational Policy Studies: Research, Measurement, and Statistics from Georgia State University and completed a 2-year postdoctoral fellowship in the Human Economy Programme at the University of Pretoria in South Africa. Camille's expertise lies in qualitative inquiry, participatory action research, visual methods, transnational feminism, international development and photovoice methodology. She has conducted research in several community-based development settings, including a multi-site photovoice study on women's empowerment in the context of microfinance in Mali. Camille's current research agenda focuses on integrating democratic research processes to explore topics related to social justice.

Craig A. Talmage serves as a Visiting Assistant Professor of Entrepreneurial Studies at Hobart and William Smith Colleges in Geneva, New York. Entrepreneurial Studies is a new, fast-growing minor that officially started in the Spring of 2016. He teaches courses on economic principles, quantitative tools, social innovation, the history of entrepreneurship theory and the senior capstone experience. He seeks to empower community members, faculty, staff and students through the development of knowledge regarding entrepreneurship and skills that match that knowledge. He completed his PhD in Community Resources and Development at Arizona State University (ASU). At ASU, he worked for the Osher Lifelong Learning Institute, Partnership for Community Development and the Southwest Interdisciplinary Research Center. He still serves as a faculty associate for ASU where he teaches Community Resilience to Emergency Management and Homeland Security students in the School of Public Affairs. He is actively involved in the International Association for Community Development, the International Society for Quality-of-Life Studies and the Association for Research on Nonprofit Organizations and Voluntary Action.

Henry Louis Taylor, Jr. is a Professor in the Department of Urban and Regional Planning at the University at Buffalo (UB). His research focuses on a historical and contemporary analysis of distressed urban neighborhoods, social isolation and race and class issues among people of color, especially African Americans and Latinos.

Talya D. Thomas is a native of Houston, Texas. She attended Clark Atlanta University, where she obtained a bachelor's in Psychology, minor in business administration and a master's in Public Administration with a concentration in community and economic development. She attended Texas Southern University where she obtained her doctoral degree in Urban Planning and Environmental Policy, with a concentration in community development and housing. She has attended and presented at various conferences throughout the United States. Some of the conferences include: National Forum for Black Public Administrators (NFBPA), Conference of Minority Public Administrators (COMPA), Planners Network, Urban Affairs Association, Community Development Society and the International Association for Community Development. She is also very active in many community service organizations: NFBPA, Top Ladies of Distinction, Inc., American Planning Association (APA), Jack and Jill of America, Inc. and Delta Sigma Theta Sorority, Inc., just to name a few. She has various publications from the spectrum of emergency management to community development. Dr Thomas has a wide range of teaching experience with Houston Community College, LoneStar College and Texas Southern University, and is currently an Assistant Professor in the Department of Urban and Regional Planning at Jackson State University.

Graciela Tonon is Doctor of Political Sciences, Magister in Political Sciences and Social Worker. She has post-doctoral studies in Qualitative Methods at University of Florence, Italy. She is the Director of the Institute of Social Research-UNICOM of the Universidad Nacional de Lomas de Zamora, Argentina. She is the Director of the Master Program in Social Sciences and CICS-UP of Universidad de Palermo, Argentina. In 2016, Graciela received the International Society for Quality of Life

Studies (ISQOLS) Distinguished Service Award for Substantial Contributing to a Better Understanding of Quality of Life Studies. She is the Editor of the International Handbooks of Quality of Life Series (Springer). She is the Vice-President of Publications of the International Society of Quality of Life Studies (2019–20). She is also the Secretary of the Human Development and Capability Association (2019–22). Graciela is Professor at Universidad Nacional de Lomas de Zamora where she teaches community social work and children at risk. She is Professor at Universidad de Palermo, Argentina, where she teaches quality of life and happiness and qualitative methodology. Her recent books published are: *Teaching Quality of Life in Different Domains* (Springer, 2019), Social Indicators Research Series, 79, Springer-International Society for Quality of Life Studies (ISQOLS); and *Quality of Life in Communities of Latin Countries* (Springer, 2017), Community Quality-of-Life and Well-Being Series. She is dedicated to the fields of: quality of life, qualitative research methodology, communities, human development, children and young people.

Leanne Townsend is a senior social scientist within the James Hutton Institute Social, Economic and Geographical Sciences group. She has expertise in the social and economic impacts of technologies and renewables in rural and international development settings. Currently, Leanne is leading research on the role of advisors and social networks in precision farming, and rural futures in relation to rapid digitization. Leanne also works on a number of projects concerned with energy in Sub-Saharan Africa, including community engagement in the development of solar energy systems in off-grid rural Kenya and Rwanda, and the impacts of blackouts in on-grid urban Nigeria. She is interested in rural and international development, agricultural innovation and digital and energy-based transformations. Leanne has worked with diverse societal and stakeholder groups with a focus on marginalization and social exclusion including Gypsy-Traveller communities, remote rural communities, traditional craftspeople in Scotland and rural and peri-urban communities in Kenya and Nigeria.

Eric Trevan is a Member of the Faculty (Tenure-Track) for The Evergreen State College with a focus on Public Administration, Non Profit Administration and Public Policy which supports his research focus of local and Tribal economies. He also serves as President of Local Solutions, a new AI market research software platform. Additionally he completed his term as Chairman of the Board of Directors for an investment/economic development corporation, Gun Lake Investments. National leadership includes the past President/Chief Executive Officer of the National Center for American Indian Enterprise Development. He is a national advocate for entrepreneurship, innovation, community and economic development. He is especially focused on working with small, minority and Native American business. He provides policy/research recommendations on a variant of community, planning, business, entrepreneurial and economic issues. His entire career has been to use strategies with advocacy, research planning and economic development in order to promote free market equitable economic growth opportunities. He recently has served over Government & Tribal relations for the Heard Museum and held past leadership positions as Chief Operations Officer-Pokagon Band of Potawatomi, CEO-Northside Economic Potential Group, Director of Planning and Development-Whiteville, North Carolina, Assistant City Manager-Port Huron Michigan and Tribal Operations Manager-Nottawaseppi Huron Band of Potawatomi. He earned his PhD at Arizona State University, master's degree in Administration from Central Michigan University and a bachelor's degree in Public Administration/Economics from Western Michigan University.

David P. Varady is Professor of Community Planning at the University of Cincinnati. He is author of six books, nine book chapters and over 60 journal articles and 75 book reviews on neighborhood development, segregation, and low-income housing policies. Professor Varady has held Visiting Scholar positions at Delft University of Technology, City of Helsinki, Rutgers University, University of Glasgow, the National Association of Realtors and the US Department of Housing and Urban Development. Since 2005, Varady has been Book Review Editor for the *Journal of Urban Affairs*.

Claire Wallace is Professor of Sociology at the University of Aberdeen. She has published widely on quality of life issues, especially in rural areas in the UK and elsewhere. She is co-author of the book *The Decent Society: Planning for Social Quality* (with Pamela Abbott and Roger Sapsford, Routledge, 2015). She was part of the dot.rural digital economy project at the University of Aberdeen together with Leanne Townsend.

Li Yin is an Associate Professor in the Department of Urban and Regional Planning at the University at Buffalo. Her research focuses on spatial models, GIS, and simulation methods applied to the analysis of urban systems for environmental planning, urban design and sustainable development.

Introduction to the *Research Handbook on Community Development*
Rhonda Phillips, Eric Trevan and Patsy Kraeger

Fundamentally, research is the process of discovery and exploration – the outcomes of which range widely from increasing understanding and finding potential solutions to gathering information that may contribute to additional inquiry. Community development as a means of improving the places we live in is a pressing issue more than ever, and further discovery and exploration of it are very much needed. It is our intent to present this volume to spur ideas and innovations in community development. At its most basic, community development is simply about making things better for the people who live there (Musikanski et al., 2019). At its most complex, it is decidedly difficult to identify the most effective or desirable approach as needs, desires, conditions, external and internal influences and confounding factors and resources can vary widely between communities. Community represents agency and solidarity (Bhattacharyya, 1995), and it is critical to understand that community is not only a destination and location but can also include a common set of ideas and values (Trevan, 2016), which inform both research and practice for the co-creation of knowledge.

By focusing on research approaches, techniques and applications, we aim to illustrate both the broad complexity of community development and its potential. We hope this will help foster greater understanding of how research contributes to scholarship and to practice, where we see the results of ideas in action.

COMMUNITY DEVELOPMENT: THEORY, RESEARCH AND PRACTICE

Community development is both an area of theoretical development, research and study, as well as a practice. Kenny, McGrath and Phillips (2017, p. xxiv) provide the following description of community development as a social movement of collective endeavor:

> it is concerned with creating better lives globally and ensuring that human beings can become agents of their own destinies. It operates on the basis of a commitment to social justice, social equality and the principles of universal human rights . . . . It is underpinned by a strong emphasis on meaningful community participation and collaboration.

It occurs in the complex site of communities which can vary widely from one to the other, even within the same regions or cultures. This wide diversity provides a fascinating platform of study and practice.

Research and practice have a complex relationship. Practice-based research and related theory are an emerging theory in the arts, education, medicine, organizational theory and sociology and other disciplines. Mohrman and Lawler (2011) recognize that institutional structures of knowledge creation, generation and application occur both within and outside the walls of academe. Recognizing and understanding that "knowledge creation, generation and application has and is changing allows both academics and practitioners alike to come together to broaden the landscape of actors who generate and develop knowledge to inform practice" (Mohrman and Lawler, 2011: 9).

This volume seeks to broaden the field of knowledge creation and generation in community development with contributions from both academe and practitioners in the community development field. Feldman and Orlikowski (2011) note that a practice-based approach to research could include empirical, theoretical and philosophical approaches. Within these three approaches there is a dualism between the actual practices being produced from "specific social actions in the social world" (p. 1241).

In this *Handbook*, readers are introduced to theoretical, philosophical, empirical analysis and case studies from around the globe to expand knowledge generation to inform social action in the field of community development. While the exact definitions surrounding practice-based research may vary, Candy and Edmonds (2018) suggest that when practice itself creates new insights and knowledge the research is practice-based. When theoretical and/or conceptual research as implemented leads to new practices in the field, we could classify that as practice-led research according to Candy and Edmonds (2018). Why the distinction? Community development is, after all, an applied discipline. Community development is to some extent practice-based research, where theory, scholarship and practice often intersect (or in some cases, collide), generating new ways of thinking about issues and potential solutions or applications.

APPROACHES TO COMMUNITY DEVELOPMENT

As mentioned, research on community development is more critical than ever. Why? First, it "helps communities to develop, evolve, and improve in a constantly changing environment"; it is thus vital to deepen understanding of community development because every policy, action, strategy, process, program, investment and so on impacts people's quality of life (Phillips and Pittman, 2015a, p. 346). Essentially, communities must be able to demonstrate the value and outcomes of their activities in order to be accountable to residents, justify and secure funding, and to assess the efficacy and outcomes of programs, policies and actions taken in the name of community development (Phillips and Pittman, 2015a). Measurement, assessment and evaluation are foundational to community development practice, especially in the context of understanding impacts and outcomes of development interventions and processes (Walzer and Blanke, quoted in Phillips and Pittman, 2015b, p. 347, Table 21.1).

One conceptual approach to thinking about measurement and assessment in community development includes Asset Based Community Development (ABCD) from Kretzmann and McKnight (1996), refocusing the field toward studying community assets based on relationships for empowered capacity as opposed to the deficit approach of the traditional needs assessment. Another approach is that of the Community Capitals Framework (Flora et al., 2004; Emery and Flora, 2006). This framework provides a method for analysing community development efforts from "systems perspectives by identifying the assets in each capital (stock), the types of capital invested (flow), the interaction among the capitals, and the resulting impacts across capitals" (Emery and Flora, 2006, p. 19). This framework has influenced subsequent thought on evaluation frameworks, and has also influenced further development of ABCD thought and application. Other conceptual approaches focusing on the relational (community members as citizens not clients), interactional sustainable community development (recognizing economic development and protection actors need to work together to achieve meaningful change for today and in the future), learning (translating theory into actual practice), participatory (people-centered, bottom-up and process oriented rather than top-down and technological), sustainability (long-term approach focusing on human and environmental welfare), sustainable livelihoods (people focused on connecting to local knowledge and assets) and networked (people working in communities recognize that changing economic and political winds impact community development) approaches have shaped the field from perspectives of research and practice (Brocklesby and Fisher, 2003; Gilchrist, 2019; Korten, 1980; Hickey and Mohan, 2004; Mathie and Cunningham, 2003; Wheeler, 2015).

Another important area of evaluation and assessment is that of community indicators which are bits of information that, when combined, paint a picture of conditions in a community – is it moving forward or away from desirable goals? (Phillips, 2003). Community indicators, when used as a system, can provide valuable information to be used in decision-making processes about investments, policies or other actions within communities (Phillips, 2005).

There are many approaches to research on community development, and its use for enhancing its efficacy. More often than not, there is not a distinct delineation between quantitative and qualitative research approaches in community development but rather mixed methods are utilized. Some research approaches are more quantitative in nature, relying on econometric and other economic-based tools or techniques (such as predictive modeling for cities or regions, cost-benefit analysis and the oft used economic impact assessment). Research methods from other disciplines and fields of study are being applied, such as evidence-based approaches from health care or scenario planning from the corporate world (we have long used strategic planning originating from the military, for another example). Qualitative aspects are often used in community development research for many reasons, not the least of which is that communities are focused on people and their interactions, and given complexities of relationships, qualitative research may yield more insight. Borrowing from sociology and other fields, social network analysis, for example, provides an analytical approach for deeply exploring relationship patterns and connections. Quality-of-life research originating in psychology is also now applied within community contexts. Appreciative inquiry, as a method of dialogic organizational development, has been adapted for use in community development and is gaining much attention as a particularly relevant method. While all these and others are not yet fully integrated or widely used within community development, they represent ways of thinking about the deep complexities within the places we live.

In this book, a myriad of approaches are included, in a variety of contexts both creating new approaches and using traditional or a new combination of traditional approaches adding depth to theory, research and practice. This edited volume brings together 27 chapters contributed by 45 authors from around the globe. It is divided into three parts: Part I: Foundations, Part II: Research Methods and Frameworks and Part III: Emerging Constructs and the Future of Community Development Research. We chose these categories to provide a way to navigate the wide terrain of community development research. In the first part, we focus on theories and essential or foundational conceptions about community development. This provides the platform on which to build and explore subsequent ideas and application. The second part presents an array of research methods and frameworks, ranging from experiential to applied. We conclude the volume with perspectives on future issues, especially those that are emerging using technology or different ways of approaching issues.

PART I: FOUNDATIONS

In the first part, a selection of eight chapters are presented. The volume begins with an exploration of the nexus or intersection of action and knowledge. Chapter 1, "Weaving reflection, action, and knowledge creation: lived experience as a catalyst into the cycle of praxis for community development" by C. Bjørn Peterson, Craig A. Talmage and Richard C. Knopf, begins our journey by calling for more critical consciousness (with a focus on personal experiences) as a basis for empowerment. It is thought-provoking by exploring collaborations across learning and action, with the request to guard the dignity of all involved in community development processes.

Elizabeth A. Dobis, Lionel J. Beaulieu and Indraneel Kumar explore a topic that holds relevance for many of the world's communities – poverty. The authors present the context of poverty in relation to the dimensions of scope, scale and space in Chapter 2, "The study of poverty in places: scope, scale, and space." A variety of disciplinary information is analysed from the place-based literature. They then examine how community development theory and practice may provide guidance for pursuing effective and evidence-based place-based poverty alleviation strategies and policy activities.

Another aspect that often arises but not always in conjunction with poverty is that of marginalized communities. Jason Reece explores this topic in Chapter 3, "In pursuit of just communities: supporting community development for marginalized communities through regional sustainability planning." During the Obama Administration the principles of sustainability and progressive regionalism made their way into the thinking of many federal agencies. The Department of Housing and Urban Development's Urban Sustainability Initiative (SCI) was designed to identify regional community solutions. This chapter analyses how SCI scales relate principles to overall processes and how solutions are identified providing greater equity for a diverse population.

Returning to strategies that identify strengths in the community, Chapter 4 focuses on "Asset-Based Community Development (ABCD): core principles" and is by Ivis García. This chapter outlines the core principles of the ABCD approach and its application within communities. Building on John Kretzmann and John McKnight's 1993 seminal work, *Building Communities from the Inside Out: A Path Toward Finding and Mobilizing a Community's Assets*, this chapter highlights engagement strategies for stakeholders across communities.

In relation to assets, institutions and organizations can serve as catalysts to energize civic engagement within communities. Chapter 5, "Stepping up the ladder: reflecting on the role of nonprofit organisations in supporting community participation" by Julia Fursova, explores how grassroots community action can support community development outcomes focused on equity and justice. This is developed through community engagement as the priority goal and having critical intrinsic values included in the process and the development of participatory action.

The next chapter deconstructs the relationship of community development and community action by viewing via a gendered lens for community decisions. Chapter 6, "Social economy, social capital, NGOs and community development: a gendered perspective" by Dyana P. Mason, explains the importance of integrating gender issues into community connections, social capital discussions and economic development. Using a feminist theory lens as well as traditional theoretical approaches, this chapter analyses the case of a US supported grassroots cooperative in Laos and community development impacts on social capital and the social economy.

Another case-focused chapter will provide a comparative analysis of community enterprises between Europe and US community-based organizations. Chapter 7, "What can Northwest European community enterprises learn from American community-based organizations?" is by David P. Varady, Reinout Kleinhans and Nuha Al Sader. The chapter discusses Community Development Corporations (CDCs) and applies a framework on Northern European Community Based Organizations (CBOs) and their impacts at the community level.

Our final comparative analysis in Part I is Chapter 8, "Community development, well-being and technology: a Kenyan village" by Claire Wallace and Leanne Townsend. The opportunities and challenges of new technology are explored and applied to a Kenyan cultural framework. How the overall usage or avoidance of the technology was adapted and adopted is discussed. Specifically, outcomes provide explanations as to how technology can enhance overall well-being for the Kenyan communities in the case.

PART II: RESEARCH METHODS AND FRAMEWORKS

This part includes 11 chapters from a range of perspectives and research methodologies and frameworks. Since community development is such a diverse and wide-ranging area of scholarship and practice, it is expected that a variety of techniques and methods are utilized.

Part II begins by using well-established research of the Grameen Bank. In Chapter 9, "Experience of group formation in Grameen Bank, Bangladesh" by Kazi Abdur Rouf, the author discusses the Grameen model and its application to community development approaches in Bangladesh. Specifically, civic engagement, the development of social capital, and community organizing provide an interesting conversation about lessons learned from a lending institution; and vice versa.

Chapter 10, "How to build an 'intentional community'" by Brenda M. Elias, uses established community development principles to provide intersectional opportunities between space and community participation. Using a "higher degree of teamwork," the 5-year research project with the Reena Community Residence, in Ontario, Canada examines how this impacts the living experiences of the residents.

Zoning provides a framework on community interaction and inclusion in communities. Chapter 11, "Inclusionary zoning and inclusionary housing in the United States: measuring inputs and outcomes" by Katrin B. Anacker, analyses the impacts of inclusive policies and how they affect socioeconomic integration. The outcomes of affordability, business returns and beneficiaries of inclusive policies are compared to the core focus of inclusive policies.

Using a case study analysing how Faith-Based Organizations (FBOs) impact international community outcomes, Chapter 12 is focused on "Enhancing evaluation capacity: lessons from faith-based community development in El Salvador," and is by James G. Huff, Jr. In Latin America, FBOs are a catalyst for community engagement, managing and responding to change, and overall stewardship. This chapter discusses the intersection of rural residents, practitioners and researchers working together to create an evaluative framework for community development outcomes.

Public participation is key for successful community development and this is explored in Chapter 13. "Managing competing interests in the public participation process: lessons from an analysis of residential displacement in Buffalo, New York's transitioning neighborhoods" by Robert Mark Silverman, Li Yin and Henry Louis Taylor, Jr. discusses public participation strategies as well as multiple competing interests that are introduced in stakeholder engagement processes. Using a case study from Buffalo, an analysis of stakeholder engagement for neighborhood revitalization using a series of focus groups is thoroughly investigated. The final results look at competing interests and how they can inform future community development activities.

M. Rezaul Islam, in Chapter 14, "Methods and framework of participatory action research for community development in Bangladesh," provides discussion on methods and participatory action frameworks as well as different contextual aspects of these methods. Using Bangladesh's high rate of participatory action research, this case study reviews community development, participation and empowerment processes and how they compare to other methods of social inquiry.

Many community developers will define success based on the health of a community. In Chapter 15, "Building a healthy community: the Coastal Georgia Indicators Coalition" by Patsy Kraeger, the literal meaning of health and how it applies to well-being in the community is explored. Using the Coastal Georgia Indicators Coalition (CGIC) and its broad reach, community development outcomes and additional resources are leveraged in this case. The chapter explores how participatory community development theories are used by the CGIC not only as a process but for informed community actions leading to quantifiable outcomes. Understanding how these apply to the CGIC may improve outcomes of community health, well-being, economic opportunity and overall quality of life.

Chapter 16, "Social indicator projects for rural communities: the case of the Northwoods Quality of Life Database" by Brandon Hofstedt, provides a framework for community indicators and their acceptance as data analytics and drivers of decisions. Using the Northwoods Quality of Life Database (NWQoL), a mix of primary and secondary data points explore overall conditions in rural America. This indicator process provides the conceptual framework, geographic scope, technical organization and data collection processes.

Food deserts is a relatively new term that has been integrated into current discussions of community development, equity and socioeconomic parity. In Chapter 17, "An exploratory study of food deserts in Utica, Mississippi," Talya D. Thomas discusses how food deserts applies to rural and urban communities. The chapter provides a framework for causes of food deserts and how communities are improving and increasing access to healthy foods and illustrates concepts by the use of a case of a small town in Mississippi.

Socioeconomic characteristics can have a causal impact on satisfaction of the overall community. Chapter 18, "Impact of socioeconomic characteristics on neighborhood environment satisfaction in deteriorated areas" by Mostafa Norouzi, Abolfazl Meshkini and Somayeh Khademi, examines the importance of how socioeconomic characteristics in the Ab-Kooh neighborhood, a deteriorated area in Mashhad, impact residents' satisfaction and perception of their community life. Factors such as density, relationships and concentrations can have a significant impact on socioeconomic characteristics. Review of recent research through the collection of questionnaires and analytics provides the basis for conversation.

Chapter 19, "Downtown revitalization, livability and quality of life in Tucson, Arizona" by Carlos J.L. Balsas, looks at the other large metropolitan area in Arizona and the overall quality of life in this urban region. Using this case study, discussion is provided on the lessons learned through the revitalization efforts in Tucson. Specifically, this chapter reflects on Tucson's urban and regional transformations and applies a quality of life and livability framework.

PART III: EMERGING CONSTRUCTS AND THE FUTURE OF COMMUNITY DEVELOPMENT RESEARCH

Our final part includes eight chapters. We seek to look at newer or current issues that impact community development research. While much is unknown about what the future may hold, these chapters represent innovative ways of thinking about community development while also recognizing that emerging issues will impact what we can or will do as we face the future.

Transitioning into Part III, Chapter 20, "Theories and concepts influencing sustainable community development" by Maria Spiliotopoulou and Mark Roseland, provides a journey of sustainable development and its intersection with community development as well as other notable areas such as ecology, economics and other social and natural sciences. The chapter also transitions into a concept of community productivity and what factors serve as productive catalysts of well-being.

Chapter 21, "Re-imagining community development: the Cocoa360 model" by Shadrack Frimpong, Allison R. Russell and Femida Handy, looks at the application of enterprising strategies enhancing community development. Using Cocoa360, a nongovernmental organization in rural Ghana, this chapter provides analyses on how innovations can support grassroots progress. A case is made to utilize this approach for other development projects in rural agriculture-based communities.

Social media has influenced connectivity with positive and negative outcomes; from the reduction of certain types of social capital to bringing empowerment to entire national movements. Chapter 22, "Community development and place attachment using an inductive social media approach" by Justin B. Hollander and Max Page, explains how software and social media can be used to strengthen community engagement. They use Holyoke, Massachusetts as a case study to provide a framework around common social media platforms of Twitter and Flickr, and explore how these platforms impact attitudes about place attachment.

Mobilizing the community is critical to the success of community development efforts. Chapter 23, "Re-imagining democratic research processes in community-based

8 *Research handbook on community development*

development: a case for photovoice" by Camille Sutton-Brown, provides analysis on how beneficiaries of community development policies must be involved in all aspects of the process. The author presents participatory models of research as providing an alternative to top-down approaches, and stresses the importance of this approach that incorporates photovoice technology. Impacts on community outcomes and increasing power dynamics for final decision-making processes are included as important features of a democratic process.

Chapter 24, "Centering aesthetics in community development: approaches from the Banff Centre for Arts and Creativity" by Jerrold McGrath, looks at how creativity is used in community development approaches. Using the case of the Banff Centre for Arts and Creativity, in Alberta, Canada, this chapter draws on multiple forms of architecture to video games design. Using an approach of positive impacts of positive interactions, this chapter analyses the subjective experiences of participants and how these dynamic behaviors can inform better approaches to work toward community solutions.

Our final selection of the following three chapters is grouped intentionally together as all explore relationships of higher education and community development. It is fitting to end the volume with discussions of higher education as the future will surely provide more opportunities to further develop curriculum, scholarship and research in community development. At the intersection of theory and practice, higher education's role in community development will continue to have a significant influence.

Anchor establishments can drive public policy and progress in connected communities. Universities plan a critical role with successful community outcomes. Chapter 25, "The new role of the university in community development" by Graciela Tonon, frames the interaction of multiple diverse groups and sharing of knowledge among these participants. This chapter looks at the role of the university, the democratization of knowledge, and respect for a diverse group of stakeholders participating in the process. Finally, arrangements between the university and community are discussed as possible approaches for future partnerships.

Chapter 26 looks at "Community innovation and small liberal arts colleges: lessons learned from local partnerships and sustainable community development" and is by Craig A. Talmage, Robin Lewis, Kathleen Flowers and Lisa Cleckner. This chapter brings forward a robust discussion of how smaller liberal arts colleges represent innovation within their communities. This includes themes of collective impact, design, innovation and entrepreneurship for community change and disruption. They utilize a case study of the efforts of a liberal arts college in upstate New York and how it approaches innovative solutions with the community.

Finally, utilizing a foundation of community development, the Community Capitals Framework (CCF) can guide relationships between a university and its host community. In Chapter 27, "Sustaining an urban education pipeline: a case study of university and community development partnership," Gloria Bonilla-Santiago explores how education and institutions can transform a neighborhood in distressed areas of a city. From the beginning, the university and community relation can be "reciprocal, collaborative and respectful of community life." This chapter examines a 25-year research project of the Rutgers/LEAP pipeline and its impact on community outcomes.

In closing, community development is a dynamic and evolving field of research, with theoretical and conceptualization dimensions, as well as applied practice and reflection.

The process is just as critical as the outcomes for overall success of community engagement, programs, projects and evaluation. Understanding the nexus of practice and research with applied research to co-create this *Handbook*, the editors recognize that applied research for developing a comprehensive theoretical base is often practice oriented. As mentioned earlier, practice oriented research contributes to the depth of understanding of the field of community development from multiple perspectives.

This *Handbook* seeks to reduce uncertainty in the field through these various perspectives. Keeping the community as the priority throughout all aspects will remain critical while researchers seek to expand to other needed aspects of exploration for community development and linkages to myriad dimensions important to collective human well-being and development. For example, continued evolution of technology in rural and urban communities will surely need much attention with the advent of artificial intelligence and as other life-altering technologies emerge. There is also much need for more research and its implications for application in the domains of subjective and other intangible feelings, preferences and values reflected in individual and collective community well-being and quality-of-life dimensions. It is important to consider both the objective (as often expressed in empirical or quantitative data and analyses) and the subjective dimensions of communities and the people who live there. In other words, it will always be instrumental to balance the variables of analysis and the tools of community decision-makers with the needs of community residents. Let us work together continuing to build on the insights within this volume for enhancing community development research both now and for the future betterment and well-being of the places we live in.

REFERENCES

Bhattacharyya, J. (1995). Solidarity and agency: rethinking community development. *Human Organization*, 54(1), 60–69.

Brocklesby, M.A., and Fisher, E. (2003). Community development in sustainable livelihoods approaches: an introduction. *Community Development Journal*, 38(3), 185–98.

Candy, L. (2006). *Practice Based Research: A Guide*. CCS Report, 1, 1–19. Sydney, Australia: University of Technology.

Candy, L., and Edmonds, E. (2018). Practice-based research in the creative arts: foundations and futures from the front line. *Leonardo*, 51(1), 63–9.

Emery, M., and Flora, C. (2006). Spiraling-up: mapping community transformation with Community Capitals Framework. *Community Development*, 37(1), 19–35.

Feldman, M.S., and Orlikowski, W.J. (2011). Theorizing practice and practicing theory. *Organization Science*, 22(5), 1240–53.

Flora, C., Flora, J., and Fey, S. (2004). *Rural Communities: Legacy and Change* (2nd ed.). Boulder, CO: Westview Press.

Gilchrist, A. (2019). *The Well-Connected Community: A Networking Approach to Community Development*. Bristol, UK: Policy Press.

Hickey, S., and Mohan, G. (2004). *Participation: From Tyranny to Transformation? Exploring New Approaches to Participation in Development*. New York: Zed Books.

Kenny, S., McGrath, B., and Phillips, R. (eds). (2017). *The Routledge Handbook of Community Development: Perspectives from Around the Globe*. Abingdon, UK and New York: Routledge.

Korten, D.C. (1980). Community organization and rural development: a learning process approach. *Public Administration Review*, 40(5), 480–511.

Kretzmann, J., and McKnight, J. (1993). *Building Communities from the Inside Out: A Path Toward Finding and Mobilizing a Community's Assets*. Evanston, IL: Institute for Policy Research.

Kretzmann, J., and McKnight, J.P. (1996). Assets-based community development. *National Civic Review*, 85(4), 23–9.

Mathie, A., and Cunningham, G. (2003). From clients to citizens: asset-based community development as a strategy for community-driven development. *Development in Practice*, 13(5), 474–86.

Mohrman, S.A., and Lawler, E.E. (2011). "Research for theory and practice." In S.A. Mohrman and E.E. Lawler, *Useful Research: Advancing Theory and Practice*, pp. 9–13. San Francisco, CA: Berrett-Koehler Publishing, Inc.

Musikanski, L., Phillips, R., and Crowder, J. (2019). *The Happiness Policy Handbook: How to Make Happiness the Purpose of Your Government*. British Columbia: New Society Press.

Phillips, R. (2003). *Community Indicators*. PAS Report No. 517. Chicago, IL: American Planning Association.

Phillips, R. (ed.) (2005). *Community Indicators Measuring Systems*. Aldershot, England: Ashgate.

Phillips, R., and Pittman, R. (2015a). *Introduction to Community Development* (2nd ed.). London: Routledge.

Phillips, R., and Pittman, R.H. (2015b). "Measuring progress." In R. Phillips and R. Pittman (eds), *Introduction to Community Development* (2nd ed.), pp. 346–62. London: Routledge.

Trevan, E.S. (2016). The influence of import substitution on community development as measured by economic wealth and quality of life (Doctoral dissertation, Arizona State University).

Wheeler, S.M. (2015). "Sustainability in community development." In R. Phillips and R. Pittman (eds), *Introduction to Community Development* (2nd ed.), pp. 72–88. London: Routledge.

PART I

FOUNDATIONS

1. Weaving reflection, action, and knowledge creation: lived experience as a catalyst into the cycle of praxis for community development
C. Bjørn Peterson, Craig A. Talmage and Richard C. Knopf

INTRODUCTION

Community Development (CD) scholars have called for a more robust engagement of the interplay of values, practices and knowledge creation (e.g., Crowley, 2018; Hustedde, 1998; Peterson and Knopf, 2016; Talmage et al., 2017; Westoby, 2016). As Bhattacharyya (2004) notes, CD is a *teleological* practice. Much like Public Education and Social Entrepreneurship, CD as a practice is based on an aspirational future social order (Bhattacharyya, 2004) and therefore requires its practitioners to articulate the future they hope to bring about (Bhattacharyya, 2004; Peterson, 2016). Any such articulation must include a discussion of values (Crowley, 2018).

Niall Crowley at the 2018 World Community Development Conference in Maynooth, Ireland conveyed the need for CD to be *values-led* (Crowley, 2018). He described values as ideals that are important to scholars and practitioners. These values can be personal and institutional, dictating what scholars and practitioners prioritize, engage and oppose. He also expressed that values are under pressure, as many scholars and practitioners must deal with challenges such as funder demands and regimes, current public discourse and overall stagnation. Still, he emphasized that values can encourage innovation and engagement in CD as a field and practice.

Community Development is the means by which adherents hope to achieve their aspirational future order. In this light, Welzel, Inglehart and Deutsch (2005) highlight "It is the values to which an activity is tied, not the activity as such that makes a society civic" (p. 140). Discussion of a specific approach to practicing CD should offer an outline of the way values are mobilized in pursuit of the envisioned future (Crowley, 2018; Peterson and Knopf, 2016; Peterson, 2016). Reflection on the values behind a given approach becomes *praxis* when that reflection intentionally interacts with action to create new knowledge – including a new understanding of a group or individual's capacity and self-story. That new knowledge informs the evolution of the teleological vision and the future intentional reflection and action taken in its pursuit (Peterson and Knopf, 2016; Peterson, 2016).

The lived experience of community members is a rich source for understanding their collective behavior as it reveals a community's collective self-perception, self-story and self-image (Bergdall, 2003). Storytelling has been shown to be a catalytic act for enhancing community participation in community building activities (Pstross et al., 2014), an emancipative expression of agency that signifies commitment to solidarity within a community (Talmage et al., 2017), and a methodological vehicle for understanding sense-making and action-taking in a CD environment (Peterson, 2016; Peterson and Knopf, 2016).

Storytelling places lived experience at the center of CD practice as key to participation and empowerment within CD (Ledwith and Springett, 2010).

Tension may be found between asset-based approaches that emphasize relatively quick action-taking (e.g. Bergdall, 2003) and critical-consciousness-raising approaches that encourage in-depth critical social analysis prior to taking action. Critical social analysis is understood here as dialogical learning done by community members in order to interpret lived experience with attention given to the role of asymmetrical power relations among individuals and groups of people and the shape and force of institutional oppression or marginalization. This chapter will demonstrate the compatibility of critical social analysis with asset-based action-taking in service to a teleological vision of agency and solidarity. However, it can be daunting, when attempted in an ad hoc fashion, to weave critical social analysis of lived experience, with action that promotes, protects and restores agency, and then incorporates the newly generated knowledge into future reflection and action. In facilitating the rhythms of this form of praxis, adopting a cyclical approach can be helpful. We suggest a cycle of praxis for facilitating catalytic CD. Therefore, a new approach to CD will be outlined in this chapter by noting the teleological vision to which the approach aspires (Dignity And Well-being For All), and the application of core values to specific practices (the Cycle of Praxis for Community Development). Before that introduction, a review of pertinent historical framing in the CD literature is in order.

THEORETICAL FOUNDATIONS

Solidarity and Agency

One of the challenges of discussing theories of CD emerges from CD's dual manifestations as both a field of study and a social practice. A theory of CD is dependent on the assumptions of its adherents. A theory can try to explain a social phenomenon, or, as Bhattacharyya (2004) observed, theories can be teleological:

> charters for action towards a goal, such as theories of democracy, freedom, equality, etc. where the purpose or the end reflexively enters the causal stream, urging, when necessary, modification of our action. The purpose of building a rocket, for instance, cannot do that; it cannot alter the laws of physics. Democratic theories are not like the laws of physics. They are not explanations but they elaborate a vision of a kind of social order. A theory of CD is of this kind. It advocates a particular kind of social order and a particular methodology for getting there. (Bhattacharyya, 2004, p. 10)

Scholarship undertaken in order to introduce and advocate for an approach to CD, as this chapter intends to do, must relay the teleological vision to which it aspires. In Bhattacharyya's words, we must name the "particular kind of social order" and "particular methodology for getting there" (2004, p. 10). The particular methodology we will explore is a process we shall refer to as the "Cycle of Praxis for Community Development." Before we can describe that methodology, we must first describe the goal towards which we are taking action. To do so, we align our vision for a "particular kind of social order" with Bhattacharyya's definitional theory of CD.

Bhattacharyya (1995) began to explore the definition of CD because he wanted to make a distinction. In the world of "helping" professionals, it can be difficult to distinguish the buzzwords from the substance. The advent of social entrepreneurship and push for corporate responsibility meant that while some businesses have actually made meaningful social impact, a number of neo-liberal endeavors have justified putting profits over people and the planet by dressing up their business in the language of value-added, empowerment and sustainability (Ihlen and Roper, 2014; Imran et al., 2014; Peterson and Knopf, 2016). Bhattacharyya (1995) recognized that a more substantive definition was required to set apart the actual work for dignity and well-being from the work that simply adopted its language.

Bhattacharyya (1995; 2004) landed on two key concepts: Solidarity and Agency. Regarding the first concept, one of his major conclusions was that communities are better identified by their relative quality of "Solidarity" than by location or other geographic descriptors. Solidarity in this context means shared values and norms for behavior, which when broken, impact community members in deeply felt ways (Bhattacharyya, 2004). This definition of community has the advantage of describing not only place-based communities, but also those that are virtual or spread out across political and geographic boundaries.

Bhattacharyya (1995; 2004) then described development in terms of "Agency." Agency, according to him, is one's capability to order one's life as one sees fit. In other words, regardless of one's social or economic status, a person should be able to act or refrain from acting in ways that maintain their dignity and well-being. This capability is missing from a person's life when they lack the basic necessities for life, experience dehumanization or are otherwise marginalized or oppressed (Bhattacharyya, 2004). Producing, protecting and restoring agency should be central to any work for dignity and well-being. However, agency at a community level is complicated by the intersecting of each member's individual makeup of identity and shared norms.

A member of a community does not experience a sense of solidarity with one group, but many (Peterson, 2016). A person may share a sense of identify with other groups based on factors such as their place, gender or race, among others. And their membership with different solidarity groups will be experienced as more or less salient across time and space (Owens et al., 2010). For example, regarding the authors' shared sense of solidarity with Arizona State University's Partnership for Community Development each author may have different feelings and experiences regarding the shared identity and norms for behavior. When on the campus of ASU, one of us may be proud of our identity and norms while the other two are ambivalent. But when we attend a conference as a group, the salience of our shared identity and norms for behavior may take on a different personal experience, resulting in a shared sense of pride in who we are and how we act (regardless of the extent to which anyone else knows who or what the Partnership is). The point of this observation is to foreground the complicating factors in operationalizing the production, protection and restoration of agency that Bhattacharyya advances as the purpose of CD.

Attending to a group's shared agency is difficult as each group member experiences the shared identity and norms for behavior (solidarity) in differing ways. Understanding the lived experience of group members is key to understanding their individual and collective behavior as "a matter of self-perception, self-story, and self-image" (Bergdall, 2003, p. 2).

Bergdall asserts, "though we would like to have community building 'from the inside out' occur spontaneously, some form of an external stimulus is usually involved" (2003, p. 2). So how do CD practitioners navigate their roles as outsiders seeking to provoke change while respecting and protecting community and individual agency?

Bhattacharyya (2004) suggests three principles of Solidarity and Agency that can inform CD methods in a manner that emphasizes an inside-out process: Self-Help, Felt Needs and Participation. The first principle of producing, promoting and restoring Solidarity and Agency is the principle of "Self-Help". Self-Help, in this context, refers to the right of a person to have agency over how their dignity and well-being is defined, created and protected. Rather than being dependent on or subject to the values and norms of some other source of power, Self-Help as a principle of Solidarity and Agency suggests an approach to CD that respects and trusts the right and capability of people to order their lives.

This sort of Self-Help is not a denial of human interdependence. Instead: "The principle rests on a concept of human beings that when healthy they are willing and able to take care of themselves, to reciprocate, to be productive, more predisposed to give than receive, are active rather than passive, and creative rather than consuming" (Bhattacharyya, 2004, p. 22). This echoes Freire's (1970/2012) insistence that people can and should be trusted to order their lives. It is a rejection of the worldview that people can be divided into makers and takers, where the takers outnumber and feed off of the production of just a few. Self-Help is not an abdication of responsibility toward one another – quite the opposite. It calls the community to partnering based on mutuality and reciprocity.

The intersection of the value of self with values in CD can be found in scholarship from the 2019 World Values Survey (Welzel, 2010; Welzel and Deutsch, 2012; Welzel and Inglehart, 2008; 2010; Welzel et al., 2005). For example, Welzel, Inglehart and Deutsch (2005) posited that self-expression values and elite-challenging action strengthen democratic institutions. Such self-expression values and elite-challenging action are termed by these researchers as emancipative social capital.

Talmage et al. (2017) extended this theory to the individual psychological. Using punk rock as a theoretical device and previous work on psychological social capital (Perkins et al., 2002), they proposed a new theory of emancipative psychological social capital containing both individual cognitions and actions. Self-expression values are found in individuation – individuals understand their uniqueness in a community – and expression – empowered individuals take action based on their uniqueness. Elite-challenging actions are found in opposition – feeling in conflict with the current status quo – and protest – taking individual action to challenge the status quo. Thus, individual action and feelings work together to strengthen communities.

The second principle of producing, promoting and restoring Solidarity and Agency is "Felt Needs" (Bhattacharyya, 2004). Felt Needs are those articulated by the community itself. They are connected to the community's sense of shared values and norms for behavior, as well as the community's sense of agency. At times people in power turn to outside expressions of the challenges a community faces when setting policy or articulating vision. In the process, the lived experiences of community members can become marginalized or forgotten altogether. Policy and vision that are constructed apart from the Felt Needs of community members potentially rob them of their agency (Bhattacharyya, 1995; Freire, 1970/2012).

Finally, the third principle of producing, promoting and restoring Solidarity and Agency is "Participation" (Bhattacharyya, 2004). Participation in this context goes far beyond the kind of limited input that is passed off as community engagement in too many public spaces. Participation as a principle of Solidarity and Agency means inclusion that protects the ability of community members to order their lives in the way they see fit.

On the psychological social capital level, Participation relates to feelings or cognitions such as sense of community and collective efficacy. Citizen participation also relates to informal neighboring behaviors (Perkins et al., 2002). Still, these models can benefit from greater foci on inclusion, inclusivity and inclusiveness. Recent evaluation and indicator models have emerged focusing on diversity, inclusion and inclusiveness (e.g., Talmage et al., 2017), but much more work is needed on the psychological level.

Inclusive participation is a difficult task for many who work in community. There are many ways to justify exclusive practices. Participation is too often reduced to a political tool for giving cover to exclusive processes (Block, 2009). Participation that protects and restores agency means adopting uncommon practices and holding oneself to high standards of inclusion. Bergdall (2003) suggests that such practices require a quality of presence more akin to an art than a science. Trust in a community's ability to make sense of their world and act as they see fit is central to any CD practice that promotes, protects and restores agency (Bergdall, 2003; Bhattacharyya, 2004; Ledwith and Springett, 2010; Peterson, 2016). The facilitation of reflection on lived experience can be a catalyst for intentional agency-promoting action on the part of a community (Bergdall, 2003; Ledwith and Springett, 2010).

Tension between Critical Consciousness-Based and Asset-Based Approaches

As the above has made clear, adopting an approach to CD that produces, promotes and restores Solidarity and Agency involves trust in the capacity of community members to understand and respond to their circumstances. Freire (1970/2012), working in Brazil in the mid-20th century, called this capacity "critical consciousness." He asserted that developing critical consciousness required a unique kind of learning, distinct from what he referred to as "Banking Education."

Banking Education relies on pedagogies that treat students like metaphorical banks. Information is "deposited" into the minds of students whose learning is based on their capability to regurgitate the information on demand, and in the form it went in. Students have little say in what or how they learn. For those in power, this form of education has the benefit of spreading some basic forms of literacy without developing widespread critical consciousness. Furthermore, Banking Education promoted people's self-image as deficient, without agency and incapable of Self-Help (Freire, 1970/2012).

Freire's critique of Banking Education went further than chastising his ideological adversaries. He saw Banking Education as a severe lack of trust in the capabilities of the average person. He decried what he called "activism" on the part of political actors because he saw in their work the same lack of trust in people that the current regime practiced, only with a different view of social order. At the same time, he chided the philosophers whose academic reflection on inequity rarely translated into meaningful policy change. He argued that true revolutionaries who wanted to work for a more equitable society needed a dynamic, robust habit of intentional reflection and mindful action, based in trust of the capacities of those experiencing oppression (1970/2012).

And yet, Freire was not simply a populist. He had strong opinions on which social systems ought to be adopted. But, he also believed that the average person who was being marginalized and oppressed did not need these opinions forced upon them. Instead, he promoted the idea of "Dialogical Education" as a means of raising critical consciousness (Ledwith and Springett, 2010).

Dialogical Education relies on methods of learning that empower the learner to lead their own education. The aim is to develop a person or group's critical awareness of the ways in which resources and power are distributed, consolidated and manipulated (Ledwith and Springett, 2010). In pursuing this knowledge, Freire believed that communities could be trusted to create a more equitable, responsible and inclusive society. The work of "revolutionaries" was to remove barriers to learning, encourage critical- and systems-thinking, convene opportunities for oppressed peoples to gather and reflect, and to insist on trusting the transformative power of Dialogical Education (1970/2012). The pursuit of this knowledge can be thought of as "critical social analysis." The quality being pursued through critical social analysis is critical consciousness. A critical-consciousness-based approach to community building is therefore deeply based in the community's lived experiences, intersecting identities, and self-image (Ledwith and Springett, 2010).

It may be argued that critical social analysis encourages passive blame-assigning rather than transformative empowerment. That would be a misunderstanding of Freire's practice and assertions. He believed Dialogical Education that engaged with multiple ways of knowing while questioning the taken-for-grantedness of the status quo would lead to new ways of acting (Ledwith and Springett, 2010). Likewise, asset-based processes could be critiqued as being insufficiently reflective on structural and institutional oppression. However, it is not necessary to ignore or deny problems and deficits in order to take an asset-based approach (Bergdall, 2003). The catalytic power of critical consciousness as an outcome of Freirean praxis should be understood as a similar process as that to which Bergdall refers to when he says, "Radical change occurs when an established image is replaced by a totally new self-understanding" (2003, p. 3). Dialogical Education becomes praxis when the reflection is married to action. In the practice of Asset-Based Community Development (ABCD) as a catalyst for inside-out community building, Bergdall (2003) suggests that practitioners see themselves as catalysts for the identification and recognition of strengths and capacities. This is exactly what Freirean praxis hoped to accomplish through transformative learning (Ledwith and Springett, 2010). Furthermore, the analysis of power need not lead asset-based approaches into a deficit mind-set. Indeed, if Bergdall (2003) is correct in his assertion that catalytic facilitators of ABCD enable a community to look at themselves realistically, then analysis of power cannot be left out. There need not be a divide between critical social analysis with its examination of power structures and political critique and democratic, deliberative, loving, hope-filled methods. Reflexive praxis is needed with any CD method if it is to resist the gravity of entrenched power dynamics.

The centrality of self-reflection and deepening understanding of one's lived experience showcases commonality, which can be found in both asset-based and critical-consciousness-raising approaches. To embrace the concerns of asset-based approaches to CD (to avoid deficit thinking, self-victimization, passive analysis) while honoring community members' lived experiences (including systematic and institutional marginalization and oppression)

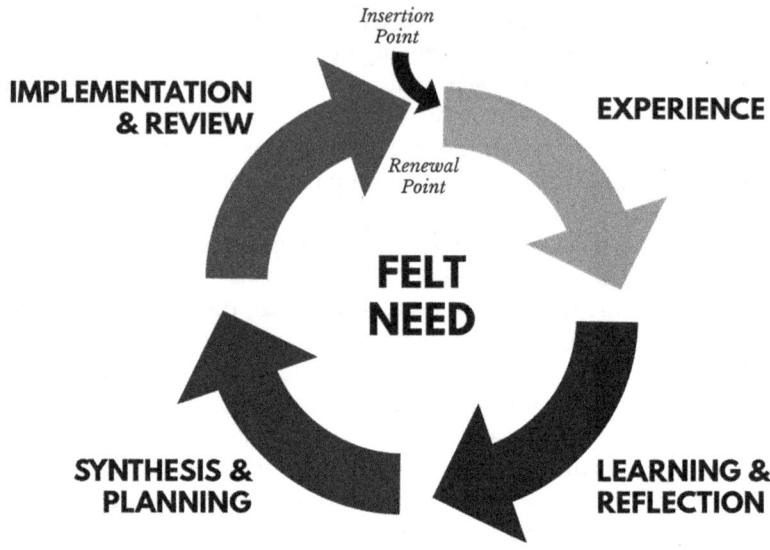

Source: Peterson (2018).

Figure 1.1 The Cycle of Praxis for Community Development

it is suggested that rhythms of action, reflection and knowledge creation be established. In order to facilitate an asset-based approach to CD that takes seriously lived experience and Dialogical Education through critical social analysis, the Cycle of Praxis for Community Development (which we sometimes shorten to Cycle of Praxis or CPCD) is presented as a new method for pursuing the promotion, protection and restoration of solidarity and agency (Figure 1.1).

INTRODUCING THE CYCLE OF PRAXIS FOR COMMUNITY DEVELOPMENT

Catalytic Action through a Cycle of Praxis

The assumption behind this approach is that an actor, to whom we will refer as the "initiator," is taking the initiative to engage a community in the production, protection and restoration of Solidarity and Agency. For the purpose of this chapter, we will assume that the initiator is a member of some community-based group (perhaps civic or nonprofit in nature) who has identified at least two to three community members (other than themselves) with whom they would like to engage on a community-relevant topic. Perhaps the initiator has heard from community members about a common complaint or perceived opportunity. Perhaps the initiator has an organizational mandate that directs their interest. In any case, for the approach we are about to lay out to work, the initiator's main concern should be to steer toward the production, protection and restoration of Solidarity and Agency. That means an abiding respect for and trust in the participants.

It is important to acknowledge an inescapable tension for the initiator; your own interests and perspectives should be secondary to those of the community *and* you are often in a position of power and influence that is not shared by the other participants. This is one reason why Dialogical Education that practices critical social analysis is an important aspect of CD: critical consciousness on the part of our community partners is a check on the power and privilege that we enjoy as initiators of community processes. Naming the inequitable power structures in our own practices does not invalidate the work of initiators, but rather invites co-ownership of future processes and structures with community members. There is much more to be said on this and we commend readers to a variety of sources for good reflection and action (see, e.g., Freire, 1970/2012; Ledwith and Springett, 2010; Ledwith, 2011; Peterson, 2016; Pstross et al., 2014; Westoby, 2016).

Praxis is characterized by intentional reflection, mindful action and the willingness to learn from our ongoing reflection and action in order to form new understandings of the world and our experiences of it. Ongoing reflection, action and knowledge creation give a rhythm to CD that can otherwise become reductive or stale. The rhythms of praxis invite us to attend to what has been called the "soul" of CD by provoking reflection on multiple ways of knowing and elements of mystery within the fabric of community (Hustedde, 1998; Ledwith and Springett, 2010; Westoby, 2016).

The approach presented here is called a *Cycle* of Praxis in order to spotlight the ongoing, iterative and seasonal nature of good praxis (Peterson, 2018). It is a prompt to facilitate rhythms of praxis in the work to produce, promote and restore Solidarity and Agency. Praxis-based approaches emphasize personal experience as the basis for reflection (Freire, 1970/2012; Ledwith, 2011; Pstross et al., 2017). In contrast to approaches that have downplayed the importance of the body or the political nature of the mundane or domestic life, the CPCD is centered on the lived experiences of people (Peterson, 2018).

At the same time, we hold in tension the paradox that our lived experiences are true even as our understanding of those experiences is dynamic. Making sense of our lived experiences and choosing our response to those interpretations is a core task of a human being. So, lived experiences are the "Insertion Point" into the CPCD, and remain a touchpoint for walking the CPCD.

In the midst of this tension, the initiator convenes a catalytic process wherein shared experience, critical social analysis and asset-based thinking link in an ongoing rhythm. We call this process walking the *Cycle of Praxis for Community Development*.

Walking the Cycle of Praxis

As we walk the CPCD for Community Development we engage in intentional "moments." Each "moment" of the CPCD is meant to facilitate some aspect of action and reflection to encourage deeper understanding of lived experiences and provoke the group toward more impactful, meaningful responses to the "Felt Need."

In the first "moment" of the CPCD, we explore a shared "Experience" in order to identify and articulate a "Felt Need." Recall that for this chapter, we are assuming that an initiator has convened two to three or more community members to participate in a process of action and reflection. At this point, the initiator reveals in greater detail to the convened community members their reason for extending an invitation. It is most important to reveal any agendas beyond convening at this point, particularly any organizational mandates, so that community members can understand what lies behind your actions. However, the most important part of this "moment" is to allow for storytelling around the shared experience or common theme.

Perhaps the initiator has brought together a group to address the growing pollution of a local waterway or because they have a mandate to engage the community around issues of transportation. Whatever the initiator suspects is deeply felt among the community members as regards to their Solidarity and Agency, the task at this moment is to convene storytelling that acts as a catalyst for identifying a shared "Felt Need" (e.g. Pstross et al., 2014). The "Felt Need" is the guide for reflection and action going forward. This emphasis helps ensure that future reflection, action and knowledge creation are meaningful in ways practical to the community's lived experience.

It helps in each moment of the CPCD to utilize focus questions in order to prompt dialogical engagement of people's lived experience and eventual response (Peterson, 2018). For the moment of "Experience," the initiator might adopt some or all of the following prompts:

1. What issue, opportunity or perceived experience led to the convening of this group?
2. Does the description given by the initiator resonate with those gathered? How is it right? What is it lacking? What does it add that you would not?
3. How deeply felt is this experience for you? Is this a new experience? An old one? Does it feel urgent or perhaps the opposite?
4. How does this experience affect you? Who else do you see it affecting?
5. If we as a group were to start a process of addressing this experience, could we start with who is here now? Why and why not? Who is missing?
6. Are there shared experiences among this group that warrant our continued meeting? How might we articulate those in a way that feels inclusive of our varied perspectives? Can we name a common "Felt Need"?

The second "moment" in the CPCD, after community members have shared their lived experience and articulated a "Felt Need," is a time of "Learning and Reflection." The task here is to deepen and broaden understanding of the lived experiences that are being addressed. The focus on Learning and Reflection represents an acknowledgment that lived experiences are true, but only part of the truth (Peterson, 2018).

Gaining deeper understanding of what has been experienced does not necessarily invalidate previous interpretations. Instead, the group is empowered in this process to critically reflect on what they have previously known. The outcomes of that reflection can range from affirmation, to amendment, to complete reform of their understanding of their lived experience and "Felt Need." The truth of lived experience can co-exist with whatever new knowledge is gained, even if it means holding in themselves the tension of competing truths.

For the moment of "Learning and Reflection" the initiator might adopt some or all of the following prompts:

1. What insights and analysis can give context and nuance to the Felt Need (Historical, Psychological, Socioeconomic, Cultural, Legal, Anthropological, Natural Scientific, and so on)?

2. What reflection and expression can give texture and depth to the Felt Need (Artistic, Religious, Philosophical, Wisdom, and so on)?
3. What technical knowledge or skills may be needed to better understand or respond to the Felt Need?
4. What interpersonal knowledge or skills may be needed to better understand or respond to the Felt Need?
5. As we gather new insights, what Learning and Reflection remain ambiguous or beyond our current scope? Can they wait or should they be addressed now?
6. As we suspect we are ready to move on to synthesis and planning, do we have a deeper understanding of the lived experiences first articulated?
7. With our new knowledge and deeper understanding, how would we outline our key Learnings and Reflections? How would we now articulate the Felt Need?
8. What assets, strengths and resources exist in our community, which might be mobilized for our purposes? (Peterson, 2018).

Notice that these prompts are designed to encourage discussion of how values, ways of knowing and knowledge creation interact contextually. The hope is to create critical consciousness regarding their epistemological and axiological assumptions.

The CPCD continues on through the next "moment" known as "Synthesis and Planning." The task here is to synthesize lived experiences with the new insights, Learning and Reflection gleaned during the CPCD thus far. The turn toward planning is the mobilization of new knowledge in the form of strategic, mindful action as a response to the Felt Need.

A few prompts that may help in this moment of synthesis and mobilization:

1. Based on our articulated need and subsequent Learning and Reflection, what are our assumptions about a potentially effective plan? What diversity exists among our assumptions? Can we continue or is more Learning and Reflection required?
2. Based on our assumptions, what tasks need effort or completion in order to effectively respond to the Felt Need? How would we prioritize these tasks?
3. Given these tasks and the priority we have assigned them, who is willing to commit what in order to ensure their completion? On what timeline?
4. Whose assistance may be required in order to complete these tasks? How can we successfully recruit/invite their participation? (Peterson, 2018).

The final moment in the CPCD is known as "Implementation and Review." The overall task here is to take action on the plan we have developed to respond to the articulated "Felt Need." Because sustainable solutions to complex challenges are rarely achieved through one set of interventions, implementation is not judged merely on the successful resolution of a problem; but, rather, on the increasing capacity of community members to effectively respond to their Felt Needs (Peterson, 2018).

Potential prompts for this "moment" include:

1. Are the agreed upon tasks being engaged in meaningful ways? Are participants choosing to be accountable to their stated commitments?
2. Are the initial lived experiences being honored in the pursuit of our planned and implemented tasks?

3. What is the pursuit of our action tasks teaching us about our previous understanding of the Felt Need? Did we ask good questions about the meaning of the initial lived experiences?
4. What are the outcomes we are seeing? How do they confirm, challenge or surprise our planning assumptions?
5. Whose experiences ought to be added to our conversation in the next iteration of the cycle? Whose voice is missing from this process?
6. What have been our strengths in this process? What have been our challenges?
7. What new assumptions do we have about this process? About our lived experiences? About our next steps? (Peterson, 2018).

At this point, the CPCD returns to the "Insertion Point" where lived experiences are once again shared. Thus, the cycle begins again. The CPCD, for that reason, not only provides progress toward ultimate goals through direct action, but also through the creation of new experiences and knowledge as the CPCD goes into its next iteration. Action now folds into further future reflection, creating new knowledge, which leads to further action, and so on.

It should also be said that the initiator may or may not be needed in second, third and continuing iterations. The CPCD can be initiated organically by simply following the prompts shared above. Indeed, when the authors have taught this method to students and community practitioners, the encouragement has always been to carry forth the process independent of the initial context as a way to self-guide through action and reflection. Adherents have remarked that the rhythms of the CPCD offer useful guidance for the application of tools and techniques not specifically mentioned here. That is precisely the hope of the authors in this writing: a platform for catalytic, participatory brilliance by our communities.

The CPCD framework presented here is meant to offer general points of emphasis to help ensure that reflection does not remain stagnant, and action is not mindless. It can be understood this way: the CPCD is a rhythm to dance to; but ultimately, it is people's own lived experience, own sense-making and own creativity that make up the actual dancing. The more a community dances, the more powerful and creative their moves become. Do not dance once and ask what good the song was. Adopt a life of dancing, and the CPCD becomes a song by which to live.

CONCLUSIONS

The CPCD is intended as a catalytic framework that encourages communities to critically engage their understanding of lived experience and mobilize their sense-making into asset-based action-taking. This approach embraces the tension between action and reflection by encouraging practitioners to find momentum in the iterative rhythms it facilitates. It is animated by Bhattacharyya's principles of CD that produce, promote and protect Solidarity and Agency. Therefore, the CPCD is offered as a tool for sense-making and action-taking in response to Felt Needs. And in order to honor the lived experience of community members, Self-Help and Participation take both asset-based and critical-consciousness-raising characteristics. Future research is needed to document the implementation of multiple iterations of the Cycle so that this approach can be further refined.

REFERENCES

Bergdall, T. (2003). Reflections on the catalytic role of an outsider in asset based community development (ABCD). Evanston, IL: ABCD Institute at Northwestern University. Retrieved 12 November 2019 from https://community-wealth.org/content/reflections-catalytic-role-outsider-asset-based-community-development-abcd.

Bhattacharyya, J. (1995). Solidarity and agency: Rethinking community development. *Human Organization*, 54(1), 60–69.

Bhattacharyya, J. (2004). Theorizing community development. *Community Development*, 34(2), 5–34.

Block, P. (2009). *Community: The Structure of Belonging*. San Francisco, CA: Berrett-Koehler.

Crowley, N. (2018). Values-led community development. Remarks given at the World Community Development Conference, Maynooth, Ireland, 25 June.

Freire, P. (1970/2012). *Pedagogy of the Oppressed* (30th Anniv. ed.). New York: Continuum.

Hustedde, R.J. (1998). On the soul of community development. *Community Development*, 29(2), 153–65.

Ihlen, Ø., and Roper, J. (2014). Corporate reports on sustainability and sustainable development: "We have arrived." *Sustainable Development*, 22(1), 42–51.

Imran, S., Alam, K., and Beaumont, N. (2014). Reinterpreting the definition of sustainable development for a more ecocentric reorientation. *Sustainable Development*, 22(2), 134–44.

Ledwith, M. (2011). *Community Development: A Critical Approach* (2nd ed.). Chicago, IL: Policy Press.

Ledwith, M., and Springett, J. (2010). *Participatory Practice: Community-Based Action for Transformative Change*. Chicago, IL: Policy Press.

Owens, T.J., Robinson, D.T., and Smith-Lovin, L. (2010). Three faces of identity. *Annual Review of Sociology*, 36, 477–99.

Perkins, D.D., Hughey, J., and Speer, P.W. (2002). Community psychology perspectives on social capital theory and community development practice. *Community Development*, 33(1), 33–52.

Peterson, C.B. (2016). The metaphors we help by. (Doctoral dissertation, Arizona State University). ProQuest.

Peterson, C.B. (2018). The cycle of praxis for community development. Retrieved 8 August 2018 from https://static1.squarespace.com/static/59ef86d080bd5e81b8c93628/t/5a9f06ab652dea8c73920685/1520371405347/The+Cycle+of+Praxis+for+Community+Development.pdf.

Peterson, C.B., and Knopf, R.C. (2016). (Re)framing sustainable development: An ecological posture and praxis. *Community Development*, 47(1), 122–35.

Pstross, M., Talmage, C.A., and Knopf, R.C. (2014). A story about storytelling: Enhancement of community participation through catalytic storytelling. *Community Development*, 45(5), 525–38.

Pstross, M., Talmage, C.A., Peterson, C.B., and Knopf, R.C. (2017). In search of transformative moments: Blending community building pursuits into lifelong learning experiences. *Journal of Education Culture and Society*, 7(1), 62–78.

Talmage, C.A., Peterson, C.B., and Knopf, R.C. (2017). Punk rock wisdom: An emancipative psychological social capital approach to community well-being. In R. Phillips and C. Wong (eds), *Handbook of Community Well-Being Research* (pp. 11–38). Dordrecht, Netherlands: Springer.

Welzel, C. (2010). How selfish are self-expression values? A civicness test. *Journal of Cross-Cultural Psychology*, 41(2), 152–74.

Welzel, C., and Deutsch, F. (2012). Emancipative values and non-violent protest: The importance of "ecological" effects. *British Journal of Political Science*, 42(2), 465–79.

Welzel, C., and Inglehart, R. (2008). The role of ordinary people in democratization. *Journal of Democracy*, 19(1), 126–40.

Welzel, C., and Inglehart, R. (2010). Agency, values, and well-being: A human development model. *Social Indicators Research*, 97(1), 43–63.

Welzel, C., Inglehart, R., and Deutsch, F. (2005). Social capital, voluntary associations and collective action: Which aspects of social capital have the greatest "civic" payoff? *Journal of Civil Society*, 1(2), 121–46.

Westoby, P. (2016). *Soul, Community and Social Change: Theorising a Soul Perspective on Community Practice*. New York: Routledge.

World Values Survey (2019). *Publications*. Retrieved 12 November 2019 from http://www.worldvaluessurvey.org/WVSContents.jsp.

2. The study of poverty in places: scope, scale, and space

Elizabeth A. Dobis, Lionel J. Beaulieu and Indraneel Kumar

INTRODUCTION

Poverty is a topic that has garnered considerable interest on the part of the academic and policy communities for many years.[1] While a variety of federal initiatives have been established in hopes of reducing poverty, the fact remains that poverty continues to be part of the fabric of many places in the United States. As Figure 2.1 shows, after an initial decrease in poverty rates during the 1960s to a low of 11.1 percent in 1973, poverty has remained a persistent problem. Over this time period, the greatest improvements have occurred among single female-headed families. While recent data reveal that poverty rates are declining, 39.7 million people (12.3 percent) remain poor in the United States (as of 2017), a number that is comparable to that found in the early 1990s (Fontenot et al., 2018). Additionally, 12.8 million of these individuals living below the poverty line are children under 18 years of age, a child poverty rate of 17.5 percent (Fontenot et al., 2018). What continues to be a challenge in implementing more effective poverty-reduction initiatives is determining the mix of factors that serve as the root causes of poverty. Is it the outcome

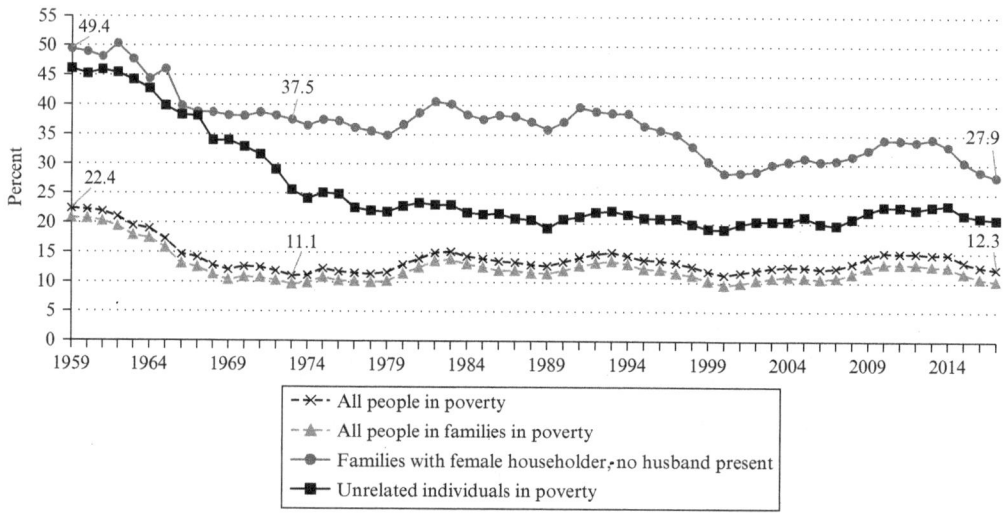

Note: Based on Haskins (2008) and authors' own calculations.

Source: Historical Poverty Tables, People and Families, 1959–2017; Table 2, US Census Bureau.

Figure 2.1 Historical poverty rates in the United States, 1959–2017

of structural shifts in the economy or the demographic composition of a community, the result of choices that individuals make in their life path, or some combination of both (Glasmeier, 2002)?

Bradshaw (2007) suggests that there are five theoretical strands in the literature on poverty. The first focuses on individual deficiencies among the poor, such as their limited educational attainment, their lower-level work-related skills, and their unhealthy lifestyle choices. The second explores the culture of poverty, the transmission of values, beliefs and norms over generations that keep individuals embedded in nonproductive subcultures. A third theory posits that poverty is the result of structural barriers with the social, economic and political systems that individuals experience, constraints that limit their opportunities to improve their well-being. A fourth theory states that poverty is the product of the spatial concentration of poverty in the United States, such as the Black Belt and Rio Grande regions of the South, or in disadvantaged cities that have suffered from the outmigration of middle-class residents to other places (Wilson, 1987). In some ways, Chetty and Hendren (2018) deliver a similar place-based message, that a child's economic mobility or capacity to escape poverty in adulthood is affected by the income level of their community. The final theory Bradshaw examines is the cyclical and cumulative interdependencies model of poverty. The central tenet of this perspective is that individuals and communities are caught in a spiral of opportunities and challenges (Bradshaw, 2007). As an example, if a community suffers from a major economic upheaval due to the loss of a major firm, the consequences are felt by employees, their families and a host of other community institutions, such as local government, the school system, social services and commercial enterprises serving the local residents.

In many respects, the five theories that Bradshaw explores can be reduced to two central arguments. One is that poverty is the result of a set of unfortunate decisions and behaviors of individuals, while a second is that poverty is the product of structural and social forces that most low-income individuals are simply unable to control or overcome (Tickamyer and Wornell, 2017). Those who embrace the first perspective are likely to push for policies and programs that target "people" while those subscribing to the second approach are inclined to seek "place-based" policies and programs for tackling poverty. The role that community development practitioners are likely to play is different in a people versus place-based policy perspective of poverty.

From a historical perspective, the dominant set of policy strategies has aligned with either the person-based or place-based orientation (Spencer, 2004). Academic research on poverty has also followed this pattern. Studies that examine the person-level factors of poverty analyse how individual and household characteristics, such as education, gender or family structure, and characteristics of the location in which the individual lives affect income levels or poverty status. The unit of analysis for these studies is either the individual or household. These studies are frequently performed using nationally representative samples (e.g., Brown and Hirschl, 1995; South et al., 2005; Thiede et al., 2015) or case studies of individuals in metropolitan areas such as Boston, Massachusetts (Katz et al., 2001).

Place-level research analyses how the environmental, economic or population characteristics of a location affect its poverty rate or level of income inequality. These locations are communities that are represented by a certain geographic unit of analysis, such as a census tract, county or metropolitan area, with various sized units embodying different definitions of community (e.g., neighborhood, city or economic region). Explanatory variables for these analyses are generally at the same level of aggregation, often employing proportions and rates to profile the features of localities. Some concepts of place, such as neighborhoods, do not have a standard definition. For example, Lynch (1960) proposed that individuals orient themselves in urban spaces via mental maps which use physical references such as paths, boundaries, districts and landmarks to distinguish cityscapes, built environment and neighborhoods. While limited literature has explored the delineation of neighborhood boundaries based on cognition and participatory research, Taylor (2012) highlights the challenges of defining "community" and "neighborhood" and how even residents can disagree on the boundaries of the neighborhood. To secure the data needed to analyse the characteristics of these places, researchers typically rely on standardized geographies, such as census block groups or tracts, as proxies for the neighborhoods.

Place-level poverty studies generally use national samples, often delineating differences across different geographic areas, such as metropolitan status (Partridge and Rickman, 2007a; Lichter et al., 2008). However, place-level analyses often restrict their data sample to certain classifications, such as rural counties (Partridge and Rickman, 2005), urban tracts (Ellen and O'Regan, 2008) or multi-state regions, such as Appalachia or the Midwest (Lichter and Campbell, 2005; Peters, 2012). Case studies also explore the spatial concentration of poverty in metropolitan areas such as Chicago, Illinois (Hu, 2014) or Columbus, Ohio (Holloway et al., 1999).

Both person-level and place-level research can be spatial or aspatial. Aspatial analyses, such as those by Holliday and Dwyer (2009) and Lichter et al. (2012), ignore the spatial relationships among units of observation, be they people or places. Spillover effects due to spatial proximity are not accounted for in these analyses. On the other hand, spatial analyses do account for the spatial relationships among observations. In these studies, the effect of relative proximity on people or places can be determined either through explanatory variables, such as adjacency, distance, or the characteristics of surrounding locations that represent the variation caused by spatial relationships (Crandall and Weber, 2004; Partridge and Rickman, 2007b) or through spatial models that consider the spatial dependence or spatial heterogeneity underlying the mechanisms affecting poverty (Peters, 2012; O'Connell and Shoff, 2014). These spatial models are important to analysing poverty given that communities are not isolated.

Poverty can be seen as intrinsically linked to the community capital assets of a community (Flora and Thiboumery, 2005). The most direct connection between poverty and community capital is through the level of human capital poor residents have and, subsequently, the jobs for which they are qualified based on their skills and knowledge (Flora et al., 2016). However, poor individuals are also affected by all other aspects of community capital (Flora and Thiboumery, 2005; Tickamyer and Wornell, 2017). For instance, they may experience the community's physical capital through the quality of the housing stock or coverage of public transportation, natural capital through the environmental quality of the neighborhoods in which they live, financial capital through their ability to obtain loans, and bonding and bridging capital through their relationships with neighbors and their integration into the larger community. Depending on each unique community's mix of various capitals, such as levels of pollution, dilapidated housing, social assistance and work opportunities, poor residents may experience both positive and negative aspects of community capitals (Tickamyer and Wornell, 2017). This link between community capital and poverty provides context for the study of poverty in places.

A clearly delineated national policy on poverty did not exist in the United States prior to the New Deal in the 1930s (Glasmeier, 2002). Poverty-reduction policies pursued from the 1930s to the 1950s tended to address efforts to stimulate the economy by investing in improvements of physical infrastructure, doing so through the establishment of quasi-governmental entities, such as the Tennessee Valley Authority, or seeking to assist the nation's poor through the creation of the Social Security and Unemployment Insurance system (Glasmeier, 2002; Spencer, 2004). During the 1960s, policymakers began to acknowledge that poverty was not distributed equally across space, that poverty was concentrated in different geographic areas of the country (Glasmeier, 2002). The result was a surge in regional economic development policies intended to tackle regional variations in poverty (Glasmeier, 2002). Furthermore, it was during this time that the War on Poverty legislation was enacted, which resulted in the formation of national policies that focused on urban and race-related poverty (Spencer, 2004).

Since the establishment of the War on Poverty, programs focusing on building human capital (e.g., Head Start and the Manpower Development Training Act), initiating community change (e.g., the Model Cities program), providing health insurance (e.g., Medicare and Medicaid), improving food access and nutrition (e.g., Food Stamps/SNAP), expanding access to affordable housing (e.g., public housing and vouchers) and increasing income (e.g., Social Security payments and the Earned Income Tax Credit) have been created or expanded in an attempt to reduce poverty (Haveman et al., 2015). Overall, these policies have been effective, particularly among elderly, disabled and black Americans (Fox et al., 2015; Haveman et al., 2015).

This chapter seeks to undertake a literature review of place-based poverty studies in the United States, focusing on those that employ a spatial approach. In particular, we explore the patterns of poverty in communities and the set of factors that may be correlated with those patterns. Such an example can be valuable in helping local decision-makers gain a better understanding of poverty and how it affects both the economy and society. We begin our review in the next section by discussing the approaches and methodologies used to study poverty, including the way in which poverty is defined and the disciplinary differences that exist with regard to the study of poverty. Next, in Space and Poverty, we examine the effect of space on poverty, exploring the geographic levels at which poverty is studied, regional and metropolitan-level variations in poverty and the role of migration, race and segregation in determining poverty rates. We conclude our chapter with a discussion of the implications of current poverty research for policymakers and practitioners, giving specific attention to possible strategies that community development practitioners may introduce in communities that are committed to alleviating poverty among their residents.

SCOPE OF PLACE-BASED POVERTY RESEARCH

This section establishes the framework for understanding place-based poverty research and the issues it explores. We start by defining poverty and the multiple ways in which researchers have defined it in their studies. Next, we discuss the different ways in which social science disciplines approach and analyse place-based poverty in the United States.

Types of Poverty

Poverty research typically utilizes the federal poverty line, the threshold at which a household is considered to be poor, to determine poverty status.[2] This measure was developed by the Social Security Administration in the 1960s, using data from the Current Population Survey, and is updated annually using the Consumer Price Index to adjust for inflation (US Census Bureau, 2017). Since 1980, the determination of whether a family household is living in poverty is dependent on total family income, family size and the number of children in the family. Prior to 1980, the sex of the family head and whether the family resided on a farm were also considered when calculating poverty thresholds.

Although there are discussions of whether a poverty measure developed in the 1960s remains both useful and accurate as a measure of poverty more than five decades later (Citro and Michaels, 1995), the metric does provide a consistent measure over time, which is ideal for studying poverty trends. A concern that is often raised with regard to the poverty line is its failure to take into account cost-of-living variations across space, possibly resulting in biased measures of poverty prevalence (Weber et al., 2005). Some argue that cost-of-living and regional price parity data are readily available through private and public sources and could be used to adjust the poverty measure.[3]

Once a family is determined to be below the federal poverty threshold, all family members are classified as poor, and it is this number that is used to calculate poverty levels and rates.[4] To study population subgroups, the total population of poor individuals is differentiated by personal characteristics such as age, race or ethnicity. Examples of research tied to population subgroups include studies of child poverty (Lichter and Johnson, 2007), elderly poverty (McLaughlin and Jensen, 1993), black poverty (Wilson, 2008) and Hispanic poverty (South et al., 2005) that frequently compare their results with those of other population groups or subgroups that are part of the non-poor.

The poverty line is the threshold used most frequently by researchers to determine whether an individual is poor or not, and geographic areas are classified as poor if a certain percentage of residents are below the poverty line. Investigators generally classify geographies with a poverty rate of 20 percent or higher as high poverty counties (Holliday and Dwyer, 2009), although other thresholds for determining poverty exist. For example, Partridge and Rickman (2005) define high poverty counties as those with poverty rates of at least 25 percent, whereas Crandall and Weber (2004) use 30 percent as their threshold.

Classifying poverty is not reliant solely on the poverty level in a given location at a specific point in time but also on changes taking place over a given time period. Peters (2009) conducts a cluster analysis of minor civil divisions (MCDs), an administrative sub-county census geography, using decennial poverty rates from 1980 to 2000. He classifies each MCD as having a low, below-average, average, above-average or very high poverty level, and whether that level is improving, worsening or relatively unchanged over time, resulting in a range of 12 different poverty classifications. Peters finds that clusters with similar poverty rates or time trends share similar economic or demographic characteristics. Ellen and O'Regan (2008) indicate, based on their analysis of urban tracts with varying poverty levels, that these differing characteristics may influence the effect that poverty policy has on a location. This could be valuable to lawmakers who are positioned to create policies that target specific locations, such as those suffering from persistent poverty. Furthermore, residents of such locations could respond in a different way to incentives and regulations than would those living in areas experiencing declining poverty rates.

When the poverty rate of a geography is high over a prolonged time period, that location is said to be persistently poor. The Economic Research Service (ERS) of the US Department of Agriculture started measuring persistent poverty in counties in the 1970s, calling them "persistently low-income" and defining them as the bottom quintile of counties by per-capita income, as calculated every decennial census since 1950[5] (Hoppe, 1985). By the mid-1990s, the current definition of persistent poverty had emerged, classifying counties with a poverty rate of at least 20 percent over the span of 30 years (as measured by four decennial censuses) as persistently poor (US Department of Agriculture, Economic Research Service, 2017b). In recent years, academic interest in the characteristics and causes of persistent poverty has increased. While researchers embrace the ERS definition as a basis for their research, they frequently expand their work beyond the ERS definition by adjusting the temporal or poverty rate thresholds (Miller and Weber, 2003; Partridge and Rickman, 2007a). Miller and Weber (2003) found that the number of counties that were classified as persistently poor was not particularly sensitive to changes in the temporal threshold but much more so to changes in the poverty rate threshold, with increases in the threshold substantially decreasing the number of persistently poor counties.

As Castle et al. (2011) discuss, persistent poverty may also refer to individuals who remain in poverty for long periods of time. For places, then, an alternate way of describing persistent poverty would be based on the number of individuals considered to be persistently poor. When analysing a longitudinal panel of single-mother families over five years in the mid-1990s, Fisher and Weber (2002) consider families in poverty for three or more years to be persistently poor and find that the highest levels of persistent poverty occur in central metropolitan and remote rural counties. These spells of poverty need not be consecutive to be meaningful. While Mattingly et al. (2011) focus on persistently high rates of child poverty in their analysis, they also classify counties by the number of decennial censuses in which they have a child poverty rate greater than 20 percent: persistent high (four consecutive censuses), frequent high (three of four censuses), intermittent high (two of four censuses) and infrequent high (one of four censuses).

Segregation indices are another measure commonly used to research poverty. These indices place a numeric value on the spatial distribution of individuals with differing poverty levels within a geography and capture the extent to which they may interact. They focus on two discrete groups of individuals, such as poor versus non-poor or poor white versus poor black. The dissimilarity index is the most commonly used measure in such studies, but the isolation, exposure, interaction and spatial proximity indices have been examined as well (Massey and Denton, 1988; Holloway et al., 1999; Lichter and Johnson, 2007; Lichter et al., 2008; Sparks et al., 2013).

Several researchers interested in poverty focus their attention on income inequality. Poverty research concentrates on the lower tail of the income distribution, utilizing the rest of the distribution only in counterpoint to the lower tail of poor individuals. Unlike the federal poverty line that differentiates between poor and non-poor and implies that all individuals within these groups are the same, income inequality accounts for the whole income distribution, indicating whether large disparities exist among a location's population. Locations with large income disparities may be considered undesirable by policymakers given that many of the same demographic and economic conditions that correlate with lower levels of income inequality are also associated with lower poverty rates (Peters, 2012). The Gini coefficient, a measure of income inequality that is frequently used in academic research (Peters, 2012), places emphasis on the variation of income within a location but does not indicate where the variation occurs within the income distribution. This tends to produce similar inequality levels for distributions skewed toward high incomes as those skewed toward low incomes.

In summary, the variety of measures we have outlined conceptualize poverty in distinctive ways. Studies using the poverty rate emphasize the proportion of the population considered to be poor within a geography, and those that use persistent poverty look at that proportion over time. Studies using segregation indices focus on the concentration of poverty within a geography, emphasizing spatial clustering of the poor rather than the level of poor within the area. Studies that focus on income inequality emphasize variation in income, rather than income levels or wealth.

Discipline Approaches and Methodologies

There is often commonality and collaboration among members of the various social science disciplines when it comes to studying poverty. While the focus of analysis may differ from one discipline to another, the social sciences have a shared interest in how society affects, and is influenced by, people and places, as well as how they influence each other (Vu, 2010; Di Fabio and Maree, 2016). In our search for articles focused on place-based poverty, we examined work from the fields of anthropology, demography, economics, geography, planning, political science, regional science, sociology and urban studies. In order to draw conclusions about common approaches and methodologies, we classified each paper by the journal in which it was published, rather than the authors' specialty. We address how social scientists from different disciplines approach their analyses and the varying methodologies they use. Table 2.1 presents a broad description of these approaches and methodologies.

Table 2.1 Approaches and methods to study place-based poverty by social science discipline

Discipline	Approach/Lens	Main Methodology*
Anthropology	culture	written analysis
Demography	population characteristics	descriptive analysis, regression analysis
Economics	economy	descriptive analysis, regression analysis
Geography	space and scale	descriptive analysis, written analysis
Planning	land use and transportation	descriptive analysis, regression analysis
Political Science	policy and institutions	descriptive analysis, written analysis
Regional Science	regions and space	descriptive analysis, regression analysis
Sociology	society and institutions	descriptive analysis, regression analysis
Urban Planning	urban focus	descriptive analysis, regression analysis

Note: * Methodology refers to the main method used for analysis and can be classified as: descriptive analysis (descriptive statistics, including mathematical indices), regression analysis (statistical regression techniques, spatial or aspatial) and written analysis (logically supported arguments). These can be applied to national studies, regional studies or case studies of places or individuals.

The disciplines of regional science and urban studies attract researchers from a variety of other disciplines that are interested in the spatial and regional analysis of socioeconomic issues, as well as urban issues and policy options. Work from these disciplines contains approaches and methodologies that are rooted in many of the member disciplines. Regional science research, especially, draws heavily from economics.

There are three distinct approaches in analysing place-based poverty: people-focused, place-focused and society-focused. These classifications are not mutually exclusive, and disciplines can fall into more than one of the categories. People-focused approaches, such as those associated with the disciplines of anthropology, sociology and economics, emphasize individuals who are poor (South et al., 2005; Frerer and Vu, 2007). This approach seeks to understand the location of individuals or households within places, along with their movement. Place-focused approaches emphasize geographic aggregations and are employed by researchers from all disciplines, with the exception of anthropology (Partridge and Rickman, 2007b; Lichter et al., 2008). This approach seeks to understand the distribution or change in the total number or proportion of poor individuals in an area. Society-focused approaches, adopted by researchers in planning, political science, sociology and urban studies, emphasize institutions (Spencer, 2004; Chapple, 2006). This approach, frequently used in combination with place-level data, seeks to understand the effects of government policies, along with services and programs offered by governmental and non-governmental institutions, on the poor.

Methodologically, analyses for each of the disciplines fall along a spectrum with two extremes. One end of the spectrum is research conducted with mathematically rigorous methods and is characterized by the use of powerful statistical techniques, particularly spatial econometric analysis (Peters, 2012; O'Connell and Shoff, 2014). In our assessment of the poverty literature, articles from economics, demography and sociology tend to cluster at this end of the spectrum. At the other end of the spectrum are techniques that use observational methods, characterized by logical arguments of hypotheses supported through observations of individuals or society (Frerer and Vu, 2007). Articles written by anthropologists represent this end of this spectrum. The most common type of study, however, falls somewhere between these two extremes. For example, regional science papers tend to use statistical techniques that involve rigorous mathematical analyses, but, in our sample of articles, these techniques are not as computationally complicated as the spatial econometric models utilized by other authors (Partridge and Rickman, 2008). Many papers in geography, political science, sociology and urban planning use indices, quadrant analysis and other mathematically descriptive methods to analyse poverty within an area (Greene, 1991; Lichter and Johnson, 2007; Hu, 2014).

The goal of researchers is to understand poverty through the lens of their specific discipline, be it a regional, spatial, societal, cultural or some other context. Researchers use a combination of analytical methods and sample or population subgroups to isolate and explore poverty through these differing lenses. For example, to analyse the spatial distribution of poverty without spatial econometric methods, authors frequently categorize places into sample subgroups based on location, such as urban-rural or central city-inner suburb-outer suburb, to understand how factors affecting poverty differ in these types of places (Holliday and Dwyer, 2009; Wang et al., 2012). Authors also isolate one of these subgroups for analysis, usually either urban or rural places, to further focus their work (Ellen and O'Regan, 2008). These methods are also used to target persistently high poverty areas or regions of the country for analysis (Lichter and Campbell, 2005; Partridge and Rickman, 2007b).

In addition to approaching the analysis of poor places using categorization or sample subgroups, some disciplines focus on the spatial concentration or incidence of poverty in population subgroups of interest. Political science and sociology often use race or ethnicity to determine their subgroups of interest (South et al., 2005; Wilson, 2008), while geography and sociology often employ age to determine their subgroups of interest (Glasmeier, 2002; Lichter and Johnson, 2007). Socioeconomic classifications, such as being homeless or a member of the working poor, are also applied to identify subgroups of interest for place-based poverty analyses in articles from geography and sociology (DeVerteuil, 2003; Thiede et al., 2015).

Another aspect that varies by discipline is the degree to which space, place and scale are explicitly included in analysis. These are concepts that are part of the core focus of fields such as geography, planning, regional science and urban studies but represent possible new areas of research by other social science disciplines, particularly anthropology and sociology. While there appears to be limited interest in an explicit place-based understanding of poverty in anthropology, it is implicitly included in research through an understanding of the culture in which poor individuals live. Similarly, place and scale are often considered in sociology to be a priori and implicit, but there is a growing literature, especially in the subfield of rural sociology, that explicitly includes these more geographic aspects of poverty (Lobao et al., 2008).

SPACE AND POVERTY

Poverty does not exist solely at the individual level but also in space because people exist and interact in space. However, it is not distributed evenly throughout space, and clusters of poverty occur within cities, counties and regions. These patterns are influenced by interactions among people and, therefore, are shaped by such characteristics as race or education. In this next section, we address the metropolitan and regional patterns of poverty, as well as how classifications and the geographic level of aggregation may affect poverty results.

Unit of Aggregation

Geographic areal units: from nation to neighborhood

Geographic areal units used to study place-based poverty are divided into three main scales: *macro* units that capture poverty averages and temporal trends, such as nations or states; *meso* units that explore spatial and temporal poverty trends at a larger level than neighborhoods, such as counties and metropolitan areas; and *micro* units that examine spatial and temporal poverty trends at the neighborhood level, such as census tracts, block groups and places (Peters, 2012). The level of geographic aggregation is important when studying socioeconomic concepts since aggregating data can cause a modifiable areal unit problem. In simple terms, aggregating individual data to create areal data at macro, meso and micro scales may obscure trends of spatial concentration that are only apparent at smaller or larger units of analysis (Lichter et al., 2008; Curtis et al., 2013). Therefore, the results of place-based poverty analyses can be different, depending on the unit of analysis that is used. As such, choosing the appropriate areal unit requires careful consideration (Peters, 2012).

Place-based poverty analyses in the United States, particularly those interested in community development, are generally not conducted at a macro scale. While states are a geopolitical unit of analysis from which many poverty policies originate, they are too large to understand more than general poverty trends. Poverty data and estimates from the macro scale work well as benchmarks from which analysts can gauge poverty levels at the regional level. To explore spatial trends in poverty, meso and micro geographic areal units are the preferred units of analysis.

Counties, a meso geographic areal unit, are most often used by researchers to study poverty. They are geographically small enough to capture trends in the spatial distribution of poverty among counties and states and serve as the building blocks of metropolitan statistical areas. There is a large amount of socioeconomic and natural geographic data available for counties from multiple governmental and academic sources, and they are directly comparable across time because they have relatively stable borders (Peters, 2012). However, as is hinted by the modifiable areal unit problem, counties can be quite large, particularly in the western United States, and may obscure spatial variation and concentration in poverty (Lichter et al., 2008). Crandall and Weber (2004) describe this phenomenon as spatial aggregation bias. Despite these drawbacks, authors such as Curtis et al. (2013) still choose to use counties as their unit of analysis because they feel that structural factors that may produce economic vulnerability may be embodied and produced at this geographic level.

At the micro scale, census tracts have been used most frequently to study sub-county poverty. This is due to the large literature exploring the spatial distribution and concentration of poverty within urban areas. These authors use tracts as a proxy to measure urban neighborhoods (Greene, 1991; Holliday and Dwyer, 2009). Tracts are geographic areas defined by the US Census Bureau that contain between 1,200 and 8,000 people, with an optimal population of around 4,000 inhabitants (US Census Bureau, 2012). While tracts are an appropriate areal unit to study the spatial variation of poverty within urban areas, they can mask sub-county spatial variation in rural areas. Rural areas have a much lower population density than urban areas, making census tracts physically much larger than in urban areas. This larger size can make them too large to conform to the boundaries of geopolitical places, such as towns, and insensitive to spatial variation in the socioeconomic and demographic composition of rural areas (Lichter et al., 2008).

Lichter et al. (2008) recommend block groups as the appropriate geographic areal unit to use when analysing spatial variation in rural poverty because they are physically smaller than tracts and socioeconomic and demographic features are available for supporting mathematical and statistical analyses. Lichter et al. (2012) use incorporated and census designated places to focus their micro analysis on communities, rather than neighborhoods. Because communities are locations in which people interact and, in the case of incorporated places, where policies are implemented, they have a large role in shaping concentrated poverty and affluence. However, places cannot be aggregated into larger geographic units because they omit areas not considered to be communities, excluding residential locations such as farms or small residential neighborhoods that do not qualify under the Census Bureau definition of a designated place. Peters (2009) considers MCDs to be a better representation of communities, especially in rural areas, because they are sociopolitical administrative units that provide mutually exclusive, exhaustive coverage of the country.

As was discussed in the previous section, understanding poverty is about temporal trends in addition to poverty levels. Micro geographic areal units present two key challenges when analysing temporal trends in poverty. The first problem is the availability of data. The United States was not fully tracted until the 1990 Census of Population and Housing. Prior to that decennial census, only a growing number of urban areas were divided into block groups and tracts. This restricts temporal research examining changes in poverty data at the tract or block group level to roughly the last 25 years. This is not an issue when it comes to using MCDs or places since data for these areal units have been collected since the 1870s and 1950s, respectively. The second problem is that the boundaries of sub-county geographies are frequently not the same from decennial census to decennial census. If authors do not correct for these changes by harmonizing the data to ensure that boundaries are consistent over time, their analyses will contain geographical biases and fail to accurately measure temporal poverty trends. A handful of authors use data harmonized by GeoLytics (Ellen and O'Regan, 2008; Peters, 2012) but others use their own process to harmonize their data (Holloway et al., 1999; Logan et al., 2014; Kumar and Kim, 2017).

The urban-rural dichotomy

Policymakers and researchers like to describe places as either urban or rural (or metropolitan or nonmetropolitan), but places, especially at the meso scale, often cannot be clearly classified into one of these two categories. Despite the fact that communities may include a combination of urban and rural spaces, the funding opportunities available to them for building community capitals are frequently tied to their classification in this dichotomy. Additionally, using a strict urban and rural dichotomy masks variation that may exist among locations within each of these categories. For example, Partridge and Rickman (2008) recognize that the socioeconomic processes affecting poverty rates may vary for metropolitan areas of different sizes. Similarly, not all rural areas are the same.

As with geographic areal units, the best urban-rural classification system to use in place-based poverty analyses is dependent on the research objective. There are six main systems that are used for academic work that differ by whether they are continuous or discrete, the amount of categorical variation they include and whether they focus on urban, suburban or rural variation. As with the modifiable areal unit problem discussed in the previous section, results can change based on the classification used (Wang et al., 2012).

The Core Based Statistical Area (CBSA) classification system developed by the US Office of Management and Budget is the most common method of differentiating urban and rural locations in research. This system divides counties into three categories based on urban population, adjacency and commuting patterns: metropolitan, micropolitan and noncore. Metropolitan and micropolitan statistical areas are functional daily activity spaces in which residents in central cities and surrounding suburbs and rural areas interact (Wang et al., 2012). These categories do not directly translate into urban and rural areas since metropolitan and micropolitan areas contain core urban settlements and surrounding rural areas of commuters, but researchers using this system to determine the rurality of a county generally classify metropolitan counties as urban and micropolitan and noncore counties as rural.

The Urban-Rural Typology (UA) classification system developed by the Census Bureau provides an alternate method of differentiating among urban and rural locations. This method uses population, density and adjacency requirements to construct a new geographic areal unit from census tracts and blocks. These units, named urban areas, capture the built urban environment of a location and are generally a micro areal unit, although they may be larger than counties for large metropolitan areas. In this typology, all locations not considered to be urban are classified as rural.

These two main classification systems were first released as part of the 1950 Census of Population and Housing and were conceptualized with a strong urban core to which workers commuted for work, which was common during the era of industrialization. Although refinements have been made to these dichotomous classifications over time as urban structure and statistical and geographic capabilities have expanded, the underlying assumptions have remained, and they cannot capture the multidimensional aspects that occur in current urban and metropolitan areas, including zones of mixed functions (Wang et al., 2012). The rest of the classification systems used by academics are based off these two main governmental classifications (Waldorf and Kim, 2015). They are designed to allow for more variations within the urban-rural and metropolitan-nonmetropolitan dichotomies.

ERS developed two county-level classification systems derived from the CBSA classification system. The first, the Rural Urban Continuum Codes (RUCC), divides metropolitan counties into three size categories by population and nonmetropolitan counties into six categories by urban population and adjacency to a metropolitan area. The second, the Urban Influence Codes (UIC), disaggregates metropolitan counties into two size categories by population and nonmetropolitan counties into ten categories by micropolitan or noncore status, adjacency to large or small metropolitan areas and whether there is an urban area in the county. Miller and Weber (2003) suggest that the RUCC should be used when researchers prefer an emphasis on suburban areas, while the UIC should be used when an emphasis on rural areas is preferred.[6]

In their evaluation of urban-rural classification systems in the United States and their ability to reveal poverty, Wang et al. (2012) collapse the RUCC codes into five categories for their analysis, as the detailed rural differentiation of the classification system can cause instability in econometric estimates. The authors suggest that, of the current urban-rural classification systems, Isserman's (2005) Rural Urban Density Code (RUDC) provides an appropriate level of variation for analysis without being too nuanced. This scheme classifies counties into four categories based on the proportion of the population that is urban and urban area size: rural, mixed rural, mixed urban and urban.

Another prominent urban-rural classification system is Waldorf's (2006) continuous Index of Relative Rurality (IRR), which uses population size, density, remoteness and the proportion of urban land to calculate a flexible measure of rurality that can be determined at any spatial scale. Waldorf (2007) then uses the IRR, population size, the CBSA metropolitan-nonmetropolitan definitions and adjacency to a metropolitan area to create the county-level, seven category Rural-Metropolitan Interface Levels classification, which groups counties into the "metropolitan sphere," "rural-metropolitan interface" and "rural sphere."

Metropolitan Patterns of Poverty: Spatial Location and Theories

Poverty became increasingly urban, particularly after 1935, as a result of the mechanization of agricultural labor, which accelerated during the 1940 to 1960 period. The displaced rural laborers migrated to urban labor markets but experienced a skills mismatch, making it difficult to find employment (Glasmeier, 2002; Michaels et al., 2012). While this structural transformation of the economy started in the late nineteenth century, it continued well into the twentieth century so that, by the 1960s, poverty had emerged as a significant and "deeply urban problem" (Kim, 2000; Glasmeier, 2002). Since the War on Poverty started in the 1960s, poverty policies have concentrated on aiding populations characterized by urbanity and race (Spencer, 2004). Place-based poverty research followed a similar path, focusing on concentrated poverty and race in urban neighborhoods, using tracts as the geographic proxy for neighborhoods.

While urban poverty served as the principal focus of academic research for many decades, research on rural poverty emerged as the focus of social scientists in the 1990s (Tickamyer and Duncan, 1990). Most of the rural poverty research to date has been conducted at the county level for the contiguous United States, using the results for urban estimates as a counterpoint to rural results (e.g., Lichter and Johnson, 2007; Partridge and Rickman, 2007b). These studies have found that persistent poverty counties are disproportionately rural and that child poverty is higher than total poverty in rural areas, especially for minorities (Miller and Weber, 2003; Lichter and Johnson, 2007; Mattingly et al., 2011). More recently, work by Lichter et al. (2008) focuses on block groups to analyse concentrated poverty in the United States, using the micro scale to extend knowledge on the characteristics and location of poverty pockets to rural counties.

Researchers frequently classify counties as urban or rural using the CBSA metropolitan-nonmetropolitan dichotomy. However, they acknowledge that this dichotomy does not translate directly to urban and rural locations (e.g., Lichter et al., 2008; Partridge and Rickman, 2008), particularly in metropolitan areas. Recent academic work has explored the spatial distribution of poverty within metropolitan areas. In the same way that the work on micro-scale rural poverty emphasizes that sub-county patterns of poverty can be drastically different than county-level patterns, due in part to the modifiable areal unit problem, these works emphasize the differences that exist among the central cities, suburbs and rural areas within metropolitan areas.

Brown and Hirschl (1995) find that household characteristics, such as employment, educational attainment and single parenthood, have the same effect on poverty status, regardless of whether the household lives in an urban or rural location. They conclude that differences in poverty between urban and rural areas are likely attributed to community characteristics and context, such as social welfare benefits or class structure. Holliday and Dwyer (2009) find that suburban poverty is not an extension of urban poverty, as poor suburban neighborhoods have a different racial and ethnic structure than urban neighborhoods, as well as their own set of advantages and disadvantages. For example, suburban neighborhoods tend to lack social and economic institutions such as hospitals, universities or large businesses and transportation connections to the central cities, which isolates the suburban poor and makes accessing social services more difficult than for the poor located in central cities.

Several theories have been proposed to explain the uneven distribution of poverty within metropolitan areas, the most prominent of which is the spatial mismatch hypothesis. This theory hypothesizes that poverty in metropolitan areas is driven by differences in the location of jobs and the workers able to fill them. This locational mismatch is usually considered to be caused by the decentralization of low-skill jobs to the suburbs in combination with economic stagnation and disinvestments within the central city, leading to a decline in job opportunities (Chapple, 2006; Wilson, 2008). Lack of access to public and personal transportation options exacerbates the spatial mismatch, as low-income and low-skilled workers concentrated in the urban center cannot access suitable jobs located in the suburbs and exurbs of the metropolitan area.

A second, closely related theory is the economic transformation or skills mismatch hypothesis. This theory hypothesizes that the restructuring of the American economy from manufacturing- to service-based industries that began in the middle of the twentieth century affects the number of low-income jobs available and the skills necessary to perform them (Hu, 2014). Wilson (2008) notes that these new low-skill service jobs are unstable, pay less and lack benefits and work protections that were commonly associated with low-skill manufacturing jobs. Researchers argue that spatial and skills mismatches are self-reinforcing mechanisms, and economic and workforce development policies need to consider both of these hypotheses (Houston, 2005; Fan et al., 2016).

Spatial mismatch is more prevalent in old, industrial regions and far less pervasive in small metropolitan areas and locations with tight labor markets (Chapple, 2006). Conversely, Hu (2014) finds that in Chicago, an old, industrial city, the poor have become more suburbanized over time, despite housing affordability and discrimination barriers. This increases their access to low-skill jobs that have also suburbanized, such that the poor are not negatively impacted by spatial mismatch but are adversely affected by the skills mismatch associated with industrial restructuring. However, Chapple (2006) finds that social services and nonprofits have yet to decentralize, creating a spatial mismatch between support services for the poor and their new location in the inner-ring suburbs.

Other theories focus on the spatial structure of metropolitan areas and where poverty is located within these models. The oldest of these is the concentric zone theory which models land use and location within a metropolitan area in concentric circles around a business-based core (a central business district surrounded by an industrial zone). Low-income residents are located in the zone adjacent to the industrial zone, followed by middle- and high-income residential zones as distance increases from the city center. Over time, academics have amended this structure by pushing the boundary separating low- and middle-income residents from between the central city and suburbs to between the inner-ring and outer-ring suburbs. This change assumes that suburban poverty is simply an extension of the poverty found in central cities (Holliday and Dwyer, 2009).

The social ecological model is more complex than the concentric zone model as the theory utilizes the entire metropolitan area's physical and social characteristics to understand the location of poverty. In this model, characteristics such as transportation systems, industrial structure, topography, residential clusters of working poor and ethnic enclaves are accounted for, but suburban poverty is still viewed as an extension of central city poverty (Holliday and Dwyer, 2009). Finally, the place stratification model includes the influence of political and industrial power as factors driving the residential location of the poor. Places are hierarchically ordered by desirability and opportunity for residents, and structural barriers and constraints exist that restrict the poor to less desirable areas (Holliday and Dwyer, 2009). Poverty is socially and racially diverse due to the effect of these barriers and constraints, and suburban poverty is no longer an extension of central city poverty, with distinctly different processes driving poverty and location.

Central city poverty has traditionally been characterized by minority populations (specifically African Americans), people with low educational attainment, presence of deteriorating and vacant housing and high rates of crime. Wilson (2013) observes that the poor black population has been pushed out to inner-ring suburbs, a situation which is consistent with the predictions of the expanded concentric zone and social ecological models. Holliday and Dwyer (2009) find that the concentric zone model still holds in many smaller Midwestern metropolitan areas. Additionally, the authors state that metropolitan areas with predominantly inner-ring suburban poverty of large black and Hispanic populations are disproportionately located in the Midwest.

However, many metropolitan areas do not follow these patterns, namely, that all suburban poverty is located in the outer-ring suburbs of many small metropolitan areas in the Northeast and South, that pervasive suburban poverty is occurring exclusively in mid-sized metropolitan areas in the South and West, and that poverty is located in both inner-ring and outer-ring suburbs of large metropolitan areas in the Northeast, South and West (Holliday and Dwyer, 2009). Holliday and Dwyer (2009) also find that poor suburban neighborhoods have a larger Hispanic population, healthier economies, lower educational attainment and more overcrowding than poor central city neighborhoods. This variety of spatial poverty patterns and diversity in the characteristics of poor neighborhoods indicates that all three theories of metropolitan spatial structures have value when it comes to understanding the location of poverty in metropolitan areas of the United States.

Finally, the depopulation hypothesis, a theory based on intra-city migration, states that areas of concentrated poverty exist, particularly in central cities, due to the outmigration of middle-income residents to more desirable neighborhoods throughout the metropolitan area. This spatial pattern emerges due to a combination of race, economics and migration, connecting it to the processes associated with spatial mismatch and industrial restructuring (Wilson, 2013). Greene (1991) uses a sample of ten large cities throughout the United States to explore whether the depopulation hypothesis holds, and finds that it cannot be generalized, but is most frequently found in the Midwest and Northeast regions of the country.

Regional Patterns of Poverty

The intra-urban and neighborhood processes that affect the spatial location of poverty cannot be examined in isolation from the rest of the urban system (Holloway et al., 1999). They must be considered within the context of patterns and changes throughout the urban and regional systems, such as migration and industrial restructuring. For example, during the twentieth century, two large waves of predominantly black migrants moved from the rural South to cities in the Midwest and Northeast in search of employment opportunities, predominantly in the manufacturing sector (Wilson, 2008; Wilson, 2013). With the industrial restructuring of the American economy from a manufacturing- to a service-based economy and the subsequent depopulation of thriving inner-city working-class communities, neighborhoods of concentrated poverty emerged.

There are several distinct clusters of concentrated, persistent poverty in the United States which can be characterized by the predominant race or ethnic attributes of their residents. These clusters have been typically identified at the level of counties but have also been detected using block group-level data, indicating that they are robust to the modifiable areal unit problem. They include Appalachia, an area of white persistent poverty; the Mississippi Delta and Black Belt crescent stretching from North Carolina to Arkansas, an area of black persistent poverty; the colonias along the US-Mexico border, a region of Hispanic persistent poverty; and the Native American reservations of the Great Plains, an area of native persistent poverty (Lichter et al., 2008).

Within these clusters, racial and ethnic disparities still persist. Despite being a predominantly white persistent poverty cluster, poverty rates in Appalachia are higher among blacks and Hispanics than whites and Asians (Lichter and Campbell, 2005). However, South et al. (2005) find that there is no racial or ethnic difference in capitalizing socioeconomic resources, such as higher education levels, to move out of poor neighborhoods; it is an individual's economic advantage that makes the difference. Similarly, the economic characteristics of a county, rather than its demographic characteristics, are what affect poverty rates and employment growth. Partridge and Rickman (2007b) find that while proximity to a metropolitan area does not affect the poverty rate of persistently poor counties, it does positively affect high poverty counties that are not persistently poor, indicating that a location's position within the metropolitan and urban system may influence poverty levels and growth. Peters (2012) finds similar variation within the urban system when analysing income inequality. He notes that high-inequality areas are located in metropolitan suburbs and Native American reservations, while high-growth areas are concentrated in micropolitan and small metropolitan areas.

Regional variations in poverty can also be characterized through the broader classification of Census regions: Northeast, Midwest, South and West. Many researchers use this approach to analyse spatial differences throughout the United States. They find that total poverty levels are highest in the South, and metropolitan poverty is lowest in the Midwest, while these same regions have higher rural and mixed rural county poverty rates than in the rest of the country (Wang et al., 2012). They also analyse regional trends in sub-county phenomena, finding that the urban resurgence of the lowest-income neighborhoods in the 1990s was most likely to occur in the Midwest and that the poverty rates and median household income of inner-ring suburbs in the South have improved relative to central cities, but have experienced a larger decline than in the rest of the country in relation to the outer-ring suburbs (Madden, 2003; Ellen and O'Regan, 2008).

DISCUSSION: IMPLICATIONS FOR POLICYMAKERS AND PRACTITIONERS

While academics and policymakers attempt to distinguish between person- and place-based policies and determine which of these two strands is more efficient or effective, this distinction is not well-defined in practice (Spencer, 2004). Policies can also be classified as demand-side or supply-side, depending on how they to attempt to mitigate poverty. Demand-side policies focus on moving jobs or workers, while supply-side policies focus on assisting people or locations (Spencer, 2004). Realistically, anti-poverty programs exist along a continuum that balances the dichotomies of policy targets (i.e., people or places) and policy mechanisms (i.e., demand-side or supply-side), and since poverty policy was first enacted in the 1930s, intentionally mixed policies have become more prevalent (Spencer, 2004).

Academic research is useful in focusing policy within this continuum to achieve policy program goals. Partridge and Rickman (2005) find that there are different mechanisms driving the poverty rates in high- and low-poverty counties, implying that a mixture of policy mechanisms must be utilized to effectively address poverty in these areas. They also find that the poverty generating process in persistently high poverty counties does not differ greatly from that of high poverty counties without a persistently elevated poverty rate, implying that many of the same policies would work in both persistent and non-persistent high poverty counties. Additionally, persistently high poverty counties are uniformly impacted by economic development, and policies do not need to be adjusted to target poverty clusters with different demographic characteristics, such as Appalachia and the Black Belt regions of the United States (Partridge and Rickman, 2007a).

While many sociologists eschew making policy recommendations based on their research and engaging in the policy-making process, other social scientists, including economists, regional scientists and political scientists, frequently make policy recommendations based on their research, working with policymakers to frame and influence program investments (Lobao et al., 2008). Partridge and Rickman (2005) suggest implementing policies to increase employment growth and education levels in high poverty nonmetropolitan counties to decrease poverty rates in these areas. Wilson (2008) recommends mitigating concentrated urban poverty through policies that decrease racial employment discrimination, revitalize poor neighborhoods, increase job training, improve public education and strengthen unions. These suggestions include demand- and supply-side mechanisms, an approach that Partridge and Rickman (2005) advise be implemented simultaneously in order to have the largest impact on reducing poverty.

The Role of Community Development Practitioners

The policy strategies above are diverse and offer community development practitioners a host of options for improving the poverty status of people and communities in urban and rural America. Practitioners frequently confront the challenge of determining which strategies, or mix thereof, make the most sense to reduce poverty and improve overall community well-being. To gain some clarity on this issue, a good starting point is Bradshaw's (2007) synopsis of the five theories of poverty and insights on programs that practitioners might pursue under each of these perspectives. In essence, the strategies community development practitioners pursue, in concert with local leaders, residents and organizations, is dependent on whether they believe poverty is: the result of poor choices made by individuals, the product of forces that keep individuals embedded in a culture of poverty, a consequence of structural barriers that limit opportunities for poor people to improve their social and economic well-being, the outcome of disparities existing across geographic space (i.e., urban, suburban, rural) or the result of interdependences between individuals and the communities and the accumulation of problems and opportunities they face over time (Bradshaw, 2007).

We assert that improving poverty conditions demands a cohesive, comprehensive set of programs and strategies that acknowledge the interconnectedness among individuals and their communities, a perspective that aligns with the fifth strand of poverty research articulated by Bradshaw. It is an approach that is consistent with Weber et al.'s (2005) critical assessment of the rural poverty literature, which notes that the odds of being poor are affected by both individual and community factors. However, the authors are unclear on what key elements should be included in comprehensive poverty-alleviation plans.

Haveman et al. (2015) assert that poverty-alleviation programs should focus on improving early childhood education, expanding parent education and investing in human capital accumulation strategies (e.g., college attendance, community college/technical education and workforce development). Tickamyer et al. (2017) note that the focus, especially in rural America, should be on enhancing job opportunities, improving access to health care, expanding housing options, strengthening local education and addressing infrastructure such as roads and transportation systems. According to Weber (2007), the emphasis should be twofold: creating jobs and building community capacity. His recommendations are based on research that reveals job creation helps reduce poverty and communities with high levels of social capital, such as the active engagement of local people and organizations, have greater success in lowering local poverty levels (Rupasingha and Goetz, 2003; Crandall and Weber, 2004).

These suggestions point to a community development framework that: (a) is comprehensive and flexible, (b) can be tailored to address the unique poverty-related challenges of each community, and (c) can build on the valuable people and organizational assets present in each place. We believe the theoretical framework that is best suited for this work is the community capitals perspective. As noted by Flora et al. (2016), the lifeblood of any community is linked to the presence and strength of seven community capitals, resources that can be invested in or tapped for promoting long-term community well-being (Jacobs, 2011). These community capitals are natural, cultural, human, social, political, financial and built.

While all community capitals are important, the one that maximizes the effective use of the various capitals is social capital. Social capital holds a community together and spurs community and economic development efforts that benefit the entire community. In communities that are thriving, be it in education, job creation, health care or community services, there is likely a broad-based corps of civic-minded people and organizations in place to undergird these important activities. This belief is reaffirmed by Phillips and Pittman (2015, p. 7) when they state, "It is difficult to imagine a community making much progress without some degree of social capital or capacity. The more social capital a community has, the more likely it can adapt to and work around deficiencies in the other types of community capital."

Though there is no universal strategy that community development practitioners can use to address all people- and place-related aspects of poverty, there are several options that may be relevant to communities seeking to take meaningful steps to tackle poverty. These strategies include:

1. *Establish an Inclusive Community Coordinating Team:* It is critical that community-driven poverty reducing efforts be undertaken by residents who reflect the demographic, economic and institutional attributes of the community. Too often, low-wealth[7] residents have little voice in guiding and shaping poverty-alleviation efforts, but their active engagement in this process is essential. Meaningful participation by

low-wealth residents must be coupled with the involvement of people and organizations having access to the mix of community capital resources that can be invested in wealth creation plans the team ultimately develops (Harvey and Beaulieu, 2010).
2. *Build Capacity:* The ability of the community team to function effectively requires constant support, coaching and training, something that community development practitioners are ideally suited to provide. The list of topics to be covered could include: identifying and studying key secondary data on poverty and related topics, exploring approaches for garnering local input, mapping community assets, understanding the nuts and bolts of building a sound plan, putting your plan into action, building strong metrics to track progress and outcomes and keeping the momentum going through sound succession planning. The investment in training not only equips the team with the skills needed to act on local priorities, it provides a mechanism for building trust among the diverse group of people who are members of the coordinating team. In essence, it helps build the type of social capital that Rupasingha and Goetz (2003) state is an essential ingredient for lowering poverty levels in communities.
3. *Expand Economic Opportunities:* As Weber (2007) and others have noted, expanding job opportunities for low-wealth individuals and families represents one of the most important ways to escape poverty. However, avenues for doing so are not always easy in places struggling with high poverty rates. Community development practitioners can guide the community team and local economic leaders in pursuing job creation strategies that build on the existing economic assets of a community. These can include well-known strategies such as business retention and expansion and economic gardening. A new strategy that practitioners can consider is exploring emerging telework opportunities. According to a 2017 report published by Global Workplace Analytics and FlexJobs, telecommuting increased 115 percent in metropolitan areas between 2005 and 2015, and nearly half of teleworkers have an associate's degree, a high school education or less. Gallardo and Whitacre (2018) find that the structure of work is shifting from a centralized work environment to a decentralized home-based work setting and that working from home has a positive impact on median household income, suggesting that a telework strategy might make sense in low-wealth places. However, expanding telework opportunities for low-income individuals requires that they have access to reliable broadband services at a reasonable cost (Gallardo and Whitacre, 2018).
4. *Strengthen Human Capital:* Improving the human capital of low-wealth individuals can expand their job opportunities and improve their economic well-being. The key is to connect them to educational and training opportunities that align with the needs of employers located in the local and regional labor market in which they are embedded (Holzer and Martinson, 2006). To aid this connection and utilize existing opportunities, it would be worthwhile to map the formal and informal training resources available to strengthen the knowledge and skills of low-wealth workers within the community. These resources could include certification programs offered by state or local workforce development agencies, community and technical colleges, land-grant universities or other state-funded universities, as well as state Manufacturing Extension Partnership programs. Developing workforce skills could also help position low-wealth individuals to qualify for telework opportunities (Gallardo, 2016).

5. *Connect to External Resources:* Poor communities often lack the full breadth of local institutions that contribute to comprehensive poverty-alleviation activities (Green and Haines, 2016). As such, community development practitioners can provide valuable guidance on the type of federal, state, private and philanthropic programs that support wealth-building activities in disadvantaged places, especially programs that align with the strategic blueprint developed by the coordinating team and the broader community.
6. *Stay Engaged over the Long-Haul:* Achieving meaningful improvements in low-wealth communities requires commitment to a five- to ten-year process, according to a critical analysis of comprehensive community initiatives (CCIs) targeting residents living in poor urban neighborhoods, small cities, towns and American Indian reservations (Kubisch, 2010). Therefore, community development practitioners must be connected to an organization that has the capacity to guide and lend assistance over an extended period of time. Consistency in the institutional support provided by community development organizations is vital to the success of the poverty-reduction initiative, even if the individual community development practitioner changes. Land-grant universities with a strong Extension community development program represent one such entity that could provide the continuity and long-term support needed to spur substantive improvements in low-wealth communities.

NOTES

1. A few examples of poverty publications produced by academics are O'Connor (2002) and Brady and Burton (2016), while examples from the policy community are Fisher (1992) and US Census Bureau (2014).
2. Under the Office of Management and Budget's Statistical Policy Directive 14, the official federal statistical measure of poverty in the United States is calculated yearly by the Census Bureau. The Department of Health and Human Services calculates a simplified version of the poverty thresholds that is frequently used for administrative purposes (Census, 2017). A relative poverty measure, the supplemental poverty measure, was introduced by the Census Bureau in 2011 to give a more comprehensive, place-based view of poverty levels in the United States. This measure, which is not intended for or used to determine eligibility for federal assistance, accounts for noncash benefits (e.g., nutritional, housing and energy assistance) and necessary expenses (e.g., income taxes, childcare expenses and health insurance premiums) and is adjusted for regional housing costs by tenure (Fox, 2017).
3. The Council for Community and Economic Research produces a for-purchase quarterly cost-of-living measure for metropolitan areas and counties. The Center for Neighborhood Technology produces a Housing and Transportation Affordability Index at multiple standard geographic levels, which measures spatial variation in cost-of-living using housing and transportation costs and is available at no cost. The Bureau of Economic Analysis produces a publicly available real personal income measure that uses consumer price index and housing data to create estimates that vary by states and metropolitan areas (US Bureau of Economic Analysis, 2016).
4. The Economic Research Service defines "deep poverty" as having cash income below half of one's poverty threshold (US Department of Agriculture, Economic Research Service, 2017a).
5. The 1950 Census of Population was the first decennial census in which total family income and individual income for individuals 14 years of age and older were collected, as opposed to just wages and salaries (US Census Bureau, 1975).
6. Miller and Weber (2003) conducted their analysis of poverty rates and comparison between the RUCC and UIC rurality classifications using 2000 Census of Population and Housing data and the 1993 classification definitions. The definitional changes in the RUCC and UIC in 2003 and 2013 may have some effect on the authors' conclusions, but the emphasis and construction of the rurality classifications have not experienced significant changes.
7. The community capitals framework focuses on "low-wealth" rather than "low-income" individuals. While the terms are similar, "low-income" refers to monetary assets, while "low-wealth" includes physical assets (e.g., real estate) and personal characteristics (e.g., education level) in addition to monetary assets.

REFERENCES

Bradshaw, T. (2007). Theories of poverty and anti-poverty programs in community development. *Community Development*, *38*(1), 7–25.

Brady, D., and Burton, L.M. (eds). (2016). *The Oxford Handbook of the Social Science of Poverty*. New York: Oxford University Press.

Brown, D.L., and Hirschl, T.A. (1995). Household poverty in rural and metropolitan-core areas of the United States. *Rural Sociology*, *60*(1), 44–66.

Castle, E.M., Wu, J.J., and Weber, B.A. (2011). Place orientation and rural-urban interdependence. *Applied Economic Perspectives and Policy*, *33*(2), 179–204.

Chapple, K. (2006). Overcoming mismatch: Beyond dispersal, mobility, and development strategies. *Journal of the American Planning Association*, *72*(3), 322–36.

Chetty, R., and Hendren, N. (2018). The impacts of neighborhoods on intergenerational mobility I: Childhood exposure effects. *The Quarterly Journal of Economics*, *133*(3), 1107–62.

Citro, C.F., and Michaels, R.T. (eds). (1995). *Measuring Poverty: A New Approach*. Washington, DC: National Academy Press.

Crandall, M.S., and Weber, B.A. (2004). Local social and economic conditions, spatial concentrations of poverty, and poverty dynamics. *American Journal of Agricultural Economics*, *86*(5), 1276–81.

Curtis, K.J., Reyes, P.E., O'Connell, H.A., and Zhu, J. (2013). Assessing the spatial concentration and temporal persistence of poverty: Industrial structure, racial/ethnic composition, and the complex links to poverty. *Spatial Demography*, *1*(2), 178–94.

DeVerteuil, G. (2003). Homeless mobility, institutional settings, and the new poverty management. *Environment and Planning A*, *35*(2), 361–79.

Di Fabio, A., and Maree, J.G. (2016). Using a transdisciplinary interpretive lens to broaden reflections on alleviating poverty and promoting decent work. *Frontiers in Psychology*, *7*, https://doi.org/10.3389/fpsyg.2016.00503.

Ellen, I.G., and O'Regan, K. (2008). Reversal of fortunes? Lower-income urban neighborhoods in the US in the 1990s. *Urban Studies*, *45*(4), 845–69.

Fan, Y., Guthrie, A., and Das, K.V. (2016). *Spatial and Skills Mismatch of Unemployment and Job Vacancies* (Report No. CTS 16-05). Minneapolis, MN: University of Minnesota, Center for Transportation Studies.

Fisher, G.M. (1992). The development of the history of the poverty thresholds. *Social Security Bulletin*, *55*(4), 3–14.

Fisher, M.G., and Weber, B.A. (2002). *The Importance of Place in Welfare Reform: Common Challenges for Central Cities and Remote-Rural Areas* (Research Report). Washington, DC: Brookings Institution Center on Urban and Metropolitan Policy.

Flora, C.B., and Thiboumery, A. (2005). Community capitals: Poverty reduction and rural development in dry areas. *Annals of Arid Zone*, *45*(3 and 4), 239–53.

Flora, C.B., Flora, J.L., and Gasteyer, S.P. (2016). *Rural Communities Legacy + Change* (5th ed.). New York: Routledge.

Fontenot, K., Semega, J., and Kollar, M. (2018). *Income and Poverty in the United States: 2017* (Current Population Reports No. P60-263). Washington, DC: United States Department of Commerce, Census Bureau.

Fox, L. (2017). *The Supplemental Poverty Measure: 2016* (Current Population Reports No. P60-261). Washington, DC: United States Department of Commerce, Census Bureau.

Fox, L., Wimer, C., Garfinkel, I., Kaushal, N., and Waldfogel, J. (2015). Waging war on poverty: Poverty trends using a historical supplemental poverty measure. *Journal of Policy Analysis and Management*, *34*(3), 567–92.

Frerer, K., and Vu, C.M. (2007). An anthropological view of poverty. *Journal of Human Behavior in the Social Environment*, *16*(1 and 2), 73–86.

Gallardo, R. (2016). *Work in Place: A Telework Friendly Policy Framework* (Publication No. 3010, POD-10-16). Mississippi State, MS: Mississippi State University Extension.

Gallardo, R., and Whitacre, B. (2018). 21st century economic development: Telework and its impact on local income. *Regional Science Policy Practice*, *10*(2), 103–23.

Glasmeier, A.K. (2002). One nation, pulling apart: The basis of persistent poverty in the USA. *Progress in Human Geography*, *26*(2), 155–73.

Global Workplace Analytics and FlexJobs. (2017). *2017 State of Telecommuting in the U.S. Employee Workforce*. Retrieved 9 January 2019, from https://globalworkplaceanalytics.com/downloads/2017-state-of-telecommuting-in-the-u-s.

Green, G.P., and Haines, A. (2016). *Asset Building & Community Development*. Thousand Oaks, CA: Sage Publications, Inc.

Greene, R.P. (1991). Poverty area diffusion: The depopulation hypothesis examined. *Urban Geography*, *12*(6), 526–41.

Harvey, M.H., and Beaulieu, L.J. (2010). Implementing community development in the Mississippi Delta. In G.P. Green and A. Goetting (eds), *Mobilizing Communities: Asset Building as a Community Development Strategy* (pp. 146–76). Philadelphia, PA: Temple University Press.
Haskins, R. (2008). *A Plan for Reducing Poverty* (Report). Washington, DC: Brookings Institution.
Haveman, R., Blank, R., Moffitt, R., Smeeding, T., and Wallace, G. (2015). The war on poverty: Measurement, trends, and policy. *Journal of Policy Analysis and Management*, 34(3), 593–638.
Holliday, A.L., and Dwyer, R.E. (2009). Suburban neighborhood poverty in U.S. metropolitan areas in 2000. *City and Community*, 8(2), 155–76.
Holloway, S.R., Bryan, D., Chabot, R., Rogers, D.M., and Rulli, J. (1999). Race, scale, and the concentration of poverty in Columbus, Ohio, 1980 to 1990. *Urban Geography*, 20(6), 534–51.
Holzer, H.J., and Martinson, K. (2006). Can we improve job retention and advancement among low income parents? *Focus*, 24(2), 31–7.
Hoppe, R.A. (1985). *Economic Structure and Change in Persistently Low-Income Nonmetro Counties* (Report No. RDRR-50). Washington, DC: United States Department of Agriculture, Economic Research Service.
Houston, D. (2005). Employability, skills mismatch and spatial mismatch in metropolitan labour markets. *Urban Studies*, 42(2), 221–43.
Hu, L. (2014). Changing job access of the poor: Effects of spatial and socioeconomic transformations in Chicago, 1990–2010. *Urban Studies*, 51(4), 675–92.
Isserman, A.M. (2005). In the national interest: Defining rural and urban correctly in research and public policy. *International Regional Science Review*, 25(4), 465–99.
Jacobs, C. (2011). *Measuring Success in Communities: Understanding the Community Capitals Framework* (Extension Extra No. 16005, Revised April). Brookings, SD: South Dakota State Cooperative Extension Service.
Katz, L.F., Kling, J.R., and Liebman, J.B. (2001). Moving to opportunity in Boston: Early results of a randomized mobility experiment. *Quarterly Journal of Economics*, 116(2), 607–54.
Kim, S. (2000). Urban development in the United States, 1690–1990. *Southern Economic Journal*, 66(4), 855–80.
Kubisch, A.C. (2010). Strengthening future work on community change. In A.C. Kubisch, P. Auspos, P. Brown, and T. Dewar (eds), *Voices from the Field III: Lessons and Challenges from Two Decades of Community Change Efforts* (pp. 138–48). Washington, DC: The Aspen Institute.
Kumar, I., and Kim, Y.J. (2017). *Spatial Data Documentation: Intergenerational Transfer of (IGT) Poverty Project 2015–2017*. West Lafayette, IN: Purdue Center for Regional Development and Department of Agricultural Economics.
Lichter, D.T., and Campbell, L.A. (2005). Changing patterns of poverty and spatial inequality in Appalachia. (Demographic and Socioeconomic Change in Appalachia Working Paper Series). Washington, DC: Appalachian Regional Commission and Population Reference Bureau.
Lichter, D.T., and Johnson, K.M. (2007). The changing spatial concentration of America's rural poor population. *Rural Sociology*, 72(3), 331–58.
Lichter, D.T., Parisi, D., and Taquino, M.C. (2012). The geography of exclusion: Race, segregation, and concentrated poverty. *Social Problems*, 59(3), 364–88.
Lichter, D.T., Parisi, D., Taquino, M.C., and Beaulieu, B. (2008). Race and the micro-scale spatial concentration of poverty. *Cambridge Journal of Regions, Economy and Society*, 1(1), 51–67.
Lobao, L.M., Hooks, G., and Tickamyer, A.R. (2008). Poverty and inequality across space: Sociological reflections on the missing-middle subnational scale. *Cambridge Journal of Regions, Economy and Society*, 1(1), 89–113.
Logan, J.R., Xu, Z., and Stults, B. (2014). Interpolating U.S. decennial census tract data from as early as 1970–2010: A longitudinal tract database. *The Professional Geographer*, 66(3), 412–20.
Lynch, K. (1960). *The Image of the City*. Cambridge, MA: The MIT Press.
Madden, J.F. (2003). The changing spatial concentration of income and poverty among suburbs of large US metropolitan areas. *Urban Studies*, 40(3), 481–503.
Massey, D., and Denton, N. (1988). The dimensions of residential segregation. *Social Forces*, 67(2), 281–315.
Mattingly, M.J., Johnson, K.M., and Schaefer, A. (2011). More poor kids in more poor places: Children increasingly live where poverty persists. (Issue Brief No. 38). Durham, NH: Carsey Institute.
McLaughlin, D.K., and Jensen, L. (1993). Poverty among older Americans: The plight of nonmetropolitan elders. *Journal of Gerontology*, 48(2), S44–S54.
Michaels, G., Rauch, F., and Redding, S.J. (2012). Urbanization and structural transformation. *Quarterly Journal of Economics*, 127(2), 535–86.
Miller, K.K., and Weber, B.A. (2003). Persistent poverty across the rural-urban continuum. (Working Paper No. 03-01). Iowa City, IA: Rural Poverty Research Center.
O'Connell, H.A., and Shoff, C. (2014). Spatial variation in the relationship between Hispanic concentration and county poverty: A migration perspective. *Spatial Demography*, 2(1), 30–54.
O'Connor, A. (2002). *Poverty Knowledge: Social Science, Social Policy, and the Poor in Twentieth-Century U.S. History*. Princeton, NJ: Princeton University Press.

Partridge, M.D., and Rickman, D.S. (2005). High-poverty nonmetropolitan counties in America: Can economic development help? *International Regional Science Review*, *28*(4), 415–40.

Partridge, M.D., and Rickman, D.S. (2007a). Persistent pockets of extreme American poverty and job growth: Is there a place-based policy role? *Journal of Agricultural and Resource Economics*, *32*(1), 201–24.

Partridge, M.D., and Rickman, D.S. (2007b). Persistent rural poverty: Is it simply remoteness and scale? *Review of Agricultural Economics*, *29*(3), 430–36.

Partridge, M.D., and Rickman, D.S. (2008). Does a rising tide lift all metropolitan boats? Assessing poverty dynamics by metropolitan size and county type. *Growth and Change*, *39*(2), 283–312.

Peters, D.J. (2009). Typology of American poverty. *International Regional Science Review*, *32*(1), 19–39.

Peters, D.J. (2012). Income inequality across micro and meso geographic scales in the Midwestern United States, 1979–2009. *Rural Sociology*, *77*(2), 171–202.

Phillips, R., and Pittman, R.H. (2015). A framework for community and economic development. In R. Phillips and R.H. Pittman (eds), *An Introduction to Community Development* (pp. 3–21). New York: Routledge.

Rupasingha, A., and Goetz, S.J. (2003). The causes of enduring poverty: An expanded spatial analysis of the structural determinants of poverty in the US. (Working Paper No. RDP22). University Park, PA: Northeast Regional Center for Rural Development.

South, S.J., Crowder, K., and Chavez, E. (2005). Exiting and entering high-poverty neighborhoods: Latinos, Blacks and Anglos compared. *Social Forces*, *84*(2), 873–900.

Sparks, P.J., Sparks, C.S., and Campbell, J.J.A. (2013). Poverty segregation in nonmetro counties: A spatial exploration of segregation patterns in the US. *Spatial Demography*, *1*(2), 162–77.

Spencer, J.H. (2004). People, places, and policy: A politically relevant framework for efforts to reduce concentrated poverty. *The Policy Studies Journal*, *32*(4), 545–68.

Taylor, R.B. (2012). Defining neighborhoods in space and time. *Cityscape*, *14*(2), 225–30.

Thiede, B.C., Lichter, D.T., and Sanders, S.R. (2015). America's working poor: Conceptualization, measurement, and new estimates. *Work and Occupations*, *42*(3), 267–312.

Tickamyer, A.R., and Duncan, C.M. (1990). Poverty and opportunity structure in rural America. *Annual Review of Sociology*, *16*(1), 67–86.

Tickamyer, A.R., and Wornell, E.J. (2017). How to explain poverty? In A.R. Tickamyer, J. Sherman, and J. Warlick (eds), *Rural Poverty in the United States* (pp. 84–114). New York: Columbia University Press.

Tickamyer, A.R., Sherman, J., and Warlick, J. (2017). Politics and policy: Barriers and opportunities for rural peoples. In A.R. Tickamyer, J. Sherman, and J. Warlick (eds), *Rural Poverty in the United States* (pp. 439–47). New York: Columbia University Press.

US Bureau of Economic Analysis. (2016). *Real Personal Income and Regional Price Parities*. Retrieved 10 March 2019 from https://www.bea.gov/regional/pdf/RPP2016_methodology.pdf.

US Census Bureau. (1975). *Historical Statistics of the United States: Colonial Times to 1970*. Washington, DC: US Government Printing Office.

US Census Bureau. (2012). *Geographic Terms and Concepts: Census Tract*. Retrieved 13 September 2017 from https://www.census.gov/geo/reference/gtc/gtc_ct.html.

US Census Bureau. (2014). *Poverty: The History of a Measure* (Measuring America Infographic Series). Retrieved 28 March 2018 from https://www.census.gov/library/visualizations/2014/demo/poverty_measure-history.html.

US Census Bureau. (2017). *History of the Official Poverty Measure*. Retrieved 29 August 2017 from https://www.census.gov/topics/income-poverty/poverty/about/history-of-the-poverty-measure.html.

US Department of Agriculture, Economic Research Service. (2017a). *Poverty Overview*. Retrieved 28 March 2018 from https://www.ers.usda.gov/topics/rural-economy-population/rural-poverty-well-being/poverty-overview.

US Department of Agriculture, Economic Research Service. (2017b). *Geography of Poverty*. Retrieved 29 March 2018 from https://www.ers.usda.gov/topics/rural-economy-population/rural-poverty-well-being/geography-of-poverty.

Vu, C.M. (2010). The influence of social science theories on the conceptualization of poverty and social welfare. *Journal of Human Behavior in the Social Environment*, *20*(8), 989–1010.

Waldorf, B. (2006). A continuous multi-dimensional measure of rurality: Moving beyond threshold measures. Selected Paper for the Annual Meetings of the Agricultural and Applied Economics Association, Long Beach, CA, 25 July.

Waldorf, B. (2007). Measuring rurality. *INContext*, *8*(1), 5–8.

Waldorf, B., and Kim, A. (2015). Defining and measuring rurality in the US: From typologies to continuous indices. Paper presented at the Workshop on Rationalizing Rural Area Classifications, National Academy of Sciences, Washington, DC, 16 April.

Wang, M., Kleit, R.G., Cover, J., and Fowler, C.S. (2012). Spatial variations in US poverty: Beyond metropolitan and non-metropolitan. *Urban Studies*, *49*(3), 563–85.

Weber, B. (2007). Rural poverty: Why should states care and what can state policy do? *The Journal of Regional Analysis and Policy*, *37*(1), 48–52.

Weber, B., Jensen, L., Miller, K., Mosley, J., and Fisher, M. (2005). A critical review of rural poverty literature: Is there truly a rural effect? *International Regional Science Review*, 28(4), 381–414.

Wilson, W.J. (1987). *The Truly Disadvantaged: The Inner City, the Underclass, and Public Policy.* Chicago, IL: University of Chicago.

Wilson, W.J. (2008). The political and economic forces shaping concentrated poverty. *Political Science Quarterly*, 123(4), 555–71.

Wilson, W.J. (2013). Combating concentrated poverty in urban neighborhoods. *Journal of Applied Social Science*, 7(2), 135–43.

3. In pursuit of just communities: supporting community development for marginalized communities through regional sustainability planning
Jason Reece

INTRODUCTION

Social and racial equity remain important yet persistent challenges in the United States. These challenges manifest within our metropolitan regions, with segregation, disinvestment and isolation from opportunity presenting profound challenges to achieving greater equity. Segregation is still highly correlated with isolation from opportunity within the geography of the contemporary American city (Acevedo-Garcia et al., 2014). These regional equity challenges have substantial importance, as a growing body of literature suggests that inequality and isolation from opportunity (particularly at the neighborhood scale) have deep impacts on a region's economic sustainability.

Regional inequities between communities can be remedied through regional strategies such as progressive regionalism (Reece, 2018). Models of progressive regionalism (sometimes referred to as the regional equity movement) seek to support community development and equity planning through collaborative interventions at the regional scale (Pezzoli et al., 2009). The regional equity movement began as a response to regional governance efforts which did not address equity concerns. As described by Angela Glover Blackwell and Rhadika Fox in their seminal paper "Regional Equity and Smart Growth: Opportunities for Advancing Social and Economic Justice in America."

> In the 1990s, many of the conversations about regionalism were driven from a smart growth perspective and rarely did these discussions lead with race and equity. Moreover, there had been limited participation from people of color in the smart growth movement. Could meaningful coalitions be built between proponents for smart growth and advocates for equity who would necessarily introduce race and the tough challenge of inner-city disinvestment into the mix? (Glover Blackwell and Fox, 2002, p. 4)

In the context of progressive regionalism, inequalities between communities are addressed through resource sharing, development of strong regional linkages (such as transportation), capacity building and concerted efforts to support the needs of challenged neighborhoods. Additionally, efforts to expand affordable housing regionally provides more housing opportunities to low-income households.

At the beginning of the Obama Administration, sustainability and progressive regionalism concepts were integrated into federal policy, particularly at the US Department of Housing and Urban Development (HUD). The agency's Sustainable Communities' Initiative (SCI), introduced by the Obama Administration as part of its economic recovery platform, sought to take equity theories and principles and translate them to practice at a scale not previously attempted in the US. The Obama Administration would launch the HUD Sustainable Communities Regional Planning and Community Challenge grant program, a program which funded nearly $250 million in grants to more than 150 local planning agencies and consortiums (US HUD, 2012). Part of the HUD SCI grant program included regional planning grants. The SCI program is grounded in principles of sustainable development and progressive regionalism, and infused with a pronounced equity mandate. The equity mandate included the intersection of equity and fair housing concerns in SCI planning principles, the development of regional equity strategies, robust community engagement with traditionally under-represented groups, and the needs of high poverty communities.

SCI regional planning grantees formed regional consortiums, usually led by a lead agency, in most cases a regional planning agency. SCI was informed by six livability principles to guide planning (US HUD, 2016).

- Provide more transportation choices.
- Promote equitable, affordable housing.
- Enhance economic competitiveness.
- Support existing communities.
- Coordinate policies and leverage investment.
- Value communities and neighborhoods.

Several of the livability principles would reflect the tenets of progressive regionalism, with an emphasis placed on providing more transportation and housing choice and targeting support for existing communities

Throughout the implementation of the HUD SCI regional planning grant program, HUD released various policy guidelines, technical assistance and other guidance to ensure better compliance with equity elements embedded within the livability principles. In addition to this assistance, a mandatory assessment was integrated into the planning process for all regional planning grantees in 2011. The Fair Housing and Equity Assessment (FHEA) was a required assessment that clearly linked an analysis of racial and social equity issues to community engagement and plan recommendations. The FHEA acted as a backstop to ensure equity elements were integrated in planning, particularly for grantees' compliance with the Fair Housing Act.

The FHEA was a three-part planning process focusing on the "three Ds" of data, deliberation and decision-making. The FHEA introduced a new geographic focus to equity analysis with a requirement that grantees analyse racial or ethnic concentrated areas of poverty (referred to as RCAP or ECAP areas). Special attention to "RCAP" or "ECAP" areas was required in the analysis for the FHEA. Additionally, the FHEA was to include direct engagement with marginalized communities, and recommendations from the FHEA analysis were to be directly included (or "bridged") into the final comprehensive plan for the region (Rose et al., 2013).

Did SCI Support Progressive Regionalism and Regional Community Development Needs?

The following chapter seeks to understand if programs like SCI and federal guidance can support regional equity. This research is a formative program evaluation of the SCI.

This research seeks to understand the impact of SCI's effort to support fair housing and equity planning, and the implications of the SCI experience. Has the SCI program presented an effective model for the federal government to support equitable regions and to affirmatively further fair housing? This formative evaluation seeks to understand if the SCI's unique federal guidance (and equity mandate) supports the implementation of a progressive regionalism model which can support regional community development needs.

CHALLENGES ADDRESSING EQUITY THROUGH SUSTAINABLE DEVELOPMENT

The SCI was also firmly rooted in sustainable development concepts and practice. Research has found many challenges to utilizing sustainable development in supporting social equity concerns. The inconsistencies and lack of clarity in defining social sustainability create a challenge to sustainable development planning and policy. Michael Gunder notes that sustainability is a "fuzzy concept," one which a common framework is interpreted and defined differently by various stakeholders (Gunder, 2006). Although this "fuzzy concept" can be more socially and politically palatable due to its vague definitions, this can also prove problematic in understanding which particular polices, programs and actions are necessary to produce sustainable communities, just cities or social justice.

Although sustainable development presents a framework to resolve development conflicts in order to balance equity, environmental goals and economic development, some scholars have expressed concern that equity falls short in sustainability planning, becoming a subordinate goal to environmental and economic concerns. Some evidence supports this concern. An early study by Warner found few cities acknowledge environmental justice as a sustainability concern (Warner, 2002). A 2006 survey of city administrations by Devashree Saha and Robert Paterson found limited evidence that cities were fully embracing equity and social justice issues in sustainability efforts (Saha and Paterson, 2008).

The ability of planners to effectively support balancing the three goals of sustainable development is challenging due to professional, fiscal, legislative and political constraints (Campbell, 1996). Studies have focused on the environmental versus social equity conflict in leading cities for sustainable development in the US, Austin, Texas and Portland, Oregon. Tretter's research suggests that sustainable smart growth planning in Austin, Texas selectively favored environmental principles over social equity concerns in the poorer and more racially segregated East Austin area (Tretter, 2013). Goodling et al. found Portland's sustainable development approach has worked to push poverty out of the core of the city, producing a form of "eco-gentrification" (Goodling et al., 2015).

APPROACH AND METHODS

This formative evaluation is a non-experimental design, using a mixed-methods approach. The research design utilizes "methods triangulation" (Denzin, 1978). Much of the data collected is qualitative and additional measures were taken to assure credibility in the analysis of qualitative data. Many strategies are recommended to address credibility concerns in qualitative research (Lincoln and Guba, 1985 and 1986). Several of these strategies, including prolonged engagement, persistent observation, triangulation and member checking were integrated into the research design.

Methods triangulation was achieved by using multiple data sources, including interviews, participant observation and content analysis or plan review. The primary methods of data analysis included an evaluation of plans produced for SCI and a thematic analysis of reporting documents from HUD, grantees and SCI capacity builders. Data sources were refined to the first cohort of grantees (awarded in 2010). These grantees are identified in Figure 3.1.

Participant Observation

Participant observation included activities pertaining to my role as a grantee capacity builder for the SCI initiative between 2011 and 2015. Participant observation included the following activities: grantee webinars, grantee "boot camps," the annual national SCI conference, capacity builder and HUD meetings, and direct one-on-one grantee capacity-building activities. Capacity-building activities were conducted in coordination with HUD, PolicyLink and the Minnesota Housing Partnership. My participation activities included more than 1,200 hours of direct engagement with the SCI program as a capacity builder over a four-year time span. Participant observation activities provided background context to assist in the design of this research.

Plan Evaluation Methods

Plan evaluation methods were utilized to analyse Grantee Final Reports, Grantee Final Regional Plans and Grantee Fair Housing Equity Assessments. The evaluation of final plans and final progress reports documents included a review of analysis, policies and recommendations identified in sustainability plans. Overarching evaluation criteria were informed by the criteria identified by Berke and Godschalk (2009).

Detailed coding criteria are modeled after protocols from the American Planning Association's (APA's) "Comprehensive Plan Standards for Sustaining Places" (American Planning Association, 2015). These criteria were narrowed from the broader set of plan criteria in the APA's "Sustaining Places" plan evaluation criteria. The plan evaluation focused on two distinct subsets of criteria, "equity strategies" and "inclusive engagement." Criteria that were selected were the most consistent with social equity goals in the planning process. The detailed evaluation criteria elements for "equity strategies" and "inclusive engagement" and scoring guidelines are provided in Table 3.1.

Thematic Analysis of Reports: Perspectives of HUD, Grantees and Capacity Builders

Thematic analysis was utilized to analyse participant observation data, interviews and HUD/grantee narrative reports. Thematic analysis involved standard methodological techniques of coding qualitative data and developing themes through six phases of analysis. Several sources of documentation were consulted for this thematic analysis. The primary data reviewed were "Final GTR [Grant Technical Representative] Performance

Grantees
Apache County
Berkshire Regional Planning Commission
California State University, Fresno Foundation
Capital Area Council of Governments
Capital Area Regional Planning Commission
Capitol Region Council of Governments
Central Florida Regional Planning Council
Chicago Metropolitan Agency for Planning
Chittenden County Regional Planning Commission
City of Knoxville, Tennessee
Des Moines Area Metropolitan Planning Organization
East Alabama Regional Planning and Development Commission
East-West Gateway Council of Governments
Evansville Metropolitan Planning Organization
Franklin Regional Council of Governments
Greater Portland Council of Governments
Gulf Regional Planning Commission
Houston-Galveston Area Council
Land-of-Sky Regional Council
Lane Council of Governments
Metropolitan Area Planning Council
Metropolitan Council (MN)
Mid-America Regional Council (RPG)

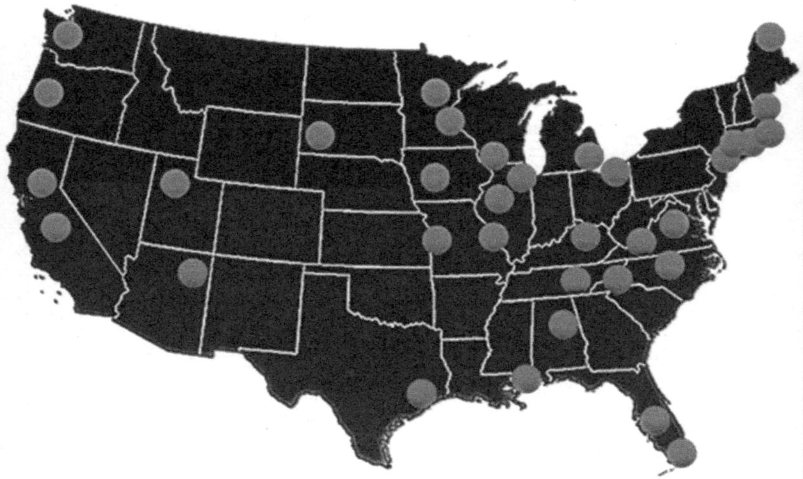

Grantees
New River Valley Planning District Commission
Northeast Ohio Areawide Coordinating Agency
Northern Maine Development Commission
Northwoods Niijii Enterprise Community, Inc.
Piedmont Authority for Regional Transportation
Puget Sound Regional Council
Region Five Development Commission
Regional Plan Association, Inc.
Roanoke Valley Alleghany Regional Commission
Rockford Metropolitan Agency for Planning
Sacramento Area Council of Governments
Salt Lake County
South Florida Regional Planning Council
Southeast Michigan Council of Governments
Southern Bancorp Capital Partners
Southwestern Wisconsin Regional Planning Commission
Thomas Jefferson Planning District Commission
Thunder Valley Community Development Corporation
Thurston Regional Planning Council (RPG)
Tri-County Regional Planning Commission (IL)
University of Kentucky Research Foundation
Windham Region Council of Governments

Figure 3.1 Location and names of the 2010 Sustainable Communities Initiative regional planning grantees

Table 3.1 Plan evaluation criteria and scoring rubric

Equity Strategies Criteria and Definitions	
Criteria	Concept Definition
Promote equitable, affordable housing	Expand location- and energy-efficient housing choices for people of all ages, incomes, races and ethnicities to increase mobility and lower the combined cost of housing and transportation
Support existing communities	Target federal funding toward existing communities—through such strategies as transit-oriented, mixed-use development and land recycling—to increase community revitalization, improve the efficiency of public works investments and safeguard rural landscapes
Provide a range of housing types	A range of housing types is characterized by the presence of residential units of different sizes, configurations, tenures and price points located in buildings of different sizes, configurations, ages and ownership structures. Providing a range of housing types accommodates varying lifestyle choices and affordability needs and makes it possible for households of different sizes and income levels to live in close proximity to one another
Plan for jobs/housing balance	A jobs/housing balance is characterized by a roughly equal number of jobs and housing units (households) within a commuter shed. A strong jobs/housing balance can also result in jobs that are better matched to the labor force living in the commuter shed, resulting in lower vehicle miles traveled, improved worker productivity and higher overall quality of life. When coordinated with multimodal transportation investments, it improves access to employment opportunities for disadvantaged populations
Plan for the physical, environmental, and economic improvement of at-risk, distressed, and disadvantaged neighborhoods	At-risk neighborhoods are experiencing falling property values, high real estate foreclosure rates, rapid depopulation or physical deterioration. Distressed neighborhoods suffer from disinvestment and physical deterioration for many reasons, including (but not limited to) the existence of cheap land on the urban fringe, the financial burdens of maintaining an aging building stock, economic restructuring, land speculation and the dissolution or relocation of anchor institutions. A disadvantaged neighborhood is a neighborhood in which residents have reduced access to resources and capital due to factors such as high levels of poverty and unemployment and low levels of educational attainment. These neighborhoods often exhibit high rates of both physical disorder (e.g., abandoned buildings, graffiti, vandalism, litter, disrepair) and social disorder (e.g., crime, violence, loitering, drinking and drug use)
Plan for improved health and safety for at-risk populations	An at-risk population is characterized by vulnerability to health or safety impacts through factors such as race or ethnicity, socioeconomic status, geography, gender, age, behavior or disability status. These populations may have additional needs before, during and after a destabilizing event such as a natural or human-made disaster or period of extreme weather, or throughout an indefinite period of localized instability related to an economic downturn or a period of social turmoil. At-risk populations include children, the elderly, persons with disabilities, those living in institutionalized settings, those with limited English proficiency and those who are transportation disadvantaged

Table 3.1 (continued)

Equity Strategies Criteria and Definitions

Criteria	Concept Definition
Provide accessible and quality public services, facilities, and health care to minority and low-income neighborhoods	A public service is a service performed for the benefit of the people who live in (and sometimes those who visit) the jurisdiction. A public facility is any building or property—such as a library, park or community center—owned, leased or funded by a public entity. Public services, facilities and health care should be located so that all members of the public have safe and convenient transportation options to reach quality services and facilities that meet or exceed industry standards for service provision. Minority and low-income neighborhoods are often underserved by public services and facilities and health care providers
Upgrade infrastructure and facilities in older and substandard areas	Infrastructure comprises the physical systems that allow societies and economies to function. These include water mains, storm and sanitary sewers, electrical grids, telecommunications facilities and transportation facilities such as bridges, tunnels, and roadways. Upgrading is the process of improving these infrastructure and facilities through the addition or replacement of existing components with newer versions. An older area is a neighborhood, corridor or district that has been developed and continuously occupied for multiple decades. A substandard area is a neighborhood, district or corridor with infrastructure that fails to meet established standards
Plan for workforce diversity and development	Workforce diversity is characterized by the employment of a wide variety of people in terms of age, cultural background, physical ability, race and ethnicity, religion and gender identity. Workforce development is an economic development strategy that focuses on people rather than businesses; it attempts to enhance a region's economic stability and prosperity by developing jobs that match existing skills within the local workforce or training workers to meet the labor needs of local industries
Protect vulnerable populations from natural hazards	A natural hazard is a natural event that threatens lives, property and other assets. Natural hazards include floods, high wind events, landslides, earthquakes and wildfires. Vulnerable neighborhoods face higher risks than others when disaster events occur. A population may be vulnerable for a variety of reasons, including location, socioeconomic status or access to resources, lack of leadership and organization and lack of planning

Inclusive Engagement Criteria and Definitions

Criteria	Concept Definition
Engage stakeholders at all stages of the planning process	Engaging stakeholders throughout the planning process—from creating a community vision to defining goals, principles, objectives and action steps, as well as in implementation and evaluation—is important to ensure that the plan accurately reflects community values and addresses community priority and needs. In addition, engagement builds public understanding and ownership of the adopted plan, leading to more effective implementation
Seek diverse participation	A robust comprehensive planning process engages a wide range of participants across generations, ethnic groups and income ranges. Especially

Table 3.1 (continued)

Criteria	Concept Definition
Inclusive Engagement Criteria and Definitions	
in the plan development process	important is reaching out to groups that might not always have a voice in community governance, including representatives of disadvantaged and minority communities
Promote leadership development in disadvantaged communities during the planning process	Leaders and respected members of disadvantaged communities can act as important contacts and liaisons for planners in order to engage and empower community members throughout the planning process. Participation in the process can encourage development of emerging leaders, especially from within communities that may not have participated in planning previously
Provide ongoing and understandable information for all participants	Information available in multiple, easily accessible formats and languages is key to communicating with all constituents, including non-English speakers. Such communication may involve translating professional terms into more common lay vocabulary
Use a variety of communications channels to inform and involve the community	Communications channels that can be used throughout the planning process include traditional media, social media and Internet-based platforms. Different constituencies may prefer to engage through different channels

Note: Scoring rubric (plans are scored from 0 to 3 on each criteria element): Not present = 0; Low achievement = 1; Medium achievement = 2; High achievement = 3.

Assessments," the 2- to 5-page documents prepared for each grantee by HUD field staff, on the grantee's performance and outcomes. Most of the assessments follow a very specific structure, beginning with basic administrative questions on performance, a rating of grantee performance (measured as good, fair or unsatisfactory) and a narrative report. Narrative reports focused on several issues, including administrative requirements and deadlines, engagement, equity, final deliverables and implementation readiness. Of the 2010 regional grantees, 36 out of 45 had accessible evaluation reports. At this time, it is unclear if reports were prepared for the other nine grantees.

FINDINGS

Plan Evaluation: Inclusion of Equity

Plan evaluation was conducted on 26 of the 45 2010 cohort grantees of the SCI. Grantees produced numerous planning documents, and all documents were reviewed, but for the purpose of this analysis only the regional comprehensive plan was scored for the plan

Table 3.2 *Average score for equity strategy and inclusive engagement criteria for each grantee*

Grantee Region	Average Equity Strategy Score	Average Engagement Score
East Alabama	1.09	1.60
Apache County	1.88	1.80
Capital Region CT	2.63	0.00
Southern Bankcorp	2.38	1.60
Windham CT	1.88	1.20
Central Valley	1.25	0.00
Central Florida	0.38	0.00
Des Moines	2.47	1.80
Sacramento	2.22	2.40
Ken Ten (KY/TN)	1.25	0.60
Franklin (MA)	1.75	2.00
Tri County (IL)	1.00	1.00
East West Gateway	2.75	1.90
Rockford Metro (IL)	2.50	1.80
Gulf Coast (MS)	2.75	2.40
Piedmont (VA)	2.28	2.65
Land of Sky (NC)	1.78	2.00
Met Council (MN)	2.25	2.40
South Florida (FL)	1.63	2.80
Roanoke Valley VA	0.50	1.20
Chittenden County VT	2.75	3.00
New River Valley VA	2.38	2.40
Northwoods NiiJii WI	1.88	1.60
Puget Sound WA	2.59	2.60
Thurston Regional WA	2.75	2.20
Thunder Valley (SD)	2.63	2.80

Note: 0 = lowest score; 3 = highest score; N = 26.

analysis. Plans were scored on a 0 to 3 scale for each criteria element. For the purpose of this analysis, scoring was focused on the inclusion of *equity strategies* and *inclusive engagement activities*. All plan evaluation scores for each grantee analysed are provided in Table 3.2. Although the discussion below analyses the cohort of grantees as a whole, it should be noted that plan evaluation outcomes vary widely between grantees (as seen in Table 3.2).

The analysis of plan evaluation criteria for equity strategies provides insight into how equity was represented in planning strategies or recommendations. The primary equity strategies identified in plan analysis were related to housing, infrastructure, workforce development and community reinvestment. Among specific equity strategy criteria, average scores (on a scale of 0 to 3) were highest for housing strategies and infrastructure strategies (Figure 3.2). "Provide a range of housing types" had the highest average score of 2.49. The criteria of "plan for jobs/housing balance" was tied with "upgrade infrastructure and facilities in older and substandard areas" for the second highest average score with 2.25. The lowest score for specific equity strategy criteria was for "protect vulnerable

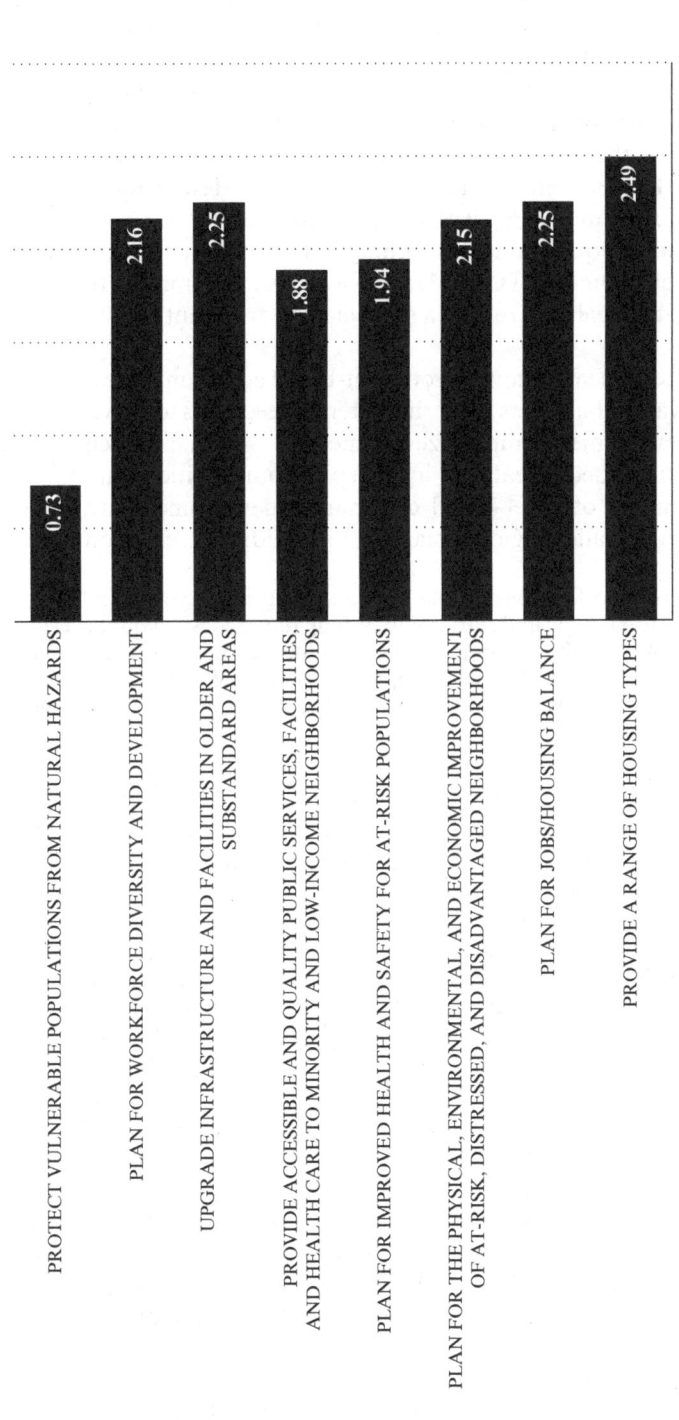

Note: 0 = lowest score; 3 = highest score; N = 26.

Figure 3.2 Average score for equity strategy criteria for all grantees

populations from natural hazards" with an average score of 0.73. The identification of affordable housing and infrastructure as the top elements for planning are not surprising given the influence of Fair Housing Equity Assessment (which was the most defined set of standards for SCI grantees) and the natural fit between metropolitan planning agencies (MPOs) and infrastructure planning, particularly transportation infrastructure.

Additional qualitative themes emerged in assessing the various equity strategies within planning documents. Larger metropolitan regions such as Metropolitan Area Planning Council (MAPC), Met Council, Puget Sound, the Regional Plan Association and the Capital Region focused on equity concerns (primarily risk of gentrification) in the context of transit-oriented development (TOD). Health was also a common theme in grantee plans, both focusing on health care but also social determinants of health impacting communities.

Tribal grantees placed a heavy emphasis on asset-based community development, looking internally to cultivate tribal assets and other internal resources for development. Asset-based community development emphasizes immediate action and self-empowerment of communities; it is intended to catalyse and inspire communities who have long been marginalized. The tenants of asset-based community development are evident in the language of the Thunder Valley regional plan for Pine Ridge Reservation:

> How long are you going to let other people decide the future for your children, are you not warriors? It's time to stop talking and start doing. A long time ago when our ancestors rode into battle they didn't know what the outcome was going to be but they did it because they knew it was in the best interest of the children and people. Don't operate from a place of fear, operate from a place of hope, anything is possible but you need to take action, the movement is here, the time is now. (Thunder Valley, 2014, p. 25)

The Arkansas grantee, Southern Bankcorp, clearly articulated the distinction between traditional economic development and equitable economic development practice. Equitable economic development efforts focused on building up existing assets and being "people" focused:

> Traditional economic development plans focus on attracting outside capital to a jurisdiction to create local jobs; however, to ensure more sustainable and equitable growth and opportunity, the economic growth strategies of the Economic Development Scenario emphasize opportunities to invest in local human capital and to maximize and build upon existing assets and entrepreneurial activity. (Southern Bankcorp, 2014, p. 25)

Based on the APA criteria inclusive engagement seeks to actively bring stakeholders into the planning process, making materials available through multiple mediums and languages, and at its highest levels, seeks to promote leadership development in the community. SCI grantees generally were very focused on equitable and inclusive engagement. Numerous grantees expressed the role of inclusive engagement as a key element to incorporating equity into the planning process. The Thunder Valley (SD) grantee described the engagement and deliberation aspects of SCI as a natural extension of Lakota culture and traditional decision-making:

> Listening is an important value shared by many of us on the Pine Ridge Reservation. It requires not just respectful silence but the ability to be open to what is being said. The work in our Oglala

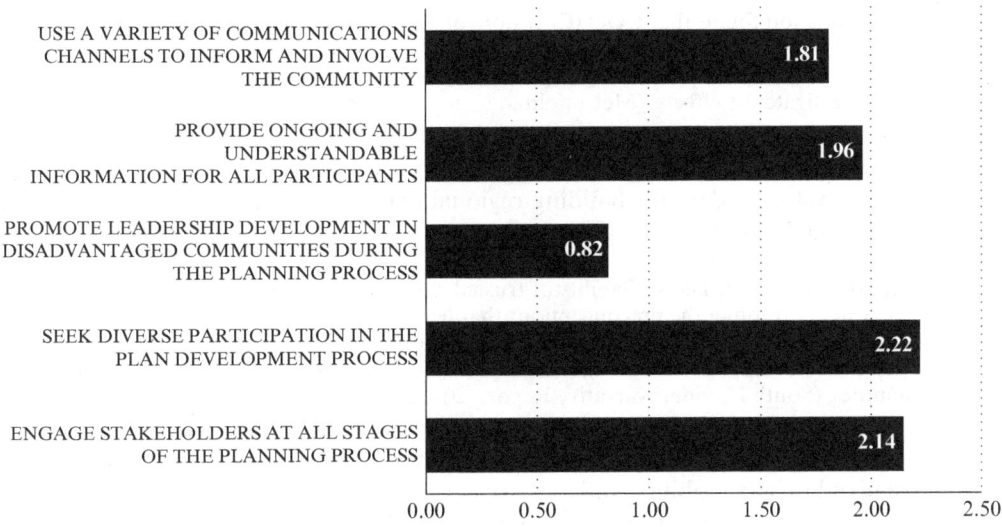

Figure 3.3 Average score for engagement criteria for all grantees

> Lakota Plan is a re-telling of what was said by elders, youth, and everyone in-between. Including as many voices as possible helped us to provide a well-rounded set of recommendations and also is important to fueling necessary changes. (Thunder Valley, 2014, p. 39)

The most common representations of inclusive engagement identified in plan analysis were meeting criteria of "seeking diverse participation" and "engaging stakeholders in all stages of the planning process" (Figure 3.3). The least common form of inclusive engagement activity was integration of efforts to "promote leadership development in the planning process in disadvantaged communities." The integration of inclusive engagement may have been directly influenced by direction from HUD (the original Notice of Funds Available (NOFA) for the program emphasized engagement with under-represented groups), and the extensive capacity builder assistance provided to grantees to ensure inclusive engagement strategies.

Thematic Analysis of Equity Concerns: Grantee and HUD Perspectives

Most grantees identified the SCI process as building community capacity, depth of analysis, shared understanding and shared vision with regards to social equity. Only a minority of grantees reported no reflections on social equity, or no reference to social equity concerns in their final reports. The Sacramento region identified the SCI process as an "inflection point to begin working with our members and low-income communities in new ways" (Sacramento, Narrative Report, 2014). Several grantees identified that the housing aspect of the SCI plan was the first time a regional vision and plan were created for housing. As described by the Metropolitan Council (Twin Cities) regional grantee:

the final evaluation found that COO [Corridors of Opportunity] has "teed up" a conversation that is critical to the region: one about the impacts of gentrification on neighborhoods and the quality of life. More than just raising the issue . . . it has developed several inter-related strands of work to mitigate the effects. (Metropolitan Council, Narrative Report, 2014)

Data was seen as a critical element for building conversations around social equity concerns. The value of data for building regional planning consensus was identified by the South Florida grantee:

The importance of data. The availability of trusted, unbiased data that is communicated well has the ability to cut through the preconceptions that people and organizations have about existing conditions and future trends and productively impact policy discussions. People are oftentimes not fully aware or appreciative of the challenges and opportunities facing the region and its communities. (South Florida, Narrative Report, 2014)

Equity issues referenced were primarily identified in the context of housing affordability, fair housing or housing mobility, and concerns regarding implementation. Other grantees focused on references to traditional community development, blight elimination, health efforts and food security projects. Small business development and minority contracting were also identified as topics of focus in equity planning. The inter-relationship between TOD and gentrification was a focus for larger urban grantees.

HUD evaluations identified political challenges and the political influence of regional planning organization boards as a barrier to stronger equity performance in some regions. These political pressures undermined consortium autonomy, fostered more conservative stances on equity concerns and created barriers to challenging the status quo. Grantees that operated primarily "top down" planning processes also did not create opportunities for community voices to impact the planning process, particularly the voices of under-represented and marginalized communities. The historical legacy of discrimination in the regions also impacted equity outcomes. These concerns are described by HUD field staff in multiple grantee evaluations:

Although SEMCOG [Southeast Michigan Council of Governments] organized and convened a consortium in accordance with the grant requirements, it never operated as an independent entity and reported directly to SEMCOG's Executive Committee that had ultimate decision-making authority. (HUD, Southeast Michigan, 2014)

That said, equity was not a major priority for this grantee. Their decision-making structure relied on a pre-existing agreement between the PDC [Planning District Commission], City, County, and University, with little room for community voices. (HUD, Thomas Jefferson Planning, 2014)

We did not always find them to be receptive to suggestions about how to increase levels of inclusion. This may have been a product of the fact that everything had to be approved by their board of directors, a conservative body composed of elected officials. (HUD, East West Gateway, 2014)

This criticism, though, should be tempered by acknowledging that the Des Moines region, while growing more diverse, is historically overwhelmingly White and much of it is deeply uncomfortable with the idea of expanding access to opportunity for low-income families and people of color, which is manifested in the clearly-stated opposition to more greater access to affordable housing and public transportation. (HUD, Des Moines, 2014)

The relationship between high capacity and high performance was a common theme in evaluations of grantees identified as strong performers. These grantees primarily were large metropolitan planning organizations (MPOs), with a track record of success and significant staff capacity. For example, the Chicago Metropolitan Agency for Planning (Chicago's MPO) received a national award for its planning process, and was also identified as a top performer of the cohort:

> CMAP (Chicago region) was an exceedingly competent and high-performing grantee. It received a prestigious first-of-its-kind award from the American Planning Association (the "National Excellence Award for a Planning Agency") for its GO TO 2040 plan, and its implementation of the LTA [Local Technical Assistance] program continues its reputation for excellence and best practices. I rate CMAP as one of the best performers among all 17 of my metropolitan area regional grantees and have nothing but praise for its overall performance. (HUD, Chicago Metropolitan, 2014)

Grantees who excelled with engagement processes were identified as strong performers by the HUD field staff in evaluations. Prolonged engagement and multi-faceted engagement processes, particularly those who could reach under-represented groups, were applauded in HUD evaluations. Illustrated in the following example from the Capital Region grantee:

> Civic engagement took place consistently through three years of the grant's period of performance. They used a combination of civic leadership labs, stakeholder interviews, web-based tools like Metroquest, Twitter, Facebook, Survey Monkey, focus groups, meetings, shared meals, workshops, visioning sessions, and traditional media outreach to engage a broad cross-section of residents in the region, including special efforts to reach traditionally marginalized populations. (HUD, Capital Region, 2014)

HUD also highlighted the work of grantees who made significant financial investments in engagement processes, particularly those that built the capacity for engagement in the region. These engagement efforts were described as transformational in building capacity for the regions.

> The Met Council committed a significant amount of its grant, $750,000, to fund community organizations working mostly in the Central, Southwest and Bottineau LRT [Light Rail Transit] corridors. The Community Engagement Team (CET) was led by Nexus Community Partners, the Alliance for Metropolitan Stability, and the Minnesota Center for Neighborhood Organizing. Twenty-three separate organizations received funding. Met Council reports that, among other things, at least 40,000 people residing in the corridors were made more aware of the projects and their potential impact, and that at least 250 people increased their capacity for leadership in their communities. (HUD, Metropolitan Council, 2014)

DISCUSSION AND CONCLUSION

The SCI was a unique, well-resourced experiment for the planning field, in which the federal government played a more robust role in supporting sustainable and equitable development. The SCI experience provides insight into planning theory, policy and practice, particularly in sharpening our understanding of equity planning. The program demonstrated that a more proactive federal role by HUD is not without complication, but can be beneficial.

Several immediate challenges would hamper the program's implementation. Some grantees did not have the organizational capacity or technical expertise to implement a complex and multi-faceted program such as SCI. Organized Tea Party resistance would challenge community engagement activities in some grantee regions. Many grantees also struggled with the complexity of addressing deep structural and societal inequalities impacting inequity in their regions.

The Limitations for Regional Equity

Dialogue and data alone can fall short in fostering implementation if they encounter strong structural and political barriers. As many grantees expressed, the equity conversation was challenging and often fraught with conflict. South Florida reflected deeply on the community and leadership divide in regards to social equity in the planning process, more explicitly detailing the divergence in perspectives that make consensus difficult:

> The equity discussion is a difficult one. As a society, our ability to address issues of equity, fairness and justice is made more difficult due to the decline of social capital and shared community identity. The ability to support, empathize, sympathize and connect to others who we many not know or who are different from ourselves is a critical aspect of this discussion. Because "equity" is a politically and emotionally charged word that means different things to different people, the equity conversation fits the frame of the "wicked problem" that is unstructured, cross-cutting, and relentless. Some work within a frame of "givers" and "takers." It will be difficult to get past that frame unless they are open-minded, intellectually curious, and willing to revisit their current impressions and beliefs. On the other hand, the equity "choir" needs to be equally open-minded and willing to listen and work with others who may not sync with their beliefs to move this discussion forward. Data will continue to be very important as it provides a factual basis for discussion. . ...It will be equally difficult for some to accept that treating everyone the same is not the same as treating everyone fairly. (South Florida, Narrative Report, 2014)

The conflicts illustrated by South Florida were not uncommon and HUD supported capacity-building resources where necessary to push along the difficult equity dialogue within grantee regions. It is challenging to imagine the SCI program producing the same equity outcomes without the addition of these capacity builders or under the guidance of another federal agency (such as the Environmental Protection Agency or Department of Transportation).

Equity efforts within sustainability are challenging because of the history of structural racism in the US. SCI illustrated the US's distinct political culture and the difficulty in remedying our long and conflicted history of racial and ethnic discrimination. The politicization of equity policies was also evident in the conservative and Tea Party resistance to SCI. Race has been a "wedge" issue utilized in political context throughout history, and has been particularly powerful in driving a wedge between groups and undermining solidarity in attempts to bolster labor and address class divides. Since the Nixon Administration's "southern strategy," race has played a substantial role in building support for conservative causes, parties and political candidates. The political organizing against SCI should be interpreted through this political history.

Scholarship and historical evidence supports the argument that localism (the antithesis of regionalism) was a reaction to provide distance from the racialized "other" and to counter desegregation efforts. The concept of "White flight" to suburban enclaves was fueled by this desire to be separate from "the other." Equity planning efforts will require significant time and energy to break down the development patterns, and policy or institutional structures created by nearly a century of pro-segregation values. Consequently, embracing equity (particularly racial equity) in the context of sustainability in the US will require effort, intentionality and persistence. As experienced with SCI, these barriers are often too much for local agencies to surmount without resources, guidance and sometimes regulatory pressure.

These challenges are exemplified by the experience of Southeast Michigan Council of Governments (SEMCOG), a grantee who received one of the worst evaluations of any grantees, and the only major metropolitan region to receive a "fair" rating in its HUD evaluation. The critique of SEMCOG is aided by additional contextual details about the conflicted history of SEMCOG in supporting equity in the Detroit region.

As one of the most racially segregated regions in the nation, Detroit has a complex and challenging history. Multiple race riots, a legacy of housing discrimination and resistance to integration has plagued Southeast Michigan (Sugrue, 2005). SEMCOG has played a critical role in this complex history. Legal scholarship has identified the agency as an example of structural racism in planning practice, because of the agency's disengagement with urban communities and communities of color, and loyalty to Detroit's predominately White suburbs. These White suburbs were forged by the White flight from Detroit and are traditionally hostile to the needs of Detroit and Wayne County. These same suburban jurisdictions drive the SEMCOG board, creating a political disincentive for the agency to advocate for the equity needs of Southeast Michigan's largest city.

Civil Rights advocates sued SEMCOG in 2004 for its imbalanced political structure. In Gary Benjamin's "SEMCOG's Business as Usual: A Failed Model," the author documents the repeated resistance of SEMCOG to supporting racial equity concerns, particularly in respect to transportation and housing in the region (Benjamin, 2011). The author notes that the imbalanced political structure of SEMCOG was created and continues to support the racialized White flight and sprawl which has dominated the region:

> Not surprisingly, the governing structure of the Southeast Michigan Council of Government (SEMCOG), as it arises out of our regional racial history, is ill-equipped to deal with the problems presented by a region where the us versus them mentality is such a strong force. The decisions made on a regular basis by SEMCOG are made through a governance structure that reflects our regional racial history. Specifically, SEMCOG's governance structure is one that relies on municipal units to participate using a "one government, one vote" philosophy. This philosophy is in itself unjust because many of the participating municipalities were created, or grew larger, because of racism. (Benjamin, 2011, p. 156)

Given the structural and political challenges to support equity in Southeast Michigan and SEMCOG's poor history, it is not surprising the grantee performed poorly. In the end, HUD's efforts to improve the outcomes of SEMCOG (and all grantees) had limitations, as expressed in the final evaluation of SEMCOG:

> Although HUD never intended regional consortiums to be subservient to an MPO board or any other governmental or quasi-governmental entity, HUD's efforts to resolve this situation were fruitless. At the end of the day, SEMCOG's initiative was largely a top down, staff- and MPO-driven effort with little significant public input or engagement. (HUD, Southeast Michigan, 2014)

Diverse Regions and Diverse Community Development Needs

The equity planning efforts from SCI also exemplify the complexity of equity planning in a diversifying US. US equity concerns vary significantly across our nation and even within regions or communities. Equity challenges are diverse due to differences in geography, population, economic conditions and local culture, as demonstrated by SCI grantees.

"Hot market" metropolitan regions, such as Seattle, Austin and Boston, focused primarily on challenges created by gentrification (displacement and affordable housing), while also attempting to ensure more economic benefits are reaching marginalized groups. "Weak market" metropolitan regions, such as St Louis or Northeast Ohio, were attempting to stem the continued disinvestment in urban areas, while also supporting regional fair housing solutions to provide access to opportunity in suburban areas.

Rural communities ranging from Appalachia to the Mississippi Delta or the Central Valley of California sought to strengthen rural economies, support human development and bring investment back to struggling small towns and cities. Tribal areas looked internally to understand how to best leverage tribal community assets to bring opportunity to reservation lands and counter intergenerational poverty. Meanwhile, across the nation, regions sought to better understand the needs of new immigrant populations and foster equitable community engagement.

Challenges but Progress

Despite these many challenges, multiple sources of data indicate that equity planning efforts did improve in most SCI regions. More importantly, the SCI planning process fostered dialogue, engaged under-represented communities and built capacity to further equity goals in the regions. Given the inherent challenges and tensions in addressing racial and social equity in our nation, and the historical difficulties in promoting equity in sustainability planning, these positive outcomes should be applauded.

The communicative aspect of SCI was also essential to the equity component of the regional planning process. The power of dialogue and engagement built a foundation for better outcomes and empowerment. Engagement and deliberation with under-represented and marginalized groups was a foundational element of SCI. Most grantees leveraged these engagements to build better institution-community relationships. The intense (rational) analytics of the regional plans were helpful in providing a base of knowledge to inform and help "make the case" for equity solutions. Equity issues such as fair housing or environment justice are harder to attack as top-down federal, state or regional intrusion when there is a strong foundation of community voices calling for them and data justifying them.

Top performing SCI grantees all noted that the emphasis on social equity and engagement made a lasting impact on organizations, with institutionalized reforms occurring as a by-product of the SCI planning process. In the case of these grantees, the federal investment for SCI was catalytic and the relationship-building and policy or institutional reforms these grantees experienced may not have happened without SCI. These particular communities illustrate the value of federal investment in regions to be transformative leaders, building models of practice for other communities to emulate.

Sustainability has had a poor track record of engaging issues of social equity in US planning practice. SCI demonstrated that sustainability can be a strong framework to engage issues of equity, but guidance and external incentive is needed to ensure this happens. Addressing equity through sustainability is not something that comes easily or naturally (given the context of the challenges in approaching equity in the US), nor is there necessarily the capacity for many planning organizations to embrace equity without robust guidance.

HUD efforts with SCI supported an innovative pilot program, demonstrating that federal funds can be catalytic in regions with adequate readiness (such as Boston, Minneapolis-St Paul and Seattle), and to build capacity in regions with limited resources and limited collaboration (such as Pine Ridge Reservation). Additionally, the enhanced attentiveness to equity illustrated that with an equity "mandate" or support, sustainability can be a framework that adequately embraces equity planning and community development.

REFERENCES

Acevedo-Garcia, Dolores, McArdle, Nancy, Hardy, Erin, Crisan, Unda Ioana, Romano, Bethany, Norris, David, Baek, Mikyung, and Reece, Jason. (2014). The Child Opportunity Index: Improving Collaboration between Community Development and Public Health. *Health Affairs.* 33(11), 1948–57.

American Planning Association. (2015). Comprehensive Plan Standards for Sustaining Places. Accessed 10 November 2019 at https://www.planning.org/sustainingplaces/compplanstandards/.

Benjamin, Gary. (2011). SEMCOG's Business as Usual: A Failed Model. *The Journal of Law in Society.* 13(1), 155–94.

Berke, P., and Godschalk, D. (2009). Searching for the Good Plan: A Meta-Analysis of Plan Quality Studies. *Journal of Planning Literature.* 23(3), 227–40.

Campbell, Scott. (1996). Green Cities, Growing Cities, Just Cities? Urban Planning and the Contradictions of Sustainable Development. *Journal of the American Planning Association.* 62(3), 296–312.

Denzin, N.K. (1978). *Sociological Methods.* New York: McGraw-Hill.

Glover Blackwell, A., and Fox, R. (2002). Regional Equity and Smart Growth: Opportunities for Advancing Social and Economic Justice in America. Funders' Network for Smart Growth and Livable Communities. Translation Paper 1. Edition 2.

Goodling, E., Green, J., and McClintock, N. (2015). Uneven Development of the Sustainable City: Shifting Capital in Portland, Oregon. *Urban Geography.* 36(4), 504–27.

Gunder, Michael. (2006). Sustainability: Planning's Saving Grace or Road to Perdition? *Journal of Planning Education and Research.* 26(2), 208–21.

Lincoln, Y.S., and Guba, E.G. (1985). *Naturalistic Inquiry.* Beverly Hills, CA: Sage Publications.

Lincoln, Y.S., and Guba, E.G. (1986). But Is It Rigorous? Trustworthiness and Authenticity in Naturalistic Evaluation. *New Directions for Program Evaluation.* (30), 73–84.

Metropolitan Council. (2014). Final Narrative Report for the HUD Sustainable Communities Initiative. Submitted to the US Department of Housing and Urban Development.

Pezzoli, Keith, Michael Hibbard, and Laura Huntoon. (2009). Introduction to Symposium: Is Progressive Regionalism an Actionable Framework for Critical Planning Theory and Practice? *Journal of Planning Education and Research.* 28(3), 336–40.

Reece, Jason. (2018). In Pursuit of a Twenty-First Century Just City: The Evolution of Equity Planning Theory and Practice. *Journal of Planning Literature.* 33(3), 299–309.

Rose, Kalima, Reece, Jason, Bergstrom, Dannielle, and Olinger, Jillian. (2013). The Fair Housing and Equity Assessment (FHEA): Potential Roles and Responsibilities for Stakeholders in the FHEA Process. A collaborative report and technical assistance guide published by PolicyLink and The Kirwan Institute for the Study of Race and Ethnicity for the US Department of Housing and Urban Development's Sustainable Communities Initiative.

Sacramento Area Council of Governments. (2014). Final Narrative Report for the HUD Sustainable Communities Initiative. Submitted to the US Department of Housing and Urban Development.

Saha, Devashree, and Paterson, Robert. (2008). Local Government Efforts to Promote the "Three E's" of

Sustainable Development: Survey in Medium to Large Cities in the United States. *Journal of Planning Education and Research.* 28(1), 21–37.

South Florida Regional Planning Council. (2014). Final Narrative Report for the HUD Sustainable Communities Initiative. Submitted to the US Department of Housing and Urban Development.

Southern Bankcorp Community Partners. (2014). Final Narrative Report for the HUD Sustainable Communities Initiative. Submitted to the US Department of Housing and Urban Development.

Sugrue, T.J. (2005). *The Origins of the Urban Crisis: Race and Inequality in Postwar Detroit.* Princeton, NJ and London: Princeton University Press.

Thunder Valley Community Development Corporation. (2014). Regional Plan: Oyate Omniciye – Oglala Lakota Plan.

Tretter, E.M. (2013). Contesting Sustainability: "SMART Growth" and the Redevelopment of Austin's Eastside. *International Journal of Urban and Regional Research.* 37(1), 297–310.

US Department of Housing and Urban Development. (2012). Initial Report to Congress. Office of Sustainable Housing and Communities. Sustainable Grant Program Evaluation.

US Department of Housing and Urban Development. (2014). Sustainable Communities Initiative Grantee Evaluation. Capital Region Council of Governments.

US Department of Housing and Urban Development. (2014). Sustainable Communities Initiative Grantee Evaluation. Chicago Metropolitan Agency for Planning.

US Department of Housing and Urban Development. (2014). Sustainable Communities Initiative Grantee Evaluation. Des Moines Area Metropolitan Planning Organization.

US Department of Housing and Urban Development. (2014). Sustainable Communities Initiative Grantee Evaluation. East West Gateway Council.

US Department of Housing and Urban Development. (2014). Sustainable Communities Initiative Grantee Evaluation. Metropolitan Council.

US Department of Housing and Urban Development. (2014). Sustainable Communities Initiative Grantee Evaluation. Southeast Michigan Council of Governments.

US Department of Housing and Urban Development. (2014). Sustainable Communities Initiative Grantee Evaluation. Thomas Jefferson Planning District.

US Department of Housing and Urban Development. (2016). Six Livability Principles. Accessed 13 July 2016 at http://portal.hud.gov/hudportal/HUD?src=/program_offices/economic_resilience/Six_Livability_Principles.

Warner, Kee. (2002). Linking Local Sustainability Initiatives with Environmental Justice. *Local Environment.* 7(1), 35–47.

4. Asset-Based Community Development (ABCD): core principles
Ivis García

INTRODUCTION

In *Building Communities from the Inside Out: A Path toward Finding and Mobilizing Community Assets* (1993), John P. Kretzmann and John L. McKnight noticed that universities and other institutions focused exclusively on the needs, deficiencies and problems of neighborhoods. In 1969, when they started their work, deindustrialization had caused massive shifts in US cities—leaving people unemployed and communities economically devastated. Instead of focusing on needs, John and Jody, as part of an Urban Research Center at Northwestern University, conducted a four-year research project that focused on residents and their gifts, talents, capacities and creativity. The basic idea was that concentrating on what was working as opposed to what was not working could help promote community development. By focusing on success stories, as told by community residents, institutions like universities, nonprofit organizations and philanthropic foundations could identify how they could support residents instead of providing them with the services they thought that residents needed. They named this way of thinking "Asset-Based Community Development."

Kretzmann and McKnight conceptualized two options, paths or dilemmas available to community leaders, institutions and others involved in neighborhood work. The first path emphasized the needs of the community addressing its deficiencies. In other words, concentrating on the half-empty glass (see Figure 4.1). Another path is committed to exploring new opportunities and discovering individuals—and by proxy communities—capabilities and assets. This path concentrated on the half-full glass. They saw this as a dilemma that institutions like Northwestern University faced, where John and Jody worked. Focusing on the deficits, however, is the more traditional path.

Figure 4.1 "Is the glass half-empty or half-full?"

THE TRADITIONAL PATH: NEEDS-BASED APPROACH

Images of South Bronx in New York City, South Central in Los Angeles and the South Side of Chicago have negative impressions on most Americans. This is because these spaces are depicted as criminal, violent, and are found to be largely associated with drugs and criminal activities. Negative conceptual images are a clear depiction of a troubled community. With the rational of deficiency oriented policies and programs, the needs of the community act as a basic fundamental principle to discuss problems among outside decision-makers. University research and funding has contributed to study the problem in detail by using funding to study community problems without offering solutions. When solutions are offered, they do not come from the residents either in terms of their voice or their capacity to make their own communities better. The value of services and policies coming from the outside becomes the only resource to solve community problems. Because of this paternalistic relationship between social service providers and residents, people have started to see themselves as clients whose needs can only be satisfied by outsiders. For example, welfare recipient's forms ask people about all of their problems such as: Are you a dropout? Are you about to be evicted? Have you abused illegal substances, and so on? These surveys make people feel worthless, that they have many problems, and cannot see their positive contributions to society.

In this way, institutions, including universities, create labeled people, where residents are identified by their deficits. Some of the labels might include school dropout, single-mother, at youth-risk, welfare recipient, among many others. Social service organizations in particular use some of these labels for their clients. In fact, one of the mistakes social service organizations made is that they treat people like clients, instead of treating people like citizens—we do not refer to legal status, but those who live, are part of, or claim their right to the city. The government has emulated human service agencies that supply funding to communities, and all this is controlled based on problem-oriented data. These problem-oriented data are evaluated as a result of surveys conducted regarding needs of communities. Residents have begun to consider themselves as being dependent, and may think of themselves as not being capable enough to possess control over their own lives. This approach breeds hopelessness. People start saying to themselves: we are deficient, we are a poor community, we are hopeless.

The needs-based or deficit-based approach to community development has become institutionalized across government, nonprofits and universities. For example, foundations request proposals with a needs statement. In a nutshell, they are asking what is wrong with the community. At times, the more deficits in a community, the more the likelihood of receiving funding. Programs usually concentrate on community needs such as poverty, crime, unemployment, poor health, slum housing and so on. Foundations tend to request a neighborhood needs map, while institutions tend to create them to obtain funding. A neighborhood needs map might include unemployment, truancy, illiteracy and a whole host of community problems (see Figure 4.2).

The community is viewed as being associated with a large number of problems and deficiencies which need to be addressed not by them but by outside institutions with power and money. There are a number of consequences of favoring neighborhood needs-based maps conceptually. For example, residents internalize that solutions come from the outside. Not only that, but that outside resources are required to cope with the identified

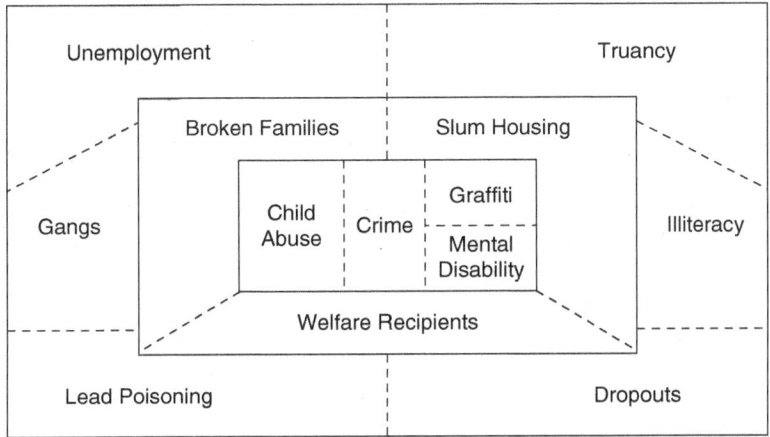

Source: Kretzmann and McKnight (1993, p. 3).

Figure 4.2 Needs-based map

needs. Since people are merely the recipients of services, the problem-solving capacities of the individual are ignored and discouraged. Resources are targeted to provide the necessary funding to service providers and professionals rather than the residents.

Moreover, the leadership of the local community is profoundly affected negatively by the availability of resources which lays its foundation on the needs map. In other words, community leaders need to speak of the community needs to advocate for resources in their communities. This is a contradiction to community organizing which presumes that people are empowered to create social change. Meanwhile, the funds provided relying on the needs map have depicted that only outside help could offer the necessary solution to the identified problem. The renewal of funding must assure that the problem should decrease in comparison to previous years. Institutions have to prove that their intervention results in less unemployment and fewer dropouts and teenage pregnancies, for example. This survival strategy, both from the dependency of funding from institutions and the dependency of residents for services, can only work based on a needs map focusing on an individual client rather than a plan that is developed for the entire community. An approach based on the needs map may never benefit individuals and their communities over the long run. This is because this way of thinking cannot bring about positive social changes at large. When institutions do not find funding for programs, efforts dwindle and may not achieve sustainability. Given the many negative consequences of favoring a needs-based map, a paradigm shift is needed.

THE ALTERNATIVE PATH: ASSET-BASED COMMUNITY DEVELOPMENT

Asset-Based Community Development (ABCD) emphasizes the creation of policies and activities involving the capacities and skills of neighborhood residents. ABCD comes from the recognition that development of an entire community can only take place if residents are able to invest their gifts and themselves in the process. Instead of depending entirely on outside resources and charity, it is better to start the process of development from within the community—that is, from the inside out. This truth has been recognized much earlier by neighborhood leaders than by researchers and social service providers. The efforts dedicated to the development of the community will be successful only if there is a clear understanding of the internal assets and capabilities of the community. Connecting all local assets of the neighborhood is one step towards rebuilding communities. This does not imply that nonprofits, foundations and universities should abandon communities, and residents need to do everything themselves. What it does say is that, if we are intentional in building communities from the inside out, power will multiply.

Before writing the green book as many community development practitioners refer to *Building Communities from the Inside Out*, Kretzmann and McKnight conducted extensive fieldwork. They visited about 20 cities and had conversations with more than 2,000 residents. Instead of asking, "What does this community need?," they asked, "Can you tell us what people who live on this block or neighborhood have done together that made things better?" After listening to thousands of stories John and Jody came up with a list of resources. They decided to call these resources assets. Initially, they identified five community assets, and about ten years later they recognize a sixth asset. A community asset map is then composed of: (1) individuals, (2) associations, (3) institutions, (4) land and the physical environment, (5) exchange and (6) culture and stories (see Figure 4.3).

The assets are the basic building blocks of a thriving neighborhood. The capacities of the individual (his or her gifts) are appropriately at the center of the asset map. There are three forms of gifts:

1. Gifts of the head (subjects residents know, e.g., law, mathematics, medicine and so on),
2. Gifts of the hands (practical skills that residents know, e.g., playing a musical instrument, fixing a car, taking care of children, videography, writing and so on) and
3. Gifts of the heart (issues residents care deeply about, e.g., childhood education, healthy eating, biking infrastructure and so on).

Associations of individuals, whether formal or informal such as block clubs, youth or older adults groups, are critical to solve problems. Groups of residents form associations because they are passionate about an everyday activity, interest or care. These associations might have no staff or staff with low pay, but what matters is that they always create the vision and engage in work with volunteers to achieve goals. Residents are found to be primarily associated with many types of religious, cultural and recreational activities. These associations act as an essential tool to offer solutions to the existing problems in neighborhoods. ABCD conceptualizes association visually as a circle (see Figure 4.3) because people come together in an organic manner. Associations usually make decisions based on consensus. There is an excellent power in associations, which often goes unrecognized. Associations can be very useful in action—they are an amplifier of gifts, creative and can reach large numbers of residents.

Private institutions such as businesses, nonprofits, hospitals, along with public institutions such as libraries, colleges and human services play a significant role in community

Figure 4.3   Six community assets

development. The distinction of an institution is that they are hierarchical in nature. A nonprofit, for example, might have an executive director, directors of different programs and employees. Decision-making is usually from the top down. Because of this, ABCD conceptualizes institutions, the second asset, in the form of a triangle or a hierarchy (see Figure 4.4).

It is worth noting the differences between associations and institutions (refer to Figure 4.4). Associations are about choice because people can come and go freely. People or citizens join associations by choice because they care about a particular issue, topic or activity. Associations thrive from individual capacities and residents collectivizing. On the other hand, institutions are about control (budgets, personnel, and so on); employees at institutions produce goods and services for consumers or clients according to their needs. However, there is power in institutions as well. They can provide resources for projects that residents have identified. With their funding, they can compensate residents for their work. They can amplify the voices of residents by recognizing their gifts and allowing them to reciprocate.

The fourth asset is physical space—that is, land, buildings, rivers, roads, and so on. Communities need to recognize local physical assets in order to mobilize them for their

▲	●
CONTROL	CHOICE
PRODUCTION GOODS SERVICES	CARE
CLIENTS CONSUMER	CITIZEN
NEEDS	CAPACITY

Source: Adapted from basic training slides, https://resources.depaul.edu/abcd-institute (accessed 9 November 2019).

Figure 4.4 Differences between associations and institutions

benefit. For example, a vacant lot could be an excellent opportunity for a community garden. There might be other new and unexpected uses for buildings and land. What seemed a glass half-empty (or unused space) can become a glass half-full, just by changing one's perspective.

The fifth asset is exchange, which could be conceptualized as the local economy and how money flows through a neighborhood. Economic development is mainly targeted here as a way to promote the success of the entire community. The development process is entirely asset-based whereby the skills and capabilities are also evaluated. An asset-based strategy is a path created to promote the development of the community. In terms of economic-based theory, it is essential to maintain the dollars of a community within its boundaries, so there is a multiplier effect of those dollars (reiterative spending with the local economy). Large corporations tend to take profits out of the area and invest them somewhere else. In that way, money is leaked out of the community, profiting outsiders but not residents. But exchange can also happen without money. For example, in "Buy Nothing" groups neighbors can borrow a lawn mower or offer free furniture. A community garden where people exchange produce can be another great way of strengthening the community.

Finally, the sixth asset is culture and stories. This asset was identified about ten years after the green book was written. By listening to many different stories of successful community building over many years, it became apparent that stories in themselves were assets. John and Jody did not come up with this list of assets themselves. Residents were aware of their local assets, and they told stories about how they mobilized these assets. Identifying assets is the first step for neighborhood action. The second step is to connect or mobilize assets. To connect assets, there must be a "connector," that is, usually an individual that knows other individuals or that recognizes associations or local institutions that can help other individuals, associations or institutions to achieve a goal. There are some characteristics of connectors:

Asset-Based Community Development (ABCD)

(1) they are gift centered, (2) well connected, (3) trusted by others and (4) believe that community is welcoming.

Using an ABCD approach, neighbors can ask three questions: (1) what can we achieve by using our own assets? (2) what can we achieve with our own assets if we get some outside help? and (3) what can we not do with our assets that must be done by outsiders? Institutions cannot do what communities can do, and communities cannot do what institutions can. This goes back to the idea that institutions can have their own role to play in community development. One of the roles of institutions is to empower residents. The "Citizen Power Progression" also known as the "Power Ladder" is a way to conceptualize how residents can move from being victims, full of needs, to producers and when in partnership with institutions to co-producers. When residents are producers or co-producers, they have control of resources for community problem-solving.

CONCLUSION

In order to give citizen power, institutions do not look at neighborhoods that are poor and need to be fixed (refer to Figure 4.5). Instead, they understand that communities have assets that can be leveraged for development. Institutions that follow the more

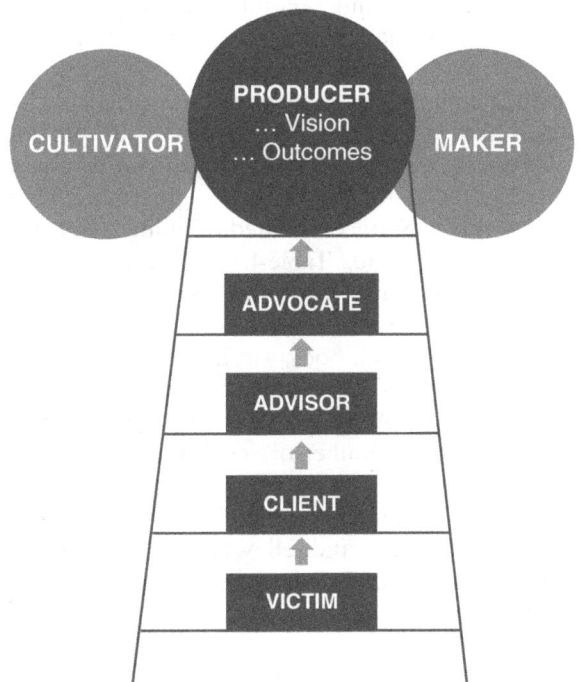

Source: Adapted from basic training slides, https://resources.depaul.edu/abcd-institute (accessed 9 November 2019).

Figure 4.5 Citizen Power Progression

traditional path of a needs-based approach often ask what is wrong with a community. Meanwhile, an asset-based approach considers what works already and how assets could be mobilized. While a needs-based map shows unemployment, truancy, illiteracy and other deficiencies, an asset map will show individuals, associations, institutions, land and the physical environment, exchange, culture and stories. A needs-based approach would argue that human capital emanates from the outside from institutions while an asset-based approach would argue that it comes from within the community. On the one hand, people are identified by their deficiencies and on the other hand, by their gifts. In a service mindset, individuals are clients while in the asset-based mindset they are citizens. In one paradigm, change is achieved by providing more resources and services. In another paradigm, change is achieved by engaging the community in their own self-determination.

Funding is spent on paying administrators and professional community developers in the needs-based approach. In the ABCD approach, we might also pay professionals, but we recognize that most of the funding should go to community residents for their work and the community projects that they envision. The model of development in a needs-based approach sadly creates dependence, as opposed to empowering people by recognizing their gifts and capacities. One leadership model is based on professional staff while the alternative is about institutions facilitating processes that are citizen-led. The most valuable resource under a needs paradigm is financial resources while ABCD is about connecting people and creating relationships. ABCD recognizes that money is not the answer but people and relationships are. Money comes after recognition and awareness of this.

ABCD tries to identify community connectors—people who connect individuals to individuals, associations and institutions. Connectors invite others to bring their gifts and they encourage others to create a solution. When people create things themselves, they own them (and are vested in them). There is a great potential to have a multiplier effect using an ABCD approach. Service-oriented thinking is not sustainable because the effort often dwindles when funding ends. Table 4.1 provides an overview of the differences between the needs-based and asset-based approaches.

ABCD is a straightforward idea: focus on what is strong and not on what is wrong. John and Jody did not expect that the green book would have sold more than 125,000 copies, used as a framework in more than 80 books and cited in more than 10,000 scholarly articles. As of 2019, there were 52 ABCD faculty and 22 fellows in the ABCD Institute at DePaul University. There are a number of regular conferences in the United States, Canada, Australia, India, the United Kingdom, to mention a few. ABCD 101 trainings are conducted twice a year in Chicago and other cities in the United States to raise funds for the Institute. A quote that faculty and fellows at the ABCD Institute frequently use to conclude their ABCD 101 training is from Lao Tzu titled "On Going to the People" and it goes like this:

> Go to the people. Live among them. Learn from them. Love them. Start with what they know. Build on what they have. But of the best leaders when their task is done the people will remark "We have done it ourselves."

This is the ABCD way. We invite you to join our movement.

Table 4.1 Needs- versus asset-based: a paradigm shift

	Needs approach	Assets approach
What institutions generally think	This community is poor and it needs to be fixed	This community has assets that can be leveraged
The data shows	What is wrong with a community	What is great already and what can be improved
Map shows	Unemployment, truancy, illiteracy and so on	Assets (individuals, associations, institutions and so on)
Human capital	Comes from the outside from institutions	Comes from within the community
People are identified	By their deficiencies	By their gifts
Individuals are	Clients	Citizens
Change is achieved	By providing more resources and services	By engaging the community in their own self-determination
Model of development	Dependence	Empowerment
Leadership model	Professional staff	Facilitated by institutions but citizen-led
The most valuable resource	Financial resources	People and relationships, but in particular "connectors"
In the long term	Dwindles	Snowballs

Source: Created by the author.

REFERENCE

Kretzmann, J.P., and McKnight, J. (1993). *Building Communities from the Inside Out: A Path toward Finding and Mobilizing a Community's Assets.* Evanston, Ill: The Asset-Based Community Development Institute, Institute for Policy Research, Northwestern University.

5. Stepping up the ladder: reflecting on the role of nonprofit organisations in supporting community participation*

Julia Fursova

INTRODUCTION

I have been a practitioner in community development and health promotion for 17 years, which includes 7 years of work outside Canada, and 10 years in Toronto (Tkaronto), one of the largest metropolitan centres in North America (Turtle Island). I have been privileged to live and work on the territory known as Tkaronto. I am cognisant of broken covenants and treaties, and aware of the profound and urgent need for a peaceful co-existence between all human and non-human inhabitants of the Earth. I am committed to honoring history, spirituality and culture of the Indigenous peoples of Turtle Island and to honoring and protecting this land in solidarity with her Indigenous stewards.[1] My research work is part of how I practice such commitment.

This chapter is informed by my doctoral research examining the role of community-based nonprofit organisations in supporting community action for health justice. I analyse nonprofits as positioned within capitalism – an institutional system based on subordination resulting from unequal access to and distribution of resources. The research uses governmentality theory, in particular its discussions of power and discourse, as an overarching analytical framework (Foucault, 1982; McHoul and Grace, 1995). To centre on the experiences of the most marginalised participants, I apply an intersectional feminist framework as an epistemological lens, and institutional ethnography and participatory action research as a methodology. The chapter describes the application of a participatory mapping tool based on Arnstein's (1969) ladder of citizen participation and provides the analysis of the findings with the application of the self-evaluation tool for action in partnership developed by Bilodeau et al. (2017). Using a case study of a neighbourhood-based network of nonprofit organisations, public agencies, grassroots groups and community residents, I discuss the roles nonprofit organisations may perform when addressing the participation of community residents and groups. Reflecting on power relations in the context of partnerships between community and nonprofit organisations, I describe three potential roles for the nonprofit organisations: managing, maintaining and sustaining participation. I further discuss such roles in connection with the levels on the ladder of community participation.

COMMUNITY DEVELOPMENT AS PRACTICE IN THE CONTEXT OF NEOLIBERAL RESTRUCTURING OF THE NONPROFIT SECTOR

My research is located at the intersection of community development and health promotion within the context of activities carried out by neighbourhood based nonprofit organisations tasked with projects and initiatives aimed at addressing social determinants of health and improving community wellbeing. Many such projects have community participation and empowerment embedded in both the goals and the values of the programmes implemented by nonprofit organisations (Rootman et al., 2001). The rhetoric of participation has been on the rise across institutions including nonprofit organisations, governments, business and public sectors, often under the labels of stakeholder participation, community partnership and community engagement, to name a few. There is a widespread concern, however, that participation is a potentially malleable concept, and can be reframed to satisfy the needs of those granted with institutional power to convene participatory processes without challenging power relations that undermine participation in the first place (Taylor, 2007; Cornwall, 2008; Katz et al., 2015). In this chapter, I examine nonprofit organisations as actors strategically positioned in the civil society and social reproduction sphere to support community participation. I discuss how nonprofits are structured by the neoliberal capitalist system to mould community participation as a process consistent with neoliberal rationale and values.

The nonprofit sector and its practices are subjected to the same political and economic pressures, trends and influences as the rest of the public sphere. Neoliberal restructuring has shaped the social and the public spheres of major capitalist and developing economies since the late 1970s and early 1980s. Brown (2003) describes the workings of neoliberalism in a radically free market as "maximized competition and free trade achieved through economic deregulation, elimination of tariffs, and a range of monetary and social policies favourable to business and indifferent toward poverty, social deracination, cultural decimation, long term resource depletion and environmental destruction" (p.1). It has been argued that neoliberal policies of deregulation and privatisation led to intensification of extractivist behavior and practices (Kidd, 2016; Gago and Mezzadra, 2017). Extractivism originated as a mode of accumulation in the capitalist system and has been a mechanism of colonial and neocolonial appropriation (Acosta, 2013). Initially, the term was used to refer to the activities associated with removal of large quantities of raw materials, but the term has been recently extended beyond extraction of natural resources to include extraction of labour, knowledge, information and land/spaces (Acosta, 2013; Kidd, 2016; Gago and Mezzadra, 2017). Extractivism is accompanied by the processes of enclosure and dispossession. The former is linked to privatisation and commodification of natural resources, public goods and services. Dispossession signifies the processes through which people and communities are displaced from the land, such as Indigenous communities during the process of colonisation, or low-income residents in urban neighbourhoods subjected to gentrification. Social service cutbacks, and austerity policies applied to nonprofit and public sectors that erase the gains of an improved social safety net, can also be considered as markers of extractivist behavior that accompany enclosure and dispossession (Kidd, 2013).

Neoliberal roll-out during the 1980s and 1990s affected the ability of the nonprofit sector to respond to the roll-back of the welfare state (Pross and Webb, 2003). The relations between the nonprofit sector and the state have been increasingly structured as contract-based relationships (Wolch, 1990; Phillips, 2003) where activities of nonprofits are increasingly streamlined to conform to a neoliberal rationale praising individual effort and responsibility while obfuscating systemic inequalities. Furthermore, neoliberalism emphasises efficiency, professionalism, and accountability to funders rather than the sector's constituents. Such priorities reshaped the practices of the nonprofit sector as more elitist, professional-led and technocratic, focused on service delivery outputs, rather than on the activities related to advocacy and public policy (Wolch, 1990; Evans et al., 2005). Within civil society neoliberalisation processes shifted the focus from collective to individual responsibility (Phillips, 2003; Pross and Webb, 2003; Evans et al., 2005; Shaw, 2011; English and Mayo, 2012). Within the community development context this shift has manifested in the proliferation of capacity building efforts aimed at community members in the context of *community empowerment* but with an emphasis on personal growth and development while displacing the focus on the communities and their relationship with institutional structures such as funders, government, public and nonprofit organisations (English and Mayo, 2012). Such trends place issues of power and participation in the centre of the critical discussions concerning research and practice in community development.

The Entanglements of Power, Participation and Empowerment

There are many typologies of participation developed and adapted to different contexts of participatory processes. Most notable are ladders of citizen participation by Arnstein (1969), a typology of participation by Pretty (1995), children and youth participation by Hart (1992) and a typology of different forms of participation by White (1996), to name only a few. These typologies are useful for understanding degrees, levels and forms of participation in different contexts. Arnstein's (1969) ladder became the most well-known and widely used tool in community development. It includes eight levels and three degrees of power, with *partnership*, *delegated power* and *citizen control* as higher levels of citizen power. Arnstein argues that real citizen participation is impossible without sharing and redistributing power, with citizen control being identified as the highest level of participation.

All typologies of participation deal with dangers of co-optation of participatory processes, with potential misuse of the concepts of empowerment or self-mobilisation in the absence of critical interrogation of power differentials among actors invited to participate. Tensions between actors, terms of involvement, and the context within which participation takes place inevitably create a *politics of participation* within which different actors exercise varying degrees of power. It is worth noting that *participation* does not necessarily guarantee *empowerment*; participation is empowering only when it is approached with an explicit focus on addressing power inequities among those who participate (Laverack, 2001).

Arnstein's (1969) ladder showed that participation at the lower levels of manipulation and therapy can be, in fact, disempowering. Yet even participation at higher levels on the ladder of participation, in an action that does not have serious implications for the process, brings less empowerment than participation at the *informing* rung in relation to an action that has more profound consequences for those who participate, as much depends on the context and scale (Cornwall, 2008). Tritter and McCallum (2006, p. 57) discussed Arnstein's ladder in the context of health care service-user participation, noting "the failure to consider the essential role of users in framing problems and not simply in designing solutions." In the context of early stakeholder involvement in framing problems, Bilodeau et al. (2017) present a similar argument on the importance of participation in developing *options for action*, rather than participation only in the implementation of action. Similarly, the depth and breadth of participation matter. Participation on the lower rungs on the ladder, for example on the *informing* rung, for a more diverse group of participants may result in more equitable participation than participation on a higher rung for a selected few, who do not reflect the full diversity of the community (Cornwall, 2008).

Cooke and Kothari (2001) discuss participation itself as a form of power that can contribute to strengthening existing power relations when approached uncritically and from a utilitarian perspective. Kesby (2005) interrogated their otherwise useful critique disputing the "unproblematic privileging of resistance" along with the premise that as a form of power participation can only be resisted (p. 2018). Instead, Kesby argued for more "spatially embedded" and "theoretically aware" participatory praxis (p. 2038). This proposition echoed Laverack's (2001) argument that participation is only empowering when it has the goal of transforming existing inequitable power relations. Power inherent in participatory processes can and should be used to counteract more domineering power structures by drawing on marginalised frameworks in order to destabilise dominant forms of power (Kesby, 2005). Power is not necessarily a negative feature; what matters is how we use power and what we mean by this rather elusive notion.

The governmentality framework developed by Foucault (1982) describes power as a less tangible force of discourse permeating social relations as a combination of language, cultural and institutional practices constantly reinforcing and reproducing itself through normalisation of what is generally considered as "common sense" (Foucault, 1982; Kesby, 2005; Felluga, 2011). Governmentality theorists argue that in its discursive form, power extends beyond the state and its institutions by permeating mass consciousness and public discourse and thus enlarging and maximising state power (Taylor, 2007). While a Foucauldian definition of power as discourse is perhaps the most comprehensive, and captures aspects of power that are not easily commodified, I strongly believe that a more concrete and tangible definition is more useful in the community development context. As an empowering practice, community development deals with negotiating and sharing power; therefore it is very important to be clear about what we seek to reclaim and share when initiating participatory processes. Power is not simply a commodity, but it does rely on the existence and inequitable distribution of commodities to reproduce itself. Intangible as it may be, it is still embedded in materiality. I approach power as a combination of institutionally valued and recognised skills, knowledge and expertise (i.e. those that constitute hegemonic discourse) coupled with access to decision-making and resources (i.e. material aspects).[2] The sharing of power involves the sharing of skills, knowledge and expertise, and it must also involve the sharing of access to resources to exercise decision-making. Otherwise, the sharing of the knowledge and skills in participatory processes may result in abuse and misuse of power by those who have greater access to resources. I will continue this discussion with the above definition of power in mind, based on a premise that the value of participatory processes lies in the ability to equalise power among actors.

Understanding the Role of Nonprofit Organisations in Enabling Community Participation

To understand how the role of nonprofit organisations when addressing community participation is structured as part of power relations, I adopted institutional ethnography as a methodology combined with participatory action research as an approach to conduct inquiry into the experiences of participation of different actors in community development work. I studied the experiences of participation in a collaborative action among community activists, frontline workers, and managers of nonprofit organisations in the context of community development work.

The research is centred on the experiences of the most marginalised actors[3] in community development processes. To make visible the experiences of such actors and the social relations that shape such experiences, I adopted an intersectional feminist lens. Such a lens afforded intellectual and spiritual space within the research inquiry for embracing different dimensions related to experiences of participation, including community history, lived experiences of diverse actors, diverse knowledge and social locations[4] and enabled focus on power differential among diverse actors in the process (Hankivsky et al., 2014). Institutional ethnography (Smith, 2006) as a methodology and participatory action research (Wallerstein and Duran, 2008) as an approach to inquiry shaped my overall research design and the choice of methods for data collection, as shown in Figure 5.1.

Analytical Framework	Governmentality Theory		
Epistemological Lens	Intersectional Feminist Framework		
Approach and Methodology	Participatory Action Research and Institutional Ethnography		
Methods	Texts Analysis	Individual Interviews	Participatory Discussions and Workshops

Figure 5.1 Analytical framework, epistemological lens, approach, methodology and methods

This chapter presents data specific to one case study selected from a broader range of research data. It focuses on the application of a participatory mapping tool and analytical framework used for understanding how institutional power relations shape participation of different actors in the context of nonprofit-community partnerships. The example of a neighbourhood based network of nonprofit organisations, public agencies[5] and community residents and groups that came together as the "Neighbourhood Table"[6] is presented.

CASE STUDY: NEIGHBOURHOOD TABLE – RESIDENTS VERSUS AGENCIES

My research examines the notions of power and participation in the context of the "Neighbourhood Table" – a semi-formal neighbourhood based network of nonprofit organisations, public agencies, community residents and emerging groups that came together as the Neighbourhood Table to address the increase of gun violence in the neighbourhood.[7] The Neighbourhood Table never had independent core funding and was supported through voluntary and in-kind contributions from its agency-members, community residents volunteering their time and skills, and through securing project-based funding through community development grants.

During its 12-year history, the Neighbourhood Table went through periods of increased funding for community engagement projects that accompanied City-led revitalisation/gentrification of the neighbourhood, and periods of austerity marked by budget cuts to public services and community-based nonprofits. The level of resident engagement with the Neighbourhood Table and its capacity to support community engagement fluctuated over the years. Participation in the Neighbourhood Table was always associated with increased responsibilities and administrative commitments for participating organisations. When faced with reduced funding, organisations struggled to respond to the increased administrative and frontline workload as they were adapting to diminished capacity. This limitation affected organisations' ability to stay involved with the Neighbourhood Table and to support community engagement projects at the neighbourhood level, as many organisations' priorities were shifting toward service delivery. Thus, the Neighbourhood Table's ability to support collaborative community development projects in partnership with community members was not consistent and was organised around short-term project-based funding. In short, the history of the Neighbourhood Table mirrors neoliberal restructuring of the nonprofit sector that took place over the course of the last 15 years.

Envisioned as a venue to bring residents, community groups, nonprofit organisations and public agencies to work together and to create a platform for community members to voice their concerns, the Neighbourhood Table was not necessarily equipped in terms of resources and internal structure to respond to concerns raised. Depending on the capacity, the scale, and internal policies and procedures of participating agencies, the opportunity for a dialogue with residents led to different outcomes. Occasionally, instead of working toward collaborative solutions, an organisation/agency withdrew from the Neighbourhood Table when frustrated by community demands. Community residents also started expressing their disappointments and frustration over cumbersome or unclear decision-making processes, the lack of cohesive action by the Neighbourhood Table members, and the lack of clarity regarding roles and responsibilities of the members. Understandably, such frustrations resulted in a diminished participation of community residents in the Neighbourhood Table. The divide and tensions between agency-members and resident-members deepened, and the level of community engagement with the Neighbourhood Table dropped. In response, the Neighbourhood Table held an extensive community consultation process and reorganised its structure to offer resident-members more opportunities to access decision-making roles. Presently, the Neighbourhood Table is organised into several action committees, according to areas of action prioritised by each committee, for example, youth development and leadership, community safety, healthy living, seniors, and so on. Each committee is co-chaired by an elected resident-member and agency-member, who also serve on the Steering Committee, the main decision-making body of the Neighbourhood Table. The Steering Committee is also co-chaired by an agency-member and resident-member.

As part of a participatory action research project, current and former resident-members of the Neighbourhood Table took part in designing an evaluation framework to assess the effectiveness of the new structure and to understand the quality of the relationships between agency-members and resident-members. Arnstein's (1969) ladder of citizen participation was chosen as a tool for participatory mapping to understand which rung on the ladder reflected the experience of participation for resident-members and agency-members, and to see if, and to what extent, members' experience of participation in the Neighbourhood Table differed depending on their roles. Arnstein's (1969) model was selected for its relative simplicity and because it is a relatable and adaptable tool with a long history of use in the context of community development. The ladder was, however, adapted to the context of nonprofit and community partnership. The adaptation of the ladder was an iterative collaborative process where I, as a researcher, presented drafts to the Neighbourhood Table evaluation working group for further work. The final version went through three editions before we ended with a product that we collectively were satisfied with and agreed to bring to the Neighbourhood Table meeting for participatory mapping activity.

The proposed ladder of community participation does not offer *citizen control* as the highest rung for participation in the original Arnstein's model. This is because this level is not necessarily pursued in the context of nonprofit community development work. The highest level on the proposed ladder becomes *delegated power* and more accurately reflects *participation* as both a goal and an intrinsic value inherent in community development activities supported by nonprofit organisations. For our purpose, the rung of *delegated power* in Arnstein's original ladder was renamed *co-production*, and the rung of *partnership* was renamed *co-design*, in order to better represent an effective and equitable partnership and *meaningful participation* as defined by community members (see Table 5.1). *Co-design* involves shared leadership and includes sharing of knowledge, skills and expertise, but resources are still controlled independently by each partner involved. The level of *co-production* involves a shared pool of resources as well as sharing of skills, knowledge and expertise and access to decision-making concerning the use of the shared resources.

The strength of the co-production concept resides in the inherent notion of partnership. Having its roots in civil rights and social justice work, co-production extends beyond consultation and involves collective development of service delivery models intended to transform wider social systems (Realpe and Wallace, 2010). The Neighbourhood Table involves nonprofit organisations providing services in the community, and therefore involves community members, who at different points of their lives are service-users of those agencies. Co-production is also one of the intended goals of the Neighbourhood Table envisioned as a venue for community members and agencies to come together to discuss and tackle issues of concern, including those related to service provision. It has been agreed that *co-production* is the rung of the ladder that is sought as a purpose of participation for the Neighbourhood Table members, including both types of partners – institutional partners represented by nonprofit organisations and agencies, and community partners, represented by individual residents and community groups.

The rung labelled *placation* in the original Arnstein model was changed to *accommodation*; for placation was perceived as too negative a term in the context of a partnership between neighbourhood based nonprofits and community. These groups enter the relationship in good faith, with nonprofits genuinely striving to support community par-

Table 5.1 Ladder of community participation

Arnstein's (1969) original rungs of participation	Neighbourhood Table's adapted levels of participation	Quality of participation for resident-members in Neighbourhood Table
delegated power in partnership	*co-production* (shared knowledge, skills, expertise, access to decision-making AND resources) *co-designing* (shared knowledge, skills and expertise)	meaningful participation
placation	*accommodation* (involvement with limitations)	symbolic participation
consultation	*consultation*	
informing	*informing*	
decoration and manipulation	*decoration and manipulation*	non-participation

ticipation. Yet many organisations found themselves structurally constrained to the level where they are conflicted between efforts to respond to community needs and restrictions imposed by funding requirements and conventional accountability models. Such a level of participation on the ladder was described as *accommodation* to reflect participants' involvement with limitations imposed by structural constraints, such as lack of staff time, space, materials resources and other shortcomings resulting from insufficient or interrupted funding.

There are three degrees of participation *quality* identified in the proposed ladder: non-participation, symbolic participation and meaningful participation. The degrees of quality and their characteristics were identified during the participatory evaluation design workshops where resident-members of the Neighbourhood Table, developed a set of indicators defining *meaningful participation*. Table 5.1 presents the ladder of community participation as developed in collaboration with the Neighbourhood Table members.

During a participatory activity to assess the quality of participation, members of the Neighbourhood Table were invited to identify the level of participation on the proposed ladder of community participation that was reflecting their experience of participation according to their roles. Resident-members and agency-members were distributed colour-coded dots according to their role, for example, resident-members, agency-members who are frontline staff and agency-members who are managers. This was done for the purpose of differentiating between different members of the group who each exercise varying degrees of power over their own and other members' participation. Sixteen people took part in the activity. Participants were invited to place a colour-coded dot at the level on the ladder that best described their experience of participation in the Neighbourhood Table. Approached this way, evaluation of participation captured not only quantitative aspects of participation by tracking the number of members at each rung of the ladder but also captured differences in the quality of participation for different members of the Neighbourhood Table according to their roles. Participants were also invited to share a story to illustrate the position of their dot on the ladder.

When responding to the mapping activity some resident-members noted that their experience differed depending on their role as action committee chairs versus their role as general members. Resident-members-at-large assessed their participation at the level between consultation and accommodation. When referring to their roles as committee co-chairs, they assessed their participation at the levels of co-production (n = 1) and co-design (n = 2). When assessing their participation as general members, residents noted that it varied between *accommodation* and *consultation* levels depending on projects. To reflect this, they chose an in-between or intermediary level (n = 5). Some resident-members assessed their participation at the informing level (n = 2).

It is notable that community partners described their experience of participation as general members only at the lower degrees of *symbolic* participation (n = 8) while institutional partners – agency-members (n = 4) and resident-members who were committee chairs (n = 3) described their experiences at the degrees of *meaningful* participation. Only two agency-members assessed their participation at the degree of symbolic, a frontline staff (n = 1) at the level of *accommodation* and an agency-member/manager (n = 1) at the in-between level of *accommodation* and *co-design*. Being in the position of the committee chair did not change the quality of participation for the agency-members. The following section of this chapter examines what factors and conditions influence the quality of participation for different members of the Neighbourhood Table.

UNDERSTANDING PARTICIPATION EXPERIENCES FOR DIFFERENT TYPES OF PARTNERS

To understand factors and conditions influencing experiences of partners in the nonprofit-community partnership, and how to advance the participation of resident-members in the Neighbourhood Table against the rungs of the ladder of community participation, I used Bilodeau et al.'s (2017) self-evaluation tool for action in partnership. This tool was chosen because action in partnership between agency-members and resident-members is the aspirational goal the Neighbourhood Table pursues. For resident-members to reach the degree of participation that they described as "meaningful" and to participate at the levels of "co-design" or "co-production," resident-members must be present as equal partners in the Neighbourhood Table.

The original tool is based on a mid-range theory that defines six requirements for effective partnership work from a series of case studies based on actor network theory (Bilodeau and Kranias, 2019). The tool makes it possible to identify enablers and barriers to effective and equitable action in partnership with reference to each of the six requirements. Each requirement comes with a set of objectives and indicators that help assess the overall strength of the requirement. In total there are 18 objectives that are unevenly spread between the six requirements. The original tool is organised as a questionnaire where the strength of each objective is assessed through "weak," "medium" and "strong" options.

The six requirements identified in the tool relate to participation dynamics, and partnership arrangements supporting equitable participation of all actors involved, and collective rather than individual action (Bilodeau and Kranias, 2019). The tool is centred on participation and each of the six requirements addresses an aspect of participation necessary for an effective and equitable action in partnership:

A) Who participates, reflecting on the range of actors involved in partnership;
B) What are the options for participation, including options for participating in developing options for action versus options for implementation only;
C) What is the extent of participation, that is how partners with the least power are engaged in negotiating and influencing decisions;
D) How is participation sustained, reflecting on the commitment of strategic and pivotal partners;
E) Is participation empowering, reflecting on partnership arrangements that equalise power among partners; and
F) Is participation collective rather than individual, reflecting on partnership arrangements that help build collective rather than individual action.
(Bilodeau and Kranias, 2019, p. 3; Fursova et al., 2018, p. 1)

To illustrate the importance of each participation requirement for achieving effective and equitable action partnership, I arranged the requirements in the shape of a wheel, where each segment represents a requirement (Figure 5.2). There is a symbolism in a wheel shape, where every segment needs to be present for the wheel to keep its shape and therefore its functionality, or in a partnership context, its effectiveness. When one or

Source: Adapted from Fursova et al. (2018).

Figure 5.2 *Partnership assessment wheel*

86 *Research handbook on community development*

more segments are underdeveloped or missing, there will be a bumpy ride, or the wheel (partnership), may fall apart.

I set up the original questionnaire from Bilodeau et al. (2017) as a framework for analysis using NVivo12 software to appraise the extent to which each requirement is met in the Neighbourhood Table. The indicators proposed in the original tool were applied to analyse stories shared by members during participatory workshops and individual interviews, in relation to their experience of participation.

Using the Neighbourhood Table as a case study of nonprofit-community partnership I analysed experiences of participation, paying attention to power differential between nonprofit and public agencies as institutional partners, and residents and grassroots groups as community partners. Institutional partners bring programs and services to the community, as well as other resources, including small-scale funding. Community partners are on the receiving end of these resources as service-users, or applicants for the small grants distributed and managed by the nonprofit organisations or agencies.

The indicators for each of the six requirements for effective action in partnership for the Neighbourhood Table varied between weak and medium, reflecting a greater degree of participation for institutional partners. Figure 5.3 illustrates the approximate strength of each requirement for partners' participation, where 1 means weaker indicators for the requirement, 2 indicates medium strength, and 3 denotes strong indicators for the requirement.

Four out of six requirements were appraised as medium strength. For two requirements, D (sustaining participation) and E (empowering participation), indicators varied

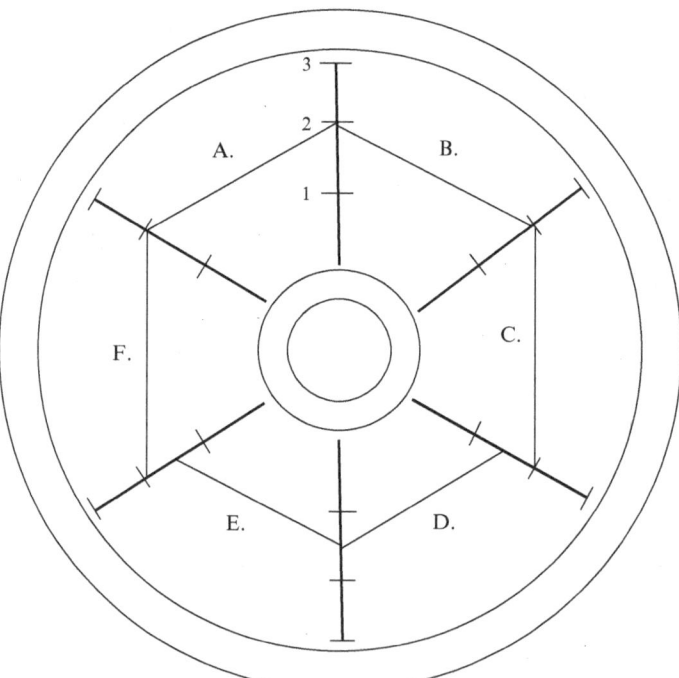

Figure 5.3   Assessment of the action in partnership at the "Neighbourhood Table"

between weak and medium, reflecting lack of access to resources and power for community partners. For example, the requirement D that reflects partners' commitment shows that resident-members engage primarily as individuals and their involvement is highly dependent on their individual circumstances, such as time availability, the need for child- or eldercare, and physical accessibility issues, to name only a few. The commitment of institutional partners depends on the strategic directions of the organisations they represent. Strategic directions and capacity of organisations change in response to funding and policy decisions, and those changes are not necessarily communicated to other partners at the Neighbourhood Table in a timely and consistent manner. In terms of resource mobilisation, some indispensable resources such as access to space for community members to meet and store their supplies/equipment are missing. For institutional partners, budget and reporting requirements in their respective organisations restrict their ability to free up and contribute funds as quickly as may be required in response to community priorities.

In relation to the requirement E that addresses partnership arrangements that help equalise power, data from community partners corresponded with weak indicators pertaining to this requirement. In particular, community partners expressed that they do not benefit from their involvement in partnership in the same way institutional partners do; in some cases, community partners' contribution to carrying out activities are neither acknowledged nor compensated equitably. In relation to both institutional and community partners, criteria and mechanisms for accountability are determined exclusively by funders, with no or little dialogue and negotiation taking place between funders and grantees.

MANAGING, MAINTAINING OR SUSTAINING COMMUNITY PARTICIPATION?

Reflecting on the requirements and indicators described in the original tool for evaluating action in partnership (Bilodeau et al., 2017), I propose three roles nonprofit organisations may perform as partners in nonprofit-community partnership:

- *Managing participation*, when the indicators are assessed as *weak*;
- *Maintaining participation*, where the indicators are assessed as *medium*;
- *Sustaining participation*, where the indicators are assessed as *strong*.

These roles influence the quality of participation for community partners, especially for those who are marginalised due to their specific social locations produced through combination of race, gender, disability, sexuality, immigration and economic status. Table 5.2 describes each role of nonprofit organisations as partners in relation to the degree of quality and levels of participation on the ladder of community participation.

At the lower levels of the ladder, the role of institutional partners is a dominant one where the nonprofit organisations perform the role of *managers* of non-institutional partners. Participation of community members is structured in ways that assist institutional partners to fulfil their organisational needs, and participatory processes at this level have little potential to shift existing inequitable power relations. Such a role is characterised

Table 5.2   The roles of nonprofit organisations as partners in community development

Degrees of quality and levels of participation		Roles of nonprofit organisations as partners in nonprofit-community partnership
Meaningful participation	Co-production Co-design	Sustaining participation
Symbolic participation	Accommodation Consultation Informing	Maintaining participation
Non-participation	Decoration and manipulation	Managing participation

by minimal participation of community members who are connected to and have lived experiences of an issue a partnership aims to address. Both institutional and community partners are involved only in implementing actions determined by their funding bodies; community partners are not included in decision-making structures. Commitment of strategic and pivotal partners is jeopardised through their limited ability to make decisions and commit resources. Inconsistent and interrupted funding affects staffing levels and access to other necessary resources to attract new and sustain existing partners. This role is also characterised by the absence of partnership arrangements aimed at equalising power among actors and supporting collective rather than individual action.

Moving up the rungs to the levels of *consultation* and *accommodation* corresponding with *symbolic participation*, the role of institutional partners changes to *maintaining participation* and reflects nonprofit organisations' efforts to reach dialogue with and increase accountability to community members. This role is characterised by an expanded range of actors, including community members with lived experience included as partners, greater involvement of partners in making decisions about options for action, opening access to community members into decision-making structures. At this stage both types of partners cope with various degrees of resource constraints undermining their continuous commitment as strategic and pivotal partners in the process. In this role institutional partners take active steps toward introducing arrangements and structures to the partnership, helping equalise power among partners and support collective action. On the community participation ladder it is reflected as participation of community partners in consultation processes, and efforts to meet in the middle at the level of *accommodation*. At this level institutional partners respond to community partners' needs albeit with limited resources and find new efficiencies through various in-kind and volunteer support among both types of partners. The levels of *co-design* and *co-production* imply collaborative decision-making and sharing of resources necessary for effective and equitable partnership. These levels require nonprofit organisations to be in the position where they are able *to sustain* their own participation and that of community members as equal partners by sharing access to decision-making and necessary resources in return for community members' knowledge, skills, time and energy. This is the hardest level to achieve within current funding conditions and accountability requirements in the nonprofit sector.

DISCUSSION AND CONCLUSION: STEPPING UP THE LADDER

To be able to move the participation of community partners to the levels of *co-design* and *co-production* that correspond with *meaningful participation* as defined by community members, the nonprofit organisations as partners must be able to perform the role of *sustaining* participation. As the case of the Neighbourhood Table illustrates, the nonprofits find themselves structurally locked between the roles of *maintaining* and *managing* participation, where they struggle to maintain community participation at the levels above the *consultation* on the ladder. Such a role is structured at the institutional level through short-term project-based funding, reduced core funding, competition for scarce resources and pressures to find new efficiencies and leverage resources consistent with the roll-out of neoliberal policies.

In their role as *maintaining participation*, nonprofits are quite literally between a rock and a hard place, on one hand they genuinely attempt to support community action but on the other they perpetually struggle with insufficient and inconsistent resources and capacities combined with pressures to meet their service delivery targets. Nonprofit organisations as institutional partners struggle with external limitations due to inadequate and inconsistent resources and accountability mechanisms that are heavily biased toward funders and higher-level policy administrators. The funding and accountability mechanisms, the political nature of strategic planning and turf wars between nonprofit agencies striving to attract and retain service-users circumvent nonprofit organisations' capacity to be reliable and dependable partners in the context of nonprofit-community partnership.

Funder-driven functional accountability models emphasising the quantitative aspect of community participation to meet short-term project needs structure nonprofits further down the ladder to the level of *managing* participation where nonprofits apply top-down approaches, relying on "technologies of participation," that is, using community participation for fulfilling organisation strategic goals. The emphasis on results-based funding and performance measurement concerning participation depoliticises partnerships and mainstreams participation as a technical process without attention to the power differential among actors in the process (Taylor, 2007). Approached as a technical exercise community participation is sought in the form of feedback and/or client/customer satisfaction surveys to satisfy reporting requirements and quality improvement protocols. In such top-down processes targets and benchmarks are established by funders and higher-level auditing institutions and do not necessarily reflect authentic community priorities and needs. Community participation is pushed down to the levels of *consultation* and *informing*. At these levels of the ladder the line between *maintaining* and *managing* participation becomes increasingly blurry. Within such roles the practices of nonprofits become consistent with extractivist behavior signified by dominance-based relationships where institutional partners extract labour and knowledge from community partners. Amidst the pressures to survive as an organisation, adapting to political climate and funding priorities, the slip from *maintaining* to *managing* may happen inadvertently in a haste to meet the targets and deliver results. It becomes tempting to use community participation in ways that mimic those of for-profit corporations, using community engagement as part of a public relations strategy rather than a genuine act of solidarity, and structure community participation at the level of *decoration and manipulation*.

Within the climate of performance measurement, practitioners are afforded little time and opportunity for reflection, for reflection does not deliver *units of service* or other measurable outcomes in the short term. Yet reflection, or the lack of thereof, almost certainly produces impact, those initially intangible outcomes that accumulate over time and result in the observable *change*. In the context of community development such change may come as a decline in community participation. In response to low engagement levels nonprofits bring yet another short-term community development initiative while not necessarily addressing the very issues that caused the engagement dip. What is interpreted as lack of engagement is often lack of trust and disillusionment. On and off participatory projects, especially those focused on needs assessment without really addressing those needs, lead to "participation fatigue" when community members refuse participation in the processes they do not trust as a pragmatic choice to avoid wasting their time and energy meeting other people's priorities (Cornwall, 2008, p. 280).

Attempting to address participation by focusing exclusively on the numbers of people without considering the extent and scale of participation the nonprofit agencies are able to support, as well as the purpose and the bounds of participation, leads to mismatched expectations on behalf of agency and community partners. Stepping up the ladder in terms of moving community participation to the upper levels of participation corresponding with greater quality of participation requires nonprofits to step up the ladder as institutional partners and to position themselves in the role of *sustaining participation*. Moving to that position requires awareness about different roles the nonprofit partners may perform when addressing community participation, and it takes courage to challenge structurally imposed roles that position nonprofits as subservient to extractivist policies of neoliberal discourse. It calls for advocating with funders and policymakers for more flexible funding and reporting mechanisms that are responsive to community needs and are based on holistic accountability models that prioritise social responsibility (Williams and Taylor, 2013). It may require the nonprofit organisations to examine internal mechanisms of reporting, communication and decision-making through the lens of equity. It also calls for developing capacity building strategies not only for community partners who are defined as marginalised, but more importantly for the partners with greater institutional power to foster arrangements that allow equalising power among partners to support collective action. Without addressing power differential between institutional and community partners, as well as within those groups, community development practitioners may inadvertently support participation of already privileged groups and by doing so undermine the very principles of community development. Critically reflecting on and assessing their role as partners will aid the nonprofits to build genuine solidarity with community groups.

NOTES

* I would like to acknowledge generous contributions of the members of Research Action Team on Evaluation and Evaluation Working Group to the development of evaluation framework and the ladder of community participation: Kaydeen Bankasingh, Zesta Kim, Ashley Nkrumah and two other members who preferred to stay anonymous. Thank you for dedicating your time, energy and ideas to this research project. I am grateful for the generous support of organisational partners, Alliance for Healthier Communities, Health Nexus and York University, who made this research project possible by providing funding, expertise and various in-kind support.

1. This territorial acknowledgment is developed with the help of resources shared on the website of Toronto Conference of the United Church of Canada (2018), accessed 15 January 2019 at https://torontoconference.ca/right-relations/territorial-acknowledgements/.
2. I would like to express my sincere gratitude to my colleague and friend Gillian Kranias for interrogating my initial definition of power by stressing that not all expertise, knowledge or skills have equal value, and thus only those that are institutionally valued or recognised gain access to decision-making and resources and subsequently become part of what constitutes "power" (Kranias, 2018).
3. I choose to use the term "actors" instead of "stakeholders" that lately permeated the nonprofit sector vocabulary. The term "stakeholders" is a for-profit, market-based term that implies individuals holding a stake, or financial share in an enterprise, rather than a group of people working toward a common goal resulting in collective benefits equitably distributed.
4. Social location is a location of individuals within systems of power defined by unique individual characteristics such as race, gender, sexuality, ability, immigration status, religion, and the like, as well as geographic location.
5. Throughout the document the terms "organisations" and "agencies" are used interchangeably assuming that both have nonprofit status and provide public programmes and services.
6. The "Neighborhood Table" is a fictional name used to preserve the confidentiality of the actual neighbourhood based network and its participants.
7. At the time the neighbourhood where the research took place was designated as a "priority" neighbourhood by the City of Toronto. Such designation denoted higher than average unemployment levels and travel distance affecting resident ability to access social and employment services. In response to the increase of gun related violence in 2005, the City of Toronto initiated a Neighbourhood Action Planning Committee, which quickly expanded to involve neighbourhood based non-profit organizations and community groups referred to in this chapter as the "Neighbourhood Table".

REFERENCES

Acosta, A. (2013). Extractivism and neoextractivism: two sides of the same curse. In Lang, M., and Mokrani, D. (eds). *Beyond Development: Alternative Visions from Latin America* (pp. 61–86). Amsterdam: Transnational Inst.

Arnstein, S.R. (1969). A ladder of citizen participation. *Journal of the American Institute of Planners*, *35*(4), 216–24.

Bilodeau, A., and Kranias, G. (2019). Self-evaluation tool for action in partnership: translation and cultural adaptation of the original Quebec French tool to Canadian English. *The Canadian Journal of Program Evaluation*, *34* (2). doi: 10.3138/cjpe.43685.

Bilodeau, A., Galarneau, M., Fournier, M., Potvin, L., Sènècal, G., and Bernier, J. (2017). Self-evaluation tool for action in partnership. Health Nexus. Accessed 14 November 2018 at https://en.healthnexus.ca/topics-tools/community-engagement/partnerships#resources.

Brown, W. (2003). Neo-liberalism and the end of liberal democracy. *Theory and Event*, *7*(1), 1–23.

Cooke, B., and Kothari, U. (2001). *The Tyranny of Participation*. London: Zed Books.

Cornwall, A. (2008). Unpacking "participation": models, meanings and practices. *Community Development Journal*, *43*(3), 269–83.

English, L., and Mayo, P. (2012). *Learning with Adults: A Critical Pedagogical Introduction*. Rotterdam, Boston, Taipei: Sense Publishers.

Evans, B., Richmond, T., and Shields, J. (2005). Structuring neoliberal governance: the nonprofit sector, emerging new modes of control and the marketisation of service delivery. *Policy and Society*, *24*(1), 73–97.

Felluga, D. (2011). Modules on Foucault: on power. *Introductory Guide to Critical Theory*. 31 January. Purdue University. Accessed 3 December 2018 at http://www.purdue.edu/guidetotheory/newhistoricism/modules/foucaultpower.html.

Foucault, M. (1982). The subject and power. *Critical Inquiry*, *8*(4), 777–95.

Fursova, J., Boulet, S., Goodluck, M., and Kranias, G. (2018). *Partnership Assessment Wheel*. Toronto, ON: Health Nexus.

Gago, V., and Mezzadra, S. (2017). A critique of the extractive operations of capital: toward an expanded concept of extractivism. *Rethinking Marxism*, *29*(4), 574–91.

Hankivsky, O. (principal author). (2014). An intersectionality-based policy analysis framework: critical reflections on a methodology for advancing equity. *International Journal for Equity in Health*, *13*(1), 119–35.

Hart, R.A. (1992). Children's participation: from tokenism to citizenship, *Innocenti Essay* no. 4, International Child Development Centre, Florence.

Katz, A.S., Cheff, R.M., and O'Campo, P. (2015). Bringing stakeholders together for urban health equity: hallmarks of a compromised process. *International Journal for Equity in Health*, *14*(1), 138–47.

Kesby, M. (2005). Retheorizing empowerment-through-participation as a performance in space: beyond tyranny to transformation. *Signs: Journal of Women in Culture and Society*, *30*(4), 2037–65.

Kidd, D. (2013). # Occupy in the San Francisco Bay. In Duxbury, N., Canto Moniz, G., and Sgueo, G. (eds), *Ces Contexto Debates, (02)*. (pp. 312–26). Coimbra: University of Coimbra, Centre for Social Studies.

Kidd, D. (2016). Extra-activism. *Peace Review*, *28*(1), 1–9.

Kranias, G. (2018), personal correspondence.

Laverack, G. (2001). An identification and interpretation of the organizational aspects of community empowerment. *Community Development Journal*, *36*(2), 134–45.

McHoul, A., and Grace, W. (1995). *A Foucault Primer: Discourse, Power and the Subject*. London: Routledge.

Phillips, S.D. (2003). Voluntary sector-government relationships in transition: learning from international experience for the Canadian context. *The Nonprofit Sector in Interesting Times: Case Studies in a Changing Sector* (pp. 17–70). Montreal: McGill-Queen's University Press.

Pretty, J.N. (1995). Participatory learning for sustainable agriculture. *World Development*, *23*(8), 1247–63.

Pross, A.P., and Webb, K.R. (2003). Embedded regulation: advocacy and the federal regulation of public interest groups. In *Delicate Dances: Public Policy and the Nonprofit Sector* (pp. 63–121). Montreal: McGill-Queen's University Press.

Realpe, A., and Wallace, L.M. (2010). What is co-production? The Health Foundation, London, pp. 1–11.

Rootman, I., Goodstadt, M., Potvin, L., and Springett, J. (2001). A framework for health promotion evaluation. *WHO Regional Publications*. European series, (92), 7–40.

Shaw, M. (2011). Stuck in the middle? Community development, community engagement and the dangerous business of learning for democracy. *Community Development Journal*, *46*(suppl 2), ii128–ii146.

Smith, D.E. (ed.). (2006). *Institutional Ethnography as Practice*. Lanham: Rowman & Littlefield.

Taylor, M. (2007). Community participation in the real world: opportunities and pitfalls in new governance spaces. *Urban Studies*, *44*(2), 297–317.

Tritter, J.Q., and McCallum, A. (2006). The snakes and ladders of user involvement: moving beyond Arnstein. *Health Policy*, *76*(2), 156–68.

Wallerstein, N., and Duran, B. (2008). The theoretical, historical and practice roots of CBPR. In *Community Based Participatory Research for Health: From Process to Outcomes* (pp. 25–46). San Francisco: Jossey-Bass.

White, S.C. (1996). Depoliticising development: the uses and abuses of participation. *Development in Practice*, *6*(1), 6–15.

Williams, A.P., and Taylor, J.A. (2013). Resolving accountability ambiguity in nonprofit organizations. *Voluntas: International Journal of Voluntary and Nonprofit Organizations*, *24*(3), 559–80.

Wolch, J.R. (1990). *The Shadow State: Government and Voluntary Sector in Transition*. New York: Foundation Center.

6. Social economy, social capital, NGOs and community development: a gendered perspective
Dyana P. Mason

INTRODUCTION

Theories of the social economy and social capital, and their relationship to community development, are distinct but overlapping frameworks that attempt to provide insights to community action. The role of nonprofit and nongovernmental organizations (NGOs) in facilitating both the social economy and social capital has also been taken for granted in community development studies. Finally, gender is often absent from many discussions on social capital and the social economy, although women and the family have always been central to community life. This chapter seeks to integrate the literature on the social economy and social capital in the context of nonprofit-led community development programs, leveraging both traditional approaches and gendered perspectives offered by feminist theories. A case illustration of a cross-national partnership between a US supported grass-roots cooperative in Laos, also provides an opportunity to study the intersection between normative views of social economy and social capital and more critical approaches to these constructs in practice. Theory and findings both provide evidence that women are often left out of consideration in a community development project, and the deliberation required to design efforts with women in mind. Specifically, NGO managers and those interested in community development must consider gender as part of their formative work in order to help set the stage for success.

The Social Economy

The social economy, sometimes called the solidarity or solidary economy, is a contested concept that can be traced back to medieval Europe through the rise of professional guilds and other associations (Moulaert and Ailenei, 2005). Beneath these early movements for associational life was a goal towards the communal – that is, organizations and activities that were formed by and for the community (Barthelemy, 2000; Hardy, 1979; Moulaert and Ailenei, 2005). Today, many define the social economy as those activities (including economic activity) that are outside both the market and state (Amin et al., 2003; Bridge et al., 2009; Kay, 2006; Quarter et al., 2001) and generate both economic and social value (Mook et al., 2010). Similarly, Levesque et al. (1999) consider the social economy as a "third sector" that combines both formal and informal dimensions at an organization level, and includes both market and non-market activities, including domestic activity. However, Mook et al. (2010) conceptualize the social economy much more broadly. Instead of claiming that the social economy and social economy organizations are outside both the market and state, they suggest that it is a multi-dimensional construct that bridges sectors – from market-based for-profit social enterprises to civil society organizations that are seeking social change.

Along these lines, Amin et al. (2003, p. vii) define the social economy from a social enterprise perspective as "non-profit activities designed to combat social exclusion through socially useful goods sold in the market and which are not provided for the state or the private sector". Chaves and Monzón (2012) argue it can include alternative business forms, including social enterprises and cooperatives. Pulling these strains together, Bridge et al. (2009) provide an analysis of the varying definitions and uses of social economy, and find that the common traits among definitions include "collective ownership, non-profit distribution, democratic governance agreements, the inclusion of multiple stakeholders in planning and control, a clear social benefit, but with recognizable systems of trade and exchange relationships . . ." (p. 104).

Fontan and Shragge (2000) recognize these conceptualizations as pragmatist or reformist in nature. That is, they are consistent with the dominant paradigm of using social economy organizations as instruments of economic change using primarily market-based strategies to create social value. However, Fontan and Shragge (2000) also consider a second perspective that links the social economy to social change, such as building new institutions of individual action, drawing from counter-cultural and/or democratic processes for change. In other words, social economy organizations are those that "give priority to people over assets (and thus to democratic processes)" (Favreau, 2000, p. 178), including cooperatives, nonprofit organizations and credit unions. Moulaert and Ailenei (2005, p. 2037) also argue that the social economy can benefit marginalized communities and reintroduce "solidary in production relations". Sattar and Mayo (1998) suggest that social economy organizations include three dimensions: *identity*, where there is a convergence of interest, *institutions* of cooperatives or mutual and voluntary associations and the *intention* of economic activities that include a social or ethical intention. Thus, the social economy can be considered to be "commercial and non-commercial activity largely in the hands of third-sector or community organizations that gives priority to meeting social (and environmental) needs before profit maximization" (Amin, 2009, p. 4).

In an international development context, the social economy, NGOs and community development are often seen as mutually reinforcing, if not outright synonymous (Eade, 2000). This is echoed by Moulaert and Ailenei (2005, p. 2050), who suggest that "The social economy is presented as a family of hybrids between market, state and civil society". Additionally, Favreau (2000, p. 184) states, "In the [Global] South, the emergence of social economy is due, to a great extent, to the development NGOs. They are its lifeline. These NGOs are grouped around what is commonly termed as community development". Complicating matters is that these conceptualizations of social economy organizations are often used interchangeably, but they have distinct implications for practice. For one, the associational and democratic aspect of social economy organizations can be seen as similar to normative and theoretical considerations of the nonprofit sector – that is, nonprofits and NGOs are usually seen as being independent organizations that are controlled by a group of individuals who commit their time voluntarily to the organization. Additionally, the nonprofit sector is often perceived as consisting of independent organizations acting outside both the state and market. Yet, many nonprofits and NGOs involved in international and community development are relatively hierarchal and non-democratic, making decisions centrally and implementing them without including local communities or clientele in the conversation. Many of them are not even based in the countries where they operate, but instead are based in the Global North, which may mitigate their impact on existing local social economies.

Social Capital

Intertwined with the concept of social economy is the theoretical framework of social capital. Social capital, or "resources embedded in a social structure which are accessed and/or mobilized in purposive action" (Lin, 1999, p. 21), was popularized by Putnam (2001) in *Bowling Alone*. Social capital can be considered the resources and assets that individuals, communities or organizations hold or can access through shared connections. Adler and Kwon (2002, p. 17) define social capital as "the goodwill that is engendered by the fabric of social relations and that can be mobilized to facilitate action", while Woolcock and Narayan (2000, p. 225) define social capital as the "norms and networks that enable people to act collectively". Yet these perspectives are at opposite ends of a spectrum. Is social capital the tangible shape of social relations, something that can be measured and mapped, such as by Burt (1995, 2000), or is it the content of those relations (Woolcock, 1998) that facilitate community cohesion? Many scholars argue it is both, although this also makes social capital difficult to study and measure (Woolcock, 1998).

Similar to ideas of the social economy, social capital has been applied to community and economic development. Focusing on the "bottom-up" versus "top-down" approaches to economic development, Woolcock (1998) mediates the benefits of challenges of leveraging social capital for economic development, finding that while poor or marginalized communities may benefit from a high level of in-group social capital, their tight-knit circles may prevent them from having access to mainstream institutions and resources. Top-down approaches, on the other hand, are influenced by specific economic, political and social contexts – each of which shape outcomes in different ways. In other words, "The structure of the state, the nature and extent of its involvement in civic and corporate life, and the organization of society together constitute the key factors determining whether a country succeeds or fails in development (Woolcock, 1998, p. 187).

Ultimately, social capital has been shown as a tool to support community development and individual action, including the social economy (Dale and Onyx, 2010; Kay, 2006; Lin, 2001; Woolcock and Narayan, 2000). Without social capital, efforts to support community and economic development may fail. For example, increasing social capital can lead to increased social and community cohesion, trust and reciprocity (Coleman, 1988; Putnam, 2001). As people or groups tap into their social capital, they are able to access the resources and assets held within that network. Or as Valente (2010, p. 13) states, "If this social capital can be mobilized, then perhaps some of the neighborhood's problems can be ameliorated".

GENDER, THE SOCIAL ECONOMY AND SOCIAL CAPITAL

One critique of the social economy and social capital is that in recent years it has related too closely to (neoliberal) assumptions about the market's ability to solve social problems. From this perspective, the market is privileged over other forms of organizing the economy in the creation of social value (Mook et al., 2010). Specifically, most research on the social economy has fallen into this instrumental approach to creating social value – where the social economy is measured by the monetary value of the organization, even attempting to monetize more informal activities, such as caring for others (Bergeron and Healy, 2013). More critical perspectives, such as feminist theories, are useful to find alternative understandings of the social economy and social capital. Feminist scholars would argue that most of the discourse on international development stems from the dominant, more "masculine" perspectives of the market (Ferber and Nelson, 2003). Yet, to this day even much of feminist economics simply seeks to monetize women's work (Cameron and Gibson-Graham, 2003). Indeed, as Floro and Willoughby (2016, p. 17) write, instead of "treating the promotion of human rights and women's well-being as ends in themselves, their labor and productivity potential are viewed mainly as means to further increase economic growth".

Thus, feminist scholars include alternative forms of social, economic and community development, including caring/giving, cooperatives, nonprofits and worker collectives (Bergeron and Healy, 2013; Cameron and Gibson-Graham, 2003) – all social economy organizations. As Bergeron and Healy (2013, p. 14) suggest, "we let go of a monolithic vision of economy [so] we can recognize the diversity of alternative enterprises (cooperatives, households), alternative systems of finance (microcredit) as well as motivations of care, interdependence, community aid, etc. as not inevitably reproducing neoliberal capitalism". Similarly, Razavi and Staab (2012, p. 6) argue that there is an undervaluation of care (both informal and formal), a skepticism towards markets, and a need for a voice for those who have traditionally been under-represented. The traditional role of women in the economy is the "social provisioning for human life, through interdependent paid and unpaid economic activities mediated by markets, households/communities, and the government" (Beneria et al., 2015, p. 42). This "social provisioning approach" (Power, 2004) identifies five areas where feminist thought diverges from market-based and traditional economic approaches: the incorporation of caring and other unpaid labor as economic activities; needing to include human well-being as a measure of economic success; the analysis of economic, political and social norms, processes and power relations, the need to include ethics and values in any analysis, and the exploration of socioeconomic differences.

Despite this, few have taken up the challenge to integrate these literatures. Razavi's (2007) report is one of the few studies that explicitly seeks to integrate the literature on social economy and feminist theories, providing a gendered approach to how care work is measured and studied. Much of this "informal economy" of care is often led by women (Razavi, 2007). This is consistent with the ideas of community development that seek to build social and community value (Amin, 2009), and suggests that "traditional economics privileged the monetized aspects of the economy, while ignoring the sphere of 'social reproduction' or 'unpaid work'" (Razavi, 2007, p. 3). Indeed human well-being as a key indicator of economic success is a core value of the social economy as well as feminist economic approaches to measuring economic value (Beneria et al., 2015). In addition, Hossein (2015; 2018) has explored the ways that women, unable to participate fully in the market, are often left to work in the informal economies. While this leaves opportunity for increased leadership roles in the social economy (Hossein, 2018), it also can leave them vulnerable to violence and exploitation (Hossein, 2015).

Conceptualizations of social capital also often lack a critical discussion of the role of gender – despite an observation by some that men and women obtain, develop and use social capital in different ways (Burt, 1998; O'Neill and Gidengil, 2013). Lowndes' (2004) work suggests that women are often left out of common measurements of social capital by focusing primarily on male-dominated activities. For example, in Hall's (1999) study of social capital, he focused entirely on sports clubs, with little effort to understand the role of activities that are traditionally seen as "women's work" including childcare or social services, although he acknowledges that women may be the key to sustaining social capital. In this way, Lowndes (2004) argues, the invisibility of women in social capital studies has led to research being primarily male-dominated instead of gender-neutral. Lowndes (2004) points out that women are more likely to know and trust their neighbors, have more contact with friends and relatives, and both men and women tend to turn to women as helpers for assistance – all measures of social capital. In other words, she finds that women and men have different "social capital profiles" (Lowndes, 2004, 52).

Perhaps the difficulty with applying a gendered approach to social capital is due to the fact that instrumental approaches to alleviating poverty may exacerbate current gender norms and political-social systems rather than challenge them (Adkins, 2005; O'Neill and Gidengil, 2013; Rankin, 2002). Portes and Landolt (2000) argue that some of the challenges in applying social capital to international development stem from the purely economic approach to social capital – leaving out the social terrain. Alternative approaches, such as by Thieme and Siegmann (2010) suggest that social networks (the infrastructure of social capital) are themselves gendered, and privilege some individuals over others. From this perspective, women are less able to access the resources held by male-dominated social networks – both in access to information as well as what is defined as appropriate work for women. In addition, while women are often asked to provide the glue that holds social networks together, they often adhere to traditional community norms, leaving women unable to receive the benefits of those networks. For example, Molyneux (2002) finds that international development projects that focus on poverty relief often rely upon women's unpaid labor, thus leaving much of the burden of the projects on women. Indeed, Rankin (2002) shows how women in Nepal are asked to perform key social obligations that increase their husbands' social standing, while not being allowed membership to the same associations their husbands belong to.

Ultimately, social capital can help support the creation of the social economy (Kay, 2006), and thus support community development. However, a lack of attention to the role of women in social relations and networks will lead to development projects that further institutionalize existing social norms and prevent women from increasing social capital and supporting the social economy (Rankin, 2002), thus mitigating the impact of the program or activity. The networks of trust and reciprocity that are necessary for social capital are also required to help support community development projects. For social capital to grow and the social economy to function at its highest level – providing social value to a community or family – considerations of gender roles and norms must not be ignored.

CASE ILLUSTRATION: A COMMUNITY DEVELOPMENT PROJECT IN LAOS

To illustrate the need for a clearer connection of the gendered perspectives of the social economy, social capital and community development, a case study was developed through the study of a US-based nonprofit active in rural Laos. In creating an organization that will primarily conduct "women's work" of care – in this case, children – the effort provides an opportunity to apply theory to practice. The program provides early childhood health and development interventions to community members who live in several rural villages. It also hopes to launch cooperative childcare centers to provide much needed health, nutritional and educational services to children under the age of five. A second, downstream goal of the organization is to develop local and community leaders (women in this specific case) to run sustainable child centers in their community and even train families in other villages on how to develop their own child center. In this way, the stated goals of the organization can be seen as helping to build a network of concerned individuals and families for community change – consistent with our broad view of both social capital and the social economy.

Laos provides a unique setting to measure the interplay of the social economy and social capital. It is a communist country where local villages are run by "chiefs" who are chosen by the villagers. Due to the central government's tight supervision and control of community projects, all development projects must be approved (and are usually monitored) by local, state and/or national departments and ministries. The country is ethnically diverse with 149 ethnic groups among 47 main ethnicities and 82 languages. Laos also struggles with poverty and childhood malnutrition. In fact, according to UNICEF, nearly half of all children (44 percent) in Laos suffer from stunting, with many more experiencing less acute, but no less serious, effects of chronic malnutrition which has been linked to cognitive delays and defects.

In the summer of 2016 and early 2017, interviews were conducted with members of all three groups active in the formation of the child centers. The executive director of the nonprofit (n = 1, male and conducted in English), the owner of the company tasked with implementing the nonprofit's plans (n = 1, also male and in English) and villagers in four rural villages (n = 55, predominately female, in Lao translated to English) within a 50km radius of Luang Prabang, Laos. Semi-structured interviews were conducted, exploring the meaning and expectations of the project to each group, as well as their experiences, challenges and values associated with working in a partnership across both cultures and organizations. With organization leaders, the questions focused on the organization and project structure, along with perceived challenges and opportunities of their current effort. With villagers, questions centered around family and village life (who handles childcare, their roles in the community) along with their views on the organizations working in the village, their hopes for the project and their willingness to support it as volunteers. Villagers were also asked who they felt should be in charge of the child centers. All interviews were analysed for consistent themes and shared meanings. With villagers, interviews were handled in focus group settings in Lao, in a setting with up to six individuals in a session and one translator.

A Gendered Perspective on the Social Economy, Social Capital and a Development Project

The planned child centers will be run on a cooperative model, where families are able to secure childcare and early childhood development services for their children in exchange for voluntary support of the center. Some individuals will be trained and have the opportunity to develop leadership roles in the center as well, although individuals will not be paid. While this is typical of a cooperative, different perspectives suggest that not paying women for their labor may continue to devalue the role of women's work in the community. Alternatively, this work can instead be valued through its contribution to social value in the community. In addition, the role of women in a village's social capital should be considered. In Laos, women are often the ones that make the economic decisions for the household. However, it is generally men that are engaged in a village's political decisions. These traditional gender norms both support and mitigate women's opportunities to be empowered members of the community, and able to access all of the recourses the community has to share.

The interviews demonstrated the social economic conceptualization of the project, which straddled both the market and more traditional forms of NGO-provided humanitarian aid. Specifically, the owner of the local, Laos-based organization was focused on the boundary-crossing (market and third sector) nature of the project, stating that at first:

> It wasn't considered work, it was just like helping out family members that were in need of the basic essentials of water or electricity or if they just wanted to clean up the school their kids were in.

While this organization evolved, and developed using a market-based approach with paying clients, the owner self-identified his organization as a social enterprise (those organizations that straddle organizational form between the market and third sector (Liu et al., 2012)). That is, he held both business and social value perspectives in regards to his organization's mission and work, with a relational perspective rather than simply a business one – consistent with definitions of social economy organizations. He stated:

> It's not a one-time-off kind of thing where we build it and then we leave. We frequently visit every village that we've ever done a project in, and we still know each other by first name, so that's really key central to get to know the villagers and have them get to know us and us with them, not just myself but the entire team. We take everybody into the field so that they know this is who you're getting money from, this is who's signing the paperwork, this is who you're going to be [helping] and who will be supervising the work.

It is interesting to note, however, that interviews with villagers, particularly in those villages that were closer to the development of the center, indicated that people were indeed thinking about whether or not they would be paid for their work in the center. When asked if they would be willing to volunteer, most said "yes", but this group, predominately women, also asked questions about who would be paid, and how much people might be paid. The notion of a cooperative with its volunteer labor in a structured environment, although not inconsistent with the values associated with a collectivist culture like Laos, seemed inconsistent with their expectations.

One of the important features of social economy organizations is that their decision-making and organizational structure is made up of democratic processes and procedures. That is, the individuals that participate have the ability to influence the direction and outcomes of the organization. Ideally, the child centers will be organized this way. As cooperatives, individuals will be encouraged to participate, and could re-shape and strengthen the community's social capital by developing new reciprocal relationships. In addition, as designed, this project will add value outside the traditional economic market. In other words, it will provide social value mostly through alternative economic forms like household support (childcare) and cooperatives. In rural villages in Laos, the care of small children is generally handled by women and occasionally their male partners (fathers) and grandparents. The child center will help to provide structure to this informal care network, and provide significant social value to those communities who choose to make use of it. The centers may also be able to provide leadership opportunities for engaged health educators and teachers, allowing empowerment opportunities for women and better access to the village's social capital – although this has not yet been developed. While childcare is perceived as "women's work" – the child center offers an opportunity to help build an important non-market economic institution – important for both social capital and social economy – that would add value to the community by supporting better child health.

Although this is the ideal structure, the existing top-down process does not seem to be meeting this standard. While working within the current political and economic systems of Laos, and seeking the general support of the community in establishing a center (communities can say "no") much of the early organization of the project has been managed exclusively by men in more traditional leadership roles. The village chiefs (men) approve the project, the health educators (so far, men) lead the initial trainings, a local field partner (male) will negotiate the permits, location and building of the centers, and the organization sponsoring this project is also run by a man. The current cultural and political context may lend itself to this arrangement, but it could make finding success difficult. For example, while success will be measured by non-economic factors (child health outcomes and women's participation), the initial approach has not been one that has been sensitive to elevating the voice of women. While women were consulted during a needs assessment during the formation of the project, the leadership to date is in direct contrast to Sobering's (2016) call for gendering organizations by ensuring that women have equal opportunities to participate in decision-making. In addition, as a male-led project (at least thus far), it remains to be seen whether the organization's ultimate culture will be able to embrace egalitarian or equity values and/or provide participants a high level of agency in decision-making – two of Sobering's (2016) other requirements to mitigate inequality.

In addition, while the role of gender is being considered, it is only thought of in the long-term, with an unstated goal of women's empowerment. The executive director stated:

> Some families will be able to afford monetary contributions. Others will be making contributions in terms of food, and things that the center will need. The thought will be that some families [particularly women] will emerge as center leadership and or staff. The thought is, longer term, that there might be some way to professionalize that role [of women] as the village figures out what and how that center is going to be sustained.

Yet, he also stated that for the short term, "Ultimately, I'm less concerned about the shape that [the structure] all takes, because I think it will be organic to the local community. It's just, is there a desire? And, is there an ability to see that there might be benefits to this?" While it goes without saying that recognizing the local culture and norms is important to project success, a lack of deliberation on the specific roles of women is inconsistent with feminist perspectives on community development.

While downstream effects of the center may be felt by the community overall, with healthier children, the challenge for the organization is to help articulate what these benefits may be and encourage village residents to allow their children to be cared for in the center. While most villagers interviewed saw value in the centers, stating they will help to provide education as well as childcare, it remains to be seen if village residents will be willing to participate in the creation and support of the center with the idea to supporting community value overall. Or, will it just be the parents of young children who are motivated to have access to the healthcare, hygiene, education and nutrition that the center promises?

Finally, voluntary participation seems to be an expectation of the project – where individuals will be offered the opportunity to participate in the cooperative, perhaps in exchange for allowing their children to stay there during the day. One consideration, however, is that in an authoritarian political environment, it is possible that individuals will be coerced to participate by local officials. When asked if they would be willing to volunteer for the child center, many villagers stated they would if they were told to (although it was not clear if this was an expected answer). It also seemed unclear to the villagers who would be asked to participate. When asked who should run the project or make decisions about the project, most villagers stated either the village chief or the village committee, which is the existing village leadership structure.

DISCUSSION AND CONCLUSION

Overall, the child centers offer an interesting approach to community development that is consistent with critiques directed towards international aid by feminist scholars and those who study the social economy and social capital. This particular community development project seeks to add value through non-economic means, a hallmark of a gendered approach to the social economy. However, it exhibited a top-down leadership structure in its formation, and is not (yet) democratically organized or run. It currently is leveraging men's participation in the formative stages of the project, and not deliberately seeking women's participation in its leadership. While women in Laos have mixed access to a community's social capital – and are responsible for both childcare and economic decisions, their role in the political sphere of the village – where development decisions are made – has thus far been absent. Indeed, organizers of international community development projects should consider nascent leadership structures and the institutional structure of social capital in the affected communities.

REFERENCES

Adkins, L. (2005). Social capital: The anatomy of a troubled concept. *Feminist Theory*, 6(2), 195–211.
Adler, P.S., and Kwon, S.-W. (2002). Social capital: Prospects for a new concept. *The Academy of Management Review*, 27(1), 17–40.
Amin, A. (2009). Locating the social economy. In A. Amin (ed.), *The Social Economy: International Perspectives on Economic Solidarity* (pp. 3–21). London: Zed Books Ltd.
Amin, A., Amin, P.G.A., Cameron, A., and Hudson, R. (2003). *Placing the Social Economy*. London: Routledge.
Barthelemy, M. (2000). *Associations: un nouvel age de la participation?* Paris: Presses de Sciences Po.
Beneria, L., Berik, G., and Floro, M. (2015). *Gender, Development and Globalization: Economics as if All People Mattered*. New York: Routledge.
Bergeron, S., and Healy, S. (2013). Beyond the business case: A community economics approach to gender, development and the social economy (Draft paper prepared for the UNRISD Conference). Geneva, Switzerland: UNRISD, 6–8 May 2013.
Bridge, S., Murtagh, B., and O'Neill, K. (2009). *Understanding the Social Economy and the Third Sector*. London: Palgrave Macmillan.

Burt, R.S. (1995). *Structural Holes: The Social Structure of Competition*. Cambridge, MA: Harvard University Press.
Burt, R.S. (1998). The gender of social capital. *Rationality and Society*, *10*(1), 5–46.
Burt, R.S. (2000). The network structure of social capital. *Research in Organizational Behavior*, *22*, 345–423.
Cameron, J., and Gibson-Graham, J.K. (2003). Feminising the economy: Metaphors, strategies, politics. *Gender, Place & Culture*, *10*(2), 145–57.
Chaves, R., and Monzón, J.L. (2012). Beyond the crisis: The social economy, prop of a new model of sustainable economic development. *Service Business*, *6*(1), 5–26.
Coleman, J.S. (1988). Social capital in the creation of human capital. *The American Journal of Sociology*, *94*(suppl), S95–S120.
Dale, A., and Onyx, J. (2010). *A Dynamic Balance: Social Capital and Sustainable Community Development*. Vancouver: UBC Press.
Eade, D. (2000). Preface. In D. Eade (ed.), *Development, NGOs, and Civil Society: Selected Essays from Development in Practice* (pp. 9–14). Oxford: Oxfam GB.
Favreau, L. (2000). Globalization and social economy: A north-south perspective. In E. Shragge and J.-M. Fontan (eds), *Social Economy* (pp. 177–91). Montréal: Black Rose Books.
Ferber, M.A., and Nelson, J.A. (2003). *Feminist Economics Today: Beyond Economic Man*. Chicago: University of Chicago Press.
Floro, M.S., and Willoughby, J. (2016). Feminist economics and the analysis of the global economy: The challenge that awaits us. *Fletcher Forum of World Affairs*, *40*(2), 15–28.
Fontan, J.-M., and Shragge, E. (2000). *Social Economy: International Debates and Perspectives*. Montreal: Black Rose Books.
Hall, P. (1999). Social capital in Britain. *British Journal of Political Science*, *29*(3), 417–61.
Hardy, D. (1979). *Alternative Communities in Nineteenth Century England*. London: Longman.
Hossein, C.S. (2015). Black women in the marketplace: Everyday gender-based risks against Haiti's *madan saras* (women traders). *Work Organisation, Labour & Globalisation*, *9*(2), 36–50.
Hossein, C.S. (2018). A Black perspective on Canada's third sector: Case studies on women leaders in the social economy. *Journal of Canadian Studies*, *51*(3), 749–81.
Kay, A. (2006). Social capital, the social economy and community development. *Community Development Journal*, *41*(2), 160–73.
Levesque, B., Malo, M.C., and Girard, J.P. (1999). L'ancienne et la nouvelle economie sociale. In P. Defourny, P. Develtere, and B. Fonteneau (eds), *L'economie sociale au Nord et au Sud* (pp. 195–216). Paris: De Boeck & Larcier.
Lin, N. (1999). Building a network theory of social capital. *Connections*, *22*(1), 28–51.
Lin, N. (2001). *Social Capital: A Theory of Social Structure and Action*. New York: Cambridge University Press.
Liu, G., Takeda, S., and Ko, W.-W. (2012). Strategic orientation and social enterprise performance. *Nonprofit and Voluntary Sector Quarterly*, *43*(3), 480–501.
Lowndes, V. (2004). Getting on or getting by? Women, social capital and political participation. *The British Journal of Politics & International Relations*, *6*(1), 45–64.
Molyneux, M. (2002). Gender and the silences of social capital: Lessons from Latin America. *Development & Change*, *33*(2), 167–88.
Mook, L., Quarter, J., and Ryan, S. (eds). (2010). What's in a name? In *Researching the Social Economy* (1st ed.). Toronto: University of Toronto Press, Scholarly Publishing Division.
Moulaert, F., and Ailenei, O. (2005). Social economy, third sector and solidarity relations: A conceptual synthesis from history to present. *Urban Studies*, *42*(11), 2037–53. Accessed 11 December 2018 at http://usj.sagepub.com/content/by/year.
O'Neill, B., and Gidengil, E. (2013). *Gender and Social Capital*. New York: Routledge.
Portes, A., and Landolt, P. (2000). Social capital: Promise and pitfalls of its role in development. *Journal of Latin American Studies*, *32*(2), 529–47.
Power, M. (2004). Social provisioning as a starting point for feminist economics. *Feminist Economics*, *10*(3), 3–19.
Putnam, R.D. (2001). *Bowling Alone: The Collapse and Revival of American Community* (1st ed.). New York: Touchstone Books by Simon & Schuster.
Quarter, J., Sousa, J., Richmond, B.J., and Carmichael, I. (2001). Comparing member-based organizations within a social economy framework. *Nonprofit and Voluntary Sector Quarterly*, *30*(2), 351–75.
Rankin, K.N. (2002). Social capital, microfinance, and the politics of development. *Feminist Economics*, *8*(1), 1–24.
Razavi, S. (2007). The political and social economy of care in a development context (Gender and Development No. Programme Paper Number 3). Geneva, Switzerland: United Nations Research Institute for Social Development.
Razavi, S., and Staab, S. (2012). Introduction: Global variations in the political and social economy of care

– Worlds apart. In S. Razavi and S. Staab (eds), *Global Variations in the Political and Social Economy of Care: Worlds Apart* (pp. 1–25). New York: Routledge.

Sattar, D., and Mayo, E. (1998). *Growth Areas of UK Social Economy*. London: UK Social Investment Forum.

Sobering, K. (2016). Producing and reducing gender inequality in a worker-recovered cooperative. *The Sociological Quarterly*, *57*(1), 129–51.

Thieme, S., and Siegmann, K.A. (2010). Coping on women's backs: Social capital-vulnerability links through a gender lens. *Current Sociology*, *58*(5), 715–37.

Valente, T.W. (2010). *Social Networks and Health: Models, Methods, and Applications*. New York: Oxford University Press.

Woolcock, M. (1998). Social capital and economic development: Toward a theoretical synthesis and policy framework. *Theory and Society*, *27*(2), 151–208.

Woolcock, M., and Narayan, D. (2000). Social capital: Implications for development theory, research, and policy. *The World Bank Research Observer*, *15*(2), 225–49.

7. What can Northwest European community enterprises learn from American community-based organizations?
David P. Varady, Reinout Kleinhans and Nuha Al Sader

INTRODUCTION

Neo-liberalism and welfare state retrenchment have shifted the economies of advanced western states and are reshaping the ways in which citizens, public, private and third sectors interact with each other. In the aftermath of the economic crisis, many European countries are implementing austerity measures and cuts in public policy, alongside longer trends of welfare retrenchment. To mitigate such challenges, entrepreneurial forms of active citizenship are considered as a new form of public management in Europe. Citizens are expected to organize to fill in gaps left by government spending cuts in health care, education and employment. There is a growing European Union-wide interest in entrepreneurial forms of active citizenship, such as social enterprises and their localized version, community enterprises (Bailey, 2012; Healey, 2015; Kleinhans, 2017).

Community enterprises (CEs) are social enterprises run by local people for the benefit of their local community. Community development corporations (CDCs) are community-based organizations that have close ties with their neighborhoods and seek to improve the physical, social and economic conditions in these areas.

In this chapter, we argue that, considering their relatively recent appearance, Northwestern European CEs may learn from longstanding examples of American community-based organizations (CBOs), in particular, but not exclusively CDCs. More specifically, the purpose of this chapter is to identify recent writings on CDCs in order to assess how the (post-)crisis context has affected their position, scope, and viability (see also Varady et al., 2015). We have used a broadened definition of CDCs to include a wide variety of CBOs including neighborhood nonprofits, worker cooperatives, community banks and so forth. A computerized literature review search using the term "community development corporations" as the search word yielded about 30 articles or book chapters published since 2010. We focused on the extent to which writers produced new knowledge on CDCs in five areas: the changing scope of CDCs, factors affecting CDC capacity, comprehensiveness and viability, citizen participation and community revitalization impacts.

We make two key arguments in this chapter. First, in terms of the hybrid nature and conflicting institutional logics of CDCs—which are themselves the product of the larger forces described above—it has become increasingly difficult but not impossible for CDC boards to reconcile economic development with social equity. Second, top-down planning is not inherently negative. Based on these key arguments and the findings of the literature, implications for CEs in Northwest Europe will be discussed.

The chapter is structured as follows. In the next section, we review the recent developments in the evolution of CDCs, especially regarding post-crisis development of this type of organization. Section three deals with the issue of CDC capacity, that is, what they are able to take on in terms of key objectives and strategies. Subsequently, section four deals with comprehensiveness and viability, discussing a key factor of CBO existence over time, that is, the "quadruple bottom line"—the necessity to simultaneously promote city-wide economic development and/or housing production, equity, environmental sustainability and institutional viability. Section 5 moves on to discuss the issue of community participation, whereas section 6 analyses current knowledge regarding CBOs' impact on community revitalization. Finally, section 7 provides the conclusions, which take the form of lessons learned from American CBOs regarding the (future) development of CEs in the Northwestern part of Europe.

WHAT ARE RECENT DEVELOPMENTS IN THE EVOLUTION OF AMERICAN CDCS?

The literature review identified four recent trends: (1) the impact of the Great Financial Crisis (GFC) on CDCs, (2) the increasing scope of work carried out by CDCs, (3) the emergence of new forms of CDCs—regional CDCs, faith-based CDCs, and (4) the advent of community enterprise-like CDCs.

The Great Financial Crisis

The GFC (roughly 2008 to 2011) had serious negative impacts on CDCs but the level of severity varied by metropolitan area. Thomson and Etienne (2017), in their study of three legacy cities, found that the GFC had the greatest impact in Detroit; the least impact in Cleveland, with Baltimore somewhere in between. By 2011, only about half of the CDCs functioning in Detroit in 2004 or 2008 were still doing so at the time they wrote their article. CDCs that had become most dependent on housing construction and rehabilitation (and who were financially dependent on developer fees) were the ones most severely affected by the GFC.

CDCs have played a role in developing strategies to address foreclosures resulting from the GFC. Turcotte et al. (2015) compare the approaches used by two CDCs in Massachusetts. The Neighborhood Developers (TND, formerly known as Chelsea Neighborhood Developers) sought to attract higher income families to purchase foreclosed properties and to raise satisfaction levels and the physical and social health of residents. Lowell's Coalition for a Better Acre (CBA) sought to acquire problematic properties that were creating a crime problem; CBA also developed housing for special needs populations and households at risk of homelessness.

Fujii (2016) highlights the value of CDCs in dealing with the foreclosure crisis in Cleveland. In general, transfer practices by private financial institutions and investors had negative impacts on neighborhoods. However, Slavic Village Development, one of the higher capacity CDCs in Cleveland, which worked for the City of Cleveland Land Bank and the Cleveland County Land Revitalization Corporation (another land bank), produced positive neighborhood benefits in terms of tax payment status by current owners as well as ownership stability.

Widened Scope

In recent years CDCs have begun to diversify away from housing development to include fields like commercial real estate, business development, community organizing and workfare development (Vidal and Keyes, 2005). But the scope is even broader than Vidal and Keyes indicate, as shown below.

Urban agriculture and environmental sustainability

Depopulation has caused some CDCs to consider urban agriculture. Burton, Bell, Carr, Development Inc., a CDC in Cleveland (cited in Thomson and Etienne, 2017, p.150) "assembled 10 acres of land that became the home of a 28-acre urban agriculture zone including an urban farming incubator, a small agriculture/aquaculture business, and an orchard, among other planned uses." Similarly, in spring, 2010 the Mary Queen of Vietnam Community Development Corporation in New Orleans planned to open a 20-acre farm in the Village de l'Est neighborhood.

> The farm would include a livestock barn, the region's first commercial composting facility, a farmers market and on an adjacent location, a seafood processing plan funded with support from the state of Louisiana. A celebrity chef agreed to buy produce from the farm to serve at a choice of New Orleans restaurants. (Cohen, 2010, n.p.)

It is unclear from the article whether "profits" from the farm will be recycled back into community projects and community services.

"The overall goal of the [Cleveland's] EcoVillage has been to develop a model urban village that will realize the potential of urban life in the most ecological way possible" (Kellogg and Keating, 2011, p.73). The four Detroit Shoreway Community Development Organization (DSCDO) housing projects include 20 new green town homes for purchase, two new green single family homes for purchase, a renovation of an existing early twentieth century home with green products and design, and four new green cottages available through a long-term lease with a nonprofit land trust organization. EcoCity (EC) Cleveland and DSCDO have jointly managed EcoVillage through most of its history. EC manages grants from the environmental policy network while DSCDO provides neighborhood planning and outreach services.

Art districts

The evolution of the Gordon Square Art District (GSAD) in Cleveland—a product of the work of DSCDO (Krumholz and Roller, 2012), occurred in three steps (1) the preservation of the Gordon Square Arcade, which contained an empty movie theatre, (2) the restoration of the Capitol Theatre building, (3) the repair of the Gordon Square Theatre and its purchase by the Cleveland Public Theatre, and (4) the decision of a third theatre company, the New West Theatre, to build a new performance center only a block from the Cleveland Public Theatre and the Capitol Theatre. Sixty new businesses, including 18 bars and restaurants and 30 artists' galleries and studios, have located in the Gordon Square Arts District between 2006 and 2012. However, DSCDO not only focused on the arts or economic development; beginning in 1979, DSCDO used a combination of their own investments; loans; and subsidies from foundations, the city, state and federal governments, to build more than 300 rental apartments mostly for low- and moderate-income families.

The Station North Arts and Entertainment District (SNAED) in Central Baltimore (Rich and Tsitsos, 2016) has used creative place making revitalization efforts successfully to attract artists and to promote cultural tourism. The Baltimore city government's 2002 designation of the area as an arts district has advanced the neighborhood's transformation. More recently, the neighborhood's commercial, residential and artistic venues have begun to flourish (Station North Arts and Entertainment District, no date).

New Forms of CDCs

Apart from "conventional" CDCs which target certain districts of parts of cities, regional CDCs and faith-based CDCs have become increasingly important in recent years.

Regional CDCs
CityWide (Dayton, Ohio) has become a major stakeholder in many economic development initiatives throughout Dayton, resulting in a strong working relationship with city officials and institutional leaders. These high-level relationships increase CityWide's access to large pools of funding and help gather support for projects. While small neighborhood-based CDCs in Dayton struggle to find adequate funding for neighborhood projects, CityWide is able to leverage their relationships with city-wide civic and political leaders resulting in decently funded comprehensive neighborhood development projects that at least have a running chance to transform neighborhoods (Varady and Badinghaus, 2013).

The Office for Community Development (OFCD, created by the Archdiocese of Philadelphia in 2001) is a regional CDC in that it serves the entire Diocese of Philadelphia. It differs from the many Catholic-sponsored community-based CDCs that have emerged since the 1960s in the tradition of neighborhood-oriented CDCs (Welch, 2013). However, as a regional CDC, the OFCD like Catholic-oriented neighborhood CDCs also has a secular orientation. Like most other CDCs (Catholic or non-Catholic) the OFCD works in areas suffering from physical blight and concentrated poverty. What sets this one apart is its work in areas with aging parishioners, which creates a need to find new uses for church properties including senior housing.

Faith-based and minority-oriented CDCs
Black churches (and historically Black colleges and universities) have played a significant role in community revitalization (Lowe and Shipp, 2014), although these efforts have generally been ignored in the general CDC literature. For example, the Canaan and Abyssinian Baptist churches in New York City strove to revitalize areas near their churches in Northeast Bronx and Central Harlem. In Columbia South Carolina, Benedict College and Allen University created the Benedict Allen CDC. The current interest of Black churches and universities in CDCs reflects the legacy of social mission, persistence of segregation, "and their use of 'Christian capitalism' which encourages members of the Black community to pool resources and invest in each other and their communities" (Lowe and Shipp, 2014, p.247). Black churches and Black colleges have chosen to use the CDC model because housing production can provide visual evidence that the churches or colleges played a leadership role. The Collective Banking Group (CBG) now called the Collective Empowerment Group (CEG) is a CDC that operates as a membership-based, service-driven organization. It was founded by Rev Jonathan Weaver, Greater Mt Nebo

African Methodist Episcopal Church in suburban Washington, DC and uses an innovative business model (Shanks et al., 2015).

To join this membership-based organization, churches pay a fee based on their membership size and their church members become card-carrying members of CEG entitled to special benefits and services. Banking partners are assessed a fee based on their asset base (fees range from $13,000 to $33,000), are responsible to support CEG community events and fundraisers and are encouraged to provide pro bono services. Both banking partners and strategic business partners receive business from participating churches and their members. The CEG now has several programs including financial education classes and counseling (Shanks et al., 2015, p.4; see also About the Collective Empowerment Group, no date). According to Shanks the CEG operates in six other cities—Baltimore, Charlotte, Miami, Austin, Newark and Cincinnati—but unfortunately does not provide detailed information on how active these chapters are. An internet search turned up no information about CEG branches.

A key assumption in writings dealing with Black CDCs is that with a shrinking federal safety net and shrinking public services, Black CDCs can step in and provide these services. This assumption is probably wrong. Black CDCs are in reality highly dependent on governmental and/or foundation support for their survival. In 2001 President Bush created the White House Office of Faith-Based and Community Initiatives to provide federal funding to religiously based CDCs. In 2009 the Obama Administration strengthened this effort by renaming it the White Housing Office of Faith-Based and Neighborhood Partnerships. Funding cutbacks for these agencies would probably have a devastating impact on Black CDCs.

Entrepreneurial and community enterprise-like CDCs
Here we are referring to CDCs that are trying to not only stimulate economic development and provide decent paying jobs but also to develop revenue streams that make them independent of government and philanthropic funding. In the Diamond neighborhoods of San Diego (Dubb, 2016), a community-based organization raised government and philanthropic funds to develop a commercial and cultural complex moored in a shopping center. A public offering enabled community residents and employees an exclusive opportunity to buy shares in the project. Eventually residents will own 50 percent of the project and a neighborhood foundation will own the remaining 50 percent. An attempt to find out more about community enterprise activity in the Diamond Neighborhood proved fruitless. Florido (2010) describes the somewhat controversial efforts of the Jacobs Foundation to implement a $500 million revitalization project to reverse neighborhood decline but fails to mention any specific CDC or CE.

Cleveland, Ohio has witnessed the emergence of a set of worker owned companies including the Evergreen Laundry, an employee owned energy company that is installing large-scale installations of solar panels for some of the city's largest nonprofit health, education and municipal buildings, and a large-scale hydroponic garden. For additional information about the Evergreen Cooperatives see (The Cleveland Model, no date; Evergreen Cooperative Laundry, no date, and Evergreen Cooperatives, no date).

Alphabet Scoop Creamery (ASC) is a for-profit social purpose business owned and operated by the Father's Heart Church and Ministry Center and is located in the East Village neighborhood of Manhattan (Johnson, 2013). Since the 1990s, gentrification has significantly transformed the community. Nevertheless, the area contains a large low-income minority population in public housing developments. ASC's business model includes a for-profit ice cream store which supports a job-training program for at-risk youths. "By appealing the desire of young professionals to do 'the right thing' [when they buy ice cream] Alphabet City balances mission and market" (Johnson, 2013, p.188).

CDC CAPACITY

Level of Capacity

Our 2015 article (Varady et al., 2015) highlighted the fact that the growth of the CDC sector in the US is counterbalanced by the fact that many CDCs are small, produce few affordable housing units and lack capacity. Our current review shows the situation to be quite nuanced. First, one has to be cautious in using staff size as a measure of capacity. The SNAED office in Baltimore only has two to three employees, yet it is able to secure enough public and private funds (mostly through private foundations, National Endowment for the Arts and university grants) to promote the Central Baltimore District through cultural events and advertising. Similarly, ASC shows that a small, underfunded and understaffed faith-based organization can, because of its ties to local universities and corporations and its utilization of volunteers, run a viable community enterprise and help many at-risk minority youths.

Secondly, even within particular cities there exists considerable variation in the capacity of CDCs. For example, in Cleveland's Slavic Village Development has a staff of 18 and is considered "one of the leading CDCs in the city" (Krumholz et al., 2006, cited in Fujii, 2016, p.308). Similarly, Cleveland's DSCDC is considered another strong CDC. In 2009 it had a full-time staff of 31 and an annual budget of $2.2 million.

Thirdly, whether CDCs can expand their scope to encompass new programmatic areas will be dependent on the extent to which they can address capacity limitations. Solitaire and Lowrie (2012) call for an expanded role for CDCs in brownfield redevelopment because greater involvement could not only facilitate the reuse of vacant sites but in addition "CDCs could (also) promote democratic decision-making by insuring local voices are incorporated into brownfield redevelopment" (ibid., p.462) However, to improve CDC capacity so as to play a role in brownfield redevelopment will require increased funding, hiring and keeping technically skilled executive directors, educating staff on contamination issues, improving network capability (e.g. interaction with other stakeholders including developers) and increasing CDC political capacity (i.e. trustful and credible relations with residents). Solitaire and Lowrie are uncertain whether CDCs will be able to address these issues.

Factors Affecting Capacity

Organizational factors
CDC scholars have devoted relatively little attention to either the motivation or the skills of the Executive Director. There are some exceptions, however. Krumholz and Roller (2012) note that Cleveland's DSCDO might not have been created (and thrived) had it

not been for Father Marino Frascati who founded DSCDO and served as its President until his death in 2002. DSCDO shows that a CDC with an entrepreneurial director and that is well-connected to a variety of funding sources can implement successful arts-based community revitalization as well as green neighborhood development. David Beach, the founder of EcoCity Cleveland, a local environmental advocacy and planning organization, played a key role in the creation of the city's EcoVillage (jointly run by EcoCity Cleveland and DSCDO). Finally, Senior Pastor Carol Vedral played a particularly important role in the development of ASC in New York City.

The importance of strong leadership is also evident in the case of Black church CDCs. Rev Jonathan Weaver was responsible for forming the CBG (see above) as part of his work as pastor at the then 700-member Greater Mt Nebo African Methodist Church in suburban Washington DC (Shanks et al., 2015). The pastor of the Metropolitan African Methodist Episcopal Zion Church located in Washington, North Carolina worked with the public and private sectors to provide improved health care through the creation of the Metropolitan Community Health Services and Agape Community Clinic and banking services via the Metropolitan Credit Union (Lowe and Shipp, 2014).

Social capital
It seems reasonable to assume that a CDC's prospects for success will be dependent on the strength of social networks (i.e. the level of social capital) which would affect the CDC's ability to recruit a network of volunteers to carry out its work (West et al., 2015). However, few, if any, scholars provide guidance for how to recruit and maintain such a network. Cohen's 2010 article on the Vietnamese community in East New Orleans is one of the rare ones to focus on social capital and how it influences CDC effectiveness in immigrant communities. This tightly knit Vietnamese community has, unlike other groups, thrived since Hurricane Katrina. "Six months after Katrina, a quarter of the community's households—1,000 in all—had returned, and 45 out of 53 businesses were back at work . . ." (Cohen, 2010, n.p.). We wonder if there are other examples of immigrant-initiated CDCs that simply have not been written up in scholarly articles. Or is it possible that immigrant groups generally do not use CDCs but rather rely on local political structures and the private economy to revitalize neighborhoods (e.g. Dominicans in New York City, see Krohn-Hansen, 2013).

McRoberts' 2003 study of collective efforts of minority churches in the Four Corners neighborhood in Boston (cited in Lowe and Shipp, 2014) shows the adverse impact of diversity (among immigrants) on social capital and, in turn, on efficacy. Most of the churches in the 0.6 square mile area were in commercial storefronts and virtually all were "commuting congregations" often made up of Caribbean, Haitian and Latino immigrants. These churches were not involved in community development underscoring the fact that many churches are not really interested in community revitalization activities.

Metropolitan networks
There exists considerable consensus that to be successful a CDC needs a variety of partners including local banks, city and state agencies, the United Way and community action agencies (West et al., 2015). In Cleveland the existence of strong intermediaries during the GFC "led to more predictable and reliable support for organizations facing otherwise turbulent environments" (Thomson and Etienne, 2017; see also Vidal and

Keyes, 2005). By 1998, Cleveland's CDCs had achieved the status of a "mature system" with an above average capacity for delivering community development in large part due to the existence of these two intermediaries (Walker and Weinheimer, 1998, as cited in Kellogg and Keating, 2011).

The Detroit Shoreway success story in Cleveland with respect to an Arts and Entertainment District was a result of the willingness of the three organizations (DSCDO, Cleveland, Public Theatre and Near West Theatre) to create the GSAD and to collaborate in fundraising. Slavic Village Development Inc. (also Cleveland) highlights the increasing number of nonprofit CDCs in Cleveland and elsewhere working with for-profit developers. In 2013, the Slavic Village Recovery Project formed a limited liability company by nonprofits and for-profits: Slavic Village Development Corporation, Neighborhood Progress Inc. (now called Cleveland Neighborhood Progress), Forest City Enterprises Inc. a large for-profit developer and RIK Enterprises LLC, a local private enterprise.

As indicated above, Cleveland's strong financial intermediaries have helped to create effective CDCs including, but not limited to, the DSCDO (the GSAD, EcoVillage) and Slavic Village Development but it is important to note that DSCDO's success has enriched the networks that DSCDO participates in (Kellogg and Keating, 2011). DSCDO has shown the neighborhood housing network that green and affordable neighborhood development projects are compatible. Furthermore, it has shifted the participation of the environmental community away from an exclusive "traditional" emphasis on open space conservation and pollution to good development patterns including housing and transportation. Finally, the supportive role of the city council representatives with respect to EcoVillage has spread green building into the organization of government: that is, Cleveland has enacted a wide variety of regulations and policies supporting green affordable development.

According to West et al. (2015) the services provided by the Massachusetts Association of Community Development Corporations (MACD) are largely responsible for the effective CDC movement in Massachusetts. This highlights the importance of vertical relations between CDCs and state and national entities as well as horizontal relations with local for-profit and governmental agencies.

SNAED's success in attracting artists and cultural tourism to Central Baltimore reflects the CBO's partnership with the city, private foundations, CDCs (the Central Baltimore Partnership, Jubilee Inc.) and local universities especially Johns Hopkins University (Rich and Tsitsos, 2016). Technically speaking, the Central Baltimore Partnership is not a CDC; rather it is an umbrella-like coordinating group. Jubilee is a CDC but one that moves around the city from one distressed area to another.

An increasing number of CDCs have tried to link up with anchor institutions (government, universities, hospitals) to create a steady source of revenue as well as a source for technical expertise. A number of these efforts are described elsewhere in this chapter. Sometimes an anchor institution plays a key role even when it is not in the same neighborhood as the CDC. As Kellogg and Keating (2011) point out, the relationship between EcoCity Cleveland and the Center for Neighborhood Development played a key function in the Detroit Shoreway EcoVillage study. Cleveland State University is on Cleveland's East Side while the EcoVillage is on the West Side.

The Philadelphia Archdiocese's OFCD relies on a variety of funding sources: "city, state or federal government grants or tax credits, private donors and foundations, corporate partners and revenue generating programs or projects (such as developer's fees, program fees, etc.)." Furthermore, OFCD has also leveraged Catholic Charities and the Catholic Campaign for Human Development resources (Welch, 2013, p.454).

Black churches have also relied on partnerships (Lowe and Shipp, 2014). Four Black churches in Columbus, Ohio through the leadership of their pastors developed a mechanism for the churches to engage with elected officials in a comprehensive revitalization strategy that would have not been possible for a single church to undertake. In New York City, two Black churches performed a key role in the implementation of a Partnership for New Homes, a New York City affordable housing program.

Urban Innovation 21, which focuses on heavily Black areas in Pittsburgh, is a consortium of higher education, government, local corporations, regional economic development organizations, lenders and charitable funds and targets current and prospective businesses (Shanks et al., 2015). Urban Innovation's strategy is to link local business development to the supply chain of anchor institutions "to meet their corporate needs and infuse capital into disinvested neighborhoods" (ibid., p.74).

Alphabet Soup Creamery (mentioned earlier) highlights the importance of vertical networks between neighborhood institutions and external entities. For example, in March 2009, ASC received a $60,000 interior design makeover, courtesy of New York University and Macy's and launched a new flavor, Macy's Cherry Jubilee. Furthermore, ASC and its employees often provide on-site catering services at the corporate headquarters of Victoria's Secret (Johnson, 2013).

COMPREHENSIVENESS AND VIABILITY

As CDCs have gone beyond housing production (to include social services and economic development) in the context of decreased government funding, they have experienced a "quadruple bottom line"; each one needs to simultaneously promote city-wide economic development/housing production, equity, environmental sustainability and institutional viability (Bratt, 2012). The recent literature on CDCs highlights the tensions between economic development and housing production on the one hand, and both community organizing and equity, on the other. Scholars wildly disagree on CDCs' ability to resolve these tensions. Randy Stoecker (1997) asserts that CDCs can focus on either community building (raising levels of social capital) or economic development (commercial real estate, market or mixed-income development), not both. When CDCs become too dependent on outside organizations, they inevitably lose control over development (see also Silverman and Patterson, 2012). Stoecker recommends two distinct types of CDCs; ones that focus on economic development/housing production, and ones that focus on community organizing.

Sarmiento and Sims (2015), drawing from their analysis of three affordable housing projects with a redevelopment plan in Santa Ana, California, are gloomy about the prospects for achieving any compromise between these two sets of goals. They state "that a narrow focus on affordable housing, as it is designed and produced within the *larger affordable housing complex*, facilitates the process of gentrification and displacement" (italics added, p.323). In other words they believe that unless planners are exclusively focused on equity they are dysfunctional. We strongly disagree. Even if a CDC was unable to stop gentrification but slowed the process down and achieved some degree of social mixing this would indeed be a notable accomplishment. And even if the affordable housing strategy involved community control we believe that low-income residents would prefer a mixed community over an exclusively low-income one.

Similarly, Varady and Raffel (1995) argue for the need for a balance vis-à- vis these two sets of goals with regard to both urban education as well as urban housing. Educators must balance off the need for helping low-income children (to promote equity) with the need to achieve and maintain high academic and disciplinary standards, in order to attract and hold middle-class parents with children in the city (to promote economic development). Sarmiento and Sims' use of the term "affordable housing complex"—which conjures the dreaded "military-industrial complex" coined by President Dwight Eisenhower—does little to promote a rational and useful debate about housing and neighborhood development policies. Cleveland practitioners use the more politically neutral and more sensible term "housing network." Housing practitioners in Santa Ana and elsewhere would be surprised and offended to learn that they are part of an affordable housing complex focused on promoting gentrification and undercutting the interests of minorities and the poor.

Black churches and Black college CDCs highlight this tension between two seemingly contradictory goals: community building (which means "fighting City Hall") and housing production. If these CDCs promote the first goal too aggressively the church and the CDC could lose its tax exempt status and also its federal funding for housing programs (Lowe and Shipp, 2014).

Turning from theory to practice we can see that various CDCs have tried to achieve a middle ground between either a focus on production or a focus on community organizing/equity. Some have been more successful than others in resolving this tension, or are at least more optimistic about the possibility of doing so.

The Archdiocese of Philadelphia is facing the challenge of what to do with churches made redundant by depopulation. In 2002, the Archdiocese of Philadelphia offered its OFCD, a regional CDC, first refusal on vacant properties. If the OFCD took up the offer, it could restructure the vacant church into a community-oriented use such as affordable housing. However, support for the policy ended when the Cardinal who had supported the policy retired. This policy change illustrates the importance of leadership in affecting CDC capacity, a point we made earlier. The first refusal policy promoted historic preservation but did not maximize income generation. In gentrifying neighborhoods, developers are willing to pay more for a property than the OFCD could for affordable housing. Thus, the economic development and equity goals were in conflict (Welch, 2013).

Rich and Tsitsos' 2016 case study of the SNAED in Baltimore questions the notion that a CDC can simultaneously promote both production and participation/equity. Although SNAED and its partner institutions such as Johns Hopkins University said that they did not desire to displace low-income residents, interviewees expressed concern about recently arrived artists as well as long-term low-income Black residents being displaced. Rich and Tsitsos argue that "the people who are most likely to be displaced from the arts and entertainment district in the future are paradoxically artists, especially those who wish to buy homes and settle in the district" (p.736). This conclusion is based on observations of community meetings. It may be that the displacement of artists seemed to be a more significant concern because the artists more articulately expressed their concerns about relocation at community events. Furthermore, the anti-displacement policies that these institutions have put into place are unlikely to prevent artists from relocating elsewhere. However, elsewhere in the article they are more optimistic about resolving the tension between economic development and equity. They note: "Many leaders explain that development in Central Baltimore is a two way process, where powerful institutions and grassroots organizations meet in the middle" (p.766).

Further adding to their pessimism, Rich and Tsitsos believe that involvement of anchor institutions leads to a top-down process, one often guided by university interests. However, this does not have to be the case. Rich's own earlier 2013 study of downtown Scranton, Pennsylvania "shows that urban revitalization efforts are not always a top-down process but can be a synthesis of grassroots, nonprofit organizations, and large institutions' efforts" (p.741).

On the other hand, Casper-Futterman and DeFilippis (2017) believe that it is possible—at least to some degree—to reconcile production with equity/community participation. They cite the example of the Bronx Community Development Initiative (BCDI) which seeks to foster growth of small businesses, especially worker owned businesses through an on-line platform which connects businesses to anchor institutions like the Bronx Zoo and Fordham University. BCDI was inspired by worker ownership models such as the Evergreen Laundry in Cleveland which is closely linked to anchors such as Case Western University. Whether Evergreen can be replicated in other cities—and for purposes other than laundry services—remains highly uncertain.

BCDI provides small businesses with technical assistance so that they can compete for contracts with these anchor institutions. The idea is to not only stimulate the creation of decent jobs for low-income Bronx residents but also to foster "a steady revenue stream from anchor institutions and transaction fees to provide independence from the local state and the mercurial and fickle mandate of private philanthropy" (Casper-Futterman and DeFilippis, 2017, p.22). It is unclear, however, from the article whether this steady revenue stream would exclusively benefit BCDI itself or whether the revenue stream would also benefit the organizations that are represented on the BCDI steering committee, that is, a number of CDCs, a labor union, a worker's cooperative and other local organizations.

Casper-Futterman and DeFilippis believe that BCDI is raising levels of social capital in this borough while at the same time promoting economic development but do not provide any statistics to back up this assertion. Thus, it is hard to say based on the evidence available from journal articles and sources whether Casper-Futterman and DeFillipis's optimism about community-enterprise-like CDCs is justified.

Steve Dubb (Dubb, 2016) provides more detail on how CDCs could utilize the community enterprise model to promote their own financial independence. CEs are a form of social enterprise and are "nonprofit organizations that operate businesses both to raise revenue and to further their social missions" (ibid., p.6)—but are ones that operate at the community level. CEs could assist nonprofits such as CDCs to increase their capacity to generate independent sources of earned income, which, unlike government or foundation grants, are typically unrestricted. This would enable CDCs to better support their operations. Unfortunately, Dubb does not provide examples of CDCs operating CEs. It remains highly uncertain whether CEs would enable CDCs to become financially independent.

Theoretically, CDCs that employ the community enterprise model could empower residents by providing them with good paying jobs and by enabling CDCs to avoid the restrictions involved with government and foundation support. Dubb (2016) notes an important caveat, however. At Evergreen Laundry (Cleveland) and similar CE-type organizations, the need for experienced management and market pressures can sometimes conflict with community wealth building values of capacity building and leadership development (Kelly and McKinley, 2015, cited in Dubb, 2016, p.10). This same type of tension exists for CDCs in general, that is, between the need for skilled executive directors and staff and the need to develop local leadership.

Owen Kirkpatrick (2011) also wants to reduce the degree to which American CDCs depend on government funding (e.g. the Low-Income Housing Tax Credit, LIHTC program) and increase their reliance on community assets. According to Kirkpatrick, community development efforts need "to 're-embed' land (community land trusts), labor (worker co-ops), and money (local community development finance institutions) back into the norms, networks and institutions of local communities" (p.1, see also Silverman and Patterson, 2012). These types of innovative community development efforts are needed, says Kirkpatrick, to cultivate a thick moral embeddedness and to emphasize values (such as trust, reciprocity, and norms) over speculation . . ." (ibid., p.17).

However, it should be noted that these three types of organizations are fundamentally similar to CDCs. That is, all four have to at least break even to survive. Silverman and Patterson (2012) mention cooperative housing as yet another alternative to CDCs. We believe that cooperative housing should be considered a tool for CDCs not an alternative. The question is how CDCs can utilize these three community asset models. Should CDCs operate CEs or should they spin them off as separate institutions and closely collaborate with them?

Krumholz and Roller's 2012 case study of Detroit Shoreway in Cleveland questions the assumption that an arts-based strategy (or any economic development strategy for that matter) inevitably drives up housing prices, leading to the displacement of existing residents. This did not happen in Detroit Shoreway. Krumholz and Roller's short article suggests that the revitalization process has *not* been controversial. Perhaps this is because Father Marino and the executive directors that succeeded him promoted both affordable housing, on the one hand (using "green building" methods), and an arts-based revitalization strategy on the other. If so, the Krumholz and Roller article offers a more optimistic perspective on arts-based neighborhood revitalization than is evident in other scholarly articles.

Finally, Johnson's case study of ASC in New York shows that whereas a faith-based community enterprise cannot solve the problem of poverty or slow down gentrification, it can through the revenue generated by an ice cream parlor help many at-risk minority youths prepare for the workforce. In addition, ASC enables the Father's Heart congregation to extend its spiritual influence outside the church building itself. As one ASC patron, who is also a volunteer, noted:

> I was part of God's creation of that ice cream parlor . . . I had let him down on so many other things that the ice cream parlor is more part of God for me than the chapel. There's a part of my creation in there that'll be part of that forever. (Schwartz, 2005, paragraph 11)

Up to now we have discussed the presumed tension between economic development and equity. Bratt (2012) also asserts that the goals of equity and environmental sustainability are often in conflict. However, Kellogg and Keating's 2011 case study of Cleveland's

EcoVillage suggests that this conflict is resolvable. The construction of affordable green housing both raised and lowered the affordability of housing; the former through higher construction costs, the latter through dramatically lower operating costs. It appears that the latter more than overcame the former.

COMMUNITY PARTICIPATION

Recent CDC literature emphasizes that it is difficult to achieve meaningful citizen participation and that a fully "bottom-up" approach toward community development is neither feasible nor desirable. The preceding section implies that worker cooperatives would not only lead to decent paying jobs, but they would also empower residents. If so, extensive citizen participation exercises via CDCs would no longer be necessary. Joyce Mandell (2010) asks: is community organizing still important if the number of successful CEs increases? Based on a 2010 case study of Lawrence Community Works (LCW), a CDC in a heavily Hispanic neighborhood in Lawrence, Massachusetts Mandell argues that community organizing remains an important prerequisite for overall community revitalization. "[Community] picnics, coupled with leadership training and civic opportunities, provide a ladder for increased activism in the community, an activism that can include collaboration or conflict with the powers that be so that true change can be won for the community" (Mandell, 2010, p. 280). Mandell's assumption—that success could be measured by effectiveness in lobbying government—is, however, questionable given cutbacks in government funding at the federal, state and local levels. Many would argue that it is important in the long run that CDCs like this one become financially independent.

Furthermore, choosing a board of directors that represents "the community" is not easy. Casper-Futterman and DeFilippis commend BCDI when they note that it is "composed predominantly [of] women of color community leaders" (ibid., p. 18). We wonder whether such a board would be representative of all Bronx residents, Whites as well as non-whites, men as well as women, non-activists as well as activists? Furthermore, anchor institutions appear to not be represented on the steering committee, thereby raising the question: will anchor institutions be willing to commit staff and money to BCDI's programs if they are not part of the decision-making process? We suspect that they will not. In sharp contrast, SNAED in Baltimore does include representatives from anchor institutions such as Johns Hopkins University.

Citizen participation has been one of the most contentious issues in SNAED in Baltimore. First, this case study highlights the questionable role of "community" in CDCs and CBOs. The basic issue is that, with such a varied area where different races, classes, ages and lifestyles interact, it is "difficult to disentangle what the community wants" (Rich and Tsitsos, 2016, p.747). Furthermore, the success of the District is influenced by "its attractiveness as a consumer-driven tourist destination, which has very little to do with the needs of the neighborhood's constituents" (ibid., p.747). Furthermore, among residents, there is a basic

> conflict of interest in the neighborhood between what some artists are interested in—a fun (and slightly dangerous) living experience with space to be creative—and the issues that concern the

bulk of residents [including established artists] ... how to defend the neighborhood against displacement, vacant and dilapidated buildings, empty lots, crime and gangs. (ibid., p.749)

Secondly, citizen participation may not be at a sufficient level of intensity to be meaningful. In reality, citizen involvement occurs indirectly through the two CDCs that partner with SNAED (Rich and Tsitsos, 2016). It would be fair to say that citizen participation takes place at a low rung in Arnstein's famous 1969 ladder of citizen participation.

Thirdly, even though the City of Baltimore and developers working in the Arts and Entertainment District have discussed the concerns and interests of all residents, some of those interviewed "remain suspicious of being excluded" (Rich and Tsitsos, 2016, pp.740–41). Finally, Rich and Tsitsos believe that larger institutions such as Johns Hopkins University had undue influence on the decision-making process.

This raises the question of what constitutes "undue influence." Do large institutions like Johns Hopkins University not have legitimate concerns that they need to fight for such as reducing the incidence of street crime? Existing literature provides little guidance to SNAED and similar types of organizations regarding the influence that anchor institutions should have on policymaking (through the board of directors or otherwise) relative to existing residents.

The OFCD, the Archdiocese of Philadelphia's CDC, illustrates the strengths and weaknesses of a regional CDC, particularly with respect to community involvement (Welch, 2013). The OFCD partners with the local Catholic parish; as a result this regional CDC has a neighborhood base. This approach has many advantages. Meetings can take place at the churches and priests can publicize programming at the Mass or in the Church newsletter. This constitutes "instrumental participation" in the citizen participation continuum developed by Silverman and Patterson (2012), and represents a low level of citizen participation in Sherri Arnstein's 1969 ladder of citizen participation. Welch recommends that more community members be included on the board of OFCD in order to help gain access to community social networks. However, we believe that convincing community residents to join the board of a regional CDC would be a challenge because residents are only likely to participate in debates directly related to their own neighborhood. In addition, the governance structure of this regional CDC makes citizen participation problematic. OFCD is accountable to the Archdiocese and its stakeholders, which include clerical authorities and professional lay staff as well as funders and donors. This is quite different than the governing structure of community-based CDCs where residents (and their leaders) serve on the board "to empower" local residents. In general, OFCD uses a top-down planning approach which makes it difficult to identify needs "on the ground."

Welch recommends that OFCD take advantage of its strengths at the regional level to adopt new cost-benefit analysis software to answer questions like: should a building be reused or should it be retained as a subsidized development? Adopting this software might help the Diocese make more "rational" decisions but such software likely would overvalue objective criteria like property values and the use of the software would likely have the unintended effect of increasing the power of OFCD staff and weakening the prospects for meaningful citizen participation.

Costa's 2014 case study of the redevelopment of Jackson Square, a low-income area in Boston, highlights the challenges and opportunities involved in collaborative neighborhood development—an approach stressing community control (at least, at the beginning of the process). An RFP (request for proposal) from the City of Boston resulted in two proposals, one from a group of CDCs and one from a group of developers. The Mayor combined the two proposals and appointed a Citizens Advisory Committee to monitor implementation. As the plan then moved forward, community-initiated recommendations either were dropped or modified. For example, the youth and family facility which represented the community's highest priority—and which aimed to solve the problem of violence among rival gangs in the area—was put on indefinite hold by the developer. Community participation, which had been extensive and inclusive in the early stages, dropped off. The Jackson Square CDCs were trapped in their effort to control the development process between their own community vision on the one hand and the financial realities of Boston's housing market, on the other.

Costa's case study indicates the need to address financial feasibility issues during the visioning phases of collaborative planning processes (p.307). "As an alternative path, instead of advocating for public support, or in addition to it, communities could direct their efforts to acquiring further economic autonomy through the creation of cooperative business and housing organizations, land trusts, community banks, etc. in order to overcome the dependence on external support" (p.307). But, is Costa's proposal for a community enterprise-type approach feasible in heavily minority low-income communities like Jackson Square?

Harvey and Beaulieu's 2010 book chapter about community development in the Mississippi Delta challenges the "common wisdom" that top-down planning is bad and that bottom-up planning is good. The authors compare two distinct community development models.

> One organization, a regional philanthropic foundation, implemented an initiative based on grassroots resident empowerment in two non-contiguous Delta counties. The other, a community development financial institution (CDFI), took a more professionally managed top-down approach in which its efforts were highly concentrated in the economic hub of one county in which it chose to work (p.147).

Counter to what might have been expected, the CDFI achieved greater success in building a foundation for long-term resident-driven community revitalization than did the foundation dedicated to community control. The reliance on a community control model enabled activists to obtain philanthropic support but did not lead to any visible projects, which eventually led to a decline in participation and a folding of the initiative. On the other hand, CDFI's effort succeeded because it allowed "their professional staff the flexibility to implement a strategic plan built on input from the community without having to include the community in deliberative processes over every step" (p.168). Thus, according to Harvey and Beaulieu, CDCs in neighborhoods with diverse interests should not necessarily seek a facilitator to obtain a consensus on the community's vision for the future. Instead, such a CDC requires someone to negotiate conflicts between groups. In some cases this may mean applying pressure to particular stakeholders. In other words, planners and scholars need to re-examine their bias against top-down planning.

Krumholz and Roller's 2012 Detroit Shoreway case study provides additional evidence that top-down planning is not necessarily bad.

Norman Krumholz is considered a "planning hero" because of his work in equity planning as Director of the Cleveland Planning Department. We are confident that if there were equity issues in the Detroit Shoreway case study he would have mentioned them. Social justice issues are noticeably absent from this article.

Father Marino Frascati created the DSCDO and the GSAD. The idea for the CDC and the District did not emerge spontaneously from "the community." Interestingly, Krumholz is silent on whether such a top-down approach is undemocratic and dysfunctional. It is conceivable that citizen involvement has increased once the products of early planning and fundraising have become apparent.

Cleveland's EcoVillage (co-run by DSCDO and EcoCity Cleveland) also shows that the top-down approach sometimes works well. As mentioned earlier, David Beach was largely responsible for the creation of the EcoVillage. It was only after an initial feasibility study was completed (involving interviews with CDCs and intermediaries) and the site, Detroit Shoreway, was selected, that citizen involvement started. "Community political support was garnered through charrettes, open design sessions and meetings with block clubs to discuss the goals and obtain feedback on designs for each of the housing projects" (Kellogg and Keating, 2011). The preceding shows that productive citizen involvement may only be possible when residents are able to react to specific designs—rather than be asked to develop a broad (but vague) vision for the community.

The genesis of ASC in Manhattan also illustrates a modified version of top-down planning. A volunteer from McKinsey and Company (a large management consulting firm) recommended that the ice cream and frozen dessert industry would produce highly positive performance and profitability. The leadership of Father's Heart used this recommendation to obtain approval of the Board of Trustees and support of the congregation. The business model first failed but a Manhattan-based marketing firm produced a strategy that ultimately worked.

The literature reviewed in this section provides evidence of how residents' interests are served by a number of initiatives brought forth by CBOs. However, we are unsure whether these initiatives truly empower residents beyond catering for certain needs that are not addressed by other actors.

IMPACTS ON COMMUNITY REVITALIZATION

Although recent work provides no conclusive empirical evidence that CDCs can stop neighborhood decline in low-income communities and promote upgrading with the same population remaining in place, recent studies do show that CDCs can attain more limited goals like coping with foreclosed and vacant housing (Fujii, 2016), attracting artists and cultural tourism to art and entertainment districts (Krumholz and Roller, 2012; Rich and Tsitsos, 2016) and preventing at-risk minority youths from dropping out of high school (Johnson, 2013).

Cleveland illustrates the limited (but important) impact of CDCs. As a result of efforts by dozens of CDCs hundreds of housing units have been built or rehabilitated by CDCs. This has stimulated private investment "so that by 2001 some neighborhood submarkets had experienced quite significant increases in population, particularly in the downtown and its adjacent neighborhoods" (Mayer and Keyes, 2005, as cited in Kellogg and Keating, 2011, p.72). Nevertheless, despite the above average CDC system, Cleveland has one of the weakest housing markets and is one of the most socially distressed cities in America (Holder, 2017; Kellogg and Keating, 2011).

Whereas the pre-2010 CDC literature focused almost exclusively on CDCs in distressed, usually heavily minority communities, an increasing number of CDCs find themselves in revitalizing ones (i.e. gentrifying communities), our literature review does not provide any conclusive answer to the question of whether CDCs can slow down gentrification where it is occurring and create stable mixed-income areas. In other words, is it possible to implement "smart gentrification"? (Grabinsky and Butler, 2015a, 2015b; Lubell, 2016). SNAED's efforts in Baltimore likely contributed to the opening of new businesses and higher property values; however displacement (especially as it relates to recently arrived artists) remains a contentious issue; and many of those interviewed do not trust either SNAED or the City of Baltimore.

Krumholz and Hexter's suggestion (2012) that CDCs go beyond brick and mortar issues to deal with broader challenges, such as "community building," education and health, is appealing. However, there is a lack of clarity about what the term "community building" means and how best to achieve it. Furthermore, urban educators differ on the best way to improve schooling and criminologists differ on how to make urban neighborhoods safer. CDCs will need to grapple with these "wicked softer problems" lacking clear-cut guidance from social scientists.

CONCLUSION: WHAT CAN NORTHWEST EUROPEAN COMMUNITY ENTERPRISES LEARN FROM AMERICAN COMMUNITY-BASED ORGANIZATIONS?

The discussion of the literature on a number of American CBOs has revealed that they are at least a decade(s) ahead of their European counterparts (see also Bailey, 2012; Varady et al., 2015). In fact, the emergence of CEs in the Northwestern part of Europe is "likely to be familiar to readers in countries such as the USA, which has long had a vigorous culture of local enterprise combined with a deep hostility to 'government'" (Healey, 2015, p. 116). While the governments of Northwest European countries are not met by such a deep hostility, trends of (welfare) state retrenchment and do-it-yourself democracies are emerging clearly, setting a context in which they respond to trends of policy reforms, austerity and budget cuts (Kleinhans, 2017). While this is a difficult entrepreneurial "ecosystem" for nascent CEs, the American experiences provide evidence that maturation of CBOs over time may solve part of tensions related to the hybrid nature of CEs in Europe—financial and economic feasibility versus social impact (Doherty et al., 2014).

In the context of a "quadruple bottom line," the need to simultaneously address economic development, housing production, equity, environmental sustainability and institutional viability (Bratt, 2012), CDCs have struggled to continuously define "community" and "target groups," neither of which are static entities. There is no reason to assume that this will be fundamentally different for their European counterparts, who have to relate to the local level and communities as well. In terms of the participation of community members and/or local residents in CBOs, the literature review has shown that productive citizen involvement (beyond the core members of CBOs) is more likely after the initial contours of the business and its objectives have been established, rather than being asked to develop a broad vision for the target community of the business. Hence, being accountable to the local community, which is a core feature of CEs in Northwest Europe (Bailey, 2012; Buckley et al., 2017; Healey, 2015) does not necessarily require strong resident participation in the initial stages of building up the business. Relatedly, the experience of CBOs has re-emphasized the importance of leadership development (cf. Selsky and Smith, 1994; Kirk and Shutte, 2004), something which needs support from beyond CEs themselves.

Like CBOs in the United States, CEs in Europe are unlikely to operate in a vacuum—coalitions with governments and other (local) stakeholders are a prerequisite to make an impact. Whereas the literature on CBOs mentions the importance of linking up with "anchor institutions" to create a more steady income (e.g. Cleveland's Evergreen Laundry), we are unsure whether such a strategy makes sense in the context of entrepreneurial "ecosystem" development in Northwest European countries, with national governments only very slowly changing their procurement systems to enable CEs to compete for public service delivery. On the other hand, the CBO experiences show that it will be extremely difficult for their European counterparts to escape the lingering dependence on government and charity funds (Bailey, 2017; Kleinhans, 2017). Traditionally, CDCs have played a fundamental role in the provision of affordable housing for poor households in American cities. While the notion of developing affordable housing may be considered as a growth area in the European institutional context, it is quite likely that realizing such an objective will be hindered by national housing systems, under which offering and managing affordable (social rented) housing is only reserved to highly regulated institutions such as housing associations. The growth of private cooperative housing is unmistakably present, but also small-scale. CEs might actually be more effective in contributing to neighborhood regeneration in general than housing production in particular (see e.g. Coatham and Martinali, 2010; Varady et al., 2015; Bailey, 2017). In the European context, this is a clear avenue for further research, considering that the number and importance of CEs is likely to increase over time.

REFERENCES

About the Collective Empowerment Group. (No date). Retrieved from http://www.collectiveempowermentgroup.org/About-Us.html, 14 August 2017.

Arnstein, S. (1969). A ladder of citizen participation. *Journal of the American Planning Association*, *35*(4), 216–24.

Bailey, N. (2012). The role, organisation and contribution of community enterprise to urban regeneration policy in the UK. *Progress in Planning*, *77*(1), 1–35.

Bailey, N. (2017). The contribution of community enterprise to British urban regeneration in a period of state retrenchment. In M. van Ham, D. Reuschke, R. Kleinhans, C. Mason, and S. Syrett (eds), *Entrepreneurial Neighbourhoods: Towards an Understanding of the Economies of Neighbourhoods and Communities* (pp. 229–48). Cheltenham, UK and Northampton, MA, USA: Edward Elgar Publishing.

Bratt, R. (2012). The quadruple bottom line and nonprofit organizations in the United States. *Housing Studies*, 27(4), 438–56.

Buckley, E., Aiken, M., Baker, L., David, H., and Usher, R. (2017). Community accountability in community businesses. Research Institute Report No. 10. London: Power to Change. Retrieved from: http://www.powertochange.org.uk/wp-content/uploads/2017/11/Research-Report-10-Digital.pdf, 21 August 2017.

Casper-Futterman, E. and DeFilippis, J. (2017). On economic democracy and community development. In M. van Ham, D. Rushke, R. Kleinhans, C. Mason, and S. Syrett (eds), *Entrepreneurial Neighbourhoods: Toward an Understanding of Neighbourhoods and Communities* (pp. 79–202). Cheltenham, UK and Northampton, MA, USA: Edward Elgar Publishing.

The Cleveland Model: How the Evergreen Cooperatives are building community wealth. (No date). *Community-Wealth.org*. Retrieved from http://community-wealth.org/content/cleveland-model-how-evergreen-cooperatives-are-building-community-wealth, 28 August 2017.

Coatham, V. and Martinali, L. (2010). The role of community-based organisations in sustaining community regeneration: An evaluation of the development and contribution of Castle Vale Community Regeneration Services (CVCRS). *International Journal of Sociology and Social Policy*, 30(1/2), 84–101.

Cohen, A. (2010). The entrepreneurial gamble. *Planning*, 76(1), 18–21.

Costa, P.M. (2014). From plan to reality: Implementing a community vision in Jackson Square, Boston. *Planning Theory and Practice*, 15(3), 293–310.

Doherty, B., Haugh, H., and Lyon, F. (2014). Social enterprises as hybrid organizations: A review and research agenda. *International Journal of Management Reviews*, 16(4), 417–36.

Dubb, S. (2016). Community wealth building forms: What they are and how to use them at the local level. *Academy of Management Perspectives*, 30(2), 1–12.

Evergreen Cooperative Laundry. (No date). Dependable commercial laundry services just down the street. Retrieved from http://www.evgoh.com/ecl/, 7 August 2017.

Evergreen Cooperatives. (No date). Cleveland People: Brett Jones President, *Evergreen Energy Solutions/Cleveland Scene Magazine*. Retrieved from: http://www.evgoh.com/2017/07/19/cleveland-people-brett-jones-president-evergreen-energy-solutions-cleveland-scene-magazine/, 14 August 2017.

Florido, A. (2010). Big plans for a struggling neighborhood. Voice of San Diego, 3 August. Retrieved from https://www.voiceofsandiego.org/neighborhoods/big-plans-for-a-struggling-neighborhood/, 14 August 2017.

Fujii, Y. (2016). Spotlight on the main actors: How land banks and community development corporations stabilize and revitalize Cleveland neighborhoods in the aftermath of the foreclosure crisis. *Housing Policy Debate*, 26(2), 296–315.

Grabinsky, J. and Butler, S.M. (2015a). The anti-poverty case for "smart" gentrification, Part 1. Washington, DC: Brookings. Retrieved from https://www.brookings.edu/2015/02/10/the-anti-poverty-case-for-smart-gentrification-part-1/, 7 August 2017.

Grabinsky, J. and Butler, S.M. (2015b). The anti-poverty case for "smart" gentrification, Part 2. Washington D: Brookings. Retrieved from https://www.brookings.edu/2015/02/11/the-anti-poverty-case-for-smart-gentrification-part-2/, 14 August 2017.

Harvey, M.H. and Beaulieu, L.J. (2010). Implementing community development in the Mississippi Delta. In G.P. Green and A. Goetting (eds), *Mobilizing Communities: Asset Building as a Community Development Strategy* (pp. 146–176). Philadelphia: Temple University Press.

Healey, P. (2015). Citizen-generated local development initiative: Recent English experience. *International Journal of Urban Sciences*, 19(2), 109–18.

Holder, S. (2017). America's most and least distressed cities. *CityLab*. 26 September. Retrieved from https://www.citylab.com/equity/2017/09/distressed-communities/541044/, 28 August 2017.

Johnson, S.P. (2013). Changing lives one scoop at a time. In R. Cimino, N.A. Mian, and W. Huang (eds), *Ecologies of Faith in New York City: The Evolution of Religious Institutions* (pp. 169–97). Bloomington, IN: University of Indiana Press.

Kellogg, W.A. and Keating, D. (2011). Cleveland's EcoVillage: Green and affordable housing through a network alliance. *Housing Policy Debate*, 21(1), 69–91.

Kelly, M. and McKinley, S. (2015). Cities building community wealth. Takoma Park MD: The Democracy Collaborative. Retrieved from https://democracycollaborative.org/, 14 August 2017.

Kirk, P. and Shutte, A.M. (2004). Community leadership development. *Community Development Journal*, 39(3), 234–51.

Kirkpatrick, L.O. (2011). Karl Polanyi and the community development movement: The local politics of embeddedness. In J.P. Rothe, L. Carroll, and D. Ozegovic (eds), *Deliberations on Community Development*. New York, Nova Science Publishers. Retrieved from https://www.researchgate.net/publication/274707030_Karl_Polanyi_and_the_Community_Development_Movement_The_Local_Politics_of_Embeddedness, 14 August 2017.

Kleinhans, R. (2017). False promises of co-production in neighbourhood regeneration: The case of Dutch community enterprises. *Public Management Review*, 19(10), 1500–18.

Krohn-Hansen, C. (2013). *Making New York Dominican: Small Business, Politics, and Everyday Life*. Philadelphia: University of Pennsylvania Press.

Krumholz, N. and Hexter, K. (2012). *Re-thinking the Future of Cleveland's Neighborhood Developers*. Cleveland: Cleveland State University Urban Publications.

Krumholz, N. and Roller, J. (2012). Here comes the neighborhood: A Cleveland success story. *Planning*, 78(3). Retrieved from http://www.planning.org/planning/2012/mar/herecomesneighborhood.htm, 14 August 2017.

Krumholz, N., Keating, W.D., Star, P.D., and Chupp, M.C. (2006). The long-term impact of CDCs on urban neighborhoods: Case studies of Cleveland's Broadway-Slavic village and Tremont neighborhoods. *Journal of the Community Development Society*, 47(4), 3–30.

Lowe, J.S. and Shipp, S.C. (2014). Black church and Black college community development corporations: Enhancing the public sector discourse. *Western Journal of Black Studies*, *38*(4), 244–59.

Lubell, J. (2016). Preserving and expanding affordability in neighborhoods experiencing rising rents. *Cityscape*, *18*(3), 131–50.

Mandell, J. (2010). Picnics, participation and power: Linking community building to social change. *Community Development*, *41*(2), 269–82.

Mayer, N. and Keyes, L.C. (2005). *City Governments' Role in the Community Development System*. Washington DC: Urban Institute.

McRoberts, O. (2003). *Streets of Glory: Church and Community in a Black Neighbourhood*. Chicago: University of Chicago Press.

Rich, M.A. (2013). "From coal to cool": The creative class, social capital, and the revitalization of Scranton. *Journal of Urban Affairs*, *35*(3), 365–84.

Rich, M.A. and Tsitsos, W. (2016). Avoiding the "SoHo effect" in Baltimore: Neighborhood revitalization and arts and entertainment. *International Journal of Urban and Regional Research*, *40*(4), 736–56.

Sarmiento, C.S. and Sims, J.R. (2015). Façades of equitable development: Santa Ana and the affordable housing complex. *Journal of Planning Education and Research*, *35*(3), 323–36.

Schwartz, Karen. (2005). Nourishment for the soul, hot fudge optional. *New York Times*. 17 April. Retrieved from https://www.nytimes.com/2005/04/17/nyregion/thecity/nourishment-for-the-soul-hot-fudge-optional.html, 14 August 2017.

Selsky, J.W. and Smith, A.E. (1994). Community entrepreneurship: A framework for social change leadership. *The Leadership Quarterly*, *5*(3/4), 277–96.

Shanks, T.R., Boddie, S.C., and Wynn, R. (2015). Wealth building in communities of color. In R. Bangs and L.E. Davis (eds), *Race and Social Problems: Restructuring Inequality*. (pp. 63–78). New York: Springer Science+Business Media.

Silverman, R.M. and Patterson, K.L. (2012). Community- and neighborhood-based organizations in the United States. In S.J. Smith (ed.), *International Encyclopedia of Housing and Home*, pp. 186–93 (pp. 186–93). Amsterdam: Elsevier.

Solitaire, L. and Lowrie, K. (2012). Increasing the capacity of community development corporations for brownfield redevelopment: An inside-outside approach. *Local Environment*, *17*(4) 461–79.

Stoecker, R. (1997). The CDC model of urban redevelopment: A critique and an alternative. *Journal of Urban Affairs*, *19*(1), 1–22.

Thomson, D.E. and Etienne, H. (2017). Fiscal crisis and community development: The great recession, support networks, and community development corporation capacity. *Housing Policy Debate*, *27*(1), 137–65.

Turcotte, D.A., Johnson, M.P., Chaves, E.J., Drew, R.B., and Sullivan, F.M. (2015). Reconstructing neighborhoods: Two case studies in foreclosed housing acquisition and redevelopment by community development corporations in Massachusetts. *Housing and Society*, *42*(1), 17–39.

Varady, D.P. and Badinghaus, A. (2013). City-wide development corporation in Dayton, Ohio (USA): Integrating economic and community development, RC 43 Conference. Amsterdam, Netherlands, 10–12 July.

Varady, D.P. and Raffel, J.A. (1995). *Selling Cities: Attracting Homebuyers through Schools and Housing Programs*. Albany, NY: State University of New York Press.

Varady, D.P., Kleinhans, R., and van Ham, M. (2015). The potential of community entrepreneurship for neighbourhood revitalization in the United Kingdom and the United States. *Journal of Enterprising Communities*, *9*(3), 253–76.

Vidal, A. and Keyes, L. (2005). *Beyond Housing: Growing Community Development Systems*. Washington DC: Urban Institute.

Walker, C. and Weinheimer, M. (1998). *Community Development in the 1990s*. Washington, DC: Urban Institute.

Welch, B.J. (2013). A dual nature: The Archdiocesan Community Development Corporation. In K.L. Patterson and R.M. Silverman (eds), *Schools and Urban Revitalization: Rethinking Institutions and Community Development*. (pp. 71–85). New York: Routledge.

West, M., Kraeger, P., and Dahlstrom, T.R. (2015). Establishing community-based organizations. In R. Phillips and R. Pittman (eds), *An Introduction to Community Development: Second Edition* (pp. 104–18). New York: Routledge.

8. Community development, well-being and technology: a Kenyan village*
Claire Wallace and Leanne Townsend

INTRODUCTION

It is now increasingly recognised that community development should include well-being (Anderson, 2017; Phillips and Wong, 2017). Generally well-being may depend upon economic development but also social development, including factors such as social cohesion, social inclusion and empowerment (Abbott et al., 2016). These elements are all important at the individual level as well as at the level of the wider society – individual and social well-being are interactive (Abbott and Wallace, 2012). Whilst at a national level, both social and economic well-being as well as individual and collective well-being have been linked, it is not clear how this applies at a community level. However, we would expect them to be also connected.

Well-being paradigms, such as that developed by the Organisation for Economic Co-operation and Development are usually generated in the context of the developed world, especially Europe and the USA (White, 2010). Yet these elements are shown to be just as important in African societies as in other parts of the world (Abbott and Wallace, 2012a; Abbott et al., 2016). In African societies people value socio-economic security (regular incomes, social and health services, cushioning against economic uncertainties), social cohesion (feeling part of a greater community, well-functioning social connections, good governance), social inclusion (involvement of all parts of the community including ethnic minorities, women, disabled) and social empowerment (a feeling that your views count, ways of being able to control the environment and the tools such as good health and education that enable this). Whilst this "Social Quality" paradigm was invented in Western Europe (Beck et al., 2001), we have shown that as a normative aspiration it applies equally in Africa, although Africans may not have the means to fully realise it (Abbott et al., 2016).

Cloutier and Pfeiffer also recognise the importance of well-being as a development goal, but bring in the neglected aspect of culture (Cloutier and Pfeiffer, 2017), something which turned out to be important in this study. Culture in this context can mean entertainment, leisure and creative activities. Indeed, cultural heritage is also an important force in rural development (Wallace and Beel, 2020). All of these elements have been shown to be enhanced by digital communications and Information and Communications Technologies (ICT) in studies in Britain (Wallace and Vincent, 2017). Indeed, ICT has become a major medium for transmitting culture.

In Britain, we have argued for looking at the role of technology in fostering social well-being at a community level (Wallace and Vincent, 2017). Whilst technology in itself is neither conducive to well-being nor otherwise, it can nevertheless be used to foster aspects of social well-being at a community level in the way that it is used. It is arguable that the extent of connectedness in modern society through communications technologies means that social well-being depends on technology as never before. The constant need to connect through social media and distress at disconnection or negative feedback in this process is a ubiquitous aspect of modern life (Hobsbawm, 2018; Turkle, 2013). In Africa, technology provides opportunities for social connectedness too, as we shall show.

Yet how people connect at a local level or into wider communities is not self-evident. Daniel Miller in his study of an English village found people connecting in various ways about both trivial and important things but changing their mediums of communication quite rapidly using Snapchat, Twitter and so on (Miller, 2016). We argue that technology can enhance well-being by improving social contacts and connections (social inclusion and social cohesion), enabling community empowerment through political representation and feedback and encouraging local civic pride and identity by creating virtual representations of the village to enhance the real one (Wallace and Vincent, 2017).

When we turn our eyes away from the well-researched West and towards Africa we find new configurations of communication. In Africa, the use of mobile phones has transformed social and economic life as most people own or have access to such a phone with use of widespread fairly cheap network providers. However, these are generally the previous generation of mobile phone technology – smart phones are still out of reach for many due to the relatively high costs. Indeed, personal experience suggests that mobile phone connectivity in Africa is often better than in rural Scotland. Africans were among the first to use mobile phones as a form of banking and money transfer in a situation where access to conventional bank accounts is limited. However, use of smart phones and internet technology is still out of reach of most Africans, especially those living in rural areas.

Mobile phones are a familiar technology in Kenya. In 2017, there were 39.7 million mobile phones in Kenya (from a population of 44 million) and it is Africa's most mobile-friendly economy (Kuo 2017) – even if the numbers had dropped between 2016 and 2017. The mobile company Safaricom (whose adverts are everywhere) set up the mobile money platform M-Pesa, the largest in the world, in 2007 and it was estimated that 43 per cent of Kenyan gross domestic product flows through it (Dahir 2016). In a country where most people do not have bank accounts and distances are large, M-Pesa is a way of transferring money from place to place and person to person. Kenyans were pioneers of this technology.

Communications technologies depend upon electricity and many places in Africa do not have regular electricity. But this did not necessarily prevent mobile communications. People have been remarkably resourceful in charging their mobile phones from car batteries or solar panels even where there was no electricity – but it might entail walking long distances to a power point, especially in rural areas. Electricity might be hard to access but it did not stop Africans from using it. Rural electrification has been an important aspect of development in Africa. Whilst the dominant paradigm until recently has been that of centralised, state controlled electric power projects, more recent development models have emphasised the importance of local off-grid solutions (Mandelli et al., 2016). This is because large scale, centralised projects have tended to neglect rural areas, where there are poorer households who are less able to afford this access, where there is less demand and where it is difficult and expensive to deliver power over long distances and dispersed populations (Mandelli et al., 2016). Therefore, community-based off-grid systems have been developed, which could use either diesel or renewable energy such as hydro, solar or wind (Kirubi et al., 2009). Kirubi and colleagues (2009) emphasised that the success or otherwise of these initiatives depends upon their sustainability, the role of the agency introducing the system, the complementary local infrastructure and the importance of involving community organisations and participants.

The way in which power is provided has implications for community development. Matarrita-Cascante and Brennan (2012) in conceptualising community development for the "twenty first century" classify the community development interventions into three types each with advantages and disadvantages (Matarrita-Cascante and Brennan, 2012). First of all the *imposed* method means that a project is set in a community with minimal community involvement. This can aid development, especially in the case of infrastructure, but does not help to develop the community because local actors are not empowered. Many technology and external aid projects take this form. The second is the *directed* model, where there is a mixture of community involvement and external direction. In these types of projects, the initiative might come from an external actor, but local participation is encouraged. The Ubuntu project that we evaluated took this form. The final example is that of *self-help* projects where the community itself initiates and controls the development. These kinds of projects involve the setting up of community broadband initiatives for example but tend to depend upon local leadership and the availability of funding, which can be fragile and change quite rapidly. Also external providers might step in when they realise that there is a demand, thus undermining self-help projects (Wallace and Vincent, 2017).

In this chapter we describe a project which brought electricity and broadband to a rural Kenyan village. We consider especially the economic, social and cultural dimensions. We look at the impact that these technologies had and how villagers adapted them in their own ways for their own purposes – not necessarily the ones predicted at the outset.

A KENYAN VILLAGE IN TRANSITION

Mutaroni Village is in rural Kenya and consists of about 2000 people located 14.5 km from the nearest town along a dusty dirt road deeply scored by gullies left over from the rainy season. A few cars but many small motorbikes operated along this road offering taxi services and bringing goods to the market. The majority of residents are subsistence farmers relying on a polyculture of cattle, sheep, goats, maize, mangoes, beans, avocados, macadamia nuts and other locally produced products. There is a local school, which forms a village hub next to the church of Christ the King. Our temporary office was located in the airy church hall with a stage at the front bearing a beribboned podium faced by rows of about 50 chairs. Chickens drifted in and out pecking crumbs from the floor. Sanitary facilities were provided by a gaily painted wooden shed in the yard over an open hole in the ground labelled "Ladies". There was no "Gents". We were frequent visitors as our stomachs adjusted to the new culinary conditions. We drank bottled water but there was a well with a windlass behind the church used by most people. Because we were occupying the church, the evening fellowship group met outside in the yard, in the shade of the battered corrugated iron fencing and chanted the evening away in chorus.

The village had a remarkable number of churches – five – which were well attended. These ranged from a newly built and magnificently gabled Presbyterian church to a huddled wooden building for Adventists. Social cohesion was provided to a great extent by this widespread religious activity. Most private houses consisted of small, square wooden shacks, although some were concrete. Women cooked over open fires using sticks of wood in an adjacent wooden shack. The houses were surrounded by mango and other trees, often with a log honey-bee hive lodged up in the branches, and also by fields with forlornly wilted maize plants, for the wet season had brought no rains this year. Most of the houses had an open-air stall nearby constructed of lashed tree branches where a Friesian cow stood munching the chopped grass brought in for it and there was often a pen for small goats – a main source of meat. The occasional skinny sheep was tethered on scraps of grass next to the road.

There were a variety of small businesses, located in one-roomed wooden shacks, just large enough for someone to stand or sit and sell different goods out of a hatch, including the ubiquitous mobile phone service, Safari.com. There were two tailors, seated in the same kind of wooden shack but containing a manual sewing machine, who offered to make or alter your clothes for you. There was also a barber shop, with just enough space for two chairs for customers, the interior tiled smartly in black and white.

Women were all well dressed in clothes obtained from Chinese traders in the town, or from bales of second-hand clothes sent from Europe. All (except the researchers and their assistants) wore dresses or skirts. Most women wrapped scarves around their heads to keep the dust out of their hair. Children roamed around barefoot dressed in a variety of outfits, including some made from the local election posters. There was a mass exodus of children to the local schools in the morning and a return flow in the evening, as with shaven heads and blue uniforms, they trekked back home through the dust carrying their school satchels. Primary education is universal in Kenya.

At the time that we started our study Mutaroni had neither electricity nor broadband. Although most people owned a mobile phone they had to walk on average 35 minutes per day to charge them with relatives or from a commercial unit. Lack of lighting meant that it was difficult to walk around the village after dark and people felt insecure outside of their homes. The working, cooking and socialising day was mostly limited to daylight hours. Lighting was mainly provided by kerosene lamps, which smoked and were liable to start fires. Children were not able to do school homework in the evening since near the equator, night falls suddenly around 6 pm. After this time, the village was plunged into darkness except for the dim glow of kerosene lamps and a weak wattage from roof top solar panels in a few houses that had installed them. Navigating around the village after dark was hazardous, given the unmarked dirt tracks and meandering locations of the various residences.

Other villages nearby were supplied electricity by the Kenya Light and Power Company, but not Mutaroni. That is why it was chosen as a location to which new technology was to be introduced. However, the central provider, the Kenya Light and Power Company, produced electricity that was unreliable and subject to frequent power cuts. Broadband was available only in towns and was likewise unreliable – as we found out when we tried to access it from our hotel. Mobile phones can also be useful as torches under these circumstances.

Into this landscape came a non-governmental organisation (NGO) called "Ubuntu" (this translates roughly as "I am because you are") with funding from a variety of international sources, including Innovate UK. Ubuntu specialises in technology projects for development. The University of Cambridge and the University of Aberdeen collaborated to research and report the social impact of the introduction of technology in this particular project. Ubuntu provided electricity through a solar minigrid, with solar panels set upon a shipping container which served as the local office. The electricity was rolled out to local houses who subscribed on a progressive basis at a lower cost than that of Kenya Power, the main electricity provider. In addition, a satellite internet installation was set up to provide free internet at a community level when first used with a small charge thereafter.

METHODOLOGY

The aim of these innovations was to foster the development of the community according to particular dimensions: social, educational, health and well-being, economic development and community empowerment through capacity building. Our job was to see if this happened. Hence, we carried out a "before" and "after" study with a baseline survey undertaken in mid-2017 and a follow up six months later in January 2018.

The survey of 117 households had a 100 per cent response rate over the two studies, and was supplemented with qualitative interviews with key informants and villagers. The research was undertaken with the help of six locally trained researchers (mostly students from nearby or from the village) who were able to communicate in Swahili and in English. The project itself was facilitated through the use of two local "*rafikis*" (which translates as "friends") who were able to communicate with villagers and help them take advantage of the new facilities. The *rafikis* were a conduit of communication between the villagers and the external Ubuntu NGO and proved invaluable for our research. The *rafikis* ensured that participation happened as they were able to explain the technology to villagers who were not familiar with it.

FINDINGS

The average age of respondents was 45.5 years with a range between 20 and 89 years. We mostly interviewed heads of households. The mean number of people in a household was 3.57 persons. Two had no formal education but most had secondary education (61 per cent) or post-secondary education (20 per cent). A separate survey was carried out for businesses which included four people from the baseline survey and six from the follow up (note, new businesses had started up as a result of the solar power). Three of the six businesses were phone-charging businesses, three were shops and two were bars. In addition there was a flour mill where grain was sent to be ground.

Mobile phones were important even before the Ubuntu project and 91 per cent of respondents had mobile phones, although only 28 per cent were smart phones. Before the Ubuntu project, internet access was mainly available only for 17 per cent of people and all of them used the mobile phone to do so. Radio was popular with 94 per cent owning a battery-operated radio, although only 14 per cent of respondents had a television.

HOW DID PEOPLE USE THE TECHNOLOGY PROVIDED?

Although the questionnaire contained questions about use of electric stoves, refrigeration, air conditioning and so on these questions proved to be irrelevant. The main way in which people used electricity was for lighting. Previously some respondents had had solar panels on their roofs or used batteries; these sources were not adequate for people's needs. In the words of one respondent "It was affected by the weather and wasn't available at all times" (male aged 22) and another male aged 60 said "my solar unit used to run low when I watched TV for many hours so it didn't meet my needs". Lighting could be used to read at night, to cook at night (although in the cooking huts over an open fire). One respondent in her 80s told us "God Bless Ubuntu" because electric lighting enabled her to read her bible at night. Furthermore, lighting at night enabled people to move around the village more safely, something that was mentioned by a number of villagers. Lighting also enabled school children to undertake homework tasks, something regarded as very important by villagers.

A second way in which electricity was used was for phone charging, which saved people walking long distances or paying a local business. In fact, those households who had electricity installed started providing a service (sometimes on a paid basis) for others to charge their phones.

Home entertainment was another important way in which their lives improved. Whilst most people listened to radio, they could now do so more conveniently and television viewing leaped from 15 per cent to 71 per cent of villagers in the six-month period. As one respondent put it "watching TV in my home makes me feel comfortable and makes me feel pride in my home". Having a television at home was seen as a major symbol of progress. They also reported an increase in listening to music and watching videos. However, this had other important spin offs beyond just leisure. Other studies have shown that access to television meant access to information which included health education, whereby women were better informed about family health issues and information about HIV/AIDS prevention could be disseminated (Kirubi et al., 2009).

In terms of more collective entertainment, the local bars were able to stay open longer and proudly showed us their new gambling machine which could run on the electricity provided! The after-hours socialising of some men was enabled by the extension of the drinking day, although their families were probably less grateful for this.

Local businesses were able to stay open for longer and electricity extended the working day both for householders and for businesses. Even if their farms did not require electricity, householders could now cook and prepare food later in the evening, which extended their working day. Respondents reported having 30 minutes more per day available for family and leisure pursuits.

The economic effects were likely to be felt over the longer term. One entrepreneur was turning from gasoline to electricity to power the flour mill and was planning the installation of a unit to make mango juice in order to take advantage of mango gluts – the mangos were all ripe at the same time of year, which was also the time when they were cheapest at the market. Therefore, the mango juicing plant would enable these products to be stored and the income to be spread more evenly over the year.

One local business was bringing in a fridge to store vaccines in order to open a local clinic for children, which would also provide a health benefit for the village. A local entrepreneur, Charles, owner of the Posho flour mill, immediately saw the advantage of this source of power to enhance his business: "Lighting – [until now], I used a battery torch. When we are using the torch the batteries are also expensive. But when you have light, you can increase the working hours." He was confident that the power would mean business growth for the mill as well as employment opportunities for the wider community.

> With the power it has enabled me to work even at night. There is also impact in terms of safety when I am around machinery – because there is light and I am able to assess where I am operating from. So I can work for longer hours than before. Because before I used to use just a torch so at 6, 6.30 pm I closed. Because now I have a longer period to work, of course there is now more money.

He is planning to use the electricity to enhance his machinery too:

> I will buy a motor for my machines, so I can switch to electricity. It will save me money in fuel, it will be cheaper. You cannot compare when you are using electricity to diesel – diesel is more expensive. Because you have to travel to get it, you have to transport it. It is also time to get it. And it is expensive fuel. . . ..with electricity I can even expand – I can install other machines, I can start animal feed production. If I have electricity, I can buy these machines and expand.

However, not all households wanted to take advantage of the electricity provided by Ubuntu power. As subsistence farmers they were used to managing with uneven incomes and the kerosene could be bought at times of surplus but cut back upon when income was short. Although Ubuntu power was also provided on a pay-as-you-go principle using M-Pesa (with which Kenyans are already familiar) people were afraid to commit themselves. The Ubuntu spokesman also pointed out that the electricity was not yet universally rolled out to all parts of the village and was seasonally variegated; in the rainy season the sky was clouded over and the solar panels would not work so well. For that reason they were seeking funding to provide a bio-digester plant to supplement the electricity generation.

The aim of the Ubuntu project was to provide sustainable, low-cost development using the environmental resources of the village. This was therefore on a different scale to the top-down provision of technologies on a national scale, such as through Kenya Power. However, in the long run it would require local ownership of the technologies and a change in the habits and lifestyles of villagers – for example in recycling their compost.

Therefore, electricity consumption mainly had the effect of enhancing existing activities (cooking by electric light, reading, charging phones) rather than being transformative at this stage of its roll out. It is interesting to note that culture was seen as a key element of improved quality of life enabled by electricity supply – listening to music, watching television and listening to radio were all important. For Kenyan villagers these aspects of connectedness and entertainment were the most important. Collective entertainment – and therefore social cohesion – was available at the local bars (mostly only for particular men) or by going round to other people's houses to watch television.

In terms of agriculture (the main occupation of our respondents), the main advantage of electricity was in chopping chaff for animal fodder (a back-breaking activity without machines) and lighting enabling them to work longer hours. Only six months after the installation of electricity, six new businesses had been set up and existing businesses were able to work longer hours due to more lighting. People also saved time by not having to walk long distances to charge their phones.

THE USE OF BROADBAND

Although we had many questions about how people used broadband, "Do you access broadband on your personal computer, tablet, laptop etc", these questions were irrelevant. They were met with blank stares and a hesitant response "But I don't have a smart phone". Smart phones were seen as the only way to access internet and most people did not have them. Indeed, the main peoples taking advantage of the new digital connections were the researchers and *rafikis* on the project!

It would seem that this technology is not so readily taken up by villagers in rural areas and enthusiasm for internet connections and usage of smart phones was to be found at this stage mainly among educated young people. Others did not really see the relevance. The Ubuntu NGO acknowledged these problems and were planning to open an internet café attached to the side of the shipping container to encourage participation in ICT communications. At the time of our study this was being built but had not yet opened. A new role for the *rafikis* would be to introduce villagers to this technology. We would predict that at least teenagers and younger generations might take advantage of this.

Some of the main beneficiaries of the new technologies were likely to be school children. Children need to study up to two hours per day after school, but this time was limited by the lack of lighting at night. This was noted in the follow up study after just six months of reliable power, when three-quarters of people claimed to have accessed educational materials through the television or internet and all respondents claimed that these technologies had improved the education for their children. Watching television was also thought to help the children's education. However, teachers were hoping for greater things from the internet, which could enable pupils as well as teachers and parents to access new learning materials. As one teacher put it:

> The power will help our children to improve their performance. Because with light, they can study at night, access the internet and even download the questions, textbooks. . ...Teachers can download new teaching materials. With the youths coming up now, the internet is something so important to them. It helps to expose them to the outside world . . . if we had internet here it would make a big difference.

The internet had therefore raised some anticipatory expectations.

REFLECTIONS ON TECHNOLOGY, WELL-BEING AND COMMUNITY DEVELOPMENT

The example of introducing technology in Mutaroni furnishes us with some lessons in community development. First of all it is important to manage this together with local people and the base for Ubuntu inside the village (donated by a local landowner) was important in making people familiar with it. The social impact survey itself must have helped to raise awareness of the presence and aims of Ubuntu. The technology was small scale and rolled out gradually so that people had a chance to adapt and learn from others. At the time that we did our research this had only been carried out partially but it was clear that there would be longer-term impacts.

The presence of the two *rafikis* was important. The *rafikis* were locally recruited people whose job it was to liaise with the local community and to provide help and assistance – to become mobile information points. They acted much like community workers. They also provided a point of contact between the researchers and the local community.

Electricity was hardly a new technology. However, the way in which it was provided by Ubuntu was new and sustainability was an important consideration (Kirubi et al., 2009). The electricity was provided at very low cost through solar panels (that require very little maintenance). However, even this required some adaptation on the part of subsistence farmers, some of whom were wary of entering contractual arrangements. Their adaptations to electricity were small scale and low level, mainly using it for lighting to enable them to extend their working and leisure days. However, this low-level usage had the advantage of bringing the community along with the project in their own evolving way.

Although there were clear economic benefits to providing electricity, the case study illustrates that one of the most important impacts was cultural, especially through providing better access for more people to television. A television – rather than a computer, a cooker or a refrigerator – was seen as the first priority. A television connected people to the outside world and the wider culture, to information, politics, sports and entertainment. Whilst televisions were individually owned, they also provided a collective resource as others could come round to watch it. This reinforces what Cloutier and Pfeiffer (2017) have described as the important but neglected role of culture in community development.

Although mobile phones were ubiquitous, smart phones were still only used by a minority. Smart phones were the main way in which internet was accessed (apart from some elderly and slow-moving computers grinding away in the internet cafes in the nearby town). Since smart phones were out of reach for most people and the internet cafes were far away in the town, internet was not seen as part of the social horizon for most people. Most people could not even imagine what they would want internet for, so the broadband provided by Ubuntu was not seen as advantageous by most (except for teachers). We would predict that this would change once the internet café was opened and people had been taught to use it. Since young people are usually the early adopters of internet, it could be assumed that they will become the technology avant-garde.

The unbridled connectedness that so pervades First World culture had yet to hit rural Kenyans in the same way. Although mobile phones are now a way of life, there is limited access to social media or broadband. However, the younger generation with enough income for a smart phone all use Whatsapp and Facebook. Therefore, connectedness involves contacts with friends and acquaintances and is to a great extent used for financial and business transactions. Social cohesion and connectedness take place mainly through older style face-to-face activities such as churches and public meetings.

Communications technology was not used directly for political participation. An interview with the local chief of this and nearby villages revealed that he did not use telecommunications to any extent – except for his personal use. It was not used to call meetings and gatherings, which were advertised and organised in more traditional ways, nor was it used for political communications and information sharing in the way that politicians use it in Britain. Although the churches were very important for the social life of the village, they did not use information technology either. Religion was a face-to-face affair. Our studies of Scottish villages revealed that email and Facebook pages were important for local economic exchange and communications (Wallace and Vincent, 2017), but this was not yet the case in rural Kenya.

In terms of our identified aspects of well-being, we can consider the economic, social and cultural impacts. In terms of economic impact, the extension of the working day could clearly be seen as an immediate economic benefit. The various initiatives by local businesses and the planned initiatives in the case of the Posho Mill would help to provide economic advantages to the community down the line as well. Whilst there were clear economic benefits from electricity, those of broadband are more difficult to foresee at this stage.

In terms of social benefits, we can include the increased possibility to socialise in the evening with extended lighting – and for some to visit the bar for more extended hours. The provision of television also allowed more socialising and we witnessed a number of people listening to music together. The improved educational facilities would certainly benefit social inclusion, but most of the collective and participatory activities in the village did not really require any technology. Nor was it used for building identity and civic pride.

In terms of empowerment, the additional information provided by television may have helped Kenyans to stay in touch and make decisions but social media, websites and so on were not used by local politicians. Kenyan society is simply not as saturated in communications as European societies.

The role of culture was important in terms of listening to music and television, although these were more passive forms of entertainment than the kinds of developments we have seen in web 2.0. However, entertainment and leisure were very important to Kenyan villagers.

The enhanced well-being of the community came about indirectly through the better educational, economic and cultural facilities that improved electricity was able to provide. Of course, it could be that the very act of providing community solar energy, bio-digesters and internet as well as doing the research on its social impact could actually help to induce community cohesion and collective self-esteem. Yet at the time that we performed our research, it was not clear how the ownership of these technologies would ultimately be passed to the village and therefore become sustainable.

Our own research role in this project as external consultants was a limited one. The project enabled us to be parachuted into the community for a week at a time (Leanne Townsend came on two occasions, Claire Wallace on only one) for a period of intensive immersion, but this meant that we had necessarily a brief acquaintance with the village. Since we were only looking at one aspect of village life, we may have failed to see more complex realities in the round. Even whilst there we were whisked away to a hotel in the nearest town before nightfall and brought back again next day as there was no accommodation available in the village, which limited further our understanding of daily (and nightly) life.

Where we were most distanced was in the way we constructed our questionnaire based upon our previous understanding of technology in rural areas in Britain and upon other development projects that we researched. Hence, we included questions about use of electricity (for refrigeration, for air conditioning, for using a cooking stove and so on) which were entirely irrelevant for these villagers. Similarly, we included questions about broadband access (on a tablet, on a personal computer, on a laptop and so on) that were beyond the comprehension of most villagers and were met with blank stares. In fact, they told us that mobile phones, a technology they were already familiar with, were seen as the way to access internet. Fortunately, we combined these questionnaires with face-to-face interviewing and intensive fieldwork, so we were able to adapt more sensitively to the cultural context of our research. People were very polite and tried to find ways to answer our questions, even if they did not understand them. This underlines the importance of qualitative fieldwork, sensitively carried out as a supplement to more survey-based research.

CONCLUSIONS

In this chapter we have looked at our experience of researching the social impact of introduction technology (in this case electricity and broadband) in a Kenyan village. Although villagers were generally delighted with the provision of low-cost electricity, they did not use it in ways we might have predicted – they used it to enhance the quality of their existing lives through more lighting and to access culture in the form of television, radio and music. Broadband was beyond their experience and although we might see take up in the next phase of the project, it was not seen as something people found particularly useful at the time of the research.

This suggests that community development models should take into account what anthropologists call the *emic* perspectives of participants on the ground as well as the *etic* perspectives of the providers and outside agencies. Community well-being needs to be understood from the ways in which people used technology to enhance their lives as well as the more "objective" measures of economic and social outcomes.

Our research emphasises the importance of working with local communities from the "bottom up" to help them with accessing new technologies and here the *rafikis* were an important element of the project work. This would fit with the directed model of community development identified by Maritta-Cascante and colleagues (ibid.). Furthermore, we would strongly endorse the use of face-to-face fieldwork rather than more distant forms of data gathering for understanding both community development and community well-being. Otherwise big mistakes could be made.

NOTE

* We would like to thank Juan Herrada, CEO UbuntuPower.org, for his help in enabling us to undertake this project.

REFERENCES

Abbott, P., and Wallace, C. (2012). Social quality: A way to measure the quality of society. *Social Indicators Research*, *108*(1), 153–67. doi:DOI:10.1007/s 11205-100-9571-0.
Abbott, P., and Wallace, C. (2012a). Happiness in post-conflict Rwanda. In H. Selin and G. Davey (eds), *Happiness across Cultures: Views of Happiness and Quality of Life in Non-Western Cultures* (pp. 361–76). London: Springer Verlag.

Abbott, P., Wallace, C., and Sapsford, R. (2016). *The Decent Society*. London and New York: Routledge.
Anderson, R.E. (2017). Community functioning that fosters sustainable social wellbeing. In R. Phillips and C. Wong (eds), *Handbook of Community Wellbeing Research* (pp. 3–10). Dordrecht: Springer.
Beck, W., van der Maesen, L., Thomes, F., and Walker, A. (2001). *Social Quality: A Vision for Europe*. The Hague: Kluwer Law International.
Cloutier, S., and Pfeiffer, D. (2017). Happiness: An alternative objective for sustainable community development. In R. Phillips and C. Wong (eds), *Handbook of Community Well-Being Research* (pp. 85–98). Dordrecht: Springer.
Dahir, A.L. (2016). The CEO of Africa's most innovative mobile company warns his "clumsy" product needs to diversify or risk dying. Quartz Africa (qz.com/africa), https://qz.com/africa/813612/bob-collymore-safaricom-ceo-warns-m-pesa-is-a-clumsy-product-that-needs-to-diversify-or-risk-dying/, accessed 11 December 2019.
Hobsbawm, J. (2018). *Fully Connected: Surviving and Thriving in the New Age of Overload*. London, New York, Oxford, New Delhi, Sydney: Bloomsbury Business.
Kirubi, C., Jacobson, A., Kammen, D.M., and Mills, A. (2009). Community-based electric micro-grids can contribute to rural development: Evidence from Kenya. *World Development*, 37(7), 1208–21.
Kuo, L. (2017). Kenya's mobile phone ownership is lower than we thought. Quartz Africa (qz.com/africa), https://qz.com/africa/900099/kenyas-mobile-phone-ownership-is-lower-than-we-thought/, accessed 11 December 2019.
Mandelli, S., Barbieri, J., Mereu, R., and Colombo, E. (2016). Off-grid systems for rural electrification in developing countries: Definitions, classification and a comprehensive literature review. *Renewable and Sustainable Energy Reviews*, 58, 1621–46.
Matarrita-Cascante, D., and Brennan, M.A. (2012). Conceptualising community development in the twenty-first century. *Community Development*, 43(3), 293–305.
Miller, D. (2016). *Social Media in an English Village*. London: UCL Press.
Phillips, R., and Wong, C. (eds). (2017). *Handbook of Community Well-Being Research*. Dordrecht: Springer.
Turkle, S. (2013). *Alone Together: Why We Expect More from Technology and Less from Each Other*. Philadelphia: Basic Books.
Wallace, C., and Beel, D. (2020). How cultural heritage can contribute to community development and wellbeing. In M. Cieslik and L. Hyman (eds), *Researching Happiness: Qualitative, Biographical and Critical Perspectives*. Bristol: Policy Press.
Wallace, C., and Vincent, K. (2017). Community well-being and information technology. In R. Phillips and C. Wong (eds), *Handbook of Community Well-Being Research* (pp. 169–88). Dordrecht: Springer.
White, S.C. (2010). Analysing wellbeing: A framework for development practice. *Development in Practice*, 20(2), 158–72.

PART II

RESEARCH METHODS AND FRAMEWORKS

9. Experience of group formation in Grameen Bank, Bangladesh
Kazi Abdur Rouf

INTRODUCTION

This chapter is about the experience of the participatory decision-making process of group forming and center organizing of Grameen Bank, Bangladesh (GB); and implementing its strategies of different savings and loan products in Bangladesh. The primary objective of the chapter is to narrate the GB group formation and center organizing experience in the villages of Bangladesh. Moreover, the chapter identifies what mistakes I made and how I could have done a better job in group forming, center designing, organizing, planning and other activities if I had had knowledge and skills on community organizing and planning ahead for my jobs in GB in the late 1970s and early 1980s. Because through my working experience and research in community-based organizations in many countries, now I have realized that I could have done a better job in my group forming and community (center) organizing assignments in the 1970s and 1980s. However, the good thing is that I learned a lot from mistakes and improved my jobs from there. Even GB learns and changes many of its loans and savings products from its field experience. The GB groups and center structure and operation system are not developed by outside experts, rather its center organizing and group forming and the GB implementation strategies have been developing from the inputs coming from the field staff of GB since 1976.

OBJECTIVES

The purposes of this chapter are

- To discuss community, civic engagement of the community, community organizing, community social capital development, and their relationships to GB center organizing and group formation.
- To explore different approaches to community organizing, designing and building and apply them to GB's specific circumstances and issues.
- To discuss different savings and credit products of GB both in Grameen Bank Phase-I and Grameen Bank Phase-II and their implementation strategies in Bangladesh.

COMMUNITY SYNONYM TERMS

The word "community" can mean different things to different people at different places. Community refers to communities of association based on religion, gender, race or geography. Cohen (1985) defines community as a system of norms, values and moral codes that provide a sense of identity for members. Fellin (2001) describes a community as a group of people who form a social unit based on common location (e.g. city or neighborhood), interest and identification (e.g. ethnicity, culture, social class, occupation or age) or some combination of these characteristics.

In Bangladesh, nongovernmental organizations (NGOs) do not use the term "community"; rather they use village organizations, group, neighborhood, center, association or society. For example, GB uses the term "Center", Bangladesh Rural Advancement Committee (BRAC) uses "Village Organization", Association of Social Advancement (ASA) uses "Association", Nejara Kari uses "Group". However, North America, Europe and many other countries use the term community. Below the chapter first discusses different thoughts and approaches to the community participatory decision-making process, community organizing, community building, community, community building and planning; secondly the chapter narrates the author's community working experience in Canada and compares his working experience with Bangladesh; thirdly, the chapter describes how GB organized, designed, formed, built and planned its groups and centers; and how GB managed its centers and groups; and its different loan and savings products in Bangladesh. Fourthly, the chapter also discusses GB's operational policies and strategies in Bangladesh at different periods.

COMMUNITY PARTICIPATORY DECISION-MAKING PROCESS

Community organizing has implications for those who express opinions about how their intended community is shaped: participatory planning and decisions means exchanging information, ideas, opinions, and process and positions with those people who live near the neighborhood. However, citizen participation and community people participation is necessary for every community people's voice and choice because the concept of participatory decision-making is that people should deliberate together over issues that affect their future and make decisions accordingly. It helps community members, group members and individuals understand one another's point of view, and it facilitates decision-making for the common good (Rouf, 2016). In professional terms, participatory decision-making means that the professional or the source of political power is not the sole decision maker, rather someone who works with the people to help them reach decisions. In community planning terms, participatory decision-making means that a community planner works with people, helps them reach decisions about community designing and planning issues, and translates these decisions into planning language throughout the decision-making process. However, the problem is if the community worker is not from the community, or he/she does not know the community's background well.

Therefore, in the community meeting and community organizing and planning, the community worker would be a facilitator, not a decision maker. Rather, he/she should identify what the community's voices and choices are. If the community people spontaneously participate and express their choices that are needed for them, the community worker tries to accommodate their voices and choices into community building and planning documents. However, it is sometimes difficult to reach common understanding among all community people. In the community participation decision-making meeting, there might be many opposing views. In that situation, a community organizing and planning worker might think the community is not homogeneous in their thinking. There should be different opinions among the community, but the community organizing worker should have the skills and capacity to translate that opposing views are also community views, but in the decision-making process, only include the majority's demands into the community design, building and action plan as a priority basis. It is important to remember that the community organizing, designing, building, and planning process is inherently democratic. Umut Toker (2012) thinks that participatory decision-making community organizing, designing and planning have been developed to achieve the following benefits:

- To let professionals become facilitators in the community decision-making process of different issues and demands of the community people.
- To efficiently design, manage and analyse community design events.
- To move toward consensus building.
- To help reach design decisions collaboratively.
- To guide parties that will contribute to and manage implementation so that the decisions made are implemented as desired by the community.

COMMUNITY ORGANIZING

Community organizing is the coordination of cooperative efforts and campaigning carried out by local residents to promote the interests of their community. Community organizing is a process where people who live in proximity to each other come together into an organization that acts in their shared self-interest. It refers to the entire process of organizing relationships, identifying issues, mobilizing around those issues, and maintaining an enduring organization. This process of building and mobilizing community is called "community organizing". It involves "the craft" of building an enduring network of central people, who identify with common ideals, and who can engage in social action on the basis of those ideals in the context of GB (Rouf, 2016). In practice, it is much more than micro-mobilization or framing strategy (Snow et al., 1986.). According to Saul Alinsky (1971) if the community already exists, someone has to help transform it to support political action. Sometimes that requires reorganizing the community by identifying individuals who can move the community to action.

According to Elizia Pan (2012) and Participedia, community organizing is a process where people who live in proximity to each other come together into an organization that acts in their shared self-interest. Unlike those who promote more-consensual community building, community organizers generally assume that social change necessarily involves conflict and social struggle in order to generate collective power for the powerless. A core goal of community organizing is to generate durable power for an organization representing the community, allowing it to influence key decision makers on a range of issues over time. Community organizers work with and develop new local leaders, facilitating coalitions and assisting in the development of campaigns. Its organizers generally seek to build groups that are democratic in governance, open and accessible to community members, and concerned with the general health of a specific interest group, rather than the community as a whole. Community organizing seeks to broadly equally empower community members.

Community organizing is the process of building power that includes people with a problem in defining their community, defining the problems that they wish to address and the solutions they wish to pursue. The organization will identify the people and structures that need to be part of these solutions, and, by persuasion or confrontation, negotiate with them to accomplish the goals of the community. In the process, "organizations will build a democratically controlled community institution that can take on further problems and embody the will and power of that community over time" (Beckwith and Lopez, 1997).

In general, community organizing is the work that occurs in local settings to empower individuals, build relationships and create action for social change (Bobo et al., 1991; Kahn, 1991; Beckwith and Lopez, 1997). Community organization is the process that builds a constituency that can go on to create a movement, and it occurs at a level between the micro-mobilization of individuals (Snow et al., 1986). Within the Alinsky model the organizing process centers on identifying and confronting public issues to be addressed in the public sphere. In the Alinsky model, the organizer is not there just to win a few issues, but to build an enduring organization that can continue to claim power and resources for the community – to represent the community in a public sphere pluralist polity. The organizer should not start from scratch but from the community's preexisting organizational base of churches, service organizations, clubs and so on. This model uses the small group to establish trust, and build "informality, respect, [and] tolerance of spontaneity" (Hamilton, 1991, p. 44).

WHAT IS COMMUNITY DESIGN?

Community design is the act of designing the physical attributes of a community. Community design is the art of making sustainable-living places that both thrive and adapt to people's needs for shelter, livelihood, commerce, recreation and social order (Hall and Porterfield, 2001, p. 3). Physical attributes of the community are location, and structure of the community organizing, building and planning. The physical structure of the community settings may affect people's ways of doing things (Lennertz, 1991). The community design approach includes (1) a community planner or designer that makes decisions with people's interests and the community planning worker's own professional background in mind, (2) people occupying the social positions and social capital and products, and (3) people adapting to the social positions and community settings while the community settings adapt to people's needs.

Community design is a process working with people, not for people (Hester, 1990). Therefore, a community worker works with people and facilitates a process during which people and community workers learn from each other; the community worker translates the wishes and aspirations of the people gathered during the process into community design and uses community workers' experience. Moreover, social capital products like social network and solidarity are the outcome of those community designs and plans, and are adapted to the people. Therefore, community design is the process of organizing community and making design decisions in collaboration with the community people who stand to be impacted by those decisions. As a community worker, we GB fieldworkers learn from center members about their day-to-day relationships with their physical settings of their neighborhood. Community design is also the documents and legal structure of the community plan. Therefore, GB center organizing, designing and planning is a participatory decision-making process with the center members of GB. GB fieldworkers advocate for participatory decision-making principles like understanding center members' needs, adopting their situations and learning from them.

Key components of community organizing are developing a timeline, identifying tasks and instruments, tools, specifying goals, identifying participatory activities of the community people, and developing outreach plans and activities. The community design and advocacy model materialized in the US in Saul Alinsky's work (Sanoff, 2000). As a community organizer, Alinsky went beyond planning and design issues and became involved in the social and organizational aspects of community building. He focused mainly on lower income communities and worked with them for social change. His well-known *Rules for Radicals* (1971) provides guidelines for proponents of this approach. However, Umut Toker (2012) believes community organizing is a continuous process with several stages in order to achieve the community's desired effective tangible outputs. According to him community design has the following stages:

- What does the community want to get out of the process?
- When and by what step does the community achieve the outcomes (timeline development)?
- What types of activities and tools are helpful for achieving the desired outcomes?
- How will community people be informed about the process and invited to provide input (identification of outreach techniques)?
- Interact with the community outreach and receive inputs from them and record them.
- The community design process is nonlinear.
- Collect feedback from community people before finalizing community design and translating the community design into tangible products.

Outcomes of community design: community designers go through systematic idea generation processes with a group of people and translate these ideas into action for the well-being of community members. Such process requires a good amount of coordination, organization and collaboration. However, the question is what are the outcomes of community design processes? The following are possible outcomes that may arise from the community design. (1) Learning from each other. (2) Community people becoming empowered by close interactions among community people and by providing inputs to the community decision-making process. (3) Fulfilling community intended goals. (4) Developing unique solutions for each community project. The community organizing process acknowledges and embraces the multiple ideas of people and develops an approach that includes multifaceted inputs.

COMMUNITY BUILDING

Another model of community organizing is community building, which encompasses elements of both locality development and social planning approaches. Community building focuses on strengthening the social and economic fabric of communities by connecting them to internal and outside resources (Smock, 2004). The goal of community building is to build the internal capacity of communities by focusing on their assets/strengths, and engaging a broad range of community stakeholders to develop high-quality and technically sound comprehensive plans (Smock, 2004). The field staff of GB let group members interact among themselves to develop their social networks and solidarity among them as well as let them identify their own problems, and solve their problems by themselves through their mutual interactions, dialogues and cooperation. To develop their bridging and bonding social capital in the group and center, fieldworkers of GB are only the facilitators and mentors of their social capital and economic capital development in their life in Bangladesh.

There is an approach called self-help and self-build. The self-help approach is also known as mutual help, mutual aid, or support groups. These are groups of people who provide mutual support for each other. Self-build is a process of the practice of creating an individual for oneself and skills development through a variety of different methods. It is in contrast to the community organizing model. The self-help approach focuses on empowering community people by helping them develop their means and skills to effect their incremental change. Turner and Fichter (1972) endorsed individuals and families' rights to self-build. John Habraken (Massachusetts Institute of Technology – MIT) is a community organizer and thinks developing a "support structure" would provide the infrastructure for a community based on their needs. However, there is a question: what happens when conflicting interests exist among community people? If the conflicting interests are not solved immediately, there would be a community crisis situation.

Role Players in the Community Design Process

The leading groups of people in the community are community members, community leaders, members of the local governments and members of the NGOs. A key informant is someone who can provide the community designer with "insider" information, but he should not be spying on community workers. A participant is an individual who is likely to be affected by the outcomes of the community design process and the project. A project champion is a person who is particularly interested in the success of the community organizing and believes in the advantage of community collective well-being. Through this process, the community designer starts translating community-based ideas into planning and design language. Muhammed Yunus is the champion of designing and streamlining the GB centers in Bangladesh.

COMMUNITY SOCIAL PLANNING

Community social planning is a form of community organizing which focuses a technical process of problem solving regarding substantive social problems that utilizes the expertise of professionals (Rothman, 2001). The goals of community social planning include the design of community formal plans and policy frameworks for delivering goods and services to community people who need them (Rothman, 2001). Community social planning facilitates bridging the social capital, economic capital and civic capital based on normative ties among community people. However, the focus is on the interests of participating agencies and the community at large, rather than the individual self-interest of neighbors. In the case of Grameen Bank center organizing and group formation planning, we fieldworkers keep the center boundary within half a kilometer of the neighborhood of the members; we form a group with 5 members and form a center with 30 members with six groups. The center meeting is conducted within the middle location (house) of the neighbors of the members.

GRAMEEN BANK GROUP FORMATION AND CENTER DESIGN STRUCTURES IN BANGLADESH

The Grameen Bank Project was a project of Bangladesh Bank, a central bank of Bangladesh, during 1976–83. The Grameen Bank Project attempted to serve those rural people in a group who were not covered by the traditional banking system. In order to provide loans to poor landless people in Bangladesh, the Grameen Bank Project formulated the following rules and regulations, center, group and village association structures and so on.

The center is for the Village Landless Association (landless community association) of GB and is organized and built by the fieldworkers of GB per the Bidimalla (Yunus, 1983). The Village Landless Association of GB is the grassroots organizational structure of landless people in GB; it is the primary and vital unit center structure of GB. In the center, the landless micro-borrowers of GB conduct their weekly meetings and annual general meetings; discuss their different issues and plans, deposit their savings, propose for loans and receive loans, and develop their future plans. A general assembly (*shadharan parishhad*) of the Village Landless Association is constituted with the Chairman of all the landless groups per GB Bidimalla (Yunus, 1983).

Grameen Bank Group of Formation Design

According to the bylaws of GB only the village landless poor are eligible to form a group. Any member from a family (i.e. a household unit) owning less than 0.4 acres of cultivable land is considered to be a landless poor person and form a group. A group can be formed with a minimum of five members. All the members of the group are the inhabitants of the same village or neighborhood. A group shall be formed with persons who are like-minded, are in a similar economic condition and enjoy mutual trust and confidence. There shall not be more than one member from the same household in any one group. There shall be a Chairman and a Secretary in each group. They shall be elected by the group members. Elections will be held at the time when a group is formed and subsequently in the month of Chaitra (last month of Bengal's calendar year) every year. Chairmen and secretaries elected in the month of Chaitra will assume their offices from the first of Baishak, the first day of Bengal's year. As of June 2018, GB has 2,568 branches in Bangladesh, its total number of groups is 1,384,180, the total number of centers is 139,314 and it is working in 81,675 villages.

The Grameen Bank center design, savings and loans transactions policies and their implementing instructions (bylaws) that follow the GB Bidimalla are described below.

Article 10 of the Bidimalla sets out the design for the center meeting structure, rules and regulations. The center of the GB borrowers is a place where borrowers organize and conduct their weekly group meetings with their neighbors. Each meeting center shall have a "Center Chief". The chairmen of all the groups in the center shall elect a Center Chief and a "Co-Center Chief" from among themselves. The Center Chief and the Co-Center Chief shall be elected in the month of Asharh (June) every year. They shall assume their offices on the first day of Shraban (July). The overall responsibility of conducting weekly meetings rests on the Center Chief (Yunus, 1983).

The savings, insurance and loan products of GB and their repayment systems were designed at different times to address GB borrowers' different needs and situations. GB executives, particularly Muhammed Yunus, always look at the center organizing, group forming, and their management dynamics, clients' socio-economic dynamics and the chain of command of GB management dynamics. For example, GB has developed its disaster loan products at different times to address the borrowers' disaster situations, but they are different from normal loan transaction systems. GB group formation and center management experience are not always linear. There have been huge conflicts among group members. This includes the breakdown of groups, the violation of rules, and bad debts since inception. Many abnormal situations have been created by the disaster suffering borrowers.

Bylaws (Bidimalla) of Grameen Bank

The bylaws of GB are called Bidimalla, and were drafted in 1978 and finalized in 1983. Groups and centers of GB and their designs, structures and functions; the conduct of group meetings; GB loans and savings products and so on are described in the Bidimalla. As part of conducting the meeting, the Center Chief shall ensure attendance of the group members at the meetings, payment of installments and overall discipline and order (Yunus, 1983). The Center Chief shall also help the bank worker present at the meeting in receiving installments and deposits and explaining bank rules. If any Center Chief absents himself from half or more of the weekly meetings held during any three consecutive months, the post of the Center Chief shall be deemed to have fallen vacant and a new Center Chief shall be elected in his place.

In article 10.6 the Bidimalla also provides that if the Center Chief becomes a "difficult loanee" at any time, that is, if he does not pay his installments for ten consecutive weeks or remains absent from the weekly meetings for ten consecutive weeks or if he has not fully repaid his loan in 52 weeks, he shall be disqualified from the post of Center Chief and the post of the Center Chief shall be deemed to have fallen vacant. In such cases, a new Center Chief shall be elected to replace him (Yunus, 1983).

Duties and Responsibilities of the Group Members of Grameen Bank

The Chairman and the Secretary of a group maintain constant contact with the Landless Association and the loan-giving bank (borrowers of GB received loans from different national commercial banks during 1978–83 (Yunus, 1983); however, the borrowers directly received loans from GB when GB became an independent bank). The Chairman and the Secretary of the group are responsible for recommending credit requirements of the individual members, ensuring proper utilization of the credit and repayment of loans. All members of the group should remain present in the weekly meetings of the group. At the weekly center meeting, each member of the group must deposit at least one taka (one taka until 2000, five taka since then) as his regular savings. This amount which is collected is his/her weekly savings deposit in the group's own account with the bank. In the weekly meeting, the Chairman of the group maintains discipline, collects weekly dues from the individual members and deposits them with the representative of the bank.

According to the Bidimalla of GB if a member of a group is found indulging in activity subversive of discipline (such as absence from weekly meetings, irregularity in payment of installments and so on) the remaining members can impose a fine on him. The money so received is deposited in the Group Fund. The Group Fund structure and its functions are discussed later in the chapter.

Moreover, in the Bidimalla of GB there is a rule for members leaving the group. A member who has no outstanding liability with the bank may leave the group voluntarily at any time. While leaving the group, he is allowed to take back the entire amount of his personal savings. If a member, who has outstanding bank loans, desires to leave the group, he must repay the entire bank loan before he leaves the group. If any member leaves the group without paying off his bank loans, the group shall be responsible for repayment of the loan of the member concerned. If the members dissolve the group without repaying bank loans, the center (association) is liable to pay off all the outstanding loans. This rule is applicable to all members of the centers. However, undisciplined centers with defaulted debts are unable to collect the outstanding loans from all the members of the center and rather create chaos in the center. In such situations, GB has had a provision for writing off bad debt loans since 1990.

Article 7.4 of the Bidimalla mentions that if the membership of any group is reduced to less than five members due to desertion by one or more of its members, the group concerned must fulfill the condition of minimum membership (i.e. five) within three months by enrolling new members. However, alternatively, two or more incomplete groups may unite to form a complete group (Yunus, 1983). This practice is continuing now to fill up the gaps of the group membership where needed. However, it is a very complex job to merge one incomplete group with other groups to make five group memberships in a center because it requires lots of chapter works, changing bookkeeping and share transfer jobs and so on. Please note that members can buy one share of GB, starting in 1983, but the Bidimalla was written in 1978. This new rule is included but it is not in the Bidimalla. Buying GB share certificates and their instructions were prepared separately in 1983. A huge amount of paperwork is needed for maintaining the members' share buying and selling transactions.

Now many GB borrowers are receiving loans from multiple agencies, who are working in their areas, by hiding information from GB. Therefore, many borrowers of GB have a huge loan burden; many of them become defaulters and they are disruptive in the center if they are expelled or forced to resign from membership of GB. These default borrowers influence other members to breach the discipline of the group or center until GB approves the provision of further loans to them. To face the center or group becoming defunct is challenging for many field staff. Even I have faced many undisciplined group members in the centers. GB field offices cannot sue aggressive borrowers. Only motivation and negation have been the tools for tackling such defunct borrowers. Therefore, it is difficult to follow the bylaws of the center/group legal design and structure because fieldworkers sometimes need to negotiate the demands and requests of members. However, fieldworkers have faced the auditor for departing from the written instructions of their job.

WORKING EXPERIENCE IN THE FIELD OF GB

At the beginning in 1976–78 GB employees worked in the Grameen Bank Project without community work experience and without manuals. For example, there were no guidelines, knowledge and ideas for how we fieldworkers could organize the centers in order to develop the center structures and functions. After working three years from scratch in 1978, a Bidimalla (bylaw) draft was streamlined from field experience and then supplied to the field officials. The improved version of the Bidimalla was circulated in 1983 (Yunus, 1983). We follow the bylaws of group formation, center organizing, designing, center structure and center management of GB. The written bylaws of GB center legal structure and functions assist us to do our jobs in a better way by following the Bidimalla. The Bidimalla has been a training manual for the trainees and the field staff of GB since 1978.

In the 1980s and onward, GB has developed written guidelines and instructions for loan transactions and savings collections rules and regulations for the fieldworkers to follow uniformly in Bangladesh. It was hectic for us (the fieldworkers) to face the disaster situations at different times and uniformly follow and apply the instructions to borrowers. However, the good thing is if we fieldworkers inform the head office of our problems and the other problematic situations to address the disaster of the borrowers, then head office is flexible and immediately changes the instructions. We adapt and respond to the ongoing situations to face and to solve the problems through interactive center meetings management processes. In this way, we field staff follow the center participatory decision-making process under the framework of the GB bylaws and other instructions (Nitimallas). We find center and group decisions are more effective for members if there are quick responses to their knowledge and inputs from them.

However, at the pilot stage of Grameen Bank Project, we field officials, including Muhammed Yunus, worked without any guidelines. Whatever our daily working experience with micro-borrowers, we fieldworkers discussed our daily experience in person with Muhammed Yunus. The venues were in different school fields, under trees or other public places. When the loan demand increased in more than one village, a guideline was requested by field staff that can help all fieldworkers work uniformly in center organizing, group forming, loan disbursing, and in savings collection and so on. Then the Bidimalla of GB (1983) consolidated all the field experience of staff and rules and regulations of forming groups and centers of GB. The different loan products and savings inventions of GB have developed at different times to address the changing environment of the borrowers of GB.

Each loan product's operation instructions (Nitimalla) and their implementation guidelines were also changed at different times to address borrowers' different situations. Borrower's weekly savings, Group Funds and Center Emergency Disaster Funds, the Grameen Pension Scheme (GPS) and other participatory innovative financial safety net savings products (weekly regular savings, individual voluntary savings, and insurance and so on) have changed at different times as well as bookkeeping and accounting jobs. However, the author finds the Bidimalla (the bylaws) of group forming, center organizing, designing and building of GB remain unchanged (1978–2018).

How I Surveyed and Organized Members of Grameen Bank in 1980

I joined GB in the GB Project Office Dhaka in April 1980. After one day's briefing, I was sent to Narandia Tangail Branch, which is 250 km away from my home village, Comilla, and 150 km away from Dhaka Capital City. The GB Narandia Branch is situated in the Narandia Bazar and all the branch staff lived in a tin sheet house beside the branch. There was no latrine in the branch or near the branch. However, luckily there was a tube well that we used for cooking, washing, baths and drinking water.

I shadowed an existing bank manager of the branch. I went to a village called Palima with him on foot on my second day in the branch. We talked together on the way; the manager described his working experience of how he has organized the landless associations, centers and formed groups; and what challenges he has faced to conduct meetings, form groups, organize centers and deal with the village elites. Before we arrived at the center meeting at 7:30 am, all center male members sit together in an open space under a tree. The Center Chief starts the meeting and then the bank worker collects the loan installments and savings from borrowers (Yunus, 1983). At the end of the meeting, three members proposed for their loans, the respective Group Chairman and Center Chief endorsed their loan proposals and then the bank worker received the loan proposals; we visited the second center meeting in another village, Pusna Pallima, at 9:00 am the same day. I was amazed to see the discipline of the members of the centers and the loan repayment and the savings collection system and their recordings and bookkeeping.

We walked to a new village adjacent to the meeting centers. There I observed 20 women waiting for us in a house. The manager and a bank worker talked with them about how to form groups and centers; how to maintain center discipline and conduct meetings; how to propose loans and get loans for themselves and deposit savings in the bank (Bidimalla, 1978). They talked with us for 45 minutes. We asked them to choose five women together to form a group and attend for seven consecutive days for group training in the same place at the same time. Afterward, we returned to the office. The bank worker completed his bookkeeping jobs and other chapter works related to the loans collection and the savings deposits. It took him about one hour. Then we went for lunch for one hour in the dormitory where we live. We again went to another village for group training at 2:30 pm, returned to the office at 5:30 pm and debriefed our whole day's work with other branch staff.

The manager told me the next day I should work independently, learning for myself about the branch geography and village names, land leasing system, sharecropping system, big trees in villages, formal and informal lending agencies existing in villages and their borrowing mechanisms and conditions and so on. Moreover, I should know about different income classes of people, and their occupations and social locations. According to the instructions of the GB Training Manual, which is different from the Bidimalla, I mapped out different physical infrastructures in the area like roads, bridges, rivers, canals, big ponds, hospitals/clinics, schools (when they were established, how many students (male and female students), student drop-out numbers in each grade). Moreover, I visited different clubs and government offices, different religious institutions like mosques, churches and temples to find out the purdah system, religious rituals, customs, people's local traditions, values, beliefs and so on of the local people within the branch area.

Moreover, I collected local market products and their prices and visited cooperatives and commercial banks, NGOs and private and public institutions that were working in the area. I explored local transportation facilities, means of agriculture, irrigation facilities for agriculture, pure drinking water facilities and cultivable lands and their frequency of annual crop productions in the area. My training also included learning about the demographics and literacy rates of each village, main manufacturing products and their processing centers. Moreover, I discerned the moneylenders economy, share economy practices in the villages, wages of different occupants, and laborers both male and female, and the poor people's household economy, different economic opportunities available for the poor people and other loan proving institutions in the villages. I collected data on food habits of the poor people, women's status in the family and in society, and village arbitration systems. In addition to these, I found out people's citizenship participation status in public decision-making practices. Moreover, I collected details of floods, cyclones, hurricanes, droughts and other natural disaster events and their records in the area and their effects on public life.

In addition, I surveyed the different social, economic, political and environmental sufferings of the people in the area. The branch manager also advised me to find out where and how many poor associations/centers could be formed for GB. He also asked me to collect details of two case histories of two poor women from two villages. I drew a map with all this information on a big sheet and submitted it to the office after two months. In addition to all these jobs, the manager asked me to organize open house meetings, conduct group training, and attend special meetings with other branch staff wherever they might be.

In order to collect all this information, I went out every day in the villages to find out and to talk with moneylenders, farmers, laborers, fishermen, carpenters, blacksmiths, formal and informal elites and village punchayets. Moreover, I visited backyard poultry, homestead gardening, identified food intake behaviors of people such as what they were eating and how many meals they took per day, and what they were buying. Besides, I attended different punchayet meetings in the villages and observed their arbitration system.

I mapped out the geographical boundary of the branch. At the same time, I visited and attended the existing centers' meetings of GB and observed members' attitudes, behaviors and discipline. Simultaneously, I learned office documentation jobs. It was very labor intensive and a hardworking job, but I enjoyed my apprenticeship in the branch. Although I was born and raised in a village in Bangladesh, I did not know about the miserable life of the poor people, particularly poor women, in villages. Usually, all GB fieldworkers do the same survey jobs that I did.

My First Experience of Organizing Groups and Center Schools in Grameen Bank

At 7:30 am on the third day of my posting in the branch Narandia Kalihati, I went to Luhuria village by walking, which is four km away from the branch. Luhuria is within the boundary of the branch. The purpose of visiting this village was to learn about the physical geography of the village and its peoples' occupations, socio-economic conditions, dress codes, house patterns, irrigation system, crop processing mechanisms, and to know what children are doing in the morning, afternoon, evening and at night. I observed there were many day laborers and farmers working the fields, plowing land with cows/buffaloes; many children (aged 6–14) were assisting their parents in the fields; many women were manually processing their crops in their houses. Poor women were working in the rich and upper middle-class farmers' houses.

At 10:00 am, I sat under a tree for a rest and to eat the biscuits that I carried with me. Many laborers came for a rest and to smoke under the tree. I talked with them informally; trying to know about their life, about their socio-economic conditions as well as to know what their wives and children are doing at home, and what they do in the off or lean season and so on. After one hour I visited a school to know about the school and its students. At 2:00 pm, I again sat under another tree where many straw cottages were located and many women did bamboo work in their houses. Some women asked me why I, a literate person, was sitting in this place. According to them, none in their life had seen a literate person sit and talk with poor women.

I introduced myself to them and told them about GB's micro-credit program service opportunity for the poor people in their area. If they were interested, I could provide them with more information about the bank. One senior lady told me that the poor village ladies are not smart enough to understand my discussion. She invited me to come to this neighborhood the next night and discuss the services of GB with people. I agreed. However, when I returned to the office at around 3:00 pm, on the way I observed many naked, malnourished children playing a marble game beside the street. I asked them why they were not going to school.

I returned to Luhuria the next day and talked with people at the street corners of Luhuria, Madras and school playgrounds and invited them to attend the open house meeting at the north west Luhuria neighborhood in the evening. At 5:00 pm, I found many poor children sitting under a tree and gossiping there. For fun, I showed my fingers one by one to teach them numbers. Some children quickly learned numbers by displaying fingers and enjoyed it. I told them, if they could come here every evening to learn literacy and numeracy skills, I can teach them voluntarily. There was a lady seated beside us; she invited me to teach the poor kids here. She could assist me to open a school here. She told me, in school, teachers do not allow poor children to be in the classes without having clothing. I started a light exercise program with the children in open spaces in their neighborhood. Twenty-five children started the basic literacy classes and did the exercise here with joy. I continued teaching in this center school and named it the Luhuria Center School.

One evening during the twilight hours, I and other staff of the branch visited another neighborhood located to the south west of Luhuria to conduct another open house meeting to inform the neighbors about the mission and vision of GB. About 50 people (male and female) attended this open house meeting. Men sat in front of us, women sat behind us and they listened to our speech. I observed many middle-class men also attended the open house meeting. We talked there for one hour that night. Many people expressed their interest in receiving loans from GB.

I collected people's names and told them I would come to this village to provide training on the GB group formation mechanism, center structure, members/borrowers' responsibilities, loan transaction system, savings deposit collections and repaying rules and regulations. They asked me to conduct training in this place in the evening. I told them they (people interested in loan receiving) should provide me with all information on their assets and liabilities during the training period. Those poor people who properly completed the training were able to register with GB and qualify for the loans.

In order to provide training to the poor people, I went to the agreed scheduled place every night at 7:00 pm. We sat together on a mat in an open space under a tree. There I taught them the GB manual for ten days. At the beginning of the training, 12 people attended the training sessions, but only five people continued to complete the training. These five trained people agreed to abide by the rules and regulations of the bank. They chose their own Group Chair and Group Secretary and filled in GB Form-1. Form-1 is a form for collecting socio-economic demographic data from the trained members. The five trained people deposited their seven days' savings (one taka for one day each). There was Tk.35.00 in the bank at the end of the training. With this savings money, they opened their Group Fund account in the bank.

Although the training started at night, I went to the center school every day at 4:30 pm to teach basic literacy and numerical skills to the children. The total number of children (both sexes and of different ages) increased to 45 in the center school. I accepted them whatever their dress codes were. In the center school, children first played for half an hour and then sat under the tree to learn the basic literacy skills.

One day it was raining during the lessons period in the school. It affects children learning. I discussed this situation with their parents; they agreed to build a house with bamboo and straw within three days. The house was built and children studied in this house. The house was named "Center House"; this Center House could be used for the children's religious and basic literacy education as well as conducting the weekly center school. This house also could be used for group training at night.

After two weeks of membership registration of the trainees, firstly two members of the group received loans, then another two members and lastly the Chairman of the group received a loan (Yunus, 1978). The loan size varied from $100–$125. I started to collect savings and installments from them every Monday morning. However, every day I walked to this neighborhood to teach the poor children. On my way to this village, I talked with other villagers, whoever I met. It was a tremendously exciting job that I have enjoyed. I use this working experience in other countries like Afghanistan, Namibia, Lesotho and Botswana when I work there to organize community schools.

On one day at noon, I met with a moneylender under a Bansai tree. He was looking at me in an angry mood and asked me why I had been giving loans to poor people in this village. I answered him with honor that the bank has mandated to provide loans to poor people here. He yelled at me and told me I should not come again to this village. The borrowers came to me and told me that this man also cautioned them not to borrow money from the bank. They were frightened. I talked to the office about this situation. Next day, all we office staff went to the village and made clear that it is their liberty and rights to receive loans from wherever they like. We also requested rural elites not to threaten the borrowers who received loans or want to receive loans from GB. The moneylender did not calm down, but he kept silent.

I had another open house meeting with poor people in another part of the village. We (the manager and I) were talking with women in a house near their locations. Three people (the moneylender, Imam and a Mattabarr (village leader)) joined us in the open house meeting and challenged us why we are talking with women that break the Purdah in the village. We had a dialogue with them and confirmed the poor women were not destroying their Purdah by attending the meeting; rather we were helping them to be involved in income generation activities like engaging in small business to alleviate their poverty. After a long conversation with them, the Mattabarr told us that women fieldworkers should work with women borrowers. However, the moneylender cautioned me I should not come to the village to give loans to poor people. I asked him why? He informed me that the day laborers who receive loans from the bank are not laboring on his lands. I found that was his conflict of interest with GB's loan program.

I told the Imam that the women attendees maintained their veiling and Purdah in the meeting; and I was respectful of their Purdah. I am also a Muslim. The women attendees of the meeting and I did not breach the Purdah there. It was a hectic situation for me to work in this village. Therefore, I immediately reported to the Union Councillor of the village; and I narrated to him of my mission, vision and activities in the village, but that the protest against my work from the elites was disturbing my job. I explained to him about Grameen Bank Project, which was recognized by the Government of Bangladesh. If rich people of this village could receive loans from the commercial banks, poor people (both men and women) also have rights to receive loans from GB.

After 40 years, now I understand that the people who are working in the community in a more sustainable way are the community change agents. However, the community work requires understanding of the environmental, economic, political and social systems and their interconnectedness and the ability to think of multiple scales in the face of very big issues. However, the community organizing work needs courage, patience and the ability to love and interact with all classes of people outside of the job description, understand disciplines and engage in big thinking. Community workers require the ability to communicate clearly and listen and collaborate with the community; it demands humility, authentic appreciation for diverse ways of serving community people and thought; and to recognize all people's mistakes and efforts in working together for their integrative positive development. It is important because the majority of people in the community struggle with issues of everyday life and economic survival in the capitalistic society.

People are willing to make changes if the changes are very easy to do or if they take on solutions that lead to a better life both at collective and individual levels. In order to have effective social diffusion, community workers must be trustworthy of the beneficiaries, meaning community workers thinking, providing services and mixing with the community people in such a way that they realize that the community worker is working for their well-being. Therefore, it is important to contact people in person on their doorsteps, which is very effective in spreading innovative work and influential sustainable behavior. Community workers' simple language communication is appreciated and important for understanding by community people.

Moreover, community workers should have the willingness to form partnerships among community people, particularly those who do not cooperate, for whom he/she is working for in the community. Community sustainability work is not a single pathway; rather sustainability has many goals, and issues are many and varied, but each of us has our own sets of skills and understanding. But these skills and understanding should be open; open to accept or deal with all issues and try to satisfy all peoples' needs and demands.

The more we are open to alternative voices, the stronger our social community design and community build that could enhance our richer choice of potentially sustainable community collective futures. These skills can all develop among community workers through hard work and loving community people. It is more likely to generate novel responses to changing conditions, and some of the innovation leads to solutions with a good fit to new conditions (Worldwatch Institute, 2013).

I told the local councillor of the village that if women members of GB do business and earn extra money in the family, they can use this money for their children's education, food intake and for a better life. The councillor attentively listened to me, but he did not give me an answer. However, I challenged myself and continued my group training work with women in the neighborhood. However, the women who were interested in loans from GB declined to receive group training and to receive loans from the bank. I asked them why they have declined to receive training and receive loans from GB. They reported to me that the moneylender, Imam and Mattabbar threatened them; they (loan receivers) would be ousted by the elites if they received loans from GB. I talked with the poor people and explained to them the conflict of interest of the elites against their loan receiving. To destroy their vicious cycle of poverty, it is necessary to take up the challenge against these unethical elites. They should challenge this threat. I am also part of their challenge. They inspire. Although my group training job was disturbed, I continued my group training and loan collection jobs in addition to the center school voluntary teaching job there. All classes of people in the village supported me to do my voluntary schooling job there.

I always look for alternatives and dialogue with the people where I am obstructed in doing my center organizing jobs. My jobs have not been linear since I worked in GB. I always adapt the situation with alternatives and follow the Bidimalla, the bylaws of GB.

There were three groups (15 members) formed in this Luhuria Center in three months. I got strength and support from the poor people of the neighborhood. It was hard work and it needed a lot of talking with the elites and with the poor people so they understood my job in the village. In the fifth month, six poor widows and three single poor mothers came to me to provide them with training and loans. They told me that their children had been suffering from lack of food, housing, clothing, education and health care for a long time. Rural elites exploited their labor. They wanted to have their own businesses by taking loans from the bank.

By the seventh month, one female group was registered and received loans from the bank. It was hurrah for these people and for me. The message spread all over the village, even neighboring villages. The situation was favorable to me for working with poor women in the village and the neighboring villages. The center members were the spokesmen of my work in the villages. After one year, I was able to form four male groups in one center and six female groups in another center in the north west Luhuria village neighborhood and two center schools in the village. One female fieldworker was recruited in the branch who took charge of the female center instead of me. I immediately saw the socio-economic change of the members of GB while I was working with them. For example, their weekly income increased, their family members ate three meals a day and members bought new clothes for their children. They were happy with their business and income; I was also excited to see the immediate positive change in the members' lives.

I also started to organize another female group and a male group separately in the neighboring village, Kurua, in the seventh month of my posting in the branch. The male group meeting was in the south east part of Kurua and the female group in the middle of the village. In the open house meeting in this village, I talked to people about my "Center School" working experience in the neighboring village of Luhuria. They knew of my previous work there. When I collected installments and savings in Luhuria village centers, three women from Kurua told us they visited the Luhuria Center and talked with the members of the Luhuria Center who provided them (the visitors from Kurua) with a positive experience of the Luhuria loan receivers' income generation, center discipline, and the solidarity in the center. The visitors from Kurua were encouraged to join GB and do business by receiving loans from GB. With this voice, I was inspired to start group training in Kurua. Ten poor women and five men were trained in Kurua to register with GB.

However, one day at noon, three boys came to the branch and accused me of being involved in a love affair with one divorced young lady, a member of a group of Kurua, although it was false. I was surprised the lady was excluded from the group by the neighbors. However, this woman was chosen as the Chair of one group. Thereafter she declined to be Group Chair, with frustration. The other group members also declined to form groups. I found three young boys from rich families of the village were observing my activities in the village; their fathers were elites of the village. I continued to conduct training for male members. After one day, at the end of the training session with male members, I could not find my cycle; it had been stolen. I was nervous, but I suspected these young boys were the thieves of my cycle. These were the boys who accused me of a love affair with the divorced young lady. I reported to a teacher of the neighborhood who was an influential person in the village.

The next morning one center schoolboy of Kurua secretly told me he had seen my cycle under the water of a pond. I gave this information to the teacher from whom I sought help to rescue my cycle. He called a meeting near the ponds with three other Mattabbars. They asked a boy to search for the cycle. After searching, the boy brought the cycle from a pond to the arbitration place. The teacher called the three young boys (who accused me of the love affair with the young divorced lady). After cross-questioning the lady, the boys and me, the teacher found it was a false conspiracy against me to stop my center organizing training in Kurua. These boys had stolen my cycle. It was a conspiracy initiated by the moneylender, the Mattabbar and the Imam. The teacher advised the villagers including these elites and the boys they should cooperate with me in my work in the village. The teacher also warned them not to make false accusations against me.

Next day I visited each of the poor women in the Kurua neighborhood and invited them to receive group training in their own neighborhood. The divorced lady was happy and she agreed to continue her group chair responsibility with GB. Later she became the Center Chief when there were 15 members registered in the center (Bidimalla, 1978). However, it took me two months to revive and to reorganize the groups and the center. Many poor males assisted me to form male groups in the Kurua neighborhood.

Again, I faced a serious, religious conflict when I started to organize another female group and center training in other parts of Kurua. Muslim people declined to be in one center if Hindu people were in the same center. Eventually, all Hindu people and Muslim people were in separate groups. My bookkeeping work at this center included these two religious peoples in one center. This was a serious, sensitive issue raised by Muslim religious leaders in the village, and it spread to neighboring villages. Thereafter, I split the center into two centers with homogenous religious faith people, but I worried in case any uncomfortable situations and issues arose among the people. However, my alternative steps of separating two centers by religion have assisted me to avoid faith-based tension among the members. I talked about this problem to the manager, and I corrected my bookkeeping work for these two centers with his permission.

I describe the above two villages' challenging group formation job experience here to inform the readers that community organizing, designing, building and planning need to be tailored and adapted to the local situation although there might be job guidelines and instructions assigned to community workers. The GB micro-credit program in the villages was not welcomed by moneylenders, elites and rich people in the 1970s and 1980s because the GB group-based micro-financing was threatening to moneylenders. The reason was GB micro-credit services hampered moneylenders' own money lending business in the villages.

The GB micro-credit lending program liberates and relieves poor people from moneylenders' bondage, exploitation and injustice. The borrowers' incomes have increased by using loans received from GB. The elites felt uneasy and jealous of the poor children's schooling. The rich people cannot use poor children as cheap labor working for them. Poor women have businesses of their own instead of doing jobs in rich peoples' houses. Poor women were saved from false blame by the Mullahs' Purdah domination. GB activities in these two villages spread to other neighboring villages. Later, I received support from the group members if I faced any difficulty in organizing, forming, designing groups and centers in villages. After one year, it was easy for me to form groups in other villages.

As male group members and borrowers work outside of their homes, I conducted the male center meetings and their loan collection sessions at night (6:00–8:00 pm), but women's center meetings were conducted in the morning (7:30–9:30 am). I was respectful of the borrowers' customs, cultures, traditions, proposals and time. I listened to their problems politely. Within one year, the general public of the village found us (GB staff) honest, sincere and hardworking with our jobs. Now all kinds of villagers welcomed our loan service jobs, although there were politics involved regarding GB at the national level in Bangladesh. Even though there were many critics and false propaganda exists in Bangladesh against Muhammed Yunus, the current Government of Bangladesh supports GB activities in Bangladesh.

If I look back to my center organizing work strategies and Grameen Bank bylaws, I find huge changes have happened now. It is because to adapt and to adjust to the changing situation of borrowers, time and context, GB introduced many new loan products and designed new savings policies and strategies to face and to address the borrowers' contemporary issues and situations. Muhammed Yunus accepted mistakes and accepted workers' decisions if they did not breach the basic discipline of the Bidimalla and the framework for organizing the groups and centers in Bangladesh.

GRAMEEN BANK PHASE-I

Group Fund Savings Products

Group Fund savings were dissolved in Grameen Bank Phase-II in 2001 (Yunus, 2002). However, Group Fund savings had been running from 1978–2000. In such accounts, 5 percent of the loan amount was deducted as a contribution to the Group Fund. This amount of money was known as "group tax" or "group saving". This money was deposited in the group's own account and the member had no personal right or claims over it. All members had equal rights to the fund. Withdrawals from the fund were made under the joint signatures of the Group Chairman and the Branch Manager. When withdrawing money from this account, the Group Chairman and the Secretary had to be present in person at the bank (Yunus, 1983).

The group tax of 5 percent of the loan money was deducted at the time of disbursement of loans from the Group Fund. Each group fixed its own rate of interest on loans from the Group Fund (the group also advanced loans without charging any interest, if it so desired). The rate so fixed applied to all loans. The group was fully responsible for the recovery of the loans given from the Group Fund. However, if this loan money was not repaid in due time according to its terms and conditions, it was considered by the bank as a breach of discipline of the group. When a member leaves the group, he is entitled to a refund of the entire amount of his personal savings deposited in the Group Fund at the rate of one taka per week. These personal savings, however, could not be withdrawn for any other reason except this one.

If any member of the group did not repay bank loans, willingly or unwillingly, the loan had to be repaid in full from the Group Fund deposits. If any loan taken from the Group Fund remained unpaid, even after the expiry of the agreed time limit, no new loan was advanced from the fund. If all the members of any group left the group willingly or if the members did not keep the group in operation, the group savings of that particular group had to be deposited in the Emergency Fund of the center/association. All the borrowers had to pay interest on all loans taken from the bank at the rate fixed by the bank. However, many campaigned against the Group Fund savings. In 2001, Group Fund savings rules were removed from GB's bylaws.

Emergency Fund

The center Emergency Fund was one of the safety net services to the members of GB in Bangladesh (Yunus, 1978). This fund was generated from borrowers' contributions. After payment of the total interest accrued on the bank loan, an amount equal to half that amount was deposited in a special fund of the center/association called an "Emergency Fund". This fund was created through compulsory contributions of all the members of the center deposited in the Emergency Fund. Money accumulated in the Emergency Fund of the center/association could be spent on the following purposes: to repay the bank loan of any member who becomes unable to repay the loan due to any accident (e.g. the death of a cow purchased with the loan money, damage of a rickshaw in accident and so on). The Emergency Fund could extend grants for repayment of the outstanding amount of loans in the case where a member of any group failed to repay his/her loan for any other reason where the total savings of the particular group were not sufficient to repay the same.

The Emergency Fund operated under the joint signatures of the Center Chief, the Co-Center Chief and the field manager (Yunus, 1983). Loans taken from the bank were repaid generally in weekly installments according to the terms and conditions of the loan. However, the loan money had to be utilized within one week of the receipt of the loan in activities for which it had been taken. Those who failed to utilize the money within one week had to keep it deposited in the bank until the opportunity for its proper utilization came. Any sort of deviation from this was considered as a serious breach of discipline (Bidimalla, 1978).

Loan Disbursement and Repayment Procedures

Credit facilities offered by the bank to the members shall primarily depend on the regular attendance of all group members in the weekly meetings, their sense of discipline and regularity in payment of loan installments (Yunus, 2002). Therefore, failure of members to attend weekly meetings in time, absence from meetings, underpayment of loan installments, non-payment and so on disqualified the group from receiving the bank's facilities.

Loans are given to borrowers only after GB recognizes group members (Bidimalla, 1983). The bank considered loan applications from the registered members (receivers of the loans) of the groups, for different economic activities. Group membership alone did not entitle a member to a loan. The members were considered qualified for loans from the bank only if they abide by the rules and regulations of the bank. Receipt of loans by the remaining members in subsequent turns depends on regular payment of installments by the members who have already received loans, and strict observance of rules and regulations by all group members.

Article 4.4 of the Bidimalla mentions that all loans taken from the bank are generally repayable in weekly installments in the weekly center meetings. In cases where the utilization of a loan generates the opportunity for daily or weekly income, the loans have to be paid off in weekly installments. However, in cases where utilization of a loan does not create opportunity for daily or weekly income, but generates a large income in a lump after the expiry of a certain period of time, a "token installment" can be paid every week. The remaining amount should be paid in one single installment immediately after the receipt of the lump income. Failure to pay this token weekly installment is considered as a breach of discipline as in cases of non-repayment of "regular" installments.

FUNCTIONS OF THE MEMBERS OF GRAMEEN BANK

The Bidimalla (bylaws) of GB mention that it is the responsibility of the center/association to motivate its members to create among them a proper attitude, a sense of discipline and a spirit of cooperation among each other and with the bank, to take full advantage of the opportunities created by GB in order to change their social and economic conditions. The center/association takes special care to create a sense of responsibility among the members and the groups who may be callous and prone to violate rules and regulations of the bank. The center/association considers it as its responsibility to ensure proper utilization and timely repayment of all loans given by the bank to its members (Yunus, 1983).

The center takes steps to create training opportunities for, and take new initiatives in, helping to increase the efficiency and skill of the members in different trades with a view to ensuring gradual improvement of their economic condition through the financial cooperation of the bank. The center/association evolves and develops within itself a permanent and effective institutional mechanism for mediation in, and settlement of, all disputes and the removal of misunderstandings among its members.

The center takes special care to create and maintain a cordial and cooperative atmosphere among its members (Rouf, 2011). All center members meet once in at least two months and, after reviewing its program, take practical steps to keep the center/association moving forward. The Center Chief maintains regular contact with the bank and extends all help and cooperation to the bank authorities for the smooth operation of this special credit program. In practice, the center itself does not usually call meetings for reviewing their performance and defaults; rather sometimes the members sit together to conspire against repayment, stop saving, increase loan sizes or claim the bank should reduce interest rates or waive interest rates – this happened mainly in the undisciplined centers.

COMMUNITY ORGANIZING WORK EXPERIENCE IN CANADA

I will not go into full detail on my community organizing in Canada. My work was centered in Toronto and I have found that many places in the US and Toronto are similar – with universal health care in Canada being the exception to both Bangladesh and the US. However, I did not study health care, but rather the role of community agencies.

There are many different types of community agencies existing in Toronto. These community agencies are very efficient in dealing with problems although many of them are very formal and structured. Here community people are more formal. People have fewer interactions with their neighbors. I found here the communities were more nationality-based, faith-based, race-based, class-based, age-based and gender/sexuality-based (Beckwith and Lopez, 1997; Kahn, 1991; Pan, 2012; and Snow et al., 1986). In Canada, disadvantaged people particularly, single mothers and newly immigrant poor women are isolated from the different community participation decision-making processes. For example, they are very busy earning income for their survival in this individualistic and expensive capitalist society.

There are no "sharing economies" or "gift economics" (Mies and Shiva, 2014) among people here in Toronto. I have been involved with many community-based agencies and socio-economic institutions in Toronto and even in the US since 1993. Here community workers usually do not walk in the community; they usually do not visit clients on their doorsteps; they do not talk with community people individually and invite people for a community meeting; rather they display written information on the notice boards in community centers, churches, street corners and other public places like community houses, community newspapers, community radio stations, weekly/monthly newspapers and so on.

Food/light refreshment is usually served in the meeting for the attendees. However, few people attended community meetings. In Toronto, community meeting schedules are tied, meaning they start and finish meetings within a particular time. Topics of community meetings are community health education, nutrition education, drug prevention education, religious education, community gardening or English as a Second Language education. Immigration community agencies brief on different community resources in Toronto; and familiarize the new immigrant with the Canadian economic, social and multicultural education via, community dance festivals and community street festivals and so on. Here community gardens and community kitchens are available where people do their gardening and cook food in these public community places and facilities.

Community street festivals are usually organized by ethnic people in summer. Open concerts, music, dance, painting, art exhibitions and other cultural activities are popular in these street festivals. Food vendors display and serve their ethnic food and do business in these festivals. Ethnic dresses are displayed and sold in the street festivals. Vendors promote their business products at the festivals. These vendors promote consumerism in the festivals. I have observed in the street festivals that food and clothes, costumes, jewelry and toys are more expensive than the usual shops and restaurants; poor people cannot afford them. However, free food samples are delivered from vendors to the festival visitors for their food business promotion.

The City of Toronto organizes many meetings in Toronto City Hall and supports many community events. However, in these community meetings and events, the City tries to disseminate its own different agendas like multiculturalism, community gardening, historic arts, promoting cultural heritage, public health education, community peace education and so on. Surprisingly, none of the agencies are involved in micro-financing activities for low income people in Toronto, but many agencies are involved in providing small business training to people. Most of these training agencies receive small and medium-sized enterprise training funds from the City of Toronto, provincial government and foundations, even from private donors.

If we look at Grameen America, it is a 501(C) – a non-profit registered micro-finance organization founded in 2008 with its head office in New York City. It follows the group-based micro-credit model in the USA. There are eight GB officials seconded to GB. It operates in 13 cities of the USA. The cities are New York City, Harlem, Omaha NE, Los Angles, Indianapolis, Boston, Oakland, San Jose, Union City, San Juan, Charlotte NC, Austin TX and Miami Florida. Its repayment rate is 99 percent and its borrowers' average credit score is 640. As of 2016, Grameen America passed an incredible milestone of investing more than half a billion dollars in more than 86,000 women entrepreneurs around the country. Grameen America's members have saved US$5.9 million deposited in different US chartered banks. Its total income is US$17.4 million and expenses are total US$15.7 million (Grameen America, 2018).

Likewise, Canada could initiate the group-based micro-credit lessons learned from Grameen America. However, here in Canada, the important thing is the micro-credit workers and community workers should be field oriented (80 percent of their work is in the field) and physically and intensively visit clients' business sites instead of doing a computer, office-based job. The fieldworker should not feel that they are a separate class or superior to their clients. The fieldworkers should not panic and be hypocritical toward clients. They ought not hide true information from the public and not be rude to clients, rather respect, love and trust clients. The community workers can learn from their community clients and be happy to see their clients' success instead of only doing routine jobs and paperwork.

In Canada, the community meetings are usually on a monthly, bi-annual and annual basis; few agencies have weekly meetings. All types of meeting agencies serve light refreshments, hot lunch or dinner to attract people to join meetings. However, after the meetings, few people interact with each other. Here in Toronto, people are busy in their paid jobs or searching for paid employment. However, people enjoy eating food at the meeting and talking to each other in the parks and recreational centers.

The professional associations' meetings are highly sophisticated. Their meetings are usually conducted in private spheres like hotels, and they have strong social networks among the professionals.

The Government of Canada is attentive in supporting and serving seniors with meditation exercises, community swimming, senior health care and safety, drug prevention education and so on. Senior homes in Toronto organize many events like indoor games competitions, music, dance, food, selling second-hand clothing, antiques and more. Here churches organize many faith lectures in the community. Seniors enjoy these events; they enjoy potluck food events where they prepare food and bring it to the events and eat together with other seniors. Many senior agencies and disability agencies arrange trips to different historical places with costs shared by the seniors or by the agencies.

There are huge employment resource centers and immigration services are available in Toronto. The Government of Canada emphasizes youth employment services; government agencies support these employment resource centers. These employment centers are fully equipped with computers, printers, internet and other employment information. These employment resource agencies offer employment development services to youths, even opening their services to all other unemployed people free of cost. Employment resource centers like COSTI, South Asian Community Support Services, YMCA, YWCA, Career Foundation Center, ACCESS Community Support Services, Cross-cultural Community Services, Storefront Community centers, Community Police and so on organize different lectures related to employment, community gardening and workplace health and safety. All ages of people join these employment resource centers and browse the computers/internet to search for jobs, read newspapers and other inquiries. Very few people find professional jobs in Toronto through employment resource counseling services. Even many senior alumni are browsing computers, reading e-news in the university libraries in Toronto and spending their time in the library instead of staying at home alone.

Many youth centers have been providing anti-drug campaigns, safe sex education and peace education. Moreover, here community agencies organize seminars on public health and nutrition education, environmental basics education and health education. However, here client-worker relationships are very formal and structured. Disadvantaged poor people would be very happy to have cordial relationships with their community workers.

In Toronto, many community centers work with people who provide basic English language training free of cost for those who are ESL, but such free language training is absent in Bangladesh. Moreover, junior schools and secondary high schools have social workers who organize parents' meetings in the schools where they discuss parental relations with their children and neighbors, and discuss how parents could deal better with their children and educate their children better.

Here many rich and middle-class families have a personal support worker who takes care of the seniors: bathing them, changing bedclothes and dressing them, cleaning clothes and houses, preparing food and assisting them to eat food and take medicines and so on. Personal support workers also bring disabled people (all ages of people) to child care centers, senior centers, schools, community centers and other recreational centers. Red Cross Canada has food services for seniors called "Meals on Wheels". This Meals on

Wheels program delivers healthy food to seniors at a minimum cost. Many churches have hot lunch services for their members. Toronto Transit Commission (TTC) has Tran Wheel vehicles that provide transportation services to seniors and physically disabled people. These are great facilities and services available in Toronto. Unfortunately, these services are unavailable in Bangladesh. Here universal health care facilities provide great services to all citizens of Canada that are absent in Bangladesh. The Canadian community agencies, hospitals and clinics are very popular for their efficient services to people in Canada. A comparison and contrasting of community organizing experience in Bangladesh and Canada can be seen in Table 9.1.

Civics is important as people re-learn skills for participating in democratic governance (Toker, 2012). The GB has bylaws for changing the Group Chair, Group Secretary, Center Chief and Co-Center Chief in a democratic way every year in order to develop all members' leadership and to maintain center discipline. This system of changing leadership is not a process of leadership development rather it is a process to empower all members of the center. Moreover, communities will become more decentralized and more resilient, with greater control over many aspects of their lives if they build networks among them. In Canada, community members connect with other community members using information technology in order to continue to build collective learning both at the community micro level, mezzo level and national level. Many young in Bangladesh connect with each other by email, Facebook and LinkedIn. Worldwatch Institute (2013) mentions that community workers find community organization is a place to stand for working for the well-being of the community people. Its community building slogan is "Choose what you love. Start from there, do what you can and connect with others".

GRAMEEN BANK LOANS PRODUCTS AND THEIR IMPLEMENTATION STRATEGIES DURING PHASE-I

GB started by providing general business loans individually to its group members and this still continues. However, in the early 1980s, GB experimented with the collective loan, a bigger loan given to a group and center instead of to the individual. The motivation for these larger loans was that they would enable members to fund more profitable activities that required greater capital, such as the installation of rice, oil and weaving mills and leasing of markets, orchards, ponds and so on. However, some members felt that they were doing most of the work while others enjoyed the benefits. Conflict among the group members in the centers resulted in the limited success of these loans, and then the bank eventually abandoned them.

In Grameen Phase-I (Classical phase (1978–2000)), GB initiated housing loans for the members for constructing their houses. This loan was introduced on a small scale in 1984, but it expanded rapidly after the devastating flood of 1987. I first disbursed this housing loan for the flood victims in my branch called Gobindashi Vhuapur Branch, during my Area Managership in Vhuapur area. The housing loan provided borrowers with a longer period, the loan to be repaid by weekly installments over several years, with an annual interest rate of 8 percent. Initially, a sum of Tk.7,000 was given to housing loan borrowers as the "Moderate Housing" loan. However, the amount of housing loan was raised to

Table 9.1 Compare and contrast community working experience in Bangladesh and in Canada

Community organizing experience in Bangladesh	Community organizing experience in Canada
Bangladesh uses the terms group, center (*kendra*), society (*samitty*) and village organization as synonyms for community	Canada uses the terms community, association and society as synonyms for formal community organizations
A majority of community organizations are in the rural area	Almost all community agencies are working in the urban area
More micro-financing institutions are in Bangladesh, both in rural and urban	Few micro-credit agencies are working in Canada despite huge demand for micro-credit among low-income people
NGOs depend on outside funding sources or Micro Finance Institutions (MFI) to cover their costs from loan investments and savings refinancing	Receive funding from local government, regional/national government and/or from foundations
Fieldworkers are not involved in fundraising events	All community organizations have fund raising programs through community dinners or fund raising campaigns throughout the year
Fieldworkers have flexibility to implement agencies guidelines to serve clients	Fieldworkers strictly follow the manual of work; they do not tailor their jobs to clients' needs
Fieldworkers are flexible about discussing various topics in their community meeting as well as being flexible about the schedule of their meetings	Community workers are very tied and rigid in their meeting topics and meeting schedule
Usually bottom-up approach	Usually top-down approach
NGOs focus on people's needs and requests	Strictly follow job assignment and follow the manual for the job
Fieldworkers are less critical	Fieldworkers are more critical and analytical
Executives connect their activities to their organizations' progress reports for the donors or for their future plan and design outcomes	Community organizers are able to connect their activities and their thinking to their organizational future planning and design outcomes
All work is done manually and documented in handwriting	All documentation is done electronically; very few jobs are done manually
Few anti-drug education, mental issue services, disability services and seniors' services	More anti-drug services, basic English language education, mental services, disability services and seniors' services are available to people
All beneficiaries are micro-borrowers	Beneficiaries are usually non-micro-borrowers, new immigrants and non-mainstream people
No small business training available to micro-borrowers	Small business training available to interested people by community organizations both free of cost and at cost
MFIs or NGOs can collect savings from micro-borrowers	MFIs or NGOs are not permitted to collect savings from their clients
Fieldworkers usually stay within their working area	Fieldworkers do not necessarily stay within the community

Table 9.1 (continued)

Community organizing experience in Bangladesh	Community organizing experience in Canada
Community workers visit groups and centers by walking and by cycle	Community workers visit or attend their meetings by driving cars or riding public transport
People can talk/contact the worker at any time	People can only talk during the meeting particularly at the end of the session as declared by the worker
Workers have primary relationships with low-income people	Community people have secondary or formal relationships with their workers
The bulk of the work is done through mutual discussion	Work done following pre-written instructions
Usually poor people are beneficiaries of NGOs	Lower middle-class, low-income unemployed, part-time workers are the beneficiaries of community organizations
Learning by doing, improve services from mistakes	Do job with pre-instructions; zero tolerance of mistakes
Fieldworkers usually do their job manually	Use computers, cell phones, power points or other electronics in meetings
Fieldworkers provide loans and services to low-income people	Mostly educational and training services to community people
Few outreach programs organized by colleges and universities can be seen	Every community college and university has community outreach programs and community outreach research centers
No food supply for attendees in the meeting	Usually every community meeting delivers food for the attendees
Dialogue-based informal discussions in meetings	Very prescribed formal discussions in meetings
Very few social safety net services are available for the clients	Social safety net services are available for beneficiaries both from public agencies and community-based agencies

Tk.25,000 to account for the rising cost of building materials. The houses built with the loans had to include four concrete pillars, a sanitary latrine and corrugated iron roofing sheets. This type of house provides protection against floods, cyclones and rain.

This housing loan was exclusively given to poor women borrowers. The land on which the house was to be built had to be in the woman's name. Grameen included this condition to make sure that a woman would not be evicted from her home in the case of dissolution of the marriage. Owning a piece of land through a housing loan from GB may be the only way for a woman to have "a house of her own". GB also introduced the "Basic Housing Loan" to rebuild the homes of borrowers that were damaged by the flood of 1987. The loan size was Tk.7,000–Tk.12,000. The maximum period for repayment of a housing loan was five years (Dowla and Barua, 2006). However, the repayment rate of housing loans was not satisfactory. As the loan repayment rate was for more than one year, borrowers were relaxed about paying the loans regularly. However, they could complete the repayment of their housing loan in their lifetime. GB introduced awards to those workers who provided more housing loans to borrowers and recovered the loans regularly. As of June 2018, GB has provided housing loans to 721,306 borrowers in Bangladesh.

During the 1980s, GB promoted children's education by helping to establish schools in the centers that I discussed earlier in the chapter. The majority of borrowers hope that their children will become educated and successful and find respectable occupations like medical doctors, engineers, lawyers or join the government services. However, the cost of higher education is high which resulted in a drop in the higher education enrolment of borrowers' children. In the year 1997, higher education student loans were introduced with easy terms and conditions for the children of GB borrowers. By receiving this loan, many children of GB borrowers completed their education and became doctors, lawyers, engineers, chemists, biologists, physicists and social scientists in Bangladesh. The graduates regularly repay their loans. As of June 2018, GB has disbursed higher education loans totaling $52.25 million (female loans $13.89 million and male students $38.36 million) to a total of 53,933 students (male student 40,970, and female students 12,963) in Bangladesh.

I was assigned to draft the GB higher education student loan manual in 1997. GB decided to provide this student loan to finance higher education for the children of GB members. Currently, all children of borrowers who are on a basic or a flexible loan and have been members of the bank for at least a year are eligible to receive higher education student loans. GB introduced scholarships for the children of GB borrowers in 2001. Scholarships are awarded every year, with priority to girls, to encourage them to get better grades in school. On average 3,000 children, at various levels of school education, receive these scholarships every year (Yunus, 2002). Now I am voluntarily conducting research on GB higher education student loan services in Bangladesh attached to the Center for Learning Social Economy and Workplace, University of Toronto.

Grameen Bank Phase-I has proved that the poor are creditworthy. The poor become self-employed and overcome their poverty by using loans from GB. Below the chapter precisely describes the Grameen Bank Phase-I activities and policies of micro-lending to poor people in Bangladesh (Dowla and Barua, 2006).

- GB targets the poor, particularly poor women, as identified by land ownership.
- It offers micro-loans to its borrowers to start income-generating activities.
- GB provides loans to the poor without collateral; it is based on trust.
- All loans must be paid back in installments (weekly or bi-weekly).
- To receive loans, poor people must form groups and belong to centers.
- A borrower can receive a new loan, often of a larger size, if the previous loan is repaid.
- Borrowers must pay into the Group Funds, weekly savings and the Emergency Funds.
- The borrowers need to memorize and follow the "Sixteen decisions", the social development charter developed by GB in 1984.

GB workers are likely to find themselves performing multiple roles, for example, marriage counselors, conflict negotiators, group trainers, civic leaders, community organizers and center managers. Every day they work either in the office or in the field to organize and form groups, mitigate conflicts, revive center discipline, advise defaulting borrowers, collect loan proposals and loan installments, and savings and so on. They work from 6:00 am until 10:00 pm. They spend 65 percent of their time in the field organizing centers, maintaining group discipline, collecting installments and savings. Currently center managers collect a huge amount of cash from the borrowers and carry this cash from the centers to the office every day. It is a risky job. They are like emergency room doctors (Woolcock, 1998). However, as it is very risky to carry cash in Bangladesh, GB should develop alternative strategies for transporting cash from the center to the branch.

GB uses the following ten indicators to assess whether members are moving out of poverty:

1. The members and their families now live in a tin-roofed house or in a house worth at least Tk.25,000; the family members sleep on cots or a bed instead of the floor.
2. They drink pure water from tube wells, boiled water or arsenic-free water or purified by the use of alum, purifying tables or pitcher filters.
3. All children of members attend school or at least have finished primary school education.
4. The members can repay a minimum weekly installment of Tk.200.
5. All family members use a sanitary latrine.
6. All family members have sufficient clothing to meet their daily needs.
7. Members have vegetable garden, fruit bearing trees and so on.
8. Members maintain an average annual balance of Tk.5,000 in their savings accounts.
9. The members are able to feed their family members three meals a day throughout the year.
10. All family members are conscious about their health; they are able to have treatment if they are suffering from illness. (Grameen Bank, 1998)

By using these poverty-free indicators, an internal survey of GB obtained data of borrowers above the poverty line (Table 9.2).

Helen Todd (1996) found that the bargaining position of GB borrowers in her sample improved as a result of participation in the bank, joining the center and attendance at the weekly center meetings. GB borrowers maintained their center discipline, regularly

Table 9.2   Annual declining poverty rates of Grameen Bank borrowers

Year	% above the poverty line
1997	15.1
1998	20.4
1999	24.1
2000	40.0
2001	42.0
2002	46.5
2003	51.1
2004	55.0
2005	58.4

Source: Grameen Bank, 2008.

deposited savings, received bigger loans, increased their businesses and repaid their loans regularly.

EVOLUTION OF GRAMEEN BANK PHASE-II – ITS PRODUCTS AND IMPLEMENTATION STRATEGIES

The GB repayment rates fell to 65 percent during 1996–2000. The *New York Times* published news of GB borrowers' default status with statistics in 2001. The reasons for increasing default rates were several consecutive disasters like floods, cyclones, droughts and hurricanes, and political turmoil from 1996 to 2001. In order to resolve the problem, GB developed a Samassha (Problematic) Cell, headed by Muhammed Yunus in Dhaka. The Samassha Cell executives analysed centers' problems, identified borrowers' sufferings and looked for possible solutions for the struggling members who were victims of natural disasters. The solutions were made case by case to revive credit discipline among defaulters and struggling members of GB, but it was like window dressing; however, it was necessary to overhaul the whole GB credit and savings design as well as re-energize the fieldworkers of GB.

In 2001–02, Muhammed Yunus himself visited many problematic centers and area offices and met with the field staff of different zones; he even stayed in the field and heard field problems directly from field staff and collected information and ideas from them to design Phase-II. He redesigned the loan products, savings products and introduced an attractive pension scheme for the borrowers in Grameen Bank Phase-II. The redesign of loan and savings products and implementation of them created an extremely important role in the rebuilding of the centers and reviving center discipline and regularizing borrowers' creditworthy behaviors. The discussion with field staff created an open floor for good ideas generation. Through deliberation, debate and extended discussion and dialogue, the winning ideas were incorporated into Grameen Bank Phase-II.

Throughout Yunus's visit to the branches, he was able to find out causes and effects of the defaulting behaviors of the borrowers. Moreover, he found the causes of problems were not only natural disasters but also bad rumours by other agencies against GB. The rumors were anti-GB propaganda. Muhammed Yunus also realized that problems were caused by structural faults within GB's system (Dowla and Barua, 2006). These problems also crystallized from the organized protest by religious leaders in Bangladesh. He found their causes also included too much on loan disbursement and mismanagement by the field staff.

The borrowers' lack of interaction with the bank was also revealed as the major cause of problems at the branches. When members stopped coming to the regular weekly meetings, the centers became mismanaged. Moreover, Muhammed Yunus explored whether lack of attendance at weekly meetings was a major reason for the crisis. Another flaw identified the inappropriate use of loan money. After an exhaustive study of the causes of the crisis of 1996–2000, various task forces started working on steps to rebuild their respective branches. Through consultation, debate and discussion, Muhammed Yunus and other executives were able to learn from each other. They realized that the loan disbursement and installment collection system was rigid under Grameen Classical Phase-I. Then they decided to redesign all the GB loan, savings and insurance products.

However, Muhammed Yunus emphatically requested the field staff to reorganize and rebuild the centers, give hope to members and tell them GB is for them; in any crisis they face, GB is with them. GB provided loans to borrowers under certain standard rules that benefit members of GB in Bangladesh. Moreover, Yunus (2001) instructs field staff to develop trust between employees and the clients of GB as well as rehabilitate borrowers to restore their businesses. He advised them to continuously motivate and give hope to borrowers with face-to-face dialogue in center meetings and special center meetings.

However, field staff reported to Muhammed Yunus that verbal motivation alone would not work. Borrowers need tangible benefits like new loan products and a system to motivate borrowers to restart repaying their loans and to revive borrowers' businesses. Muhammed Yunus also found that to ensure effective participation and execution of necessary changes, there needed to be special attention paid to improving staff morale and performance. Through a process of trial and error and small-scale pilot testing during field study periods in two years, 2000–01, the main elements of Grameen Bank Phase-II began to take shape. There are many new loan products tested with the new loan repayment system; however, not all tested products are included in the regularized new products in Phase-II. However, through deliberation, debate and an open process of dissemination and critical evaluation of results, only the winning ideas survived and were incorporated into the new system of Grameen Bank Phase-II (Dowla and Barua, 2006). Now GB had 8.9 million borrowers in Bangladesh. The chapter below briefly narrates the Grameen Bank Phase-II system.

Grameen Bank Phase-II Rules and Procedures

Grameen Bank Phase-II came up with a number of fundamental changes in the elements of GB loan transactions including the flexible loan, the basic loan and the six-month repayment schedule and six-month quality control check. The flexible loan has a limit in that whatever amount a borrower repays, he or she can borrow twice the amount back after a six-month interval, and then every six months thereafter an amount equivalent to the amount repaid (Dowla and Barua, 2006). However, there is a credit limit to the flexible loan where borrowers have to pay off half the loan in six months in order to stay current.

At the beginning of the initiation of Grameen Bank Phase-II, field staff faced difficulty tracing borrowers, and they could not even find them. Borrowers did not come forward to reorganize themselves because they thought there was no incentive to repay the loan. On the other hand, borrowers felt that, because they had defaulted, the bank would not give them another loan even if they repaid their old loans. In order to reach out to these borrowers, the field staff established contact with many of the defaulters and urged them to restart their installment payments. However, there was distrust between defaulting borrowers and fieldworkers. Moreover, fieldworkers found that while many borrowers would promise to come back and repay, they would not keep the promise. Then the bank introduced the written contract (*chukti*) for the borrowers from the bank. Then came the idea of "*chukti rin*", a contractual loan called a "flexible loan" (Yunus, 2002).

Another outcome of the development was the dismantling of the Group Funds and Emergency Funds that were essential elements of Grameen Phase-I. Then after, in 2000–02, GB developed the new loan products (mentioned earlier in the chapter) and an attractive insurance policy (the GPS) that motivated defaulting borrowers of GB to revive the center discipline and receive new loans (flexible loans and easy loans) to revive their business capital. Borrowers started to deposit their weekly savings, voluntary savings and premiums of their pensions (GPS). Although the Emergency Fund safety net stopped, families of deceased borrowers of GB received a total of Tk.8 million to 10 million in life insurance benefits each year. Each family received Emergency Funds up to a maximum of Tk.2,000 from GB. A total of 54,469 borrowers of GB died during 1979–2004. Their families collectively received a total amount of Tk.114.0 million (Dowla and Barua, 2006). Borrowers were not required to pay any premium for this life insurance. Borrowers came under this insurance coverage by being a shareholder of the bank (Yunus, 2002). As of June 2018, GB life insurance had accumulated $194.039 million and the total amount paid out from life insurance to borrowers of GB was $5.91 million in Bangladesh.

Field staff were re-energized through new financial incentives and promotional incentives designed for them. Now the GB repayment rate has recovered to 97 percent. The premium deposits of GPS are 100 percent although all members of the center are not present on time in their weekly center meetings. Even now members do not come at the same time; some come early and others wait for the latecomers which is a waste of time for the early attendees and disturbs their business and urgent jobs. However, bank workers go to the weekly center meetings in the morning and start loan and savings collections, whoever is repaying their loans and savings. Now GB is more flexible about the attendance of the borrowers in the weekly meetings than before. If for any reason all members need to sit together or need to resolve any problems, they organize special meetings at their own convenient places at night.

Grameen Bank has been adding to and changing its different loan products and savings products, at different times since 1976, as is mentioned in Table 9.3.

The newly designed contractual savings plans have enabled the poor to break free from the trap of low savings in Grameen Bank Phase-II. GB cumulative loans disbursed as of June 2018 were $25 billion and members' savings deposited were $1.6 billion and non-members savings had deposited $751 million in Bangladesh.

Savings mobilization among members and non-members was the most important change instituted under Grameen Bank Phase-II. Now center managers were needed to explain all these new products in their group training and center meetings as well as in general meetings in order to inform borrowers about the Grameen Bank Phase-II products and services that were for them. Although Muhammed Yunus has left GB, Grameen Bank Phase-II is running very well because the design of Grameen Bank Phase-II is more suitable for the members than before; they feel more comfortable with the present loan operation system and savings deposit collection procedures. The bank has developed trust with its borrowers through the new loan operation and savings collections system.

GB took a long time to develop a new flexible system and field-tested it over months. GB finally introduced the new system in September of 2000. It was a simplified and generalized GB loan transaction system. Now this works equally well both in normal and disaster situations. It allows enterprising borrowers to move ahead faster. Muhammed Yunus comments that everybody has fallen in love with it. Borrowers love it and staff love it because it is so simple because it offers tailor-made loans rather than the previous one-size-fits-all type of loans. The new system basically has introduced two types of loans: (a) basic loans and (b) flexible loans. A borrower can take a basic loan for any income-generating purpose with mutual agreement between the borrower and the bank,

Table 9.3 *Grameen Bank's different loan and savings products developed at different times*

Name of the loan products and savings products	Date of commencement	Status
Unwritten piloted general loan	August 1976	Discontinued in Grameen Bank Phase-II
Collective enterprise loan	November 1982	Ditto
Housing loan	May 1984	Continues
Basic housing loan	September 1987	Continues
Capital recovery loan	September 1990	Discontinued after Grameen Phase-II
Family loan	July 1992	Discontinued
Food stock loan	March 1992	Discontinued
Installation of tube-well loan	November 1992	Discontinued
Building sanitary loan	February 1993	Discontinued
Leasing	October 1993	Continues
Supplementary loan	October 1994	Discontinued
Cattle rearing	December 12994	Continues
Loan for homestead purchase	March 1996	Continues
Pre-basic housing loan	October 1996	Continues
Special general loan (scaling up)	June 1997	Discontinued
Seasonal loan 2	September 1997	Discontinued
Seasonal loan 3	September 1997	Discontinued
Intermediate loan	October 1997	Discontinued
Higher education loan for members' children	October 1997	Continues
New flexible loan	29 December 1999 for Jamalpur special project	Continues with improvement
New flexible loan	24 April 2000, for three projects (Jamalpur, Habijong and Nilphamari)	Continues with improvement
Flexible loan (September 2000 edition)	22 May 2000, for all zones	13 February 2001, for all zones and projects
Basic loan for regular members	31 May 2000, for all zones and projects	–
Basic loan for regular members (September 2000 edition)	14 September 2000, for all zones and projects	Continues
Basic loan for regular members (February 2001 edition)	13 February 2001, for all zones and projects	Continues
Group Fund	1980	Discontinued
Center Emergency Fund	1980	Discontinued
Individual savings	1978	Continued but saving amount increased to Tk.5 from Tk.1
Individual voluntary savings	2001	Continues
Center Emergency Fund	2001	Continues
Grameen Pension Scheme	2000	Continues
Buy Grameen Bank share	1983	Continues

Table 9.3 (continued)

Name of the loan products and savings products	Date of commencement	Status
Beggars loan	2003	Discontinued
Beggars voluntary savings	2003	Continues
Beggars join GB group members	2005	Continues
Housing accessories buying loan	1999	Continues
Center School savings	1982	Discontinued
Center Emergency Fund	2001	Continues
Deposits/savings for "double in seven years"	2001	Continues
General public deposit savings	2003	Continues
Monthly Income Plan (usually staff retirement pension money), but borrowers can save more than Tk.20,000 under this plan for at least five years and receive monthly benefits	2002	Continues

Source: Dowla and Barua (2006).

unlike the old system where all loans were for one year. Basic loans can be for only six months (Yunus, 2002).

Grameen Bank Phase-II Operation Incentives to Field Staff

Both regular borrowers and defaulting borrowers were happy with these new products and procedures. However, fieldworkers paid more attention to the flexible loan in order to regularize defaulters into basic loans where they can receive more basic loans than flexible loans. In Grameen Bank Phase-II, the respective fieldworkers and branch managers received awards from Muhammed Yunus and got quick promotions and incentives from the bank. Through this process, GB defaulting borrowers became regularized and revived and center discipline was reorganized within two years.

The new system brought excitement and inter-branch competition in GB to reorganize the centers and to revive the discipline between the borrowers and the centers (Dowla and Barua, 2006). This system introduced a grading system for branches. The grading system created competition among field staff and the centers, and it awarded color-coded "stars" to indicate the quality of performance of a branch and the centers. If a branch has 100 percent repayment and 100 percent center discipline records for two consecutive years it is awarded a "green star". If the repayment falls below that during any two successive years, the star is lost. Likewise, GB introduced different colored stars like "blue stars", "violet stars" and "red stars" for the various performances of the branch. Branch staff can wear the stars as a badge of honor and display their stars on the branch stationery to show their achievement.

CONCLUSIONS

Grameen Bank's group forming, center organizing and designing of different types of loans and savings products were based on GB members' needs and their feedback on GB products and services. GB has streamlined its bylaws and its different products after testing in centers at different places in Bangladesh. In the beginning, we GB fieldworkers did our jobs from scratch, without guidelines, but we improved our jobs from learning by doing and from mistakes. We made lots of mistakes in our jobs. Muhammed Yunus, our employer, accepted our mistakes and gave us an opportunity to correct mistakes with innovation. He gave us liberty to experiment with new ideas and let it shape a program for the well-being of poor people in Bangladesh. Now the rules of group formation, center organizing and designing structures of GB are streamlined. However, we faced huge difficulty from the moneylenders, rural elites and community agencies during the 1970s and 1980s. However, our persistent persuasion, sincerity, the cooperation of borrowers and proper leadership of Muhammed Yunus led to our job success in Grameen Bank, Bangladesh. This bank continues to be flexible in accepting new ideas from the field and developing suitable services for its borrowers. GB builds its structure and develops its different products throughout its internal innovations.

REFERENCES

Alinsky, Saul (1971). *Rules for Radicals*. New York: Vintage Books.
Beckwith, David with Cristina Lopez (1997). Community organizing: People power from the grassroots. COMM-ORG: The On-Line Conference on Community Organizing and Development. Accessed January 2010 at http://sasweb.utoledo.edu/comm-org/papers.htm.
Bidimalla (1978). *Grameen Bank Bidimalla*. Dhaka: Grameen Bank Project.
Bidimalla (1983). *Grameen Bank Bidimalla* (revised version). Dhaka: Subarna Publishing.
Bobo, K., J. Kendall and S. Max (1991). *Organizing for Social Change: A Manual for Activists in the 1990s*. Cabin John, MD: Seven Locks Press.
Cohen, A.P. (1985). *The Symbolic Construction of Community*. New York: Tavistock Publications and Ellis Horwood Limited.
Dowla, Asif and Dipal Barua (2006). *The Poor Always Pay Back: The Grameen-II Story*. Bloomfield, CT: Kumarian Press Inc.
Fellin, P. (2001). *The Community and the Social Worker* (3rd edition). Itasca, IL: F.E. Peacock Publishers.
Grameen America (2018). *Grameen America Updates*. New York: Grameen America.
Grameen Bank (1998). Grameen Bank Training Institute poverty reduction survey parameters. Grameen Bank Training Institute, Dhaka.
Grameen Bank (2008). Monitoring and Evaluation Department of Gramen Bank. Grameen Bank, Dhaka.
Hall, K.B. and G.A. Porterfield (2001). *Community by Design: New Urbanism for Suburbs and Small Communities*. New York: McGraw-Hill.
Hamilton, Cynthia (1991). Women, home, and community. *Women of Power*, 20(Spring), 42–5.
Hester, R.T. (1990). *Community Design Primer*. Mendocino, CA: Ridge Times Press.
Kahn, Si (1991). *Organizing: A Guide for Grassroots Leaders*. Silver Springs, MD: NASW Press.
Lennertz, W. (1991). Town-making fundamentals. In Andres Duany and Elizabeth Plater-Zyberk (eds), *Towns and Town-Making Principles*, pp. 21–4. New York: Rizzoli.
Mies, Maria and Vandana Shiva (2014). *Ecofeminism* (Kindle edition).
Pan, Elizia (2012). Community organizing. Participedia. Accessed 6 July 6, 2012 at https://participedia.net/en/methods/community-organizing.
Rothman, J. (2001). Approaches to community intervention. In J. Rothman, J. Erlich and J. Tropman (eds), *Strategies of Community Intervention: Macro Practice*, pp. 27–64 (6th edition). Itasca, IL: F.E. Peacock Publishers, Inc.

Rouf, K.A. (2011). Grameen bank women borrowers' family space and community space development in patriarchal Bangladesh. *International Journals of Research Studies in Psychology*, 1(1), 17–26. ISSN 2243-7681.

Rouf, K.A. (2016). Eradication of poverty through community economic development using micro financing: Lessons learned from Grameen Bank Bangladesh. *International Journal of Research Studies in Management*, 5(2), 1–10.

Sanoff, H. (2000). *Community Participation Methods in Design and Planning.* New York: Wiley.

Smock, K. (2004). *Democracy in Action: Community Organizing and Urban Change.* New York: Columbia University Press.

Snow, David A., E. Burke Rochford, Jr, Steven K. Worden and Robert D. Benford (1986). Frame alignment processes, micro-mobilization, and movement participation. *American Sociological Review*, 51(4), 464–81.

Sobhan, Rahman (2005). A macro policy for poverty eradication through structural change. Discussion paper no. 2005/3, World Institute for Development Economics Research, Helsinki, Finland.

Todd, Helen (1996). *Women at the Center: Grameen Bank Borrowers after One Decade.* Boulder, CO: Westview Press.

Toker, Umut (2012). *Making Community Design Work.* Chicago: Planners Press.

Turner, J.F.C. and R. Fichter (1972). *Freedom to Build: Dweller Control of the Housing Process.* New York: Macmillan.

Woolcock, Michael (1998). Social theory, and poverty alleviation: A comparative historical analysis of group-based banking in developing economies. Unpublished PhD dissertation, Brown University, Providence, RI.

Worldwatch Institute (2013). *State of the World: Is Sustainability Still Possible?* Washington, DC: Island Press.

Yunus, Muhammed (1978). *Bidimalla.* Dhaka: Oslo Printers.

Yunus, Muhammed (1983). *Bidimalla.* Dhaka: Grameen Bank.

Yunus, Muhammed (1998). Alleviating poverty through technology. *Science*, 2282(5388), 409–10.

Yunus, Muhammed (2001). Expanding microcredit outreach to reach the Millennium Development Goal: Some issues for attention. Paper presented at the International Seminar on Attacking Poverty with Microcredit organized by Palli Karma Shuhauk Foundation (PKSF) in Dhaka, 8–9 January.

Yunus, Muhammed (2002). *Grameen Bank II: Designed to Open New Possibilities.* Dhaka: Grameen Bank.

10. How to build an "intentional community"*
Brenda M. Elias

An innovative housing model that provides integrated support services to a mixed community of adults with physical, developmental and mental health needs demonstrates how the use of social and personal space intersects with social participation levels and the creation of an "intentional community." Intentional communities are not a recent phenomenon (Brown, 2012) and there are many diverse examples around the world. As early as the 6th century, Buddha's followers rejected wealth, turned to meditation, and joined together in ashrams to model an orderly, productive and spiritual way of life. Common themes in intentional communities are living cooperatively, solving problems non-violently and sharing experiences with other members. This chapter will explore whether these themes are present at the Reena Community Residence (RCR).

DEFINING COMMUNITY

Phillips and Graham (2000) and Phillips et al. (2003) describe the "language of 'community' to be sufficiently elastic that its meaning remains contested," and yet governments at all levels are looking at how to involve "community" both in the process of policy development and in service delivery. Nonprofit agencies in Canada have formed to represent community across many sectors including housing, health and social services. One way to look at community, as articulated by Christenson, Fendley and Robinson (1994) is: "A community is defined as people that live within a geographically bounded area who are involved in social interaction and who have one or more psychological ties with each other and with the place in which they live." The definition emphasized in this research is community as a place or geographical community with common interests (Drier, 2001; Forchuk et al., 2006).

Housing stability is linked closely with health, emotional wellness and quality of life. This has been demonstrated by Robert D. Putnam (Putnam, 2000) as he identified a connection between strong community engagement and lower levels of illness. In a groundbreaking book based on vast data, Putnam shows how we have become increasingly disconnected from family, friends, neighbors and our democratic structures – and how we may reconnect. He warns that our stock of social capital – the very fabric of our connections with each other – has plummeted, impoverishing our lives and communities. Therefore, if the presence of social capital is emerging within an "intentional community" this would serve as a measure of improved health and quality of life.

Along this same line of thought, Leon Pastalan studied the linkages between spatial behaviour and housing to encourage urban planners to think about human factors such as adopting a compassionate lens when designing personal and social spaces. M. Powell Lawton (1977) developed the concept of person-environment fit, defined as the degree to which individual and environmental characteristics match. Person characteristics may include an individual's biological or psychological needs, values, goals, abilities or personality, while environmental characteristics could include intrinsic and extrinsic rewards, demands of a job or role, cultural values, or characteristics of other individuals and collectives in the person's social environment.

Person-environment fit can be understood as a specific type of person-situation interaction that involves the match between corresponding person and environmental dimensions. Even though person-situation interactions as they relate to fit have been discussed in the scientific literature for decades, the field has yet to reach consensus on how to conceptualize and operationalize person-environment fit. This is due partly to the fact that person-environment fit encompasses a number of subsets, such as person-staff fit. Nevertheless, it is generally assumed that person-environment fit leads to positive outcomes, such as satisfaction, performance and overall well-being.

Susanne Iwarsson (2012) conducted a review of the literature on person-environment fit over a 20-year period to determine how research strategies interact with health aspects and how practitioners make use of and evaluate the effects of this dynamic in housing projects. The practice applications and findings of this theoretical approach have significant relevance for health and social service providers actively contributing to building community and participant engagement.

Defined as a planned residential community RCR provides services with a higher degree of staff teamwork than other settings. Tenants hold a common social vision, focusing on the benefits of living together. Tenants will be asked to look out for their neighbours and can expect the same help from others. For the purposes of this study, community is understood first and foremost as a geographical place.

From an ecological and sociological perspective the emerging levels of social participation and the building of social capital (Putnam, 2000) in this "intentional community" could be seen by the tenants as life changing events. This research will outline how one type of intentional community has evolved in the Region of York, Ontario, Canada and whether it has become life changing.

Findings from this five-year research project (2012 to 2017) at the new RCR, opened in 2012, report on the lived-experience of a diverse population with special needs as they transition from various residential settings and connect to a new "Intentional Community with Supports." Tenant responses may align with the person-environment fit theoretical framework (Lawton, 1977) which has been adopted to look at aspects of personal and social spaces.

WHAT IS COMMUNITY DEVELOPMENT?

This leads us to a definition of "community development" as "a group of people in a locality initiating a social action process (i.e., planned intervention) to change their economic, social, cultural and/or environmental situation" (Christenson et al., 1994). Where initiatives are taken to tackle social problems, a contemporary approach is to form partnerships with those who share the same goals.

The sense of community instilled through community groups is very powerful and significant skills are acquired in learning about organizing, lobbying, advocacy and bringing attention to issues which are all part of awareness raising and creating sustainability

(Minkler et al., 2002; Sommer, 1969; Tang and Pickard, 2008; Torjman, 2007). Evidence of how this sense of community is defined will be documented at RCR.

ABOUT REENA

Established in 1973 to serve individuals with intellectual/developmental disabilities, Reena has pioneered several innovative supportive housing models including: group homes, triplexes, condo clusters, clustered apartment models, residential/respite models and family partnership models. In 2012, Reena built the new RCR, an 80,000 square foot, four-storey apartment building with:

- Sixty apartments on upper three floors for up to 84 tenants;
- One-, two- and three-bedroom units, some of which are accessible;
- Two Day Programs on the main floor for young people with intellectual disabilities living elsewhere in the region; and
- Holocaust Remembrance Garden and Education Centre honouring the over 200,000 people with disabilities who were targeted and killed by the Nazi Regime.

RCR is an innovative housing model that provides integrated support services to a mixed community of adults with physical, developmental and mental health needs. Inclusion is one of Reena's core values and in practice this means that tenants with varied abilities come together, interact with their neighbours and find meaning in their lives that they may not have had in previous settings. Beyond providing a home and supports, life at RCR builds community where people can live, learn, play and work.

Reena opened the first elderhome in North America, in 2000, and two others have since opened. This nonprofit agency is supporting over 1,500 participants in day programs, respite, outreach and residential services to children, adults and seniors.

Reena's statement of values is CLEAR:

C for Care – compassion and justice;
L for Leadership – to repair the world;
E for Empowerment – value others as you value yourself;
A for Accessibility – remove barriers;
R for Respect – honour our humanity.

The person-environment fit theoretical framework (Lawton, 1977) has been adopted to look at aspects of successful aging in an intentional community and ascertain what aging in place looks like in the RCR. Successful aging is a socially constructed concept (Baltes and Baltes, 1990) that has proven difficult to define. A number of studies have examined genetic, lifestyle and social determinants. These determinants have coincided with fundamental aspects of aging and recent clinical trials suggest that caloric restriction, physical activity, cognitive intervention, stress reduction and social programs may enhance cognitive and emotional health in people as they age (Depp et al., 2010; Bedney et al., 2010; Bigonnesse et al., 2014; Bookman, 2008; Lien, 2013; Hirdes, 2007).

Over the last two generations persons with intellectual/developmental disabilities have a longer life expectancy but frequently age prematurely so that services and supports may be required as early as age 40–50 with similar needs to a much older person (Shaw et al., 2011; Courtenay et al., 2010; Statistics Canada, 2011). This is a consideration when Reena is concerned with tenants who were not expected to live long lives and have survived to age in place in this new setting.

PLANNING WITH A COLLABORATIVE APPROACH

Reena spent seven years planning collaboratively with partner agencies to construct this innovative housing model. This reflects one of the basic principles of a deliberate community development approach with extensive planning and discussion of the project and prospective tenants. A total of six nonprofit corporations have participated in this collaboration and provide evidence of capacity building, efforts at engaging the larger community and recognizing the social capital (Putnam, 2000; Gilmor, 2012; Simpson, 2005) this generates. The new building opened in September 2012 and has now been fully occupied for the past six years.

With this unique opportunity in mind, a research project was designed to tap into the lived-experiences of the individual tenants as they transitioned from various community settings to the new RCR. The benefits of this study will demonstrate how individuals adapt to the new housing and make use of the social and personal space, and, in addition, how new models of support service provided by partnering organizations impact groups with diverse needs and demonstrate the basic principles of community development at a very practical level.

This opportunity offered a chance for agencies to work together, maximize capacity and share expertise so that tenants are supported by a seamless delivery of service (Walker, 2011; Williams et al., 2009; Young, 1998). An environmental scan and interviews with key informants were included in this collaborative approach to inform the development of the model. Program and steering committees were established with agency staff and community members to review the progress of this initiative. A protocol manual was developed to address financial matters, coordinated intake and applications from prospective tenants. A care-planning team was struck to fine-tune the integrated service system and adaptations were ongoing as needed. Using the person-environment fit theoretical model was another consideration when selecting prospective tenants from a long waiting list.

Administrative integration is another important feature of this housing model that includes nonprofit partners: Circle of Care, St Elizabeth Health Care and the March of Dimes Canada who are actively engaged in providing ongoing and integrated support and gathering evidence-based learning about building intentional communities.

RESEARCH PLAN

To explore the ecological and sociological perspective of social participation levels in this case study research methods included tenant interviews and participant observations and both qualitative and quantitative measures to gather data and monitor change over time.

A major focus examines whether tenants embrace a common social vision that places importance on living and sharing life together and whether these activities impact their perceptions of their own health, levels of social participation and whether a new sense of belonging is created.

The RCR is adjacent to the new Schwartz/Reisman Center and tenants have access to a variety of recreational, cultural and educational programs, a wellness center and Mount Sinai's ambulatory health care facility. Many tenants have a family doctor for the first time in their lives. This campus location has easy access to transportation/parking (for visitors, tenants and service staff) and provides numerous opportunities for tenants to enjoy the amenities on campus, while inviting the entire community to use the many resources of the RCR. These include a multi-purpose room, large computer room, meeting rooms, Holocaust Remembrance Garden, education center, life-skills suite and greenhouse.

THE RESEARCH PROJECT

When a new building opens a unique opportunity presents itself to conduct a case study designed to tap into the lived-experiences of the individual tenants as they transitioned from various community settings to the new RCR. The benefits of this study will demonstrate how individuals adapt to the new housing and make use of the social and personal space.

Scope and Objectives of the Reena Community Residence Research Project

The most important indicator of success for this research project is to examine the lived-experience of the individuals who relocate to and adapt to their new home as part of an intentionally built community. This longitudinal study (2012 to 2017) observed and captured both qualitative and quantitative data before and after the tenants moved and monitored change over time.

It will ascertain what aging in place looks like in the RCR. How research strategies interact with health aspects and how practitioners make use of and evaluate the effects of this person-environment dynamic in housing projects will be reviewed (Iwarsson, 2012). Successful aging is a socially constructed concept (Baltes and Baltes, 1990) that will be examined by asking the tenants about their activities of daily living.

The literature states that there are distinct differences between men and women with intellectual disabilities with regard to rates of disability, with women requiring more services and the older age groups, referred to as more frail, needing more help to maintain independence (Shaw et al., 2011). Both men and women have consented to participate, and any differences will be recorded.

Methodology

This research project offers a rare opportunity to study a group of up to 84 individuals ranging in age from 18 to 82 years, who have transitioned from apartments, group homes, hospital settings or family homes to an "intentional community" with integrated services. A description of the methodology follows:

- Individual interviews with tenants of the new building and participant observation by trained volunteer researchers are methods adopted for this case study. Qualitative interviews will explore the social, psychological, cognitive and physical responses to the transitional change in residential status of the individuals.
- Social participation will be measured by examining emerging levels of activity within the building and outside including on the grounds where there is a lot of opportunity for activity, gatherings and sports, recreation and leisure, volunteering and supported paid employment.
- These new tenants will be interviewed individually before they move, and then at routine times over a five-year period after they move in. By asking the tenants about their subjective perceptions of their experiences, an informed understanding of the meaning of "intentional community" will emerge over time along with a snapshot of the person-environment fit and how space is being used post occupancy.
- This approach to conducting an in-depth, primarily qualitative case study along with site observations will yield a tremendous amount of information about life in this new building.
- The confidentiality and privacy of individuals participating in these interviews has been protected and appropriate consent protocols have been developed and implemented.

Expected Project Outcomes

There is potential for an "intentional community" to be created that is welcoming, inclusive and promotes a healthy aging process and a chance to explore self-care through social and recreational pursuits to improve perceived health. Additionally, learning on many levels about what is expected of neighbours and making new friends and what it means to belong to a community will raise awareness of issues in the new building. Collaborative community partnerships may show improvements to the service system and provide a cost-effective alternative to long-term or institutional type of care.

An examination of the integrated service model delivered by the four nonprofit agencies on-site will be conducted in relation to the characteristics of tenant subpopulations best served under the model and its potential impact for clients and the health system.

Tenant Interviews

Every participant was interviewed by a trained research team member at least eight times over the course of the five-year project to determine how the use of social and personal space intersects with activities of daily living and social participation levels.

The interview used a semi-structured guide with five questions that cover activities of daily living and feelings about the living situations (before and after move). One final open-ended question allows the participant to include any other thoughts and perceptions about the lived-experience and services offered at the RCR.

Quantitative and qualitative methods were adopted in this longitudinal five-year case study. Interview questions about the person-environment fit were administered eight times: pre and post move, at year one to year five in September 2017 with 65 tenants. Initial intake of tenants was a positive outcome of joint planning activities and provided easier access for all Activities of Daily Living services including such in-home support services as: cleaning and laundry, shopping and meal planning, personal care, caregiver relief and respite. These services are critical to the success of this intentional community.

The final sample size was 65 tenants who agreed to participate and are described as follows:

- The age range of the participants is 18 to 82 years.
- There were 23 females and 42 males. It is interesting to note that there are twice as many males as females in this study.
- There have been 12 tenants who have moved out providing evidence of an attrition rate of negative person-environment fit; however, no follow-up provision to interview those former tenants has been made for those who no longer live in the building.
- One study participant died before the scheduled move and the first participant died three years after the move with two other participants dying in the final months of 2017 for a total of four deaths of participants over the five-year period.

All data has been collected and coded as individual responses to the six questions, then aggregated and coded according to the six questions. Emerging themes were recorded and quantified from the aggregated data analysis and actual qualitative quotations have been compiled to display the type and unique characteristics of expressions used by respondents. This opportunity offered a chance for agencies to work together, maximize capacity and share expertise so that tenants are supported by a seamless delivery of service. The service partners have been fully engaged in providing ongoing and integrated support and gathering evidence-based learning about building intentional communities.

Observations

As the research teams enter the residence, a descriptive log was kept for each visit along with general comments about the residential spaces, both personal and social, to capture the effects of housing design in a more humanistic manner and comment on the person-environment fit.

All participants were interviewed in their home setting prior to their move to the new building to establish a comparator and baseline for their self-perceived health status and identify individual service requirements. There have been 12 tenants who have moved out early in the five-year study period indicating a mismatch of person-environment fit. During the research time period four study participants died, all of natural causes. Wave 1 set of interviews was completed for all participants post move in 2012, Waves 2 to 5 sets of interviews were completed in the summer/fall of 2013 to 2017.

Interview results

Results of interviews are summarized in the following six themes: tenant experiences, the building, perceived health improvements, support services, sports and recreation and pets. The pace of move-ins was deliberately slow allowing for ample preparation time to accommodate individual needs and unanticipated health care delays and months of preparation in some cases. The transition plans have been comprehensive and implemented very smoothly with the involvement of family and caregivers.

Comments about the building included the fact that no balconies were part of the design and would have added an extra outdoor space to each apartment. Some of the neighbours are noisy and frequently pets also make noise on some floors. Issues related to garbage accumulation and smoking have arisen similar to other types of apartment buildings and staff are working on helping tenants to resolve these issues. Consistent with earlier reported findings this study has documented a successful transition for most of the individual tenants and an increase in feelings of overall health and well-being as reported by the participants.

What the tenants are saying
The overwhelming majority can be described as "ecstatic" with their new apartments and the experience of buying items such as toasters, big screen televisions and microwaves is very exciting if you have never owned such items before. There is a special unit set aside for groups of tenants to learn how to use these new items, learn how to cook and also how to take care of their apartment. For other tenants they say it is the freedom to come and go as they please. They are independent and learning many new things about their surrounding neighbourhood, about grocery shopping, banking, cooking, keeping their units clean, doing their own laundry, gardening, and living on their own.

> I check my blood sugar and it's done amazing, I take meds 3x day, make my bed and learned how to take care of my apartment.

> I love it here and want to stay forever and ever . . . it's nice . . . very quiet and peaceful.

> I really like my new room and I am very happy . . . the building is beautiful.

> I take care of my health now, like joining the aquafit and yoga classes, feel as healthy as a horse.

One young man said that he "hated it in the group home" and is trying very hard to keep up with cleaning his apartment.

> When I moved in I was so tired, found it very stressful and I slept a lot at the beginning, feeling better now.

Tenants are learning what it is like to live with other people, learning how to cook simple meals and share common space and also to share service providers. Some have pets to care for and at last count there are 11 cats and dogs living in this intentional community offering a new chance to share pet care, which is a new learning experience and responsibility for many.

> I have never lived by myself before and I am working on socializing at evening programs like karaoke.

Four tenants say they "hate" living in York Region and want to move back to Toronto to be closer to the "action in the city," only five tenants have cars and double bus fares for crossing jurisdictional boundaries of York Region and City of Toronto transit are proving to be too costly.

Some concerns have been raised regarding safety and security in the building and tenants have requested a Tenant Council. Smoking and fire drills are other issues that have been identified as a worry to tenants. Workshops with emergency responders: police and fire and paramedics have been arranged to inform tenants of their responsibilities such as when it is appropriate to call 911 and what constitutes a real emergency.

Initial expressions of feelings of increased independence, freedom, safety and feelings of inclusiveness are themes that have emerged from this research design. Self-directed care has taken on a new meaning for tenants dealing with service providers on a daily basis. Researchers' observations reflect a fantastic and supportive environment to relieve pre-move levels of worry and anxiety about what life would be like at the RCR.

Unexpected outcome – Tenant Council
As the residence opened in 2012, the staff and tenants started a journey that has clearly defined its open and participatory nature. Ballot elections for the Tenant's Council were held and the first four-person Council was formed in December 2013, only one year after full occupancy of the building. Together with staff, Council members defined their roles and began a process to articulate their responsibilities, make decisions and provide action for the challenges that came to them. This represented a significant learning curve for the tenants and the acquisition of new skills to understand the role of a council and how to serve on behalf of all RCR tenants. In addition, a handbook was developed that provides some background and further context to those creative processes while also articulating principles, roles and responsibilities for those elected to the Tenant's Council. The handbook will no doubt further develop as this intentional supportive community evolves. In keeping with building an intentional community a clear statement of Tenant Council Principles was adopted as follows:

- Cooperate and communicate with each other
- Treat each other fairly, with dignity and respect
- Promote harmony and teamwork in all relationships
- Strive to understand each other's perspectives
- Encourage and consider opinions of other members and invite their participation in decisions
- Recognize that Council roles are different, but each is equal
- Understand that to keep Council membership you are required to be living as a good neighbour
- Recognize that fellow tenants in their personal lives may experience crisis – show compassion and understanding
- Understand that as a Council member you are a role model
- Understand that some issues you deal with on Council may be confidential
- Roles are defined as Chairperson (1) the meeting organizer; Note Taker (1) minute taker/to do list; Event Planner (1) community builder; Treasurer (1) budget/money

The Tenant Council exists to give a voice to tenant and resident concerns on the issues of the building. Tenant Council members have described themselves as advocates for the tenant population as a whole. As one Tenant Council member so rightly said:

we are able to do things here that we haven't been able to do other places we have lived . . . the Tenant Council is one of those things that we have a say in . . .

The primary roles of a Council member are to bring issues of health and safety forward immediately, to work with staff to face building challenges as they arise and to plan for activities and events as a community. The three roles identified for the new Tenant Council at RCR are as follows:

1. Advocates – pathway to staff.
2. Health and Safety monitors.
3. Event/Community planning.

Originally, a consultant was hired to assist the Tenant Council to develop this handbook and a simple problem-solving tree to explore issues as they arrive is another innovation for this group of tenants learning how to solve problems. This is an important and significant outcome illustrating the creation of social capital through participating and defining a community much earlier than expected. Person space has been adjusted to meet individual needs and provision for use of social spaces, such as the library, program rooms and computer room, is encouraging tenants to move about the building. These post-move findings are consistent with the literature that reports longer life expectancies when accompanied by a strong supportive environment such as multiple ages living together in an "intentional community."

FINDINGS

Findings from Year 5, the final phase of this Reena research project, report on the lived-experience of a diverse population with special needs as they transition from various residential settings and connect to a new "Intentional Community with Supports." Tenant responses are defining person-environment fit and emerging levels of social participation as life changing events. Many have reported improved self-perceived health, made new friends and arranged to attend social events together.

In the fall of 2017, after living about five years in the new building, results of interviews are summarized in six themes that documented the tenant experience, comments about the building and recreational opportunities. The transition plans have been comprehensive and implemented very smoothly with the involvement of family and caregivers. The research interview used a semi-structured guide with questions that cover activities of daily living and feelings about the living situations (before and after move). The interviews were administered eight times: pre and post move, at years one through five with 65 tenants. The age range of the participants is 18 to 82 years and reflects a diverse multigenerational population. Aside from positive assessments of their apartment and the RCR building, the local and on-site access to health, support and recreation services and the possibility of having pets scored also very highly. This unique model of support is one major outcome of this community development effort which has already had a huge impact in building social capital and reducing local and regional service costs and demand. The RCR tailor-made sense of community is very powerful and significant skills have been acquired by the tenants in learning about organizing, lobbying, advocacy and bringing attention to issues which are all part of awareness raising and creating sustainability (Minkler et al., 2002).

Residential Space

A beautiful modern building with wide hallways and an abundance of natural light and amenity spaces are features of the RCR. Since September 2012, all units have been occupied. Great care has been exercised in selecting and supporting each and every new tenant to ensure that their settlement process is of exceptional sensitivity and quality.

Person-Environment Fit

The importance of good design and functionality cannot be emphasized enough as it defines personal space, common spaces, flexibility and accessibility. One observation at RCR in the early move-in stage was how quickly the tenants rallied to defend a pet hamster that had been taken outside and was being physically abused by another tenant. That tenant has since moved out of the building partially due to peer pressure regarding his disruptive behaviours which did not fit the new environment and emerging definition of what boundaries are envisioned by the tenants and the acquired power of collective actions from living among a group of tenants in an intentional community.

Defensible space is a term coined by Oscar Newman in the 1970s and describes a model for residential environments which inhibits crime by creating the physical expression of a social fabric that defends itself. Early findings provide some indication that person-environment fit in the RCR is finding a balance for tenants between the creation of new rules and self-regulation. An interesting phenomenon observed during the settling in period was staff revisions to the "open" concept through creation of a "fortress mentality" where the program rooms are locked at 4:30 pm, denying access to tenants after hours and in the evening. A curfew was declared that doors were to be locked by 11:00 pm thereby planting the seeds in minds for behavioral control although the doors were not in fact locked. Tenants are beginning to voice concerns about limited access to social spaces such as the computer room after hours without staff supervision and these issues are to be raised at the new Tenants' Council, which indicates growing levels of confidence and social capital.

An important lesson learned for planners and architects is to evaluate the use of space after building occupancy and take a second look at how the space is being used and what is working or not. The author, as practitioner, participated in an evaluation of a home for the aged with the architect Jerome Markson, who was awarded the Ontario Association of Architects Design Award for "sensitively planning and detailing the place to a gentle scale. This home was described as a sweet haven for the elderly" (Markson, 2014). Another architect practitioner alludes to housing form and housing design using a "more humanistic manner" (Simon, 1993).

An intentional community such as RCR presents a real chance to evaluate how this modern space is being used. Once the environment is in place and tenants have settled in, it is still the person that determines the fit in the new space. The reality is that difficult behavioural issues at RCR persist that disrupt and disturb other tenants. Creative ways to deal with these issues need to be explored on an ongoing basis to allow all tenants to take steps toward total independence, developing a sense of belonging to a community and variations of the use of residential spaces.

Strengths and Weaknesses

For the tenants at the RCR, one clear strength is their living environment which embodies inclusion, one of Reena's core values and practices that has now been embraced by the community partners and is reflected in the intentional community. Tenants with varied abilities come together to live a meaningful life, interacting with their neighbours. Some other strengths and weaknesses are as follows:

- Weakness: to accommodate special needs, significant funds not included in the original plan were invested in the development of three units for those who required 24/7 congregate care. One unanticipated budgetary consequence of embracing this subpopulation, after the building was opened, was the additional investment required to create accessible roll in showers, install ceiling track lifts, apply wall protection and other equipment which was time-consuming and delayed tenant moves.
- Weakness: expertise available from community partners that are "age-friendly" has not yet been fully transmitted to Reena, possibly due to unfamiliarity with the language and approach to responding to needs of older tenants. The concept of service integration to address aging in place must be sharpened through a thorough and deep screening of prospective tenants based on trust with referral agencies and ongoing discussions for selection for the right person-environment fit.
- Strength: self-directed care has taken on a new meaning for these tenants as they gain confidence in deciding what care and when it is to be offered and deal with these service providers on a daily basis. It is seen as a strength that fosters independence. Beyond providing a home and supports, this innovative housing model builds a community where people can live, learn, play and work.
- Strength: community partnerships draw on significant "pooled" resources, enhance service capacity, social capital and allow for extensive staff training. A higher degree of staff teamwork is evident as many tenants were known to several agencies from the past and now the services are integrated on-site. This administrative streamlining of services is a significant outcome observed by the researchers.

Great care has been exercised in selecting and supporting each and every new tenant to ensure that their settlement process is of exceptional sensitivity and quality. Extensive staff training has been put in place allowing all partners to learn about their respective approaches to care. What has emerged is a new definition of what an "intentional community" starts to look like at the RCR for the tenants and visitors, partners, researchers, staff and family members, friends, caregivers and volunteers are all part of that defining process. Almost everyone observed makes comments freely on this phenomenon and describes it in ways that have meaning for them. A sense of belonging to a very special community at RCR has been captured and documented. It is fascinating to try to capture these observations as they evolve.

RCR is still in the early stages of establishing an intentional community. Patience is a true virtue as each individual adapts to the new setting and the personal and social spaces become occupied. It has been a learning experience for everyone involved and for the most part it has certainly been a positive one and we will see what the next years will bring as this intentional community matures.

DISCUSSION

This supportive housing model creates a residence for an intentional community with built-in supports in a mixed community of adults with developmental, physical, cognitive or mental health needs. This client population ages at a faster rate than the general population and requires unique services and supports. Two key components of the innovative housing model at RCR are the provision of highly integrated services with extraordinary teamwork and the sense of community being constructed through social capital and social participation of the tenants. As the tenants are encouraged to hold a common vision and shared values and norms, they have focused on the importance of living, learning and sharing life together.

Thus, social engagement in the community is being nurtured through a conscious construction of community (Putnam, 2000; Manzo and Perkins, 2006). Through the cultivation of community, good neighbours and reciprocity are being fostered. This is a manifestation of social capital as described by Putnam (2000).

Using the lens of person-environment fit (Lawton, 1977) to examine tenant adaptation during this relocation indicates that these adults are thriving with the independence and sense of agency that this model of supportive housing has facilitated. The increase in social participation that this model affords provides the opportunity for personal growth and has reduced social isolation and dependence that these tenants experienced before their move. Benefits and implications for this intervention related to successful aging are evident in this housing model as one barrier to access to service has been removed. Access to information about available services is provided on-site thus providing a "bridging" to social capital and aging in place (Putnam, 2000). Active engagement in social activity is a key component of successful aging theory (Baltes and Baltes, 1990). Social participation through volunteerism, social interaction and neighbourly reciprocity enables tenants to live an enhanced and more meaningful life.

LIMITATIONS

This is a unique setting specifically built to accommodate an "intentional community" of tenants with a variety of special needs. The case study findings are not readily replicated and the sample size is small; however, the documentation of the age-related changes and transitions experienced by these tenants is noteworthy as the environment serves as an incubator of social participation that warrants ongoing learning applicable to other research efforts in person-environment and aging in place. The results are evidence of active community development.

SUMMARY AND CONCLUSIONS

Administrative integration is another significant service feature on-site at RCR including nonprofit partners: Circle of Care, St Elizabeth Health Care and the March of Dimes Canada. Tenant responses are defining person-environment fit at RCR as there is an obvious intersection between social and personal space most notably observed in the reception/lobby area where a high level of group activity and social interaction takes place on a daily basis.

This opportunity at RCR offered a chance for agencies to work together, maximize capacity and share expertise so that tenants are supported by a seamless delivery of service. Ongoing training and support for staff is essential as this on-site work is very demanding. Partners will come and go as each organization tests out appropriate services and costs for their clients.

Both internal and external communication is extremely important with so many stakeholders involved in this complex project with integrated support services. A newsletter/website for the building would be a simple way to keep tenants, staff, volunteers, friends, staff and family members informed about what is going on in the intentional community. In 2018, a first newsletter, another community milestone, was published for RCR.

Six themes from the self-reported tenants' aggregated data have been described as follows: tenant experiences, the building, perceived health improvements, support services, sports and recreation and pets. Initial expression of feelings of increased independence, safety and feelings of inclusiveness are themes that reinforce the notion of person-environment for this group of tenants. The findings also demonstrate:

- Social participation levels are beyond expectations: tenants are embracing newfound independence – there are new jobs paid and unpaid – and innovative approaches reflected in the creation of a volunteer receptionist position as a training opportunity for future employment of tenants. There is an obvious intersection between social and personal space most notably in the reception/lobby area as well as in the library and program rooms which are used in the evening for movies and Karaoke and other social events.
- RCR offers the opportunity to foster, build and maintain relationships as there is evidence of tenants making new friends, socializing and making plans to spend time together both inside and outside the RCR. Tenants are developing social capital. The definition of sense of community at RCR is very powerful in this setting with the active participation of the tenants in building their own support networks and improving their own health.
- Self-perceived health has been reported as improved with increased levels of fitness and recreation with visits to the Schwartz/Reisman centre and medical clinic located next door.
- Important components have been identified for consideration in future "intentional community" and supported housing design.

This approach to conducting an in-depth, primarily qualitative case study is intended to document the potential for an "intentional community" to be created that is a welcoming and learning environment, inclusive, supporting personal independence and growth, reciprocity and engagement in social life, and promotes a holistic understanding of a healthy aging process.

RECOMMENDATIONS

Based on five years of observation by the research team of the RCR several ways to enhance the quality of life and social participation levels in this environment going forward are recommended as follows:

1. Reena needs to develop a plan to become a more "age-friendly" community as the organization must become familiarized with basic principles of gerontological practice distinct from developmental service approach including the importance of person-environment fit and aging in place.
2. Cross training for health and social service support needs is necessary, on-site, to support total independence of tenant decision making for self-directed care.
3. Policies and procedures need to be developed related to screening for new tenants and to identify those interested in possible involvement on the Tenant Council and as editors of the RCR newsletter and other on-site volunteer opportunities.
4. Great strides have been made to establish a Tenant Council, which must have more representatives of all subgroups in the building including youth, seniors and those with medically complex needs, and offer problem-solving training programs for all participants.

Overall, this case study captures qualitative and quantitative information about the role that person-environment fit plays in healthy aging. Never before has tenant life experience been documented in this way as the survival rate and longevity of this group of tenants is unprecedented in the history of Canada. It is our privilege to observe the evolution of this fascinating intentional community that demonstrates the basic principles of community development in living cooperatively, solving problems in non-violent ways such as the creation of a Tenant Council and sharing experiences with other members of the RCR and in the neighbourhoods beyond this geographical place. A strong emerging sense of community is undeniable evidence of how to build a successful "intentional community."

This case study confirms that life at RCR does in fact embody inclusion, a sense of belonging to a unique community, healthy aging and social participation as evidenced by tenants with varied abilities interacting with their neighbours and reporting improved quality of life.

NOTE

* I would like to acknowledge research team members: Susan Roher, Chair Research Advisory Committee, Reena; Dr Suzanne Cook, PhD, York University and BA student, Dorothy During, York University; Mindy Ginsler, University of Toronto; MSW students, Amanda Neves and Christina Lanteigne, MSW, University of Toronto; and Annelise Callisto, Vengayi Kanyere, Idah White, BASc, University of Guelph Humber and Erin Hill, Humber College and the ongoing support of March of Dimes Canada, St Elizabeth Health Care and Reena.

I would also like to acknowledge a total of $50,000 of research funds received from the University of Guelph, Humber College and New Horizons Canada for this project.

This chapter is dedicated to the memory of Leon Pastalan, PhD environmental gerontology (1930–2018), founder and first editor of the *Journal of Housing for the Elderly* and a mentor of this author at the graduate school, University of Michigan.

REFERENCES

Baltes, P.B., and Baltes, M.M. (1990). Psychological perspectives on successful aging: The model of selective optimization with compensation. In P.B. Baltes and M.M. Baltes (eds), *Successful Aging: Perspectives from the Behavioural Sciences* (pp. 1–34). New York: Cambridge University Press.

Bedney, B.J., Goldberg, R.B., and Josephson, K. (2010). Aging in place in naturally occurring retirement communities: Transforming aging through supportive service programs. *Journal of Housing for the Elderly*, 24 (3/4), DOI: 10.1080/02763893.2010.522455.

Bigonnesse, C., Beaulieu, M., and Garon, S. (2014). Meaning of home in later life as a concept to understand older adults' housing needs: Results from the 7 age-friendly cities pilot project in Quebec. *Journal of Housing for the Elderly*, 28(4), DOI: 10.1080/02763893.2014.930367.

Bookman, A. (2008). Innovative models of aging in place: Transforming our communities for an aging population. *Community, Work & Family*, 11(4), DOI: 10.1080/13668800802362334.

Brown, J., and Hannis, D. (2012). *Community Development in Canada*. Toronto: Pearson Canada, Inc.

Christenson, J., Fendley, K., and Robinson, J. (1994). Community development. In J. Christenson and J. Robinson (eds), *Community Development in Perspective* (pp. 9–12). Ames, IA: Iowa State University.

Courtenay, K., Jokinen, N., and Strydom, A. (2010). Caregiving and adults with intellectual disabilities affected by dementia. *Journal of Policy and Practice in Intellectual Disabilities*, 7(1), 26–33. 3

Depp, C., Vahia, I., and Jeste, D. (2010). Successful aging: Focus on cognitive and emotional health. *Annual Review of Clinical Psychology*, (6), 527–50.

Drier, P. (2001). *Place Matters: Metropolitics for the 21st Century*. Lawrence: University Press of Kansas.

Forchuk, C., Nelson, G., and Hall, G. (2006). It's important to be proud of the place you live in: Housing problems and preferences of psychiatric survivors. *Perspectives in Psychiatric Care*, 42(1), 42–52.

Gilmor, H. (2012). Social participation and the health and well-being of Canadian seniors. *Health Reports*, 23(4), Catalogue No. 82-003-x, Ottawa, Canada: Statistics Canada. Retrieved 11 October 2018 from: http://www.statcan.gc.ca/pub/82-003-x/2012004/article/11720-eng.htm.

Hirdes, J. (2007). Aging at home: Balancing evidence (interRAI) with needs and resources home care research and knowledge, Dept of Health Studies and Gerontology, University of Waterloo.

Iwarsson, S. (2012). Implementation of research-based strategies to foster person-environment fit in housing environments: Challenges and experiences during 20 years. *Journal of Housing for the Elderly*, 26(1), DOI: 10.1080/02763893.2012.651378.

Lawton, M.P. (1977). The impact of the environment on ageing and behavior. In J.E. Birren and K.W. Shaie (eds), *Handbook of the Psychology of Ageing* (pp. 276–301). New York: Van Nostrand Reinhold.

Lien, L. (2013). Person-environment fit and adaptation: Exploring the interaction between person and environment in older age. Doctoral thesis, Design and Human Environment, Department of Health, Oregon State University.

Manzo, L.C., and Perkins, D.D. (2006). Finding common ground: The importance of place attachment to community participation and planning. *Journal of Planning Literature*, 20, 335–50.

Markson, J. (2014). Tribute to six decades of design. *Modern Canadian Architecture*. Retrieved 11 October 2018 from: www.jeromemarksonarchitect.com.

Minkler, M., Wallerstein, N., and Hall, B. (2002). *Community-Based Participatory Research for Health*. New York: Thousand Oaks, Sage Publications.

Newman, O. (1973). *Defensible Space: Crime Prevention through Urban Design*. Toronto: Macmillan Publishing Company Canada Ltd.

Pastalan, L.A. (1930–2018). Obituary by Benyamin Schwarz, 19 November 2018. Retrieved 19 November 2018 from: https://doi.org/10.1080/02763893.2018.1505457.

Phillips, S.D., and Graham, K.A. (2000). Hand-in-hand: When accountability meets collaboration in the voluntary sector. In K.G. Banting (ed.), *The Nonprofit Sector in Canada: Roles and Relationships* (pp. 22–8). Montréal, Québec, Canada: McGill-Queen's University Press.

Phillips, S., Graham, K., and Ker, A. (2003). The new trilateralism: Experiments in federal–municipal–community relationships. Paper presented to the Association for Research on Nonprofit Organizations and Voluntary Action (ARNOVA), Denver, CO.

Putnam, R.D. (1993). The prosperous community: Social capital and public life. *The American Prospect Inc.* 13 (Spring). Retrieved 11 October 2018 from: epn.org/prospect/13/13putn.html.

Putnam, R.D. (2000). *Bowling Alone.* New York: Simon and Schuster.

Shaw, K., Cartwright, C., and Craig, J. (2011). The housing and support needs of people with an intellectual disability into older age. *Journal of Intellectual Disability Research*, 55(9), DOI: 10.1111/j.1365-2788.2011.01449.x.

Simon, J. (1993). Housing form and use of domestic space. In J.R. Mirron (ed.), *House, Home and Community: Progress in Housing Canadians* (pp. 188–202). Montreal, Quebec, Canada: McGill Queen's University Press.

Simpson, L. (2005). Community informatics and sustainability: Why social capital matters. *Journal of Community Informatics*, 1(2), 102–19.

Sommer, R. (1969). *Personal Space: The Behavioral Basis of Design.* Englewood Cliffs, New Jersey: Prentice-Hall, Inc.

Statistics Canada (2011). The 2011 Census. Government of Canada. Retrieved 11 October 2018 from: www.gc.ca.

Tang, F., and Pickard, J.G. (2008). Aging in place or relocation: Perceived awareness of community-based long-term care and services. *Journal of Housing for the Elderly*, 22(4), DOI: 10.1080/02763890802458429.

Torjman, S. (2007). Shared space: The communities agenda. Report, Caledon Institute of Social Policy. Ottawa, Ontario, Canada.

Walker, D. (2011). *Caring for Our Aging Population and Addressing Alternate Levels of Care.* Report to the MOHLTC. Toronto, Ontario, Canada.

Williams, P., Peckham, A., and Rudoler, D. (2009). Mapping the state of the art: Integrating care for vulnerable older populations. Canadian Research Network for Care in the Community Report. Retrieved 11 October 2018 from: http://www.CRNCC.ca.

Young, H. (1998). Moving to congregate housing: The last chosen home. *Journal of Aging Studies*, 12(2), 149–65.

11. Inclusionary zoning and inclusionary housing in the United States: measuring inputs and outcomes*
Katrin B. Anacker

INTRODUCTION

Housing is a critical part of community development. Over the past few decades, both housing affordability,[1] with its focus on households, and affordable housing, with its focus on the housing stock, have gradually declined for many low-, very low- and extremely low-income people in many communities in the US for multiple reasons (Anacker, 2019). At the household level, average household incomes have nominally declined or stagnated until just recently (US Bureau of the Census, n.d.b). At the local level, many plans have focused on job growth without taking the workforce and multifamily rental housing into account (Myerson, 2016; National Association of Home Builders, 2015). This chapter provides insights into research on inclusionary zoning and inclusionary housing.

Housing affordability and affordable housing challenges may be caused by many factors, including housing market forces and exclusionary zoning (Babcock and Siemon, 1985; Davidoff and Davidoff, 1971; Dietderich, 1996; Downs, 1973, Ellickson, 1981; Fischel, 1985; Tiebout 1956). These housing affordability and affordable housing challenges may lead to a housing and jobs imbalance that could have several consequences. First, the imbalance may lead to a concentration of low-income households in low-opportunity neighborhoods, contributing to racial and ethnic residential segregation (Anderson, 2003; Brown, 2001; Jacobus, 2015). Second, it may lead some workers who are unwilling or unable to live in unaffordable and/or low-opportunity neighborhoods to move to distant neighborhoods (Anderson, 2003; Brown, 2001; Jacobus 2015). Third, it contributes to traffic congestion, air pollution and high expenditures on transportation by workers who have to commute to distant jobs as well as all taxpayers who have to pay for the infrastructure (Anderson, 2003; Brown, 2001; Jacobus, 2015). Fourth, it limits the pool of essential workers living in the neighborhood, including first responders such as paramedics, emergency medical technicians, police officers, firefighters and rescuers, as well as other trained and certified professionals, such as public school teachers (Anderson, 2003; Fox and Rose, 2003; Myerson, 2016; Regional Inclusionary Housing Initiative, n.d.). Because of all the consequences discussed above, a housing/jobs imbalance may cause residential segregation, sprawl, commuting, air pollution, taxpayer expenditures and local worker shortages.

Over the past decades, states have pursued four types of local, state and regional housing planning requirements. First, inclusionary zoning and inclusionary housing (IZ/IH), practiced, for example, in Maryland and Virginia, is regulated by local governments whose respective states grant them the power to adopt IZ/IH policies that require developers to include a certain number or proportion of affordable housing units in new residential developments with a certain number of units (Cowan, 2006). Second, affirmative measures, practiced, for example, in Florida and Oregon, require local governments to plan for affordable housing as part of a state's comprehensive planning and growth management programs, although there is no monitoring or enforcement system that tracks whether affordable housing is actually built (Cowan, 2006; Smith et al., 1996). Third, fair share allocation systems, practiced, for example, in California and New Jersey, have the state or the regional authority determine the regional need for affordable housing and then allocate a fair share of affordable housing to municipalities (Cowan, 2006). Fourth, "anti-snob" land use laws, practiced, for example, in Massachusetts, Connecticut and Rhode Island, limit the ability of local governments to use their zoning power to exclude the construction of affordable housing, facilitating the work of affordable housing developers (Cowan, 2006).

In summary, IZ/IH and "anti-snob" land use laws are on the opposite ends of the spectrum when considering the four types of local, state and regional housing planning requirements. IZ/IH is an active approach, while "anti-snob" is a passive approach, as it does not actively initiate housing (Breagy, 1976). Whereas the former requires affordable housing in new developments over a certain threshold in those *municipalities* that want affordable housing, the latter facilitates affordable housing for those *developers* that want affordable housing, even if the municipality does not (Cowan, 2006).

While there is no federal database, California has a state-wide database of IZ/IH (Hickey et al., 2014).[2] Interestingly, New Jersey has a state-wide mandate for local data collection (Hickey et al., 2014), although there is no requirement for data analysis and program evaluation. The questions of whether many local databases are public or accessible to researchers and whether these databases are comparable remain. Given the lack of national data and the piecemeal nature of local data, this chapter provides a national literature review of general discussions and a brief overview of case studies of IZ/IH. Future research efforts should focus on building and analysing a national database and evaluating state and local IZ/IH programs.

PUBLIC POLICY

IZ/IH is a regional, state and local housing policy that has the goal of expanding the local or regional affordable housing supply to achieve long-term housing affordability and facilitate socioeconomic integration, which often results in mixed-income communities (Anderson, 2003; Brown, 2001; Burchell and Galley, 2000; Calavita and Grimes, 1998; Calavita et al., 1997; Ellickson, 1981; Fox and Rose, 2003; Hickey, 2013, 2014; Hickey et al., 2014; Madar, 2015; Madar and Willis, 2015; National Housing Conference, n.d.; Schuetz et al., 2008; Schwartz et al., 2012). IZ/IH units may be developed, purchased, rented, managed, renovated, rehabilitated or resold by municipalities, developers, local housing authorities, designated housing development agencies or nonprofit organizations (Anderson, 2003; Regional Inclusionary Housing Initiative, n.d.). IZ/IH typically applies to newly constructed developments but also to substantial rehabilitations or condominium conversions (Anderson, 2003; Schuetz et al., 2008). The minimum threshold for IZ/IH that necessitates including affordable units may range from 5 to 50 units or be a certain percentage (Anderson, 2003; Ellickson, 1981; Schwartz et al., 2012). Many municipalities require that IZ/IH units be similar in size and design to market-rate units, avoiding stigmatization and the fear of a negative impact on property values (Anderson,

2003; Regional Inclusionary Housing Initiative, n.d.; Schwartz et al., 2012). Also, IZ/IH units may be dispersed throughout jurisdictions (Schwartz et al., 2012). However, some IZ/IH units may be smaller, have lower-cost interior finishes and have fewer amenities compared to market-rate units in order to support housing affordability (Brown, 2001; Ellickson, 1981; Fox and Rose, 2003; Schwartz et al., 2012).

IZ/IH may be a voluntary local or regional planning tool under prescribed conditions, a voluntary tool through ad hoc negotiated agreements or a mandatory land use ordinance with or without incentives (Anderson, 2003; Burchell and Galley, 2000; Fox and Rose, 2003; Porter, 2004; Regional Inclusionary Housing Initiative, n.d.; Smith et al., 1996; Schuetz et al., 2009, 2011; Schwartz et al., 2012).[3]

IZ/IH may apply state-wide, jurisdiction-wide or to designated neighborhoods (Anderson, 2003; Hickey, 2014). Some jurisdictions have multiple inclusionary housing policies, such as one at a municipal level and a second at a district, corridor or neighborhood level (Madar, 2015). Whereas some IZ/IH policies may be stand-alone, others may supplement existing programs (Hickey, 2014). Some IZ/IH policies may broadly apply to any residential development, while others may only apply to certain types, based on type, size, tenure or structure (Schuetz et al., 2009, 2011).

IZ/IH may be directly funded by local and federal policies. At the local level, IZ/IH may be funded by Community Development Corporations (CDCs), housing trust funds, developer impact fees, real estate transaction fees, property tax levies, dedicated sales and use taxes, and tax increment financing (TIF),[4] as well as bond financing (Burchell and Galley, 2000; Calavita and Mallach, 2010; Fox and Rose, 2003; Freeman and Schuetz, 2017; Myerson, 2016; Schwartz, 2015; Smith et al., 1996). IZ/IH is typically indirectly funded by developers by cross-subsidizing affordable and market-rate units, accepting lower financial returns, spending less on projects, building fewer units or building somewhere else (Burchell and Galley, 2000; Ellickson, 1981; Madar, 2015; Madar and Willis, 2015; National Association of Home Builders, 2015; Schuetz et al., 2008, 2009, 2011; Schwartz et al., 2012; Wheeler, 1990). A developer may be more likely to participate if the direct and indirect subsidies are higher than the expenditures for constructing the additional affordable housing units (Madar, 2015; Madar and Willis, 2015). He or she may be more likely to increase the price of market-rate units if the relative elasticity of supply and demand is in his/her favor and if alternative land uses face similar burdens (Schuetz et al., 2009, 2011). Thus, IZ/IH may not necessarily lead to inclusion (Schwartz et al., 2012).

At the federal level, IZ/IH may be promoted by the following six programs:

1. The Low Income Housing Tax Credit (LIHTC) program, jointly administered by the US Department of Housing and Urban Development (HUD) and the US Department of the Treasury;
2. The Community Development Block Grants (CDBG) program, administered locally;
3. The HOME Investment Partnership program, administered locally;
4. The Federal Rehabilitation Tax Credit, also known as the Historic Tax Credit, administered by the National Park Service, the Internal Revenue Service and the respective state Historic Preservation Office;
5. The New Market Tax Credit (NMTC), administered by the Community Development Financial Institutions (CDFI) Fund, housed under the US Department of the Treasury; and

6. The (somewhat limited) Sustainable Communities Regional Planning Grant program, whose most recent award was made in FY 2011, housed under the Partnership for Sustainable Communities, an interagency partnership among HUD, the Department of Transportation (DOT) and the Environmental Protection Agency (EPA) (Calavita and Grimes, 1998; Freeman and Schuetz, 2017; Massey et al., 2013; Partnership for Sustainable Communities, n.d.; US Department of Housing and Urban Development, n.d.c; US Department of the Interior, n.d.; US Department of the Treasury, n.d.).

IZ/IH may also be linked to the Housing Choice Voucher Program of the US Department of Housing and Urban Development (Fox and Rose, 2003).

Since the early 1990s, IZ/IH has been adopted in 507 inclusionary housing programs in 27 states and the District of Columbia and 482 municipalities, resulting in more than 150,000 affordable housing units nationwide as of 2014, or about 0.1 percent of the existing housing stock in the US (Freeman and Schuetz, 2017; Hickey et al., 2014; Jacobus, 2015; National Housing Conference, n.d.; Schuetz et al., 2009, 2011; Schwartz et al., 2012).[5] However, IZ/IH programs are distributed unevenly across the US (Schuetz et al., 2008). Most IZ/IH programs are established in housing markets with high house prices and high house price growth, where affordable housing opportunities are scarce and difficult to develop and implement, or in places with strong local constituencies (Fox and Rose, 2003; Hickey, 2013, 2014; Schuetz et al., 2009, 2011). At the state level, of the 507 identified inclusionary housing programs, 36 percent were located in New Jersey and 29 percent in California (Calavita and Grimes, 1998; Hickey et al., 2014). For example, about 30,000 inclusionary housing units were produced between 1999 and 2006 in California (Hickey et al., 2014). At the local level, Montgomery County, Maryland has relatively high numbers of inclusionary housing units (Brown, 2001; Calavita and Mallach, 2010; Hickey et al., 2014; Montgomery County n.d.; Schwartz et al., 2012).

MEASURING INPUTS FOR INCLUSIONARY ZONING AND INCLUSIONARY HOUSING

There are many aspects of inputs for inclusionary zoning and inclusionary housing that may be measured in different ways. The sections below differentiate between developer requirements and incentives on the one hand and the period of affordability on the other.

Developer Requirements and Incentives

In the case of IZ/IH, local or state planning departments shift some or all of the direct costs of building and operating affordable inclusionary housing units from the taxpayer to the developer. The cost may be partially or fully absorbed by either the developer or passed on to the homebuyer (Burchell and Galley, 2000; Freeman and Schuetz, 2017; Hickey, 2014; Knaap et al., 2008; Madar and Willis, 2015; National Association of Home Builders, 2015; Schuetz et al., 2008, 2009, 2011). Some argue that IZ/IH programs increase the affordable housing supply without using taxes, others believe that mandatory IZ/IH programs are essentially a tax on new residential developments, and others state that IZ/IH increases housing affordability issues, as the costs are passed on to homebuyers (Burchell and Galley, 2000; Ellickson, 1981; Knaap et al., 2008; Schuetz et al., 2009, 2011). Local or state planning departments may implicitly subsidize developers by allowing them to build more units than allowed in a master plan, often resulting in higher density (Anderson, 2003; Brown, 2001; Calavita and Mallach, 2010; Hickey et al., 2014; Knaap et al., 2008; National Association of Home Builders, 2015; National Housing Conference, n.d.). In some areas and neighborhoods, this density bonus may allow a developer to produce more housing units at a relatively low cost per additionally allowed unit, possibly resulting in increased revenues and profits (Anderson, 2003; Calavita and Grimes, 1998; Ellickson, 1981; Fox and Rose, 2003; Hickey, 2013, 2014; Mukhija et al., 2010; Regional Inclusionary Housing Initiative, n.d.; Schuetz et al., 2008, 2009, 2011). Local or state planning departments work with developers of owner-occupied or rental, market-rate, multifamily housing to provide a minimum number or proportion of units that are eligible for low- or moderate-income households at prices or rents below market rate, which are sometimes called "set-asides" (Anderson, 2003; Brown, 2001; Burchell and Galley, 2000; Calavita and Grimes, 1998; Calavita and Mallach, 2010; Fox and Rose, 2003; Madar, 2015; Meltzer and Schuetz, 2010; National Housing Conference, n.d.; Regional Inclusionary Housing Initiative, n.d.; Schwartz et al., 2012; Shaw, 2018; Smith et al., 1996). The proportion of affordable units within market-rate developments depends on the local, regional or state program but typically ranges between 10 and 20 percent per development (National Association of Home Builders, 2015; National Housing Conference, n.d.).

In some areas and neighborhoods, and for larger structures allowed by density bonuses, a local building code or structural engineering may require a developer to switch from wood frame to steel/concrete construction or to add an elevator and other safety features, such as sophisticated fire alarm systems, pressurized exit stairs and other fire safety provisions at a relatively high cost per additionally allowed unit, possibly resulting in decreased revenues and profits (Fox and Rose, 2003; Hickey, 2013; Schuetz et al., 2008). However, in some neighborhoods with intermediate density, including some parts of Brooklyn in New York City, developers may not take advantage of the density bonus because of height limits that may impede adding additional housing units, resulting in only 15 percent of affordable units permitted in designated areas (Hickey, 2013, 2014). Also, developers may not take advantage of the density bonus in some neighborhoods with medium to low rents, as profits and thus incentives will not be high enough (Madar, 2015).

Local or state planning departments may also offer in-lieu fees that are transferred to local affordable housing trust funds as an alternative to providing affordable housing units (Anderson, 2003; Brown, 2001; Calavita and Mallach, 2010; Cray, 2011; Ellickson, 1981; Fox and Rose, 2003; Hickey, 2013; Hickey et al., 2014; Madar, 2015; National Association of Home Builders, 2015; Regional Inclusionary Housing Initiative, n.d.; Schuetz et al., 2008, 2009, 2011; Schwartz et al., 2012). Alternatively, they may accept land dedications to land trusts in lieu of a certain number of affordable housing units (Brown, 2001; Ellickson, 1981; Hickey, 2013; Myerson, 2016; National Association of Home Builders, 2015; Schuetz et al., 2008, 2009, 2011).

Furthermore, planning departments may allow off-site development through linkages, typically in the community of the inclusionary housing unit and subject to distance constraints (Anderson, 2003; Brown, 2001; Calavita and Grimes, 1998; Calavita and Mallach, 2010; Fox and Rose, 2003; Madar and Willis, 2015; Porter, 2004; Regional Inclusionary Housing Initiative, n.d.; Schuetz et al., 2009, 2011; Schwartz et al., 2012; Shaw, 2018).

In some cases, developers may be allowed to contract with for-profit or nonprofit third parties to provide the required units (Madar, 2015).

Moreover, local or state planning departments may even offer cash subsidies or defer or waive a variety of development fees or taxes, including application fees for submittals, development approval fees, building application fees, utility connection fees, property taxes or real estate transfer taxes, especially when the fees or taxes are expected to cause severe financial hardships for developers or potential future occupants (Anderson, 2003; Brown, 2001; Fox and Rose, 2003; Hickey, 2013; Mukhija et al., 2010; Myerson, 2016; Regional Inclusionary Housing Initiative, n.d.; Shaw, 2018). Local or state planning departments may also allow special accommodations, such as expedited reviews during the development review process, relaxed planning or design standards for something like parking, or certain thresholds to exempt small developments from IZ/IH requirements (Anderson, 2003; Fox and Rose, 2003; Hickey et al., 2014; Madar, 2015; Mukhija et al., 2010; National Association of Home Builders, 2015; National Housing Conference, n.d.; Schuetz et al., 2008, 2009, 2011; Schwartz et al., 2012).

Critics of IZ/IH have made arguments that some requirements may amount to a taking, discussed in what is known as the "Takings Clause" in the Fifth Amendment of the Constitution, which specifies that no private property shall be taken for public use without just compensation (Anderson, 2003; Ellickson, 1981). First, they argue that the required set-asides reduce the economic value of private land (Anderson, 2003). A counterargument is that the set-aside requirement affects a certain proportion but not all units, making it possible to obtain a just and reasonable return on investment while taking advantage of the benefits that many municipalities offer (Anderson, 2003). Second, critics believe that set-asides do not sufficiently fulfill the required nexus between a legitimate community interest and the requirement (Anderson, 2003; Ellickson, 1981). However, there may be a legitimate community interest in providing affordable housing for residents with low and moderate incomes, which may include essential workers, as discussed above (Anderson, 2003). Third, critics state that set-asides force private landowners to bear the public burden of the lack of affordable housing (Anderson, 2003). A counterargument is that any new development that does not include affordable housing will increase community challenges that IZ/IH tries to address (Anderson, 2003).

Recently, in a number of states IZ/IH programs have been struck down as violating the state's rent control law. For example, an appellate court in California ruled in 2009 that requiring inclusionary zoning to create affordable rental units was inconsistent with state law, which prohibits rent control (Cray, 2011; Hickey, 2013, 2014). The *Palmer/ Sixth Street Properties, L.P. vs. the City of Los Angeles* decision of 2009 was followed by the Budget Act of 2011, which approved the statewide elimination of more than 400 Redevelopment Agencies (RDAs) in California, effective 1 February 2012, along with the elimination of the only inclusionary housing policy (i.e., the requirement to have 15 percent of all new housing units in designated redevelopment areas affordable to low- and moderate-income households), as well as the loss of $1 billion in local funding for affordable housing (Hickey, 2013; Hickey et al., 2014; State of California Department of Finance, n.d.).

The above legal decision and the Budget Act caused most jurisdictions in California to stop pursuing inclusionary rental housing altogether (Hickey, 2013). However, a few municipalities currently ask for voluntary compliance with inclusionary rental requirements in cases where developers request zoning modifications or upzonings (Hickey, 2013). Other municipalities, including San Francisco and San Diego, either assess a fee on affordable housing developments or offer developers the option to produce additional affordable housing units on site instead of paying a fee (Hickey, 2013). As most new developments in California are multifamily rental units built in housing markets with very high and increasing rent levels, many households will continue facing difficulties in finding affordable rental housing (Hickey, 2013).

Period of Affordability

The length of the affordability period for IZ/IH housing may range from 10 years to perpetuity, although the vast majority of IZ/IH programs remain affordable for at least 30 years (Anderson, 2003; Brown, 2001; Fox and Rose, 2003; Hickey et al., 2014; Meltzer and Schuetz, 2010; Schuetz et al., 2009, 2011). The longer the affordability period, the higher the likelihood that an inclusionary unit will remain affordable and that restrictions will be more stringent (Schuetz et al., 2009, 2011).

The longevity of inclusionary housing units may be based on several factors. First, it depends on local rules that determine the length of affordability or local legal mechanisms, such as deed covenants and deeds of trust, that notify jurisdictions about illegal sales, improper refinancing, over-encumbrances with second loans, defaults and foreclosures (Fox and Rose, 2003; Hickey et al., 2014; Myerson, 2016; Regional Inclusionary Housing Initiative, n.d.). Some IZ/IH programs require beneficiaries to work in the community and occupy their respective units full time (Myerson, 2016).

Second, the longevity depends on resale restrictions, which may be rights of first refusal and price controls (Smith et al., 1996). The former gives a municipality or local planning department the right to decide whether or not to purchase a unit before it is placed on the market (Anderson, 2003; Ellickson, 1981). The latter may cap future equity gains in home values, allowing current homeowners to retain only part of the appreciated value while also retaining affordable homeownership for subsequent homeowners (Brown, 2001; Ellickson, 1981; Fox and Rose, 2003; Hickey, 2013; Hickey et al., 2014; Myerson, 2016). However, some IZ/IH programs may only require units to remain at below-market rates after the first occupants have moved out (Schwartz et al., 2012).

If units are locked into an affordable price for a very short time, sellers may sell them quickly (Brown, 2001). In turn, if units are locked into an affordable price for a very long time, sellers may not realize a good return on their investments (Brown, 2001). Resale formulas may be based on set percentages, appraisals, mortgages, the growth in area median income (AMI), cost of living, inflation rates, transaction costs, capital improvements, closing costs not paid for by the initial seller or a hybrid of any or all of these metrics (Anderson, 2003; Hickey et al., 2014; Regional Inclusionary Housing Initiative, n.d.).

These resale restrictions may also deter interested homebuyers from purchasing inclusionary units in the first place or may cause some current homeowners or developers to sell units at a loss or face foreclosure (Hickey, 2013). Also, the Federal Housing Administration (FHA) may not insure the mortgages of those housing units whose price restrictions will survive foreclosure (Hickey, 2013). Thus, some jurisdictions allow affordability restrictions to expire upon foreclosure, resulting in an FHA waiver (Hickey, 2013).

After resale, IZ/IH units may revert back to the original grantor or a third party, such as a municipality, a public housing authority, a nonprofit agency or a housing or land trust (Anderson, 2003).

Third, the longevity of inclusionary housing units depends on pre-purchase and post-purchase stewardship practices. Examples of pre-purchase practices include classes for future homebuyers that provide oversight of the leasing and tenant selection process, review and training of property managers, and management of a centralized waiting list and the tenant selection process (Hickey et al., 2014). Examples of post-purchase practices include classes for homeowners and monitoring illegal sales, foreclosures or rental management practices (Hickey et al., 2014).

Fourth, the longevity depends on strategic partnerships with nonprofit housing developers and community land trusts (CLTs), limited equity cooperatives (LECs), for-profit administrative agents, local housing authorities, and cities or counties (Freeman and Schuetz, 2017; Hickey et al., 2014; Myerson, 2016).

Permanent affordability represents the highest value for public investment in affordable housing production (Johnstone, 2009). It may also counter the anticipated decrease of affordable housing stock in the near future (Ray and Roset-Zuppa, 2008). For example, the National Low Income Housing Coalition (NLIHC) calculated that almost 450,000 of the nation's 1.4 million federally assisted rental units may leave the affordable housing stock as their owners opt out of existing programs, as the mortgages of units mature or as units no longer comply to code (National Low Income Housing Coalition, n.d.). Khadduri et al. (2012) estimate that more than one million affordable units developed under the LIHTC program may leave the affordable housing stock by 2020 due to expiring affordability terms. Assessing the risk of units potentially leaving the affordable housing stock may be of critical importance, as preserving an existing affordable housing unit costs 40 percent less than building a new one (National Low Income Housing Coalition, n.d.; Ray and Roset-Zuppa, 2008).

A past and current challenge is IZ/IH units aging out of the program, resulting in a net loss of affordable units (Anderson, 2003). This often occurs when the length of the affordability period is relatively short or before new affordable units have been built (Anderson, 2003). Solutions are to extend the length of the affordability period, possibly to perpetuity, or construct more affordable housing (Anderson, 2003).

MEASURING OUTCOMES OF INCLUSIONARY ZONING AND INCLUSIONARY HOUSING

The magnitude of the outcomes of IZ/IH depends on several factors, as there is much diversity in the goals, structures and stringency of IZ/IH and other land use policies in IZ/IH neighborhoods and surrounding areas (Anderson, 2003; Schuetz et al., 2009, 2011). First, the outcomes of IZ/IH are determined by whether local or state policies are mandatory or voluntary. Mandatory policies typically result in a higher number of affordable housing units, although they may suppress development when some developers decide to shift the location of their activities (Brunick et al., 2003; Hickey et al., 2014; Madar and Willis, 2015; Mukhija et al., 2010; see Schuetz et al., 2009, 2011 for an alternative finding). Also, mandatory policies may result in a higher number of affordable housing units for a wider range of income levels (Brunick et al., 2003). The trend seems to be the adoption of mandatory rather than voluntary programs (Brunick et al., 2003).

Second, the outcomes are determined by the sizes of the density bonuses and cost offsets, whether there is an option for developers to opt out or cash out of a program and whether there are other types of residential or nonresidential developments in the neighborhood (Calavita and Mallach, 2010; Hickey et al., 2014; Schuetz et al., 2009, 2011). Communities with more generous density bonuses and less stringent IZ/IH requirements may achieve a higher number of affordable housing units (Schuetz et al., 2008, 2009, 2011).

Third, the outcomes are determined by the nature and the dynamics of local or regional housing market conditions, including the elasticity of supply and demand (Hickey et al., 2014; Meltzer and Schuetz, 2010; Schuetz et al., 2008, 2009, 2011). When IZ/IH is mandatory and a neighborhood is characterized by relatively high property values, increasing density may spur additional developments without any additional subsidy, as in the case of New York City, but may also increase property values and thus negatively impact demand, as some homebuyers may buy elsewhere (Madar and Willis, 2015; Meltzer and Schuetz, 2010; Schuetz et al., 2009, 2011).

Fourth, the outcomes depend on the local or regional economy and the response of jurisdictions and developers. For example, during and after the Great Recession, many communities increased the threshold of income for eligible households or gave developers the option to rent instead of selling their affordable housing units (Hickey, 2013). On the other hand, many developers stopped building homes, including IH units (National Association of Home Builders, 2015).

Fifth, the outcomes are determined by whether there is local or regional political will, successful public engagement, and nonprofits that support the design and implementation of IZ/IH in the first place (Meltzer and Schuetz, 2010; Myerson, 2016). Communities with a higher proportion of politically liberal residents and a higher number of established affordable housing nonprofits tend to support policies and programs that result in a higher number of affordable housing units (Hickey, 2014; Meltzer and Schuetz, 2010). One criticism of IZ/IH is that it creates distortion and unfairness in the market by placing the entire burden of the IZ/IH units on developers and homebuyers (National Association of Home Builders, 2015).

Outcomes for Beneficiaries

Most IZ/IH programs target low- and moderate-income households through income tiering or, less commonly, income averaging (Anderson, 2003). With regard to income tiering, municipalities may require that a certain number or proportion of units be set aside for households earning less than 80 percent or between 80 and 120 percent of the AMI (Anderson, 2003; Brown, 2001; Burchell and Galley, 2000; Cray, 2011; Ellickson, 1981; Fox and Rose, 2003; Madar, 2015; Meltzer and Schuetz, 2010; Myerson, 2016; Smith et al., 1996; Schwartz et al., 2012). With regard to income averaging, municipalities may require developers to set the average price of IZ/IH housing units at a level so that households earning 50 percent of the AMI do not spend more than 30 percent of their gross income on rent (Anderson, 2003). Despite the focus on low- and moderate-income households, most IZ/IH programs tend to serve somewhat advantaged households compared to other affordable housing programs (Schwartz et al., 2012). Also, most IZ/

IH programs serve owners rather than renters, as they require that IZ/IH units have the same tenure as non-IZ market-rate units within developments (Schwartz et al., 2012). Since IZ/IH policies were first established in large, affluent, progressive communities, most IZ/IH programs are located in suburbs (Calavita and Mallach, 2010; Schuetz et al., 2008; Schwartz et al., 2012).

In the IZ/IH literature, economic, racial and ethnic integration is typically analysed through the school attendance and test performance of children who reside in IH units (Schuetz et al., 2008). IZ/IH programs successfully promote racial and ethnic integration, as shown by Orfield (2006) for Montgomery County, Maryland, and Holmqvist (2011) for Davis, California. In the case of Montgomery County, Maryland, IH is widely dispersed in 11 jurisdictions and disproportionately located in neighborhoods with lower poverty rates than neighborhoods with public housing units (Schwartz et al., 2012). Children who live in IH typically attend schools with lower poverty rates and have higher test scores compared to children who live in public housing units (Schwartz et al., 2012).

Socioeconomic, racial and ethnic integration was confirmed, clarified and strengthened in July 2015 by HUD's Affirmatively Furthering Fair Housing (AFFH) rule, through which the approximately 1,200 jurisdictions that received CDBG funding can submit their comprehensive Assessments of Fair Housing (AFHs), which set fair housing priorities and goals, assisted by publicly open data and mapping tools provided by HUD (Freeman and Schuetz, 2017; Hickey, 2013; National Association of Home Builders, 2015; US Department of Housing and Urban Development, n.d.b). However, in January 2018 the Trump administration postponed the imminent deadline of AFFH to 31 October 2020, allowing the CDBG grantees more time to complete their AFHs (Capps, 2018).

Outcomes for the Housing Stock

Researchers have found positive, negative, mixed and no outcomes of IZ/IH on the housing stock, based on analyses, evaluations and nexus studies (Anderson, 2003; Clapp, 1981). In terms of positive impacts, Brunick and Maier (2010), Hughen and Read (2014) and Rosen (2004) found that IZ/IH increased housing production, and Hughen and Read (2014) found that IZ/IH increased house prices. Knaap et al. (2008) found that IZ/IH programs in California, studied from 1988 to 2005, marginally but significantly increased the proportion of multifamily rental homes. Chakraborty et al. (2010) found that for the case of six metropolitan areas, suburban jurisdictions with a higher number of units zoned or designated for high-density development, a shorter distance to downtown and a higher proportion of Whites in 1960 had a higher number of multifamily rental housing units. On the other hand, in the case of the San Francisco metropolitan area, Schuetz et al. (2011) found that IZ/IH decreased the price of market-rate housing "during cooler regional markets" (p. 297).

In terms of negative outcomes, Ellickson (1981) found that in the case of California, IZ/IH increased the price of single-family housing about 2 to 3 percent faster than in non-IZ/IH communities, as developers needed to cross-subsidize market-rate units to achieve the desired profit. Powell and Stringham (2004; see also Basolo and Calavita, 2004) analysed 50 cities in the San Francisco metropolitan area and found that, on average, IZ/IH policies have resulted in 228 units per year since 1973 and that they increased house prices of new homes by $22,000 to $44,000 in the median city. In the case of suburban Boston, Schuetz et al. (2009, 2011) found that IZ/IH increased the price of market-rate housing, especially during periods of regional house price appreciation. Schuetz et al. (2008) found that IZ/IH decreased housing production rates because developers may have decreased or even stopped building in IZ/IH communities, switching activities to non-IZ/IH communities. Finally, Mukhija et al. (2010) found that IZ/IH did not have any impact on house prices and production rates.

Outcomes for Communities

Similar to the outcomes of IZ/IH on the housing stock, researchers have found that IZ/IH has positive, negative and mixed outcomes on communities. Many scholars point out that IZ/IH provides beneficiaries access to high-opportunity neighborhoods with relatively low poverty rates (i.e., less than 10 percent) and relatively high-performing schools (i.e., schools with average test scores at the fiftieth percentile or above among schools within the state) (Fox and Rose, 2003, Schwartz et al., 2012). Madar and Willis (2015) point out that neighborhood infrastructure may be negatively impacted by the increased density that often follows IZ/IH. Also, there may be resistance (sometimes called Not in My Back Yard (NIMBY)) to new IZ/IH units (Massey et al., 2013; National Association of Home Builders, 2015; Porter, 2004).

Schwartz et al. (2012) analysed the impact of IZ/IH programs in terms of school outcomes in Montgomery County, Maryland, where the Housing Opportunities Commission (HOC) randomly assigned households to public housing apartments in the vicinity of schools with low poverty rates. Schwartz et al. (2012) then compared elementary school performance in math and reading between children in an IZ/IH program and children who attended schools with high poverty rates. They found that children who attended schools with low poverty rates had statistically significant gains in math but statistically insignificant gains in reading. It would be interesting to analyse other communities with IZ/IH programs.

CONCLUSION

The large variations among the designs and scopes of local IZ/IH programs result in different outcomes (Schuetz et al., 2008). IZ/IH is not a one-size-fits-all tool (Schuetz et al., 2008). As Tip O'Neill, Speaker of the House between 1977 and 1987, stated, "All politics is local.' When it comes to inclusionary housing, it should be said, 'All success is local'" (quoted in Brunick and Maier, 2010, p. 174).

Whereas IZ/IH may address affordable housing production, it will not necessarily solve a local affordable housing crisis because of its limitations in scope. IZ/IH seems to work best in housing markets with high house prices and high house price growth, as density bonuses are most valuable in markets with high property values and high demand (Fox and Rose, 2003; Hickey, 2014; Schuetz et al., 2009, 2011; Solé, 2010). IZ/IH also seems to work well in markets with relatively low base zoning, where IZ/IH creates large incentives (Hickey, 2014).

There are several future challenges for IZ/IH policies. First, some current IZ/IH policies were designed for certain housing market conditions that may have since changed (Hickey,

2013; Madar, 2015). Some IZ/IH policies may lack the flexibility to adjust to different housing market situations or different homebuyer, homeowner or resident circumstances (Madar, 2015). This lack of flexibility may have been especially unfortunate during and after the Great Recession, which resulted in job losses, involuntary part-time work and foreclosures, thus negatively impacting incomes, wealth, credit scores and homeownership opportunities. One example of inflexibility is that some jurisdictions may not allow existing homeowners or developers to rent out inclusionary units. Other jurisdictions, especially those in strong markets, may have difficulties finding eligible future homebuyers who have incomes that are low enough to qualify for an IZ/IH program yet high enough to quality for mortgage financing (Hickey, 2013; National Association of Home Builders, 2015).

Second, most IZ/IH policies have expiring affordable price controls. Thus, IZ/IH units will leave the program, resulting in increased affordable housing challenges (Brown, 2001). These challenges have been exacerbated by decreased housing construction after 2006. For example, 1.979 million privately owned housing units were completed in 2006, reaching a peak, while only about 584,900 units were completed in 2011, resulting in a trough (US Bureau of the Census, n.d.a).

Finally, future IZ/IH policies will have to creatively address how higher costs per unit will be financed, as many housing trust funds have relatively low balances and, as in most metropolitan areas, land will become scarce (Brown, 2001). In turn, developers will increasingly pursue more infill developments with higher land prices closer to the metropolitan core, structured parking, steel and concrete construction with elevators and other safety features to accommodate for heights, and the higher risk of building high rises, which cannot be built incrementally (Hickey, 2013). Sources of funding could be existing or additional cross subsidies from market-rate units, public subsidies or mortgages (Madar and Willis, 2015). Jurisdictions could offset these higher development costs per unit by:

(a) providing direct and indirect subsidies (Hickey, 2013);
(b) providing public land at a discounted rate (Hickey, 2013; Schuetz et al., 2008);
(c) providing first-right-of-refusal for public housing authorities and nonprofits that purchase inclusionary, for-sale homes (Hickey, 2013);
(d) reducing the number of required units (Hickey, 2013);
(e) streamlining the entitlement process (Hickey, 2013);
(f) relaxing lot coverage, public space and parking requirements (Hickey, 2013);
(g) facilitating off-site construction of inclusionary units (Hickey, 2013);
(h) increasing allowed rent levels (Hickey, 2013);
(i) increasing allowed household income thresholds (Hickey, 2013); or
(j) decreasing inclusionary requirements for high rises (Hickey, 2013).

NOTES

* This material is based upon work supported by the National Science Foundation under Grant No. 1636520. Any opinions, findings and conclusions or recommendations expressed in this material are those of the author and do not necessarily reflect the views of the National Science Foundation.

1. Housing affordability can be measured in multiple ways, including the housing expenditure-to-income ratio, which specifies that a household with housing costs higher than 30 percent of its income is cost burdened and a household with housing costs higher than 50 percent of its income is severely cost burdened (Anacker and Li, 2016; US Department of Housing and Urban Development, n.d.a). Another measure of housing affordability is based on the hourly wage required to rent a two-bedroom unit without paying more than 30 percent of one's income on housing (National Low Income Housing Coalition, 2017). A third measure is the residual income approach, which is the amount of income that an individual can spend on housing after taking other necessary expenditures of living into account (Stone, 2006; 2009a; 2009b).
2. Hickey et al. (2014) discuss the online database of local inclusionary housing programs maintained by the California Coalition for Rural Housing (www.calruralhousing.org). However, this database does not seem to be publicly accessible. An e-mail sent to the California Coalition for Rural Housing was not answered. See also Grounded Solutions Network (n.d.) for a list of state inclusionary housing programs.
3. Oregon and Texas prohibit mandatory IZ/IH, as they wish to protect private property rights (Hickey, 2014; Schwartz et al., 2012).
4. TIF can also be used to help subsidize the cost of building affordable units in a mandatory IZ/IH program.
5. Hickey (2014) compares the 150,000 affordable IZ/IH units to the estimated 2.5 million units built through the LIHTC program.

REFERENCES

Anacker, K.B. (2019). Editorial introduction. *International Journal of Housing and Society*, 19(1), 1–16.
Anacker, K.B., and Li, Y. (2016). Rental housing affordability of U.S. renters during the Great Recession, 2007 to 2009. *Housing and Society*, 43(1), 1–17.
Anderson, M. (2003). *Opening the Door to Inclusionary Housing*. Chicago, IL: Business and Professional People for the Public Interest.
Babcock, R.F., and Siemon, C.L. (1985). *The Zoning Game Revisited*. Boston: Oelgeschlager, Gunn, and Hain.
Basolo, V., and Calavita, N. (2004). Policy claims with weak evidence: A critique of the Reason Foundation study on inclusionary housing policy in the San Francisco Bay Area. Working Paper. June. University of California, Irvine and San Diego State University.
Breagy, J. (1976). *Overriding the Suburbs: State Intervention for Housing through the Massachusetts Appeals Process*. Boston: Citizens Housing and Planning Association of Metropolitan Boston.
Brown, K.D. (2001). *Expanding Affordable Housing through Inclusionary Zoning: Lessons from the Washington Metropolitan Area*. Washington, DC: The Brookings Institution.
Brunick, N., and Maier, P.O'B. (2010). Renewing the land of opportunity. *Journal of Affordable Housing and Community Development Law*, 19(2), 161–90.
Brunick, N., Goldberg, L., and Levine, S. (2003). *Voluntary or Mandatory Inclusionary Housing? Production, Predictability, and Enforcement*. Chicago: Business and Professional People for the Public Interest.
Burchell, R.W., and Galley, C.C. (2000). Inclusionary zoning: A viable solution to the affordable housing crisis? Inclusionary zoning: Pros and cons. *New Century Housing*, 1(2), n.p.
Calavita, N.K., and Grimes, K. (1998). Inclusionary housing in California: The experience of two decades. *Journal of the American Planning Association*, 64(2), 150–69.
Calavita, N., and Mallach, A., eds (2010). *Inclusionary Housing in International Perspective: Affordable Housing, Social Inclusion, and Land Value Recapture*. Cambridge, MA: Lincoln Institute of Land Policy.
Calavita, N., Grimes, K., and Mallach, A. (1997). Inclusionary housing in California and New Jersey: A comparative analysis. *Housing Policy Debate*, 8(1), 109–42.
Capps, K. (2018). The Trump administration just derailed a key Obama rule on housing segregation. *CityLAB*, 4 January.
Chakraborty, A., Knaap, G.-J., Nguyen, D., and Shin, J.H. (2010). The effects of high-density zoning on multifamily housing construction in the suburbs of six U.S. metropolitan areas. *Urban Studies*, 47(2), 437–51.
Clapp, J.M. (1981). The impact of inclusionary zoning on the location and type of construction activity. *Real Estate Economics*, 9(4), 436–56.
Cowan, S.P. (2006). Anti-snob land use laws, suburban exclusion, and housing opportunity. *Journal of Urban Affairs*, 28(3), 295–313.
Cray, A.F. (2011). *The Use of Residential Nexus Analysis in Support of California's Inclusionary Housing Ordinances: A Critical Evaluation*. Sacramento, CA: California Homebuilding Foundation.
Davidoff, P., and Davidoff, L. (1971). Opening the suburbs: Toward inclusionary land use controls. *Syracuse Law Review*, 22(2), 509–36.

Dietderich, A.G. (1996). An egalitarian's market: The economics of inclusionary zoning reclaimed. *Fordham Urban Law Journal*, *24*(1), 23–104.

Downs, A. (1973). *Opening up the Suburbs: An Urban Strategy for America*. New Haven, CT: Yale University Press.

Ellickson, R.C. (1981). The irony of inclusionary zoning. *Southern California Law Review*, *54*(6), 1167–216.

Fischel, W. (1985). *The Economics of Zoning Laws: A Property Rights Approach to American Land Use Controls*. Baltimore: Johns Hopkins University Press.

Fox, R.K., and Rose, K. (2003). *Expanding Housing Opportunity in Washington, DC: The Case for Inclusionary Zoning*. Washington, DC: PolicyLink.

Freeman, L., and Schuetz, J. (2017). Producing affordable housing in rising markets: What works? *Cityscape*, *19*(1), 217–36.

Grounded Solutions Network (n.d.). *Is Inclusionary Housing Legal?* Portland, OR: Grounded Solutions Network.

Hickey, R. (2013). *After the Downturn: New Challenges and Opportunities for Inclusionary Housing*. Washington, DC: Center for Housing Policy.

Hickey, R. (2014). *Inclusionary Upzoning: Tying Growth to Affordability*. Washington, DC: Center for Housing Policy.

Hickey, R., Sturtevant, L., and Thaden, E. (2014). *Achieving Lasting Affordability through Inclusionary Housing*. Cambridge, MA: Lincoln Institute of Land Policy.

Holmqvist, A. (2011). The effect of inclusionary zoning on racial integration, economic integration, and access to social services: A Davis case study. Master's thesis. University of California, Davis.

Hughen, W.K., and Read, D.C. (2014). Inclusionary housing policies, stigma effects and strategic production decisions. *The Journal of Real Estate Finance and Economics*, *48*(4), 589–610.

Jacobus, R. (2015). *Inclusionary Housing: Creating and Maintaining Equitable Communities*. Cambridge, MA: Lincoln Institute of Land Policy.

Johnstone, K.A. (2009). *Permanent Affordability: A National Conversation*. New York: The Association for Neighborhood and Housing Development.

Khadduri, J., Climaco, C., Burnett, K., Gould, L., and Elving, E. (2012). *What Happens to Low-Income Housing Tax Credit Properties at Year 15 and Beyond?* Washington, DC: US Department of Housing and Urban Development.

Knaap, G.-J., Bento, A., and Lowe, S. (2008). *Housing Market Impacts of Inclusionary Zoning*. College Park, MD: National Center for Smart Growth Research.

Madar, J. (2015). *Inclusionary Housing Policy in New York City: Assessing New Opportunities, Constraints, and Trade-Off*. New York: Furman Center for Real Estate and Urban Policy, New York University.

Madar, J., and Willis, M. (2015). *Creating Affordable Housing out of Thin Air: The Economics of Mandatory Inclusionary Zoning in New York City*. New York: Furman Center for Real Estate and Urban Policy, New York University.

Massey, D.S., Albright, L., Casciano, R., Erickson, E., and Kinsey, D.N. (2013). *Climbing Mount Laurel: The Struggle for Affordable Housing and Social Mobility in an American Suburb*. Princeton, NJ: Princeton University Press.

Meltzer, R., and Schuetz, J. (2010). What drives the diffusion of inclusionary zoning? *Journal of Policy Analysis and Management*, *29*(3), 578–602.

Montgomery County (n.d.). *Moderately Priced Dwelling Unit (MPDU) Program*. Rockville, MD: Montgomery County Department of Housing and Community Affairs.

Mukhija, V., Regus, L., Slovin, S., and Das, A. (2010). Can inclusionary zoning be an effective and efficient housing policy? Evidence from Los Angeles and Orange Counties. *Journal of Urban Affairs*, 32(2), 229–52.

Myerson, D.L. (2016). *How Did They Do It? Discovering New Opportunities for Affordable Housing*. Washington, DC: National Association of Home Builders.

National Association of Home Builders (2015). Inclusionary zoning primer. Washington, DC: National Housing Conference.

National Housing Conference (n.d.). *Inclusionary Housing: The Basics*. Washington, DC: National Housing Conference.

National Low Income Housing Coalition (n.d.). *Project-Based Housing*. Washington, DC: National Low Income Housing Coalition.

National Low Income Housing Coalition (2017). *Out of Reach: The High Costs of Housing*. Washington, DC: National Low Income Housing Coalition.

Orfield, M. (2006). Land use and housing policies to reduce concentrated poverty and racial segregation. *Fordham Urban Law Journal*, *33*(3), 101–59.

Partnership for Sustainable Communities (n.d.). *About Us*. Washington, DC: US Department of Housing and Urban Development, Department of Transportation, and Environmental Protection Agency.

Porter, D. (2004). *Inclusionary Zoning for Affordable Housing*. Washington, DC: Urban Land Institute.

Powell, B., and Stringham, E. (2004). *Housing Supply and Affordability: Do Affordable Housing Mandates Work?* Los Angeles: Reason Foundation.
Ray, A., and Roset-Zuppa, P. (2008). *A Risk Assessment Method for Preservation of Assisted Rental Housing.* Gainsville, FL: Shimberg Center for Affordable Housing.
Regional Inclusionary Housing Initiative (n.d.). *Policy Tool #1: Developing an Inclusionary Zoning Ordinance.* Chicago: Business and Professional People for the Public Interest.
Rosen, D. (2004). Inclusionary housing and its impact on housing and land market. *National Housing Conference Affordable Housing Policy Review*, 3(1), 38–47.
Schuetz, J., Meltzer, R., and Been, V. (2008). *The Effects of Inclusionary Zoning on Local Housing Markets: Lessons from the San Francisco, Washington DC, and the Suburban Boston Areas.* New York: Furman Center for Real Estate and Urban Policy, New York University.
Schuetz, J., Meltzer, R., and Been, V. (2009). 31 flavors of inclusionary zoning: Comparing policies from San Francisco, Washington, DC, and Suburban Boston. *Journal of the American Planning Association*, 75(4), 441–56.
Schuetz, J., Meltzer, R., and Been, V. (2011). Silver bullet or Trojan horse? The effect of inclusionary zoning on local housing markets in the United States. *Urban Studies*, 48(2), 297–329.
Schwartz, A.F. (2015). *Housing Policy in the United States.* New York: Routledge.
Schwartz, H.L., Ecola, L., Leuschner, K.J., and Kofner, A. (2012). *Is Inclusionary Zoning Inclusionary? A Guide for Practitioners.* Santa Monica, CA: RAND.
Shaw, R. (2018). *Generation Priced Out: Who Gets to Live in the New Urban America.* Berkeley, CA: University of California Press.
Smith, M.T., Delaney, C.J., and Liou, T. (1996). Inclusionary housing programs: Issues and outcomes. *Real Estate Law Journal*, 25(2), 155–71.
Solé, J.P. (2010). Foreword. In N. Calavita and A. Mallach (eds), *Inclusionary Housing in International Perspective: Affordable Housing, Social Inclusion, and Land Value Recapture* (pp. ix–xiii). Cambridge, MA: Lincoln Institute of Land Policy.
State of California Department of Finance (n.d.). *Redevelopment Agency Dissolution.* Sacramento, CA: State.
Stone, M.E. (2006). What is housing affordability? The case for the residual income approach. *Housing Policy Debate*, 17(1), 151–84.
Stone, M.E. (2009a). *Renter Affordability in the City of Boston.* Boston, MA: Center for Social Policy.
Stone, M.E. (2009b). Unaffordable "affordable housing": Challenging the U.S. Department of Housing and Urban Development's area median income. *Progressive Planning*, 180(Summer), 36–9.
Tiebout, C.M. (1956). A pure theory of local expenditures. *Journal of Political Economy*, 64(5), 416–24.
US Bureau of the Census (n.d.a). *New Privately Owned Housing Units Completed: Annual Data.* Washington, DC: US Bureau of the Census.
US Bureau of the Census (n.d.b). *Table A-1: Households by Total Money Income, Race, and Hispanic Origin of Householder – 1967 to 2016.* Washington, DC: US Bureau of the Census.
US Department of Housing and Urban Development (n.d.a). *Affordable Housing.* Washington, DC: US Department of Housing and Urban Development.
US Department of Housing and Urban Development (n.d.b). *HUD Role on Affirmatively Furthering Fair Housing.* Washington, DC: US Department of Housing and Urban Development.
US Department of Housing and Urban Development (n.d.c). *Six Livability Principles.* Washington, DC: US Department of Housing and Urban Development.
US Department of the Interior (n.d.). *Tax Incentives for Preserving Historic Properties.* Washington, DC: US Department of the Interior.
US Department of the Treasury (n.d.). *Investing for the Future: Empowering America's Economically Distressed Communities.* Washington, DC: US Department of the Treasury.
Wheeler, M. (1990). Resolving local regulatory disputes and building consensus for affordable housing. In D. DiPasquale and L.C. Keyes (eds), *Building Foundations: Housing and Federal Policy* (pp. 209–39). Philadelphia, PA: University of Pennsylvania Press.

12. Enhancing evaluation capacity: lessons from faith-based community development in El Salvador
James G. Huff, Jr.

How can the evaluation research capacities of small-scale organizations working to promote community-based development be strengthened and sustained? And how can these capacities be developed in such a way so that community members have increased opportunities to participate in the processes of evaluation research and to make use of research products to strengthen their own capacities to direct beneficial change? To address these questions, I briefly summarize some of the key lessons I have learned from a decade of evaluation research and practice with a Salvadoran non-government organization (NGO) that partners with local religious and government organizations in the design and implementation of community-based initiatives. The chapter begins by describing the theory of change that guides the efforts of ENLACE, a faith-based NGO that promotes community-based development in rural communities throughout El Salvador. This is followed by a short description of some of the challenges that organization staff and community partners have faced in trying to generate and sustain an effective program of evaluation. I conclude the discussion by describing key adaptations and changes that my Salvadoran colleagues and I have implemented along the way in order to enhance the capacities of all stakeholders to carry out evaluation research and to make good use of the information it generates.

STRENGTHENING COMMUNITY CAPACITY IN RURAL EL SALVADOR

ENLACE (Entidad Natural Latinoamericana de Cooperación Estratégica) is a Salvadoran NGO founded in 1993. As a faith-based organization, ENLACE, which is the Spanish word for "link," works primarily with local churches that are engaged in different community-based projects of change. Most of ENLACE's church partners are small congregations made up of fewer than 100 members and all are affiliated with evangelical and Pentecostal forms of Christianity. In 2017 ENLACE worked with approximately 70 different churches located in rural and semi-urban communities across El Salvador. The organization expanded its partnerships in 2013 to include several churches in Nepal and it initiated church partnerships in neighboring Guatemala in 2015.

ENLACE's approach to community development prioritizes the building of the leadership and organizational capacities of local religious and civic leaders and strengthening the relationships among local organizations so that they can implement community change initiatives. The organization's core programs are guided by a theory of change which posits that the relational and interactional patterns of community-based organizations are fundamental to achieving long-term outcomes of community transformation.

Such a model, which has been called a network theory of change, focuses on how "the relationships, networks, and connections, among [community] entities, and not just the characteristics of the entities themselves, affect outcomes" (Funnell and Rogers, 2011).

For ENLACE, a local church congregation is the primary actor for initiating and building new social and organizational networks in a given community of place. The organization's main strategy, known as the Church and Community Program, invests in training church leaders and members to become active agents of community-building. It is a strategy that focuses on "strengthening the ability of residents, organizations, and institutions to foster and sustain . . . change, both individually and collectively" (Kubisch et al., 2002, p. 26).[1] ENLACE's core programming incorporates many of the basic principles and practices of Comprehensive Community Initiatives (CCIs), which have been described carefully by researchers working in communities across the US (Kubisch et al., 2002; Kubisch et al., 2010).

The training carried out by ENLACE staff anticipates several interconnected outcomes of change. First and foremost are the changes that are to occur within a church partner. ENLACE's asset-based approach recognizes each congregation to be a somewhat unique local organization that contains varying stocks of human, cultural, social, financial and built capital (Emery and Flora, 2006). A primary aim of the Church and Community Program, therefore, is to involve congregants in a process of identifying their internal resources and to motivate the church body to invest these assets into working with other local, non-church entities in the design and implementation of small-scale development initiatives. Participation in the program generates changes in the internal organization of the church and particularly in the functioning of the groups (e.g. *ministerios* or internal ministries) that carry out the day-to-day activities of congregational life. Many of ENLACE's church partners, for example, have created *ministerios* that create new relational ties with non-Pentecostal neighbors and mobilize church members' participation in the design, fund-raising and implementation of local development initiatives (Huff, in review).

Another related outcome of change is the formation of new linkages between a local church and other community organizations. While these organizational ties vary from one community to the next, they usually include connections between a church and local entities like the mayor's office, government health clinics, public and private schools, community development associations (known as ADESCOs) and other churches. As noted, a core assumption of ENLACE's model is that the formation of new, local networks will facilitate the flow of assets across different community capitals, which in turn will initiate "an ongoing process of assets building on assets" (Emery and Flora, 2006, p. 22).

A final set of changes is related to the local development initiatives that are generated by the collaborative work of a church and its community partners. Such initiatives vary widely in terms of their size, scale and project cycle timeline. Approximately 50 percent of the funding for each initiative is generated by local church members, community residents and by the local mayor's office, and the remaining funds come from contributions made by US-based churches and from donations by individual households, family foundations and businesses in the US and Canada (Huff, in review). Initiatives typically consist of local infrastructure (e.g. roads and foot bridges) and school improvement projects, home gardens, aquaculture (e.g. tilapia ponds) and microenterprise development, potable water and sanitation (e.g. pit latrine) initiatives, improved stoves and home construction projects, and various community health initiatives. In 2016, ENLACE reported that local partner organizations completed over 600 different initiatives, including 130 pit latrines, 140 eco-stoves, 60 small-scale poultry farms and two potable water systems, among others (ENLACE Annual Report, 2016, p. 8).

Since 2008 I have worked voluntarily in a consultative role with two full-time, Salvadoran staff at ENLACE to initiate and maintain a department of research and evaluation. For the past decade we have identified outcomes and indicators of change for core programs, developed research instruments that are used in community baseline studies, and, most recently, initiated the process of analysing and reporting on household data (based on the global Multidimensional Poverty Index) collected from the communities where local church and organizational partners work. Along the way we have also conducted ethnographic process evaluations of different aspects of a local project cycle (Huff, 2010).

THE CHALLENGES OF BUILDING EVALUATION CAPACITY

Resourcing and sustaining a program of evaluation research at ENLACE have presented challenges that are common to many organizations working to facilitate comprehensive models of community change. The most fundamental task has been to conceptualize and create a strategy of evaluation that is logically appropriate to the model of change that ENLACE promotes. The process of designing such a strategy has been a very gradual and iterative one. A slow and steady approach has been required in part because of the professional capacities of the team itself. Most were formally trained in methods of social scientific investigation that did not consider how both the processes and products of research might be used to empower communities and strengthen community capacity. In the early stages, our team focused on designing an outcomes-based approach to help ENLACE staff determine the extent to which various distal outcomes of community transformation were being achieved as a result of the initiatives that local partners implemented. As I note below, this approach has been adapted and revised since then. This early work, however, enabled the team to better understand ENLACE's overarching theory of change and to recognize some of the shortcomings of conventional evaluation practice for understanding the effectiveness of the community-based initiatives it promotes.

Related challenges have concerned the constraints of time and budget, and of the inaccessibility of certain data, all of which are problems frequently confronted by evaluators working in Majority World contexts (Bamberger et al., 2004). The team conducted its first community baseline study at the end of 2009 in a remote, rural community located in northeastern El Salvador. Since that time the team has worked with local church and organizational leaders to complete baseline studies of nearly 30 different communities. The survey instrument we created makes possible the collection of individual- and household-level data for over 100 different socio-economic, health, educational and standard of living (e.g. sanitation, water, electricity, housing, assets, and so on) indicators. Participants are selected through simple random sampling and the data is aggregated from a representative sample of households in each study community. However, due to time and budget constraints the team did not create a study design that includes control or comparison groups.[2] In many communities, the process of identifying households has also been very challenging given that local census data is sometimes out of date or simply unavailable. Moreover, cleaning and inputting the data has proven to be an especially time-consuming task for staff and has caused considerable delays in sharing what we are learning about household patterns of multidimensional poverty. Consequently, it has been difficult to find adequate time to report on key study findings with ENLACE staff and community participants.

A final challenge continues to be how to strike a good balance between two of ENLACE's key goals, namely to build community capacity and to understand the changes that are (or are not) occurring at the community level (Center for Community Health and Development, 2017). Until recently, the involvement of community partners in the research process has been somewhat limited. Local leaders have been an invaluable resource for mobilizing community participants to assist in the collection of household data for baseline studies. ENLACE staff regularly train a group of local participants – which have included scores of churchgoers, ADESCO members and young people from youth clubs and schools – in carrying out survey research for each baseline. Their involvement has greatly sped up the process of collecting household data; moreover, there is good evidence that their participation has enabled the team to improve survey question quality and, consequently, the overall quality of the data collected. Nevertheless, the important work of learning about community change is still generally understood by ENLACE staff and community partners as a task that is carried out primarily by the two "technical" staff in the Department of Evaluation and Research. Our challenge remains to develop an evaluation design that enables all stakeholders to understand the relationship between changes in local organizational capacity and the broader effects of the development initiatives carried out by church and community leaders.

REEVALUATING EVALUATION: SOME LESSONS LEARNED

The decade of steady work undertaken by ENLACE's research team has provided ample opportunity to rethink and enhance our evaluation strategy. Perhaps the greatest lesson we have learned along the way concerns the fundamental importance of regularly setting aside time to review and assess the assumptions and processes that comprise our evaluation practice. Sustaining regular habits of reflection, feedback and adaptation are absolutely essential to creating a supportive learning environment. Such habits have made it possible for the team to revise and strengthen the evaluation capacities of both ENLACE and its church and community partners. I highlight a few of those revisions here to conclude.

Repurpose Instruments for Learning

Since the beginning, most of our efforts have focused on maintaining a research strategy oriented towards the implementation of community baseline studies. Findings from the baseline research have been invaluable for developing a broad picture of the patterns of multidimensional poverty that exist within a given community of place. Yet the breadth and complexity of the data included in the studies make them unwieldy as a tool for learning in the short and medium term. The baseline studies will remain an important reference point for tracking outcome changes in the long term, and especially as we work to develop a research design that more effectively evaluates the progress that community partners are making towards "moving the 'bottom-line' indicators" (Center for Community Health and Development, 2017). More recently, the team has used baseline findings to develop and disseminate community *informes* (reports). These reports are shared with ENLACE coaches and with local church and organizational partners (in a community forum) who use them to better understand and identify critical needs in the community. In this way, the work of the evaluation team reinforces the decision-making and planning capacities of community partners as they design initiatives that best address local concerns.

Redirect Outcome Assessment Energies

The time spent on managing a strategy focused on the collection of data related to distal outcomes also means that we have given less attention to learning systematically about very important intermediate outcomes. As noted, ENLACE's core programming trains churches and other community organizations to become active agents of community-building. A present learning focus, therefore, is to understand the processes by which churches and local partner organizations become "catalysts for change" (Center for Community Health and Development, 2017). An example of an intermediate outcome that is generated by the training that ENLACE church coaches facilitate is the formation of church and community committees. Such committees ostensibly signal the expression of new leadership capacities and novel organizational forms in the local community. In the near term, the team will give increased attention to documenting the processes by which such intermediate results emerge "downstream" from the training activities that church coaches direct and "upstream from longer-term economic, environmental, political, or demographic changes" (Better Evaluation, 2018). We will use evaluation and monitoring tools associated with an "outcome mapping" approach, for example, in order to better assess the capacities that ENLACE's Church and Community Program aims to build in local organizational leaders "who will ultimately be responsible for improving the well-being of their communities" (Earl et al., 2001, p. 3).

Prioritize Strategic Time-Savers

Determining how to make optimal use of the time and energy of the research team has also been a very iterative process. A relatively straightforward change has been the adoption of mobile data collection technologies for the community baselines. A pilot project is currently underway that will allow staff and community participants to use mobile devices (including cell and smartphones) to carry out household surveys for the baseline studies. The use of such tools promises to greatly reduce the time that the team allocates to the inputting and cleaning of survey data. Correspondingly, the team is working to create additional efficiencies by building the research capacities of their staff colleagues. This involves training church coaches in the implementation of baseline studies and in the mobilization of local participants in community-based research. Such adaptations, then, are not simply about making efficient use of time. They also aim to increase the opportunities for ENLACE's community partners to participate directly in the research process.

More importantly, they seek to facilitate more consistent patterns of information-sharing and feedback with community partners as they try to better understand the local initiatives they direct.

Build Reciprocal Relationships (and Maintain Methodological Rigor)

Our efforts to enhance evaluation research capacities will continue to have both an internal and external focus. We hope that the internal focus will enable the ENLACE staff to better understand the efficacy and impact of their efforts to help local organizations become catalysts of positive change in their communities of place. At the same time, we continue to look for ways to increase the participation of community partners in evaluation research activities, and especially in the processes of decision-making related to how the products of research are to be used for adapting and improving community-based initiatives. This external dimension of our work has reminded us of the fundamental importance of sustaining healthy relationships among all stakeholders – relationships that facilitate trust, mutuality and interdependence. As researchers, this has required us to regularly assess the degree to which our own research practice "both protects and promotes the interests of participants and ensures that methodological rigor can be maintained" (Pittaway et al., 2010, p. 242). We recognize that our own professional training inclines us to critically reflect on the rigor of the research methodology we adopt; however, we also recognize that the same training was sorely lacking in preparing us to facilitate and sustain a more participatory and reciprocal practice of evaluation research. Learning to strike a good balance has proven to be a very challenging yet worthwhile endeavor.

NOTES

1. See Huff (2017); and Huff (in review) for a fuller discussion of why ENLACE identifies local churches as key actors in generating community change and for more detail on how staff carry out the training that church members receive as participants in the Church and Community Program.
2. The indicators included in the survey instrument were operational measures of the various distal outcomes that ENLACE staff anticipated achieving as a result of the local initiatives that community partners implement each year.

REFERENCES

Bamberger, M., Rugh, J., Church, M., and Fort, L. (2004). Shoestring Evaluation: Designing Impact Evaluations under Budget, Time and Data Constraints. *American Journal of Evaluation*, 25(1), 5–37. https://doi.org/10.1177/109821400402500102.

Better Evaluation. (2018). Outcome Mapping. Retrieved 15 December 2018 from Better Evaluation: https://www.betterevaluation.org/en/plan/approach/outcome_mapping.

Center for Community Health and Development. (2017). Chapter 1, Section 5: Our Evaluation Model: Evaluating Comprehensive Community Initiatives. Lawrence, KS: University of Kansas. Retrieved 1 January 2018 from the Community Tool Box: https://ctb.ku.edu/en/table-of-contents/overview/model-for-community-change-and-improvement/evaluation-model/main.

Earl, S., Carden, F., and Smutylo, Q. (2001). *Outcome Mapping: Building Learning and Reflection into Development Programs*. Ottawa: International Development Research Centre.

Emery, M. and Flora, C.B. (2006). Spiraling Up: Mapping Community Transformation with Community Capitals Framework. *Journal of the Community Development Society*, 37(1), 19–35.

ENLACE Annual Report. (2016). Retrieved 19 November 2019 from ENLACE: https://www.enlace.link/wp-content/uploads/2017/04/AR2016_Final_Web_R4.pdf.
Funnell, S.C. and Rogers, P.J. (2011). Purposeful Program Theory: Effective Use of Theories of Change and Logic Models [VitalSource Bookshelf version]. Retrieved 14 December 2019 from vbk://9780470939895.
Huff Jr., J.G. (2010). The Evangelical Church as Community Asset? An Ethnographic Analysis of Asset-Based Development in Rural El Salvador. *NAPA Bulletin*, *31*(1), 105–25.
Huff Jr., J.G. (2017). Of Specters and Spirit: Neoliberal Entanglements of Faith-Based Development in El Salvador. *Urban Anthropology and Studies of Cultural Systems and World Economic Development. Special Issue: The Impact of State-Level and Global-Level Neoliberal Agendas on NGOs in Latin America*, *46*(3/4), pp. 173–220.
Huff, Jr., J.G. (in press). The Role of Local Churches in Building Community Capacity: Insights from Rural El Salvador. In *Building Community Capacity and Resilience through Rural Development: Perspectives from Latin America*, P. Lachapelle, I. Gutierrez-Montes, and C. Butler Flora, eds. New York: Routledge.
Kubisch, A.C., Auspos P., Brown, P., and Dewar T. (2010). Community Change Initiatives from 1990–2010: Accomplishments and Implications for Future Work. *Community Investments*, *22*(1), 8–12.
Kubisch, A.C., Auspos P., Brown, P., Chaskin, R., Fulbright-Anderson, K., and Hamilton, R. (2002). *Voices from the Field II: Reflections on Comprehensive Community Change*. Washington, DC: The Aspen Institute.
Pittaway, E., Bartolomei, L. and Hugman, R. (2010). Stop Stealing Our Stories: The Ethics of Research with Vulnerable Groups. *Journal of Human Rights Practice*, *2*(2), 229–51.

13. Managing competing interests in the public participation process: lessons from an analysis of residential displacement in Buffalo, New York's transitioning neighborhoods*
Robert Mark Silverman, Li Yin and Henry Louis Taylor, Jr.

A TYPOLOGY OF COMPETING INTERESTS IN PUBLIC PARTICIPATION

This chapter begins with the premise that public participation is a contested process and involves a variety of competing interests that aim to shape public policy. Moreover, each interest that engages in dialogue about public policy exhibits a different degree of fidelity to the public participation process. Our premise, that public participation is a contested process, does not discount the literature on deliberative planning and other techniques used by equity planners and community development practitioners to augment citizen input (Jones, 1990; Forester, 1999; Creighton, 2005; Shipley and Utz, 2012). However, we stress that despite genuine efforts to design tools and techniques to enhance public participation, equity planners and community development practitioners continue to face barriers to fostering full participation due to the contested nature of the policy process. Among these obstacles is the presence of institutional stakeholders and vested interests that have an undue influence on the content, shape and implementation of the public participation process. In order to parry the effects of these interests and protect citizens' access to the participation process, we argue that equity planners and community development practitioners must assume a strong advocacy role.

The rationale for advocacy in planning and community development is grounded in Davidoff's (1965) seminal work on pluralism in planning. In this article, Davidoff argues that professionals engaged in planning and community development should advocate for the interests of their clients, particularly those who are disenfranchised from policy and decision-making processes. Since the appearance of Davidoff's article, others have expanded on the role of advocacy in community development. For instance, Needleman and Needleman (1974) expanded on the notion of advocacy in professional practice and argued that community development practitioners are critical to disseminating information to grassroots groups and empowering them. This work was elaborated upon by Krumholz (1990) where the parameters of equity planning were defined in a manner that positioned city planners and community development professionals as advocates for redistributive policies benefiting the poor and disenfranchised. The role of advocacy in community development was further expanded upon by Silverman et al. (2008) where the use of advocacy planning to enhance citizen control was applied to discussions of work done by university-based consultants. There has been continued discussion and empirical analysis of the role of advocacy in community development since these foundations were set. In their recent works, Brenman and Sanchez (2012) and Walls (2015) apply these lessons and reframe the discussion of advocacy and equity in community development to the contemporary context.

Wedding values rooted in advocacy and equity to community development practice has been an impetus behind efforts to expand the scope of public participation in the policy process. The development of these streams of thought have paralleled one another. When Arnstein (1969) introduced the ladder of citizen participation, her argument was strongly influenced by contemporaneous literature on advocacy, black power and civil rights. In that work, she argued for community control in the public participation process facilitated by the empowerment of community-based organizations. One of the enduring contributions of Arnstein's work is her ladder of citizen participation which divides the scope of participation across varying degrees of nonparticipation, tokenism and citizen power. This typology and the themes it is built upon have influenced subsequent literature on community development and empowerment. In her review of the literature on public participation, Roberts (2004: 343) revisited these conceptual foundations reaffirming that citizen participation "intentionally seeks to level the playing field among the participating social actors" and that it is necessary for community development practitioners to invite more grassroots interests to the table as part of any citizen participation process. Silverman (2005) reached similar conclusions in his analysis of community development corporations, and he offered extensions to Arnstein's ladder in his articulation of a citizen participation continuum.

Despite calls for greater advocacy and expanded citizen control, the public participation process remains somewhat inaccessible to grassroots groups. Roberts (2004: 343) argues that it is dominated by stakeholders whose privileged status is based on expertise, money or position, and that the public participation process is primarily driven by top-down directives from legislative and executive authority. Others, like Chaskin (2005) and Silverman (2009), echo these views, arguing that the public participation process is compromised by tension between democratic interests at the grassroots level and public officials operating from rational planning frameworks or pursuing traditional patronage politics. Similarly, McGovern (2013: 310) found that even progressive local regimes lacked fidelity to public participation, "scaling back outreach efforts, co-opting citizen advocacy, and managing public forums in ways that dampened critical reflection and deliberation."

Given the relative inaccessibility of the public participation process to grassroots interests, there is a need to pay added attention to expanding advocacy and citizen control in community development work. In particular, this need continues to be acute in working class and minority communities where inner-city revitalization is underway – the same types of communities that were the focus of Davidoff's and Arnstein's critiques from over a half-century ago. Although the context of inner-city revitalization has shifted since it was hotly contested during the urban renewal period, it has remained a focal point in debates concerning the need for community control in the public participation process. During urban renewal, working class and minority residents faced displacement as neighborhoods were cleared to make way for freeways, civic centers and other large-scale development projects (Anderson, 1967; Teaford, 1990). During the waning years of urban renewal, Worthy (1977) chronicled how grassroots interests asserted themselves to challenge hospitals, universities and other institutional investors that were pursuing neighborhood development projects. Today, urban revitalization no longer occurs against

Table 13.1 Typology of competing interests in public participation

Stakeholder Group	Degree of Participation	Grassroots Advocacy	Equity Goals	Outcome Orientation
Local Government	Nonparticipation → Tokenism	Weak	Low → Moderate	Status Quo
Institutional Stakeholders	Nonparticipation → Tokenism	Weak	Low → Moderate	Status Quo
Private Consultants	Nonparticipation → Tokenism	Weak	Low → Moderate	Status Quo
Community Development Professionals	Tokenism → Citizen Power	Moderate	Moderate → High	*Wildcard*
Advocacy Group	Citizen Power	Strong	High	Redistributive
Residents	Citizen Power	Strong	High	Redistributive

the backdrop of urban renewal; instead it is discussed in the context of gentrification. Still, neighborhood development projects are driven by institutional investments made by hospitals, universities, local government and large cultural institutions (Adams, 2003; Birch 2009; Silverman et al., 2014). Now, as before, calls for greater citizen control in the public participation process and more equitable inner-city revitalization remain resonant.

The typology presented in Table 13.1 offers greater insight into the dynamics of how competing interests affect the public participation process. We argue that this typology is applicable in a variety of contexts, but our analysis applies it to urban revitalization in an inner-city context. The typology illustrates how stakeholder groups are anticipated to engage in the public participation process across four dimensions. The first dimension represents the degree of participation that individual stakeholder groups are expected to gravitate toward. This dimension conforms to Arnstein's ladder of citizen participation which classifies public participation as occurring in various degrees of nonparticipation, tokenism and citizen power. The second dimension represents the level of commitment that individual stakeholder groups are expected to exhibit toward grassroots advocacy in the public participation process. Commitment to grassroots advocacy ranges from strong to weak in the typology. The third dimension represents how individual stakeholder groups are expected to prioritize equity goals during the participation process. The prioritization of equity goals ranges from high to low in the typology. The fourth dimension represents the expected outcome orientation of individual stakeholder groups. The outcome orientation of stakeholder groups is binary, leaning toward policies that support status quo outcomes or redistributive outcomes.

In Table 13.1, we apply the typology to stakeholder groups that are often engaged in public participation processes related to urban revitalization projects in inner-city contexts. At the individual stakeholder group level, the typology is used to predict the positions participants in the process will assume across the four dimensions outlined above. However, the typology also allows us to predict where coalitions may form across stakeholder groups. For instance, the typology predicts a high degree of congruence between local government, institutional stakeholders and private consultants. Likewise, the typology predicts a high degree of similarity between advocacy groups and residents.

Using the typology, we predict that two coalitions are expected to emerge from the public participation process: One that is predisposed toward status quo policies and the second seeking redistributive policies. However, the typology also identifies community development professionals as occupying a middle-ground, or wildcard, position, in the public participation process. This is a critical position in the policy process because the leanings of community development professionals in the public participation process can result in outcomes that either solidify the status quo or strengthen arguments for redistributive policies. It should also be noted that in the latter case, where community development practitioners lean toward redistributive outcomes, a relative stalemate may result between the two coalitions.

In the next section of this chapter we apply the typology of competing interest in public participation to a project we conducted in Buffalo, New York as a team of students and faculty from the University at Buffalo. The project involved a multi-stage public participation process designed to analyse residential displacement and generate policy recommendations to promote equitable neighborhood revitalization. The research was conducted as part of a multi-city analysis of neighborhood change focused on identifying ways to curb displacement in gentrifying neighborhoods. The multi-city analysis was part of a larger research project called Turning the Corner (TtC) which was done in collaboration with the Urban Institute and partners from four other cities across the United States: Detroit, Milwaukee, Phoenix and the Twin Cities (Taylor et al., 2018). The focus of that project was to identify planning strategies to address negative externalities caused by neighborhood change and heightened risks of displacement due to inner-city revitalization. Data used in this analysis were collected through a series of stakeholder meetings and focus groups with renters, homeowners, city planners and other stakeholders. We were part of a team of university-based researchers leading the project in Buffalo, and effectively assumed the role of community development professionals.

APPLYING THE TYPOLOGY TO INNER-CITY REVITALIZATION IN BUFFALO

The research conducted in Buffalo for the TtC project involved three tiers of public participation. Combined, the three tiers were used to create sustained interaction with stakeholders in order to generate community development recommendations that promoted equitable inner-city revitalization. This purpose was reiterated throughout the research process to participants in order to maintain a focus on addressing inequalities in the urban revitalization process. Also, the researchers were positioned as advocates for minority and low-income residents disenfranchised from local decision-making in the public policy process.

In the first tier of public participation, a citywide group of stakeholders from local government, philanthropic organizations, advocacy groups, the business community and local anchor institutions was assembled to review data collected by the research team. The data measured trends in population and housing characteristics at the census tract level. The analysis applied an adaptation of the methodology developed by Lisa Bates (2013) to identify neighborhoods at risk of gentrification and displacement. Using that methodology, neighborhood indicators were summarized for the participants who analysed them collaboratively. Those deliberations led to the ranking of neighborhoods at risk of gentrification and the identification of three neighborhoods for more extensive analysis during the next stages of the project.

The second tier of public participation involved leaders from neighborhood-based organizations and other nonprofit service providers based in the three neighborhoods identified as transitioning. These neighborhood-based stakeholders participated in a workshop where prior work done on the project was reviewed. The neighborhood-based stakeholders were asked to provide feedback on the project, as well as participate in the design and planning of subsequent stages of the research. This was an intentional step which aimed to solidify the perspectives of neighborhood-based stakeholders into the analysis. The elevation of neighborhood-based stakeholders at this stage of the work allowed for them to serve as a ballast against citywide stakeholders who participated in the neighborhood selection process for the study. This reinforced our focus on addressing inequalities in the urban revitalization process and emphasized our role as advocates.

During the third tier of public participation, focus groups were held in each neighborhood with renters, homeowners and other neighborhood-based stakeholders to discuss perceptions of displacement and gentrification. A total of nine focus groups were held. In each neighborhood, one focus group was held with renters, one was held with homeowners and one was held with other neighborhood-based stakeholders. Data from the focus groups were used to construct a narrative informed by standpoint theory to generate policy recommendations promoting more equitable neighborhood revitalization (Adler and Jermier, 2005; Anderson, 2017). Standpoint theory focuses on amplifying the voices of groups traditionally disenfranchised from the planning and policy processes. Researchers who apply standpoint theory make a conscious decision to reference the perspective of disenfranchised groups at each stage of data collection and analysis. By emphasizing the perspectives of these stakeholders over others in the policy process, we empowered them through the participation process. This approach to community development research was compatible with our efforts to pursue advocacy-driven research.

At the end of the process, all the stakeholders involved in the study participated in a peer review of the draft project report. The draft report was distributed to all the individuals who participated in the process; they were given opportunities to submit written comments, as well as attend a workshop where the report was summarized and further discussed. Those comments were used to inform the final report and to inform the policy recommendations that grew out of it.

The next three subsections of this chapter discuss how competing interests shaped the project during different stages of the participation process. The first subsection examines how we managed competing interests that were most pronounced during the first tier of the public participation process. This subsection focuses on how disputes over the definition of gentrification, competing goals of participants and disagreements about data and its analysis shaped subsequent steps in the research process. The second subsection examines how we managed competing interests during the second and third tiers of the public participation process. This subsection focuses on disputes over neighborhood boundaries, discontent about prior studies of participants' neighborhoods, and how suspicions about collusion between us and local government shaped the degree to which we built trust with participants at the grassroots level. The third subsection examines how we managed pressures from city planners who disputed the finding and recommendations of the research, and who devalued perceptions of grassroots participants. This subsection discusses how decisions related to managing resistance from local government ultimately shaped the design and outcome of the research.

Definitions and Data

In the first tier of the public participation process, a citywide group of stakeholders was assembled to review data collected by the research team. The group was composed of representatives from local government, philanthropic organizations, advocacy groups, the business community and local anchor institutions. The group was evenly balanced in terms of membership. The group was asked to collaboratively analyse data measuring trends in population and housing characteristics at the census tract and block group levels. After two separate workshops, those deliberations produced a ranking of neighborhoods at risk of gentrification and the identification of three neighborhoods for more extensive analysis during subsequent stages of the project.

Although the group eventually reached a consensus on the three neighborhoods to be examined in greater detail, the deliberations leading up to this decision revealed cleavages, or differences, between stakeholder groups. These cleavages were consistent with characteristics identified in the typology of competing interests in public participation. During the examination of the neighborhood indicators, members of local government and institutional stakeholders were most apt to resist the selection of neighborhoods where their institutional investments and interests were prominent. In part, this resistance was a reaction to our position as advocates for grassroots interests in the processes. Consequently, these stakeholders bared concerns that greater attention to neighborhoods they were engaged in would generate demands for additional community benefits and linkages associated with revitalization projects in which they were invested. In contrast, members of advocacy groups were more apt to gravitate toward selecting neighborhoods where underrepresented groups were most at risk of displacement due to institutionally-driven urban revitalization. These groups were very receptive to the data and the methodology adopted to analyse it.

As the process unfolded and a subset of neighborhoods that were at risk of gentrification and displacement began to emerge, local government and institutional stakeholders stepped up their challenges. These challenges crystalized around three issues that had repercussions for subsequent stages of the project. First, these stakeholders disputed the underlying definitions used in the deliberations. For example, the definition of the term gentrification and its applicability to the city of Buffalo was disputed by them. It was argued that there were multiple definitions of gentrification and that no existing definition of gentrification applied to neighborhood change and revitalization efforts in Buffalo. Stakeholders from local government were particularly adamant about this point, and argued that there was no evidence of gentrification occurring in the city. In essence, conflicting definitions of gentrification were brought into the discussion of neighborhood change to obfuscate the analysis. Although the definition of gentrification was a focal point of the objections from representatives of local government and institutional stakeholders, they also disputed whether displacement was occurring in the city. When confronted with data showing population decline in neighborhoods undergoing revitalization, these stakeholders argued that the term displacement was being used in a pejorative manner. They argued that displacement was not detrimental to residents. Instead, they viewed it as a largely voluntary process in Buffalo where many residents simply chose to leave revitalizing areas due to personal preferences and tastes.

In addition to disputing how underlying definitions were used in deliberations with other stakeholders, representatives of local government and institutional stakeholders lodged a number of challenges to the appropriateness and accuracy of the data used in the analysis of neighborhood trajectories. A primary source of data used to examine neighborhood demographic trends came from the US Census' American Community Survey (ACS). Representatives from local government and institutional stakeholders disputed the accuracy of these data and argued that confidence intervals reported with ACS data rendered them an unreliable source of information about demographic and housing trends. Objections to the data also involved other issues. For instance, questions were raised about whether census tract level or block group analysis should be emphasized. Moreover, the 10 to 15-year timeframe for trend analysis was objected to, with city planners arguing that a time horizon dating back almost a half-century was necessary to examine contemporary patterns of neighborhood change. Just as the definitions of gentrification and displacement were obfuscated by these stakeholders in the deliberations, challenges to the baseline data used in the analysis served to block grassroots advocacy and the project's goal of promoting equitable outcomes.

Despite the objections of city planners and institutional stakeholders, advocacy groups engaged in this tier of the public participation process remained resolute as they referenced the data and made arguments in support of further analysis of the three neighborhoods that emerged from it. However, the consensus to move forward involved some concessions in response to objections raised by representatives of city government and institutional stakeholders. Most notable, we agreed to use the term gentrification sparingly in the future, since it was identified as a contentious issue by city planners and some institutional stakeholders. In its place, more neutral terminology, like neighborhood change, was adopted. Although seemingly semantic in nature, this decision had repercussions, particularly during the second tier of the public participation process.

Navigating Grassroots Distrust

During the second tier of the public participation process, leaders from grassroots organizations and other nonprofit service providers based in the three neighborhoods identified as at risk of residential displacement participated in a community workshop. The workshop had dual purposes. First, the grassroots stakeholders were provided with an overview of the prior work done on the project and a description of how their neighborhoods were identified for further analysis. The stakeholders were asked to give feedback to the research team and guidance as the project moved forward. Second, the neighborhood-based stakeholders assisted in the design of a recruitment strategy for the third tier of the public participation process. This assistance included the identification of participants for the resident and neighborhood-based stakeholder focus groups and a refinement of topics to be discussed in those meetings. The role of grassroots stakeholders at this juncture of the project was critical, since it enhanced community control in this aspect of the research design.

Despite our efforts to create a grassroots-driven process, grassroots stakeholders attending the workshop were initially distrustful. This was the result of two factors. First, there were no references to gentrification in the materials presented to grassroots stakeholders at the workshop. Second, in addition to our research team, members of the City's planning department attended the workshop. In reaction, grassroots stakeholders were livid, and voiced concerns that the absence of any discussion of gentrification showed that the project was flawed and being done in bad faith. From the perspective of the grassroots stakeholders, gentrification was a highly salient issue in the three neighborhoods selected for the analysis. The absence of this topic from the discussion was viewed as dishonest and a threat.

The identification of this concern by grassroots stakeholders represented a critical moment in the project. Although prior work had focused on empowering stakeholders through participatory research, conflicting stakeholder interests threatened to derail the project. A substantial part of the workshop was dedicated to a frank discussion about our intentions and autonomy from the City. In those discussions, the grassroots stakeholders were assured that we were committed to a community-driven process where stakeholders defined problems. We clarified that the City was not in control of the project and that our goal was to develop community-driven recommendations to present to policy-makers at the end of the process. Within that context, we reaffirmed our commitment to serve as honest brokers in the process and advocate for grassroots interests. These assurances were important and the grassroots stakeholders found them credible. In part, this was the case because members of our team had an extensive track record of working with some of the grassroots stakeholders, and the city planners attending the workshop acknowledged our autonomy.

Once trust was re-established with the community, we were able to continue with the planned workshop activities. Grassroots stakeholders provided feedback about the neighborhoods selected, their internal dynamics, and offered referrals and contact information for the neighborhood-based focus groups. Although the workshop ended on a positive note, we learned a valuable lesson from the grassroots stakeholders. It was essential to reaffirm our autonomy and commitment to empowering the community at every stage of the research process. At each stage, trust has to be re-established and the advocacy role of community development professionals has to be clearly articulated. In subsequent stages of the project, this was the first order of business. Although maintaining trust and autonomy is critical to any community development project, it is particularly important in the context of inner-city neighborhoods undergoing revitalization. At one level, grassroots and neighborhood-based stakeholders in this context are often fatigued due to the repeated studies of their neighborhoods by city government and consultants. In addition to being overstudied, grassroots stakeholders are understandably distrustful and disillusioned by the lack of tangible outcomes that these studies have yielded.

The issue of being overstudied and the paucity of community benefits growing out of past studies were expressed during both the second and third tiers of the public participation process. For example, across the focus groups held with neighborhood-based stakeholders, participants commented on how reports generated from past studies collected dust on shelves in City Hall, and how development and neighborhood improvements promised by the City never materialized. Against this backdrop, we made concerted efforts to distinguish our role in the public participation process from other outside groups. As the typology of competing interests in public participation shows, community development professionals occupy a middle ground between local government and institutional stakeholders with a status quo orientation and advocacy groups and residents with a redistributive orientation. Throughout the process, it was important for us to dissociate ourselves from other outside groups, align ourselves with grassroots interests and articulate a strong advocacy stance.

Resistance from Local Government

Following the third tier of the public participation process, we had assembled both quantitative and qualitative data measuring a spectrum of neighborhood characteristics and perceptions of neighborhood-based stakeholders. These data were analysed and a draft project report was written. Following the stakeholder-driven process adopted for the project, we circulated the draft report for peer review. The peer review process encompassed all of the stakeholder groups who were engaged in the three tiers of public participation. Copies of the draft report were circulated to approximately 70 individuals and they were given the opportunity to submit written comments, as well as attend a workshop where the draft report could be critiqued by all of the stakeholders. The feedback session was well attended by a cross-section of stakeholders who participated in each stage of the project. In addition to attending the feedback session, some institutional stakeholders submitted written comments about the draft report. Most notable among this subgroup were planning staff from the City.

On balance, most of the feedback on the draft report was supportive of its findings and recommendations. Suggestions from peer reviewers focused on the need for clarification about the analysis and recommendations, as well as suggestions related to policy implementation strategies. Unlike other stakeholders engaged in the process, planners from the City articulated strong opposition to a number of the draft report's facets. In hindsight, opposition from city planners was not surprising, since we perceived an undercurrent of City-based opposition to the participatory framework that was adopted early on in the project. The first sign that there was local government resistance to the project surfaced when it was initiated. Originally, the research team planned to hold a public event to kick-off, or launch, the project. At this event, the Urban Institute's TtC initiative would be described followed by a question-and-answer session about the research to be done in Buffalo. Local officials, community groups and the media were invited to attend. However, the kick-off event was canceled a few days before it was scheduled to occur at the request of the City. The cancelation of the event was due to the City's concerns about wine being served to attendees and fears that the topic of gentrification might fuel heated public debate. The cancelation of the kick-off event meant that there was limited public awareness of the project outside of the stakeholder groups invited to participate.

As described, local government and institutional stakeholders raised objections to the use of the term gentrification and the reliability of US Census data that formed a foundation for neighborhood selection during the first tier of the public participation process. Although we thought these issues were amicably resolved at that stage in the project, they resurfaced when the draft report was circulated. At this point, city planners submitted extensive written comments to the project team. Those comments re-articulated earlier objections about the reliability of using US Census data to describe neighborhood characteristics, as well as how trends were mapped, displayed and interpreted in the analysis. City planners also raised questions about the presentation and analysis of data from other federal sources, as well as proprietary data provided by the City. In addition to objections to the measurement and analysis of quantitative data, city planners disagreed with how perceptions of neighborhood trends were reported by focus group participants and argued that they should be removed from the analysis. The qualitative data reporting residents' perceptions received some of the harshest rebukes, with city planners arguing that the perceptions of neighborhood-based stakeholders were ill-advised, unrepresentative or not based in fact. In addition to challenging the quantitative and qualitative data, city planners challenged any language in the report that remotely suggested residents perceived gentrification occurring, or that City policies influenced neighborhood change. These objections led to extensive, and sometimes animated, dialogue between the research team and the city planners.

After taking the city planners' comments into account, revisions were made to the draft report. However, those revisions did not satisfy the representatives from the City. What ensued was an extensive process of redrafting the report with intermittent in-person meetings to discuss the report's text and provide clarification about data sources and methodology. In total, there were five sets of written comments submitted to the project team about the report over a five-month period of time, as well as multiple email exchanges, telephone calls and in-person meetings. In the end, the impasse was resolved with an agreement that the full TtC report would be released to the Urban Institute so it could be used in their multi-site analysis while a policy brief based on the report would be delivered to the City. As a result of this agreement, the full report was disseminated as part of the national study, but its local impact was curtailed.

This outcome is quintessential of the type of result predicted by the typology of competing interests in public participation. In this case, community development professionals sided with advocacy groups and residents, but they were not able to garner support from other stakeholders predisposed to oppose redistributive outcomes. The result was a stalemate. Although local government was relatively unsuccessful in its efforts to alter the content and thrust of the final report, it did put a damper on its dissemination. In retrospect, the curtailment of the final report's dissemination began with the cancelation of the kick-off event which removed the TtC project from public view. Successively, city planners challenged data, the scope of stakeholder input, analysis and recommendations at each stage of the participation process. This series of sustained challenges led to a low probability that any recommendations coming out of the study would be adopted by the City. However, there were three unexpected positives associated with this outcome.

First, much of the community-driven character of the final report was preserved. The city planners' efforts to change the analysis in the report and filter out its focus on grassroots and residents' perceptions were repelled to some extent due to the research team's adoption of an advocacy position at the initiation of the project. Our fidelity to a collaborative, community-driven research model was reinforced by our partnership with the Urban Institute, since the general approach adopted in the larger multi-site analysis called for community engagement and mixed methods analysis. Second, the existence of an external partner meant that the project had a life outside of Buffalo. Our partnership with the Urban Institute created a path for disseminating the final report to a national audience, regardless of the suppression of its visibility locally. This alternative path for dissemination also created a way to bypass the City and make the report available to local advocacy groups interested in drawing from it to influence local policy. Third, our ability to protect the community-driven character of the final report prevented it from being watered down and disseminated through the City's channels where it could be used to thwart grassroots opposition to displacement and gentrification. This meant that local debate over the merits of status quo versus redistributive community development policy would continue.

LESSONS LEARNED

The case examined in this chapter is instructive for community development professionals. It reminds us that public participation is a contested process where interests compete to influence the scope and outcomes of public policy. These interests can be observed at every stage of the community development process, and how they influence the process during early stages has ripple effects moving forward. This was exemplified in our discussion of how city planners challenged definitions used in the dialogue about neighborhood change and data used to analyse it at each stage of the participation process. This type of sustained opposition incrementally chipped away at efforts to advocate for redistributive outcomes in the neighborhood revitalization process and ultimately reduced the impact of public participation. More broadly, the case examined in this chapter serves as a reminder to community development professionals that we cannot assume that all stakeholders who engage in public participation have fidelity to the process.

The case examined in this chapter illustrates how the typology of competing interests in public participation can serve as a planning tool for community development professionals. The typology predicts how individual stakeholder groups fit into the participation process, and it predicts the degree to which they will emphasize equity goals and outcomes. The typology also helps community development professionals anticipate where coalitions will form across stakeholder groups. This foresight improves the ability of community development professionals to leverage their wildcard (middle-ground) role in the public participation process. In the case examined in this chapter, the typology assisted us in reaffirming our advocacy role at each stage of the public participation process. This helped buffer us from pressures by local government and institutional stakeholders to abandon our role as advocates for redistributive outcomes. Understanding the wildcard role that community development professionals fill in public participation made us more effective advocates for groups traditionally disempowered in the process.

The case examined in this chapter also serves as a reminder that it is essential for community development professionals to maintain a high level of public visibility throughout the public participation process. In some respects, we fell short of achieving this in Buffalo. Our project was obscured from public view early on with the cancelation of the kick-off event. Removing the project from the public eye early in the process helped to clear the path for the curtailment of the dissemination of the final report. However, local efforts to suppress the dissemination of the final report were counteracted by our relationship with an outside partner, the Urban Institute, which opened up an alternative avenue for dissemination.

Finally, the case examined in this chapter highlights that the default position in the public participation process is typically to reach a stalemate. This should not be discouraging to community development professionals. As our analysis of public participation in Buffalo illustrates, a stalemate in the process at the local level does not preclude the adoption of asymmetrical strategies to advocate for redistributive policies. We accomplished this by disseminating our final report through outside partners, as well as sharing our experiences and insights with other community development professionals through publication like this one. In retrospect, community development professionals must appreciate that a stalemate is not a defeat. In contrast, it is an impetus for continued advocacy.

NOTE

* The project that is reflected upon in this chapter was supported by a research grant from the Ralph C. Wilson Jr. Foundation. This chapter was based on reflections from a larger national effort titled "Turning the Corner: Monitoring Neighborhood Change for Action," a project guided by the Urban Institute's National Neighborhood Indicators Partnership and the Funders' Network Federal Reserve-Philanthropy Initiative. Launched in January 2016, the project piloted a research model that monitors neighborhood change, drives informed government action, and supports displacement prevention and inclusive revitalization. Local teams in Buffalo, Detroit, Milwaukee, Phoenix and the Twin Cities conducted independent research to understand neighborhood change and displacement risk in their communities. The Urban Institute, funded by the Kresge Foundation, is synthesizing lessons across the five cities. For more information, see http://www.neighborhoodindicators.org/turningthecorner (accessed 11 October 2018).

REFERENCES

Adams, C. (2003). The meds and eds in urban economic development. *Journal of Urban Affairs*, 25(5): 571–88. doi: 10.1111/j.1467-9906.2003.00003.x.
Adler, P. and Jermier, J. (2005). Developing a field with more soul: Standpoint theory and public policy research for management scholars. *The Academy of Management Journal*, 48(6): 941–4. doi: 10.5465/amj.2005.19573091.
Anderson, E. (2017). Feminist epistemology and philosophy of science, *The Stanford Encyclopedia of Philosophy*, Edward N. Zalta (ed.), accessed 11 October 2018 at https://plato.stanford.edu/archives/spr2017/entries/feminism-epistemology/.
Anderson, M. (1967). *The Federal Bulldozer*. New York: McGraw-Hill.
Arnstein, S.R. (1969). A ladder of citizen participation. *Journal of the American Institute of Planners*, 35(4): 216–24. doi: 10.1080/01944366908977225.
Bates, L.K. (2013). *Gentrification and Displacement Study: Implementing an Equitable Inclusive Development Strategy in the Context of Gentrification*. Portland: City of Portland, Bureau of Planning and Sustainability.
Birch, E.L. (2009). Downtown in the new American city. *ANNALS of the American Academy of Political and Social Science*, 626(1): 134–53. doi: 10.1177/0002716209344169.
Brenman, M. and Sanchez, T.W. (2012). *Planning as if People Matter: Governing for Social Equity*. Washington, DC: Island Press.
Chaskin, R.J. (2005). Democracy and bureaucracy in a community planning process. *Journal of Planning Education and Research*, 24(4): 408–19. doi: 10.1177/0739456X04270467.
Creighton, J.L. (2005). *The Public Participation Handbook: Making Better Decisions through Citizen Involvement*. San Francisco: Jossey-Bass.
Davidoff, P. (1965). Advocacy and pluralism in planning. *Journal of the American Institute of Planners*, 31(4), 331–8. doi: 10.1080/01944366508978187.
Forester, J. (1999). *The Deliberative Practitioner: Encouraging Participatory Planning Processes*. Cambridge: MIT Press.
Jones, B. (1990). *Neighborhood Planning: A Guide for Citizens and Planners*. New York, Routledge.
Krumholz, N. (1990). *Making Equity Planning Work: Leadership in the Public Sector*. Philadelphia: Temple University Press.

McGovern, S.J. (2013). Ambivalence over participatory planning within a progressive regime: Waterfront planning in Philadelphia. *Journal of Planning Education and Research*, 33(3): 310–24. doi: 10.1177/0739456X13481246.

Needleman, M.L. and Needleman, C.E. (1974). *Guerrillas in the Bureaucracy: The Community Planning Experiment in the United States*. New York: John Wiley & Sons.

Roberts, N. (2004). Public deliberation in an age of direct citizen participation. *American Review of Public Administration*, 34(4): 315–53. doi: 10.1177/0275074004269288.

Shipley, R. and Utz, S. (2012). Making it count: A review of the value and techniques for public consultation. *Journal of Planning Literature*, 27(1): 22–42. doi: 10.1177/0885412211413133.

Silverman, R.M. (2005). Caught in the middle: Community development corporations (CDCs) and the conflict between grassroots and instrumental forms of citizen participation. *Community Development*, 36(2): 35–51. doi: 10.1080/15575330509490174.

Silverman, R.M. (2009). Sandwiched between patronage and bureaucracy: The plight of citizen participation in community-based housing organizations in the US. *Urban Studies*, 46(1): 3–25. doi: 10.1177/0042098008098634.

Silverman, R.M., Lewis, J. and Patterson, K.L. (2014). William Worthy's concept of "institutional rape" revisited: Anchor institutions and residential displacement in Buffalo, NY. *Humanity & Society*, 38(2): 158–81. doi: 10.1177/0160597614529114.

Silverman, R.M., Taylor, H.T. and Crawford, C. (2008). The role of citizen participation and action research principles in main street revitalization: An analysis of a local planning project. *Action Research*, 6(1): 69–93. doi: 10.1177/1476750307083725.

Taylor, H.T, Silverman, R. and Yin, L. (2018). *Buffalo Turning the Corner*. Buffalo: University at Buffalo, Center for Urban Studies.

Teaford, J.C. (1990). *The Rough Road to Renaissance: Urban Revitalization in America, 1940–1985*. Baltimore: Johns Hopkins University Press.

Walls, D. (2015). *Community Organizing: Fanning the Flames of Democracy*. Cambridge: Polity Books.

Worthy, W. (1977). *The Rape of Our Neighborhoods: And how Communities Are Resisting Takeovers by Colleges, Hospitals, Churches, Businesses, and Public Agencies*. New York: William Morrow.

14. Methods and framework of participatory action research for community development in Bangladesh
M. Rezaul Islam

INTRODUCTION

Participatory action research (PAR) is one of the important tools in social sciences research. This tool is particularly pertinent in the community development field though it is not widely examined and innovated in Bangladesh. PAR has been developed as a set of tools that help to enable community people to be actively involved in generating knowledge about their cultures, livelihoods and conditions and how these can best be transformed (Fals-Borda and Rahman, 1991). PAR is basically the methodology that was originally proposed by Kurt Lewin in 1940. PAR is a subset of action research of qualitative methods. The focal point of this qualitative method is the entire human experience and the meanings ascribed by individuals' living experience; broader understanding and deeper insight into complex human behaviors thus occur as a result (Lincoln, 1992; Mason, 2006). Lincoln (1992) argued that this kind of qualitative method is a naturalistic, participatory mode of inquiry that discloses the lived experiences of individuals. As multiple realities this method is based on subjective experience and circumstance (Wuest, 1995), and the prime goal is to interpret and document an entire phenomenon from an individual's viewpoint (Creswell, 1998; Mason, 2006). This approach attempts to uncover the world through another's eyes, in a discovery and exploratory process that is deeply experienced (Gilbert, 2001). This is systematic and orientated around analysis of data whose answers require the gathering and analysis of data and the generation of interpretations directly tested in the field of action (Greenwood and Levin, 1998). This kind of research is termed as a number of subheadings such as participatory action research, participatory research, community-based participatory research, and other forms of participative inquiry (Greenwood and Levin, 1998; Gibson et al., 2001).

The purpose of all action research is to impart social change, with a specific action (or actions) as the ultimate goal (Greenwood and Levin, 1998; McNiff and Whitehead, 2006). This type of research is successfully used with an agenda for social change in community development that embodies the belief of pooling knowledge to define a problem in order for it to be resolved (Greenwood and Levin, 2006). The epistemological assumptions of such kind of research hold knowledge creation as an active process (McNiff and Whitehead, 2006). This kind of research is called community-based participatory action research (CBPAR) which is a collaborative approach to research that involves all stakeholders throughout the research process, from establishing the research question, to developing data collection tools, to analysis and dissemination of findings. It is a research framework that aims to address the practical concerns of people in a community and fundamentally changes the roles of researcher and who is being researched. It is applied research; it seeks to change issues that are critical to communities and focuses on engaging community members in research directed at addressing their social concerns. Islam and his associates (e.g. Islam, 2018, 2017, 2014a and b; Islam and Shamsuddoha, 2017; Islam and Hasan, 2016; Paul and Islam, 2015; Islam and Hossain, 2014; Islam and Morgan, 2012a and b) conducted a number of studies with the nongovernmental organizations' (NGOs') community empowerment projects with the blacksmiths, goldsmiths, and climate change and disasters affected char land and coastal people in Bangladesh, where the PAR method was partially used. The main limitation of those studies was that the methodological section was not fully based on PAR framework. Based on the experiences of those studies this chapter aims to develop methods and framework of PAR in Bangladesh.

CONCEPTUAL AND THEORETICAL CONTEXT: COMMUNITY DEVELOPMENT VERSUS PARTICIPATORY ACTION RESEARCH (PAR)

Community Development

The definition of community development is not clear in the literature as both terms "community" and "development" have been problematic in the literature. A community is a small or large social unit that has something in common, such as norms, religion, values or identity. Communities often share a sense of place that is situated in a given geographical area or in virtual space through communication platforms. Brint (2001) suggests that communities are connected primarily through common emotions and personal interests in one another. He defines communities as aggregates of people who share common activities and/or beliefs and who are bound together principally by relations of affect, loyalty, common values and/or personal concern (i.e. interest in the personalities and life events of one another). Each community establishes traditions and patterns of behavior which may be implied or written as rules. Members of a community share some kind of a bond such as location, interests, background or identity, situations or experiences. Thus a community is a social institution, that is, a stable structure and agreed set of procedures and conventions that provides social order and meaning (Scott, 2001). On the other hand, development is the process in which someone or something grows or changes and becomes more advanced (Cambridge Dictionary, 2019). Development is again a wide and multidimensional concept. It has different meanings and scope in different disciplines and there is no consensus on its universal meaning and definition. In general, development means the growth of an individual, society, a nation or entire world in terms of both economic and non-economic activities. One of the simplest definitions of development is provided by Robert Chambers. Chambers (2004) mentioned development as the notion of good change. Sen (1999) found development as a freedom where attention is thus paid particularly to the expansion of the "capabilities" of persons to lead the kind of lives they value and have reason to value.

The concept of community development has been muddled with some associated words such as "community participation" (Hickey and Mohan, 2004), "community empowerment" and "community cohesion" (Ratcliffe and Newman, 2011) along with "community education" towards "active citizenship" (Annette and Mayo, 2010). Community development has been and continues to be a contested field, characterized by varying definitions and competing theoretical perspectives, aims and objectives (Mayo, 2008). Craig et al. (2011) argue that community development has always had an ambiguous nature. They add that this is a goal, self-evidently the development of communities, in the context of social justice agendas, notwithstanding the different interests at work in defining what this is all about and why it matters. It has been promoted top-down to facilitate self-help in order to legitimize reductions in service provision, shifting responsibilities from the public sector to the voluntary and community sectors, whilst opening up new spaces for the private market.

In many ways, the concept of community development has been linked with community actions such as social needs, promoting cooperation, equalities and social and environmental justice from the bottom-up. In this regard, Taylor (2011) argues that community development may be promoted in ways that move beyond the *top-down/bottom-up* dichotomy, working both sides of the equation. The underlying aims are to strengthen democratic processes and promote social justice agendas, enabling the voices of the most disadvantaged to be heard more effectively, whilst setting out to transform rather than to bypass or even undermine the structures of public service provision (Craig et al., 2011). Like Taylor, Paulo Freire's approach to community education and experiential learning aimed to enable oppressed people to develop a critical understanding of their situation, questioning previously accepted ideas in order to develop strategies for social change, actively engaging with others, collectively, to transform oppressive social relationships.

Participatory Action Research

PAR is a course of action of working collaboratively with those most involved or impacted by the issue. Freirian is possibly the first person who used the PAR initiatives (Le Grange, 2009). Over the past quarter-century or more, participatory research methods have been applied in Africa, Asia and Latin America, as well as in the Global North (Tandon, 2005). Through the application of innovative methods, including the use of popular theater (Boal, 1979), experts' monopoly of knowledge has been challenged, and alternative approaches to research have been developed. PAR attempts to create new knowledge that incorporates multiple perspectives by systematic inquiry with the collaboration of those affected by the issue being studied, for the purposes of education and taking action on effecting policy change (Macaulay et al., 1999).

Reason and Bradbury (2001) mention that PAR is a participatory, democratic process concerned with developing practical knowledge in the pursuit of worthwhile human purposes, grounded in a participatory worldview which we believe is emerging at this historical moment. It seeks to bring together action and reflection, theory and practice, in participation with others, in the pursuit of practical solutions to issues of pressing concern to people, and more generally the flourishing of individual persons and their communities. Community-based participatory research is defined as a collaborative research approach that is designed to ensure and establish structures for participation by communities affected by the issue being studied, representatives of organizations, and researchers in all aspects of the research process to improve health and well-being through taking action, including social change (Viswanathan et al., 2004). As Downey et al. (2010) explain, community-based participatory research diminishes the traditionally heightened power of the researchers through a community-driven approach to change with the goal of strengthening a community's problem-solving capacity through collective engagement in the research process (Viswanathan et al., 2004). Macaulay et al. (1999) mention this as a process of creating new knowledge that incorporates multiple perspectives by systematic inquiry with the collaboration of those affected by the issue being studied, for the purposes of education and taking action on effecting policy change. PAR is a systematic collection and analysis of data for the purpose of taking action and making change by generating practical knowledge (Gillis and Jackson, 2002). Koch et al. (2002) define PAR as a qualitative research methodology option that requires further understanding and consideration. PAR is considered a democratic, equitable, liberating and life-enhancing qualitative inquiry that remains distinct from other qualitative methodologies. MacDonald (2012) also considers PAR as qualitative features of an individual's feelings, views, and patterns which are revealed without control or manipulation from the researcher. The participant is active in making informed decisions throughout all aspects of the research process for the primary purpose of imparting social change; a specific action (or actions) is the ultimate goal.

PAR comprises a family of research methodologies which aim to pursue action and research outcomes at the same time. It therefore has some components which resemble consultancy or a change agency, and some which resemble field research. The focus is action to improve a situation and the research is the conscious effort, as part of the process, to formulate public knowledge that adds to theories of action that promote or inhibit learning in behavioral systems. In this sense the participatory action researcher is a practitioner, an interventionist seeking to help improve client systems. However, lasting improvement requires that the participatory action researcher helps clients to change themselves so that their interactions will create these conditions for inquiry and learning. Hence to the aims of contributing to the practical improvement of problem situations and to the goals of developing public knowledge we can add a third aim of PAR, to develop the self-help competencies of people facing problems. Valuing everyone's strengths and ensuring that all voices are heard results in a unique capacity-building approach. Working together allows us to engage with collaborators from various backgrounds, perspectives and disciplines, including community members, practitioners and producers. Through participatory action research, the conditions are created to support innovation and change around issues of food security. The guiding principles of PAR are (MacDonald, 2012):

- Meaningful relationships;
- Sharing power;
- Building individual, organizational, community and systems capacity;
- Participatory methods and leadership approaches;
- Transformative ways of understanding and taking action;
- The unique contributions and perspectives of all team members and participants;
- Responsive and accountable leadership;
- Clear and transparent decision-making processes;
- Accessibility of opportunity to participate; and
- Activities rooted in real community needs.

POSSIBLE RESEARCH FRAMEWORK AND METHODS: ANSWER OF WHAT, WHY AND HOW

A CBPAR framework conceives many aspects and perspectives. Advancement Project – Healthy City (2011) mentions that CBPAR recognizes that:

- Community knowledge is one-off and provides key insights that ground-truths administrative data;
- Complex social issues often cannot be well understood or resolved by "expert" research;
- Interventions from outside of the community have often had disappointing results;
- Communities should have equal inclusion and collaboration in the identification, research and resolution of community issues;
- There is value and legitimacy in the knowledge of individuals, families and others in the community.

CBPAR has a specific goal of collaborative research that engages stakeholders in an iterative research and action process. Figure 14.1 shows that CBPAR links research and action, recognizing that social action requires further research and social research requires further action.

This research framework is highly recommended for the application of methods of data collection and analysis to generate findings that have highly practical results. The key stakeholders for such findings are typically made up of community members, practitioners and local policymakers who wish to design an intervention that benefits a geography-based community. It thrives with community engagement in the research process, especially through primary data collection methods such as interviews, focus groups and community-engaged mapping. This community engagement ensures that those who represent that place, particularly those who reside there, ground this research with their unique perspectives and experiences.

The core principles and values of the CBPAR framework ensure that the community members from all levels will participate in the research process and develop outcomes that they can use to make changes in their own communities (Ahmed and Palermo, 2010). However, it requires a high level of contact and interaction between researchers and the community to engage in "a time-intensive process of collaboration over research design, implementation, and analysis". Burns et al. (2011) used a participation continuum (Figure 14.2) of how a researcher involves his/her community in research and where they fall on the continuum.

Various methods for data collection are used in PAR. For each specific issue or situation, the researcher and participants collaborate to establish the appropriate methods of data collection (Gillis and Jackson, 2002; McNiff and Whitehead, 2006). However, it is recommended that triangulation is an important feature in this research framework to increase the reliability and validity of this research framework (Streubert and Carpenter, 1995). Triangulation helps looking at a vibrant and complex situation in many ways which helps to bring out the reality of the complex situation of a community. Focus groups, participant observation and field notes, interviews, diary and personal logs, questionnaires and surveys are effective methods of data generation employed in PAR.

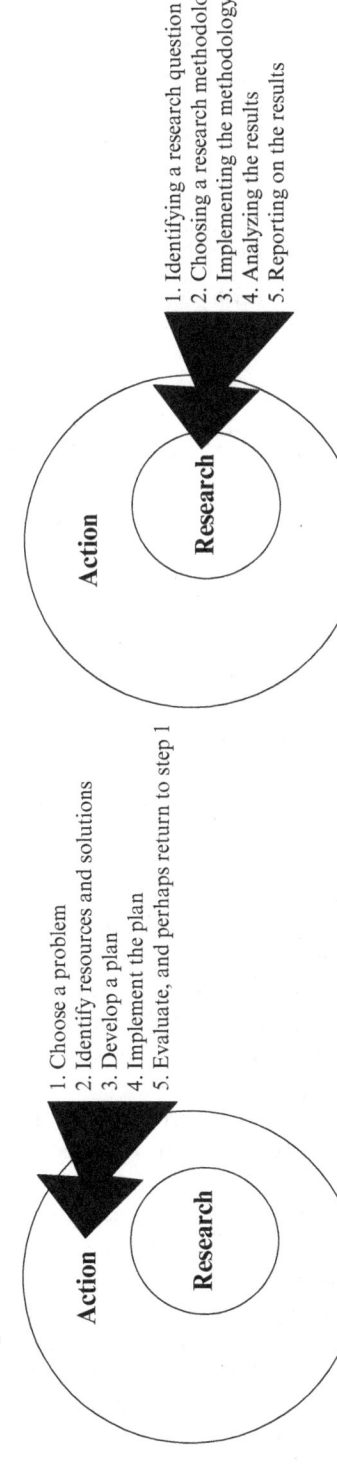

Source: Burns et al. (2011).

Figure 14.1 CBPAR links research and action

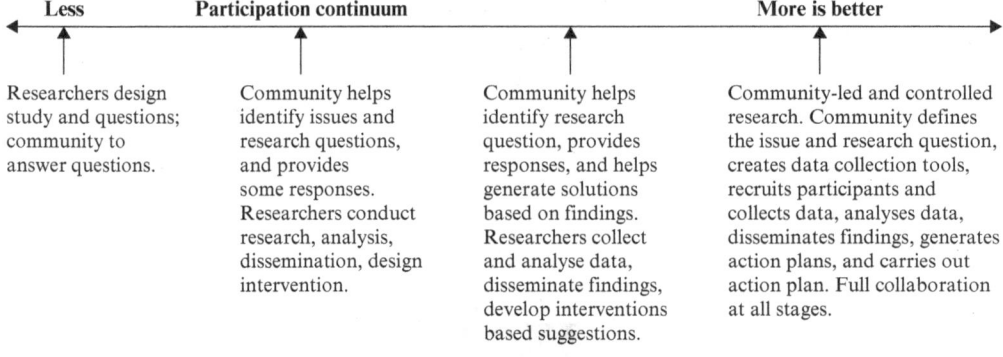

Source: Burns et al. (2011).

Figure 14.2 Participation continuum of CBPAR

Focus Group Discussions (FGD)

FGD is one of the important data collection tools that can visualize the real voices of a community. Kitzinger (1995) mentions that this method is considered a socially orientated process and a form of group interview that capitalizes on communication between the researcher and participants in order to generate data. A focus group generally consists of 7 to 12 individuals who share certain characteristics relevant to the focus of the study (Marshall and Rossman, 2006). The core emphasis is given to issues where the facts can be explored through a group discussion, as an individual usually does not provide such kind of information using other types of data collection methods such as key informants' interview, face-to-face interview, and the like. FGD is a useful method to collect data on issues such as corruption, sexual abuse, NGOs development debate, corporate responsibility, role of the family members in child protection, child labor, and so on. Here, the small number of individuals in a focus group facilitates an atmosphere for optimal communication among all participants towards generating useful data. During a focus group, the researcher creates a supportive environment in which discussion and differing points of view are encouraged (Marshall and Rossman, 2006). Ideally, in PAR, all participant viewpoints are recognized and valued, as all participants have an opportunity to communicate (McTaggart, 1991).

This kind of discussion is held in a cordial environment which is free from all sorts of aggravations and concerns so that participants can feel free and safe about expressing their desires and feelings (Islam, 2009). The discussion session is conducted with two members of the research team such as a moderator and a note taker. It is crucial that the moderator is fully impartial and free from all sorts of biases. A moderator will facilitate the whole discussion considering the time, quality of discussion and field management. The moderator will facilitate the participants in such a way so that each and every participant can have sufficient room and opportunity to talk. The moderator will manage the flow of the discussion so that a meaningful discussion can end properly. On the other hand, the note taker will take notes in writing and as a recording (as well as video recording if possible) throughout the discussion. A note taker's role is most crucial.

S/he will not take notes; rather s/he will memorize the main themes and points of this fruitful discussion. S/he will take such important points which are very relevant to the research objectives. S/he will also take the verbatim and point out the nature and types of data in order to restore the contextualization of the topic and community. In PAR, all involved in the research process are active participants throughout the entire research process (Greenwood and Levin, 1998; McNiff and Whitehead, 2006). Gillis and Jackson (2002) noted that even though the topic of discussion is left up to the focus group "the facilitator typically provides some structure". According to Morgan (1997), combining participant observation with focus groups is useful in gaining access to the group, focusing on sampling and site selection, while also useful for checking tentative conclusions and possible changes to be implemented.

Participant Observation

Participant observation is an innovative qualitative research method of inquiry and a rich source of data collection that is commonly employed in PAR (Marshall and Rossman, 2006). In many cases this data collection tool is the only method to collect concealed and complex contextual and cultural aspects of a community that may not be possible using other data collection tools. Gillis and Jackson (2002) and Mulhall (2003) argue that this kind of data collection tool provides the researcher with privileged access to research subjects in a social situation and captures the context of the social setting in which individuals function by recording subjective and objective human behaviors. The researcher becomes part of the process being observed and immersed in the setting, hearing, seeing and experiencing the reality of the social situation with the participants (Marshall and Rossman, 2006). In this aspect, Spradley (1980) comments that here the researcher is not only a participant-observer, not only observes activities, participants and physical aspects of the situation, but also engages in activities appropriate to the social situation. It is also important that the researcher should be careful that s/he should be alert to the context of the particular community that can significantly vary from one community to another. Marshall and Rossman (2006) further note that participant observation entails the systematic noting and recording of events, behaviors and objects in the social setting through the use of detailed and comprehensive field notes. The field note is a pen-picture of that community which should be written in such a way to visualize the real story of that community. This data collection method provides huge opportunities to a researcher so they can attain first-hand knowledge of social behavior as it unfolds over time in the social situation (Gillis and Jackson, 2002). As a result, the researcher obtains a broader view of what is occurring and has the opportunity to detail what is communicated and what is implicit in the situation (Streubert and Carpenter, 1995).

Interviews and In-Depth Interviews

PAR uses both face-to-face interviews and in-depth case interviews. Stringer (1999) mentions that interview is a method used in PAR which enables participants to describe their situation in such a way so that they can tell a researcher what is happening in the community and of what they are doing in their community. This method is mostly significant to capture the daily human activities and experiences (Kaufman, 1992;

Kvale, 1996). This interviewing offers researchers access to people's ideas, thoughts and memories in their own words, rather than the words of the researcher (Reinhartz, 1992). The researcher explores a few general topics to assist in uncovering the participant's perspectives, but demonstrates the utmost respect for how the participant frames and structures the responses (Marshall and Rossman, 2006). Ultimately, an interview is "a face-to-face verbal interaction in which the researcher attempts to elicit information from the respondent, usually through direct questioning" (Gillis and Jackson, 2002). Both the researcher and the participant share and learn throughout the interviewing process in a reciprocal manner. Again, throughout the PAR process all participants are active in the development of the interview guide, as well as data analysis. It is essential that interview questions be carefully formulated to ensure that participants are given maximum opportunity to present events and phenomena in their own terms and to follow agendas of their own choosing (Stringer, 1999).

STEPS OF PAR DESIGN FOR COMMUNITY DEVELOPMENT IN BANGLADESH

As I mentioned earlier, PAR as a research framework and method is complex and it needs to consider many aspects that include contextual perspectives and different types of local stakeholders. However, it is more inclusive and proper coordination is needed. As a result, there is always a tension between a number of forces that lead to personal, professional and social change. My own thinking is that this kind of action research needs a deep inquiry into one's practices in service of moving towards an envisioned future, aligned with values. However, an organized context is very essential to conduct this research. It is also a collaborative process as it is done in a social context and understanding the change means probing multiple understanding of complex social systems (Riel, 2019). German et al. (2012) provide a process that guides the various steps in PAR as shown in Figure 14.3.

German et al. (2012) argue that some of these steps are carried out by the facilitators or action research team, and others by the community (often with facilitation). These steps follow a logical sequence over time, and have functional linkages among them.

Figure 14.4 shows the illustration of the research design that encompasses four steps that include planning, preparation, data collection including research ethics and data analysis and evaluation.

Planning Step

This step includes rapport building with the community and stakeholders, team and partnership building with stakeholders, research design, resource allocation (time, manpower and funding) and research management. As I mentioned earlier, a successful PAR method is possible when different stakeholders, especially the researcher and community people, will come together. In a community there are different levels of partners such as community leader, members of civil society and general members of the community or the members of institutions or beneficiaries. It needs an inclusive partnership where both parties become more familiar with one another in terms of motivating factors, strengths, weaknesses and complementarities, and to build rapport. It helps to build trust, minimize

Methods and framework of participatory action research 233

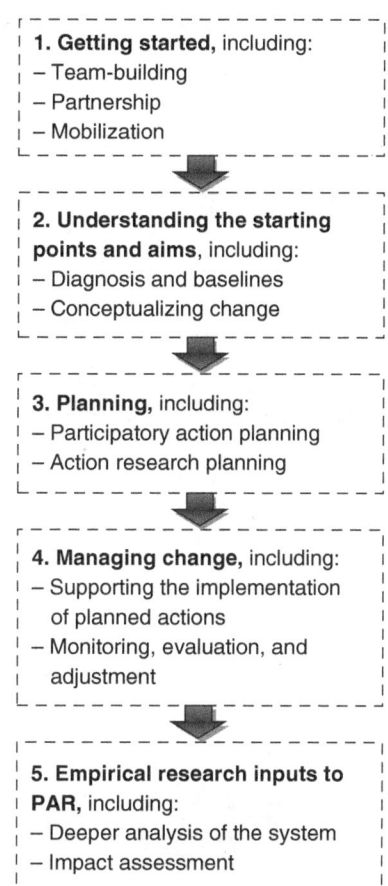

Source: Developed by the author based on German et al. (2012).

Figure 14.3   Process in the steps of PAR

suspicions by clarifying aims and clearing any doubts, and increases the chances that team members or partners will come to the table with a positive attitude. This facilitates developing a common understanding of the background of team members, ensuring that people are at the same level of understanding, exploring differences in work style, understanding the importance of working as a team, finding possible challenges, agreeing on roles and responsibilities for team members and partners. Research design is another important aspect in this planning step to conduct such qualitative research. A case study research design is usually used for this study. However, filed selection (such as community and respondents), sampling and possible data collection methods are important considerations in this stage. This is important as this kind of research needs much time where time and funding allocation and proper research management are components that need proper coordination and integration.

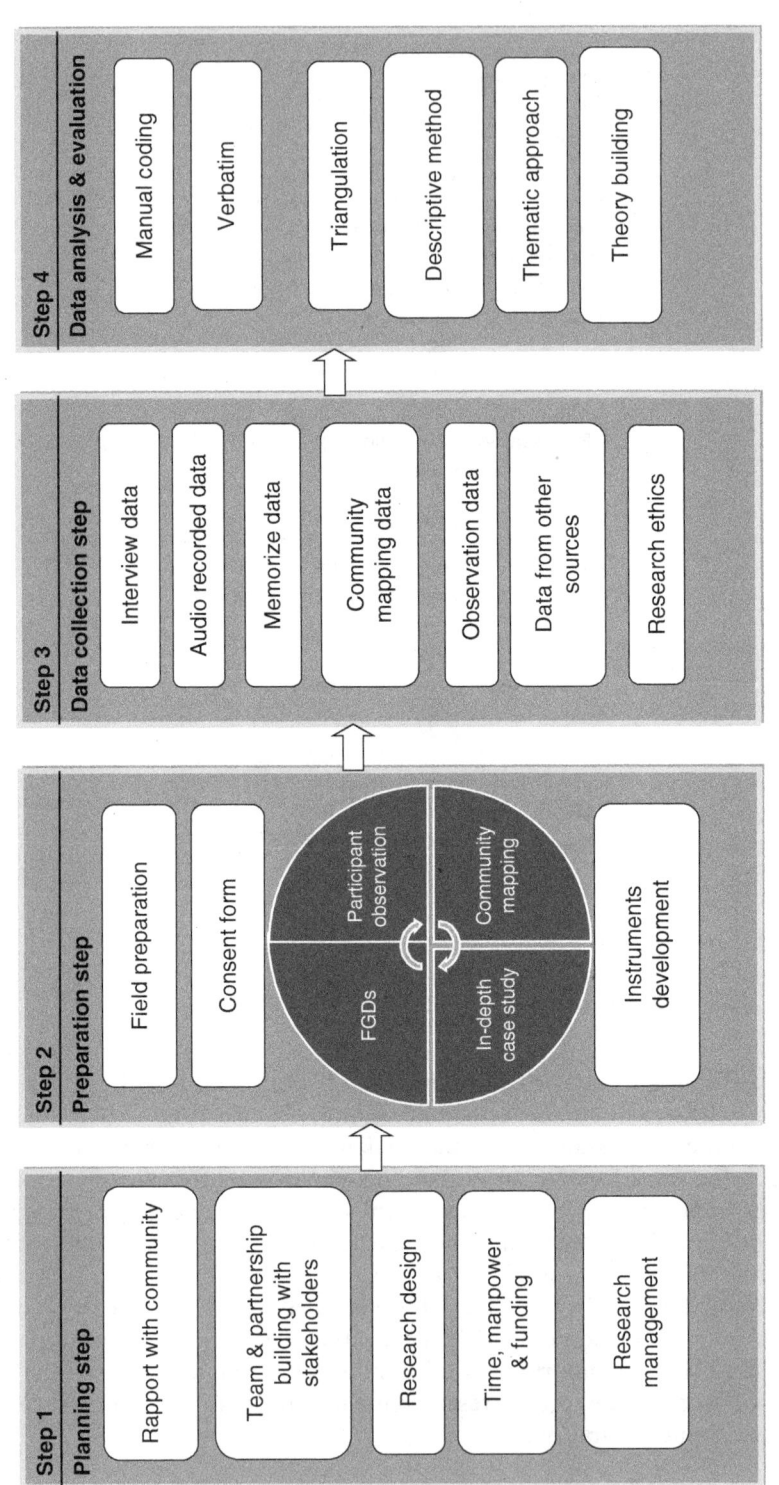

Figure 14.4 Steps of research design

Preparation Step

This step needs to consider four aspects namely field preparation, consent, data collection methods and development of data collection instruments. Initially, it is important to carry out proper field preparation; field preparation includes communication, mental and physical preparation, transport and readiness. Permission from the community to conduct the study is most crucial. Usually, there is power-relation in every community and it is important to consider local power structure, culture, community language and community settings for such study. Then, a researcher needs to think about proper data collection methods in the light of the research objectives, local context and culture, nature of respondents and timing. In this connection, to develop data collection methods is also important. Usually, an unstructured and open-ended data collection procedure is followed here.

Data Collection Step

This is a very crucial step where the qualitative data through multi-method data collection methods should be collected. In general, a researcher gathers data by interview, observation, audio-visual, memorizing data, and data from community mapping. Much data can be gathered through other sources such as local library, museum, local heritage, individual diary and visiting historical places. It should be very clear that these data collection methods will be employed in such a way that they can be suitable to collect specific types of data from the specific type of data collection methods. Some important data should be collected through one more data collection method for validation (Islam, 2009). Research ethics are an important consideration for such study (research ethics are discussed in a different section).

Sample of an in-depth case interview

In-depth Case Study (ICS) Guideline
Vulnerability, Social Dignity & Livelihood Choices of the River Bank Erosion Victims in Bhola District of Bangladesh

[The main objective of this research is to explore the vulnerability, social dignity and livelihood choices of the river bank erosion victims in Bhola District of Bangladesh. This research is funded by Islamic Relief Bangladesh. I will be grateful if you share your opinions regarding this topic. Please note that the information given by you will be used only for research purpose, and I will not disclose your opinions and I will strictly maintain the confidentiality of your information. You have every right to withdraw yourself from the interview at any time. After finishing the interview, I will let you know what I have written and recorded from you, and you have opportunity to add or delete any of your opinions/information that you will provide. Your cooperation would be highly appreciated]

Guideline Code # ....   Name of the Interviewer: ....
Type of respondent: .... Date of interview: ....
Union name: ....   Cell phone number ....
Upazilla Name ....

Your life history (born, brought-up, education, family life and employment)
Briefly tell me what you do during river erosion in your community
Tell me about one extreme suffering that you face by river erosion
Impacts of river erosion on the following aspects:
Your individual/household income and employment
Your food
Social life
Education
Cultural activities
Changes of your lives, livelihoods and practices due to river erosion
Living house and living place
Employment
Food habits
Social lives
Displacement/migration
Your observation about the government and nongovernment initiatives
Before river erosion
During river erosion
After river erosion such as rehabilitation

Sample of an FGD guideline

Focus Group Discussions (FGDs) Guideline
Vulnerability, Social Dignity & Livelihood Choices of the River Bank Erosion Victims in Bhola District of Bangladesh

[The main objective of this research is to explore the vulnerability, social dignity and livelihood choices of the river bank erosion victims in Bhola District of Bangladesh. This research is funded by Islamic Relief Bangladesh. We will be grateful if you share your opinions regarding this topic. Please note that the information given by you will be used only for research purpose, and we will not disclose any of your opinions and we will strictly maintain the confidentiality of your information. You have every right to withdraw yourself from the interview at any time. After finishing the interview, we will let you know what we have written and recorded from you, and you all have opportunity to add or delete any of your opinions/information that you will provide. All of your cooperation would be highly appreciated]

Guideline Code # ....   Name of the Interviewer: ....
Union name: ....   Date of interview: ....
Upazilla Name ....   Cell phone number ....

Origin of this community
River erosion in the community last 10 years
Causes of river erosion
Impacts of river erosion on the following aspects:
Your income and employment
Your food
Social life
Education
Cultural activities
Older people

Disabled people
Children
Widow
Pregnant mothers
Changes of your lives, livelihoods and practices due to river erosion
Living house and living place
Employment
Food habits
Social livings
Coping strategies
Resilience
Displacement/migration
Government and nongovernment initiatives:
Before river erosion
During river erosion
After river erosion such as rehabilitation

Sample of a key informants' interviews (KIIs) guideline

Key Informants' Interviews (KIIs) Guideline
Vulnerability, Social Dignity & Livelihood Choices of the River Bank Erosion Victims in Bhola District of Bangladesh

[The main objective of this research is to explore the vulnerability, social dignity and livelihood choices of the river bank erosion victims in Bhola District of Bangladesh. This research is funded by Islamic Relief Bangladesh. I will be grateful if you share your opinions regarding this topic. Please note that the information given by you will be used only for research purpose, and I will not disclose your opinions and I will strictly maintain the confidentiality of your information. You have every right to withdraw yourself from the interview at any time. After finishing the interview, I will let you know what I have written and recorded from you, and you have opportunity to add or delete any of your opinions/information that you will provide. Your cooperation would be highly appreciated]

Guideline Code #	Name of the Interviewer: ....
Type of respondent:	Date of interview: ....
Union name:	Cell phone number ....
Upazilla Name	

Your experience about river erosion in this region in the last 10 years
Impacts of river erosion on the following aspects:
Your income and employment
Your food
Social life
Education
Cultural activities
Older people
Disabled people
Children
Widow
Pregnant mothers
Changes of your lives, livelihoods and practices due to river erosion
Living house and living place

> Employment
> Food habits
> Social lives
> Displacement/migration
> Coping strategies of the river erosion
> Resilience of the river erosion victims
> Local human and social capitals to face river erosion
> Government and nongovernment initiatives for the river erosion victims
> Before river erosion
> During river erosion
> After river erosion such as rehabilitation
> Your suggestions to overcome river erosion problem

Data Analysis and Evaluation

This is again an important step where a researcher needs to formulate an appropriate data analysis plan and then data evaluation. There are a number of aspects such as coding, verbatim, data analysis framework (triangulation, thematic approach and descriptive method) and theory building. Usually a PAR method follows manual coding where the analytical text should be written in descriptive method using proper themes where a triangulation approach can provide solid data analysis. This is called a pen-picture of the community. Text should be rich, using many contextual and local examples. Verbatim either in quoting or dialogue form is mostly appropriate in order to visualize the real scenario and daily livelihoods of the community. Data analysis should run in such a way so that a reader can understand that this description is providing the real scenario of a particular community. This text will give the real feelings and real composition of the community so that s/he understands that this is happening in front of him/her. This is a live picture of the community. In the whole process, evaluation will guide the researcher in order to keep himself/herself on the right track. This evaluation also helps a researcher to avoid the biases and cultural and local conflicts, and then helps him/her to develop new theory.

BENEFITS OF PAR IN COMMUNITY DEVELOPMENT

PAR is an important tool which is commonly used for community development projects. In fact, without using the PAR method, it is quite impossible to implement community development projects at community level. According to Stringer (1999) and Selenger (1997), PAR is democratic, thus enabling the participation of all people; equitable, as it acknowledges equity of people's worth; liberating, in that it provides freedom from oppressive, debilitating conditions; and life-enhancing, which enables the expression of people's full human potential. McTaggart (1989) adds some other benefits such as it is an active approach to improving social practice through change; congruence on authentic participation; collaboration; establishing self-critical communities; and involving people in theorizing about their practices. The PAR method facilitates the community people to put the practices, ideas and assumptions about institutions to the test, involves record-keeping, requires participants to objectify their own experiences and involves making critical analysis. McTaggart (1989) articulated that PAR starts with small cycles and groups, and allows participants to build records while allowing and requiring participants to give a reasoned justification of their social (educational) work to others.

To consider the above discussion, we can summarize the benefits of PAR as follows. PAR:

- Facilitates collaborative, equitable partnerships in all phases of research;
- Balances research and action for the benefit of all;
- Recognizes community as a unit of analysis;
- Builds on community strengths and resources (assets);
- Promotes joint learning, skill-sharing and capacity building among all partners;
- Engages in a long-term process and commitment;
- Emphasizes and engages in addressing the often complex causes of local problems;
- Disseminates findings and knowledge gained to all partners and involves partners in the process of taking action or next steps;
- Involves systems development through a cyclical and iterative process.

ETHICS IN PAR PRACTICE

Research ethics are an important aspect in the PAR method. Ethics are the principles and rules that guide how people should be treated, when they are participants in a research process. In Bangladesh, there is no ethical body who can approve this ethical guideline. In most of the cases, we follow the Western ethical guideline provided by either different ethical institutions or authors. Here, the researcher should always be aware of, and aim to address, inequalities within the group, and to provide opportunities for the less powerful to express their opinions and have them heard. In line with this, the partners should understand that they can withdraw from the research. A researcher should ensure that participants do not have unrealistic expectations regarding the outcomes of the research process. If the community people are educated then obtain signed consent forms from them, if they are illiterate their verbal consent should be given (Islam, 2009). Guaranteeing anonymity and confidentiality should be strictly maintained. In addition to the above, Kwan and Walsh (2018) mention some important points for ethical guideline in the PAR method:

- Think of community as a unit of identity that contributes to essentialism and identity politics;
- Follow an approach for the vulnerable and marginalized towards addressing the risk of re-stigmatization;
- Build up collaboration and equal partnership throughout the entire research for underestimating the complexity of power dynamics and relations;
- Use an emergent, flexible and iterative process;
- Be geared towards social action as well as culturally inappropriate expressions.

CHALLENGES OF PAR IN COMMUNITY DEVELOPMENT ACTIVITIES

Despite many strengths there are some challenges that might be considered while conducting PAR at a community setting. These challenges can be classified as commune and procedural level challenges (Islam and Siti Hajar, 2013). The commune level challenges were very much associated with the existing conditions of the community. They include low socio-economical conditions, cultural barriers, lack of research knowledge, and non-cooperation from the funding as well as the operating organization. On the other hand, the procedural level challenges are those challenges that are considered as the limitations of a qualitative research approach in the data collection process. These include reliability and validity of research findings, complexity and diversification of human behavior, research ethics, unavailability and inaccessibility of data, and power-relation. Each challenge is inter-related and inter-influenced, which later contributes to its complexity. In many cases, this would be confusing for novice researchers and others first learning this type of research approach. There is generally a lack of access to a sufficiently comprehensive and balanced way to learn about the diverse origins, theories, methods, motives and problems associated with this complex field (Greenwood and Levin, 1998). In many cases, researchers fail to include community members in the research team (Gillis and Jackson, 2002).

German et al. (2012) mention the following challenges of the PAR method:

- Motivating and sustaining interest: this includes motivating people to take action; sustaining interest over long periods or during prolonged diagnostic and planning phases; motivating partners while also managing their expectations.
- Power dynamics: another common challenge faced by PAR facilitators is the complex web of power relations that shape patterns of participation, communication, decision-making and, ultimately, benefits capture.
- Strengthening local and external institutions.
- Institutional weaknesses.
- Managing the research-development tension: this includes challenges faced in implementing PAR, as well as challenges faced in validating action research findings to a larger audience.
- Project funding and sustainability: over the long time which the processes often play out and given the limited duration of donor funding cycles. This is most applicable to climate change adaptation, given the prolonged time scale over which climate change and human "adaptive capacity" play out.

CONCLUSION

The main objective of this chapter was to develop the methods and framework of PAR for community development in Bangladesh. Drawing from a number of community empowerment and community development field-based research examples, this chapter composed a number of aspects of PAR that included the conceptual framework of PAR and community development, possible research framework and methods of PAR, steps of PAR, ethics, benefits and challenges. This will help researchers to conduct the PAR method in the community development field in Bangladesh. The main limitation of this chapter is still the lack of contextual perspectives and examples. Despite this limitation, this chapter provides a template of the methods and framework of PAR with its definitions, principles, strengths and challenges in the community development field in Bangladesh.

REFERENCES

Advancement Project – Healthy City (2011). *A Short Guide to Community Based Participatory Action Research*. Los Angeles: Advancement Project – Healthy City.
Ahmed, S.M., and Palermo, A.G.S. (2010). Community engagement in research: frameworks for education and peer review. *American Journal of Public Health*, *100*(8), 1380–87.
Annette, J., and Mayo, M. (2010). *Taking Part? Active Learning for Active Citizenship and Beyond*. Leicester: NIACE.
Boal, A. (1979). *Theatre of the Oppressed* (C. McBride and M. McBride, trans.). London: Pluto Press.
Brint, S. (2001). Gemeinschaft revisited: a critique and reconstruction of the community concept. *Sociological Theory*, *19*(1), 1–23.
Burns, J.C., Cooke, D.Y., and Schweidler, C. (2011). A short guide to community based participatory action research. *Advancement Project – Healthy City*. Available online: www.advancementprojectca.org (accessed on 17 October 2018).
Cambridge Dictionary (2019). Community development. Cambridge: Cambridge University Press. Available online: https://dictionary.cambridge.org/dictionary/english/community-development.
Chambers, R. (2004). Notes for participants in PRA-PLA familiarisation workshops in 2004. Institute of Development Studies, U.K.
Craig, G., Mayo, M., Popple, K., Taylor, M., and Shaw, M. (eds). (2011). *The Community Development Reader: History, Themes and Issues*. Bristol: Policy Press.
Creswell, J.W. (1998). *Qualitative Inquiry and Research Design: Choosing among Five Traditions*. Thousand Oaks, CA: Sage Publications.
Downey, L.H. (principal author) (2010). Capacity building for health through community-based participatory nutrition intervention research in rural communities. *Family & Community Health*, *33*(3), 175–85.
Fals-Borda, O., and Rahman, M.A. (eds). (1991). *Action and Knowledge: Breaking the Monopoly with Participatory Action-Research*. New York: Apex Press.
German, L.A., Tiani, A.M., Daoudi, A., Maravanyika, T.M., and Chuma, E. (2012). *The Application of Participatory Action Research to Climate Change Adaptation in Africa: A Reference Guide*. Ottawa, Canada and Bogor, Indonesia: International Development Research Centre and Center for International Forestry Research.
Gibson, N., Gibson, G. and MacAulay, A.C. (2001). Community-based research. In J. Morse, J. Swanson and A. Kuzel (eds). *The Nature of Qualitative Evidence* (pp. 161–82). Thousand Oaks, CA: Sage Publications.
Gilbert, K.R. (2001). Collateral damage? Indirect exposure of staff members to the emotions of qualitative research. In K.R. Gilbert (ed.). *The Emotional Nature of Qualitative Research* (pp. 147–61). Boca Raton: CRC Press.
Gillis, A., and Jackson, W. (2002). *Research Methods for Nurses: Methods and Interpretation*. Philadelphia: F.A. Davis Company.
Greenwood, D.J., and Levin, M. (1998). Action research, science, and the co-optation of social research. *Studies in Cultures, Organizations and Societies*, *4*(2), 237–61.
Greenwood, D.J., and Levin, M. (2006). *Introduction to Action Research: Social Research for Social Change*. Thousand Oaks, CA: Sage Publications.
Hickey, S., and Mohan, G. (2004). *Participation: From Tyranny to Transformation? Exploring New Approaches to Participation in Development*. London and New York: Zed Books.
Islam, M.R. (2009). Local knowledge and globalization in Bangladesh: NGOs' role for social capital and community development. Published PhD dissertation, University of Nottingham, Nottingham, United Kingdom.
Islam, M.R. (2014a). Improving development ownership among the vulnerable people: challenges of NGOs' community empowerment projects in Bangladesh. *Asian Social Work and Policy Review*, *8*(3), 193–209.
Islam, M.R. (2014b). NGOs' role for social capital and community empowerment in community development: experience from Bangladesh. *Asian Social Work and Policy Review*, *8*(3), 261–74.
Islam, M.R. (2017). NGO community empowerment projects in Bangladesh: how do these fit the local context? *Local Economy*, *32*(7), 763–77.

Islam, M.R. (2018). Climate change, natural disasters and socioeconomic livelihood vulnerabilities: migration decision among the char land people in Bangladesh. *Social Indicators Research*, *136*(2), 575–93.

Islam, M.R., and Hasan, M. (2016). Climate induced human displacement: a case study of Cyclone Aila in the southwest coastal region of Bangladesh. *Natural Hazards*, *81*(2), 1051–71.

Islam, M.R., and Hossain, D. (2014). Island *char* resources mobilization (ICRM): changes of livelihoods of vulnerable people in Bangladesh. *Social Indicators Research*, *117*(3), 1033–54.

Islam, M.R., and Morgan, W.J. (2012a). Agents of community empowerment? The possibilities and limitations of non-governmental organizations in Bangladesh. *Journal of Community Positive Practices*, *12*(4), 703–25.

Islam, M.R., and Morgan, W.J. (2012b). Non-governmental organizations in Bangladesh: their contribution to social capital development and community empowerment. *Community Development Journal*, *47*(3), 369–85.

Islam, M.R., and Shamsuddoha, M. (2017). Socioeconomic consequences of climate induced human displacement and migration in Bangladesh. *International Sociology*, *32*(3), 277–98.

Islam, M.R., and Siti Hajar, A.B. (2013). Methodological challenges on community safe motherhood: a case study on community level health monitoring and advocacy programme Bangladesh. *Revista de Cercetaresi Interventie Sociala*, *42*(September), 101–19.

Kaufman, B.A. (1992). In pursuit of aesthetic research provocations. *The Qualitative Report*, *1*(4), 1–8.

Kitzinger, J. (1995). Qualitative research: introducing focus groups. *British Medical Journal*, *311*(July), 299–302.

Koch, T., Selim, P., and Kralik, D. (2002). Enhancing lives through the development of a community based participatory action research program. *Journal of Clinical Nursing*, *11*(1), 109–17.

Kvale, S. (1996). The interview situation. In S. Kvale, *Interviews: An Introduction to Qualitative Research Interviewing* (pp. 124–43). Thousand Oaks, CA: Sage Publications.

Kwan, C., and Walsh, C.A. (2018). Ethical issues in conducting community-based participatory research: a narrative review of the literature. *The Qualitative Report*, *23*(2), 369–86.

Le Grange, L. (2009). Participation and Participatory Action Research (PAR) in environmental education processes: for what are people empowered? *Australian Journal of Environmental Education*, *25*(January), 3–14.

Lincoln, Y.S. (1992). Sympathetic connections between qualitative methods and health research. *Qualitative Health Research*, *2*(4), 375–91.

Macaulay, A.C., Commanda, L.E., Freeman, W.L., Gibson, N., McCabe, M.L., Robbins, C.M., and Twohig, P.L. (1999). Participatory research maximises community and lay involvement. *British Medical Journal*, *319*(7212), 774–8.

MacDonald, C. (2012). Understanding participatory action research: a qualitative research methodology option. *The Canadian Journal of Action Research*, *13*(2), 34–50.

Marshall, C., and Rossman, G. (2006). *Designing Qualitative Research* (4th ed.). Thousand Oaks, CA: Sage Publications.

Mason, L. (2006). Mixing methods in a qualitatively driven way. *Qualitative Research*, *6*(1), 9–25.

Mayo, M.C. (2008). Community development, contestations, continuities and change. In G. Craig, K. Popple and M. Shaw (eds), *Community Development in Theory and Practice* (pp. 13–27). Nottingham: Spokesman.

McNiff, J., and Whitehead, J. (2006). *All You Need to Know about Action Research*. Thousand Oaks, CA: Sage Publications.

McTaggart, R. (1989). 16 tenets of participatory action research. Available online: http://www.caledonia.org.uk/par.htm (accessed 15 November 2006).

McTaggart, R. (1991). Principles for participatory action research. *Adult Education Quarterly*, *41*(3), 168–87.

Morgan, D.L. (1997). *Focus Groups as Qualitative Research*. Newbury Park: Sage Publications.

Mulhall, A. (2003). In the field: notes on observation in qualitative research. *Journal of Advanced Nursing*, *41*(3), 1–19.

Paul, S., and Islam, M.R. (2015). Ultra-poor char people's rights to development and accessibility to public services: a case of Bangladesh. *Habitat International*, *48*, 113–21.

Ratcliffe, P., and Newman, I. (eds). (2011). *Promoting Social Cohesion: Implications for Policy and Evaluation*. Bristol: Policy Press.

Reason, P., and Bradbury, H. (eds). (2001). *Handbook of Action Research: Participative Inquiry and Practice*. London: Sage Publications.

Reinhartz, S. (1992). *Feminist Methods in Social Research*. New York: Oxford University Press.

Riel, M. (2010). Understanding action research. *Research Methods in the Social Sciences*, *17*(1), 89–96.

Riel, M. (2019). Understanding collaborative action research. Center for Collaborative Action Research, Pepperdine University.

Scott, W.R. (2001). *Institutions and Organizations* (2nd ed.). Thousand Oaks, CA: Sage Publications.

Selenger, D. (1997). *Participatory Action Research and Social Change*. New York: Cornell University.

Sen, A. (1999). *Development as Freedom*. New York: Random House, Inc.

Spradley, J. (1980). *Participant Observation*. New York; Toronto: Holt, Rinehart and Winston.

Streubert, H.J., and Carpenter, D.R. (1995). *Qualitative Research in Nursing: Advancing the Humanistic Imperative*. Philadelphia: J.B. Lippincott Company.
Stringer, E.T. (1999). *Action Research* (2nd ed.). Thousand Oaks, CA: Sage Publications.
Tandon, R. (2005). *Participatory Research: Revisiting the Roots*. New Delhi: Mosaic Books.
Taylor, M. (2011). Community organising and the Big Society: is Saul Alinsky turning in his grave? *Voluntary Sector Review*, *2*(2), 257–64.
Viswanathan, M. (principal author) (2004). Community-based participatory research: assessing the evidence – summary. Evidence report/technology assessment no. 99. Agency for Healthcare Research and Quality, U.S. Department of Health and Human Services, USA.
Wuest, J. (1995). Breaking the barriers to nursing research. *The Canadian Nurse*, *91*(4), 29–33.

15. Building a healthy community: the Coastal Georgia Indicators Coalition*
Patsy Kraeger

INTRODUCTION

The Coastal Georgia Indicators Coalition ("CGIC") formed in 2008 as a partnership and coalition between the City of Savannah and Chatham County governments and the local United Way affiliate, the United Way of the Coastal Empire. I often refer to these groups as *founding stakeholders* or *sponsors and partners* in this chapter. The Coastal Georgia Indicators Coalition contracted with a local university research partner, Armstrong Atlantic State University, to examine community needs by considering the feasibility of an indicators project to aid in decision-making for efficiently and effectively distributing scarce resources to improve the community overall.[1] Weffer, Mullooly, Sylvester, DeLugan and Hernandez (2014) recognize that "researchers, community stakeholders and policy makers are increasingly working together to tackle deeply entrenched social problems" (p.120).

COASTAL GEORGIA INDICATORS COALITION FORMATION

The Coastal Georgia Indicators Coalition was formed during a period of great economic unrest in the United States, known as the Great Recession. Importantly, this case examines how community partners formed an alliance to facilitate dialogue into action for improved outcomes in health and well-being and economic opportunity. This case explores improvements in overall quality-of-life needs in the Savannah-Chatham area given the scarcity of resources allocated by the founding institutional coalition stakeholders. The founding sponsoring stakeholders were seeking to shape a shared vision for change that alleviated community problems. Kania and Kramer (2011) suggest:

> that large-scale social change comes from better cross-sector coordination rather than from the isolated intervention of individual organizations. . . .[T]hat substantially greater progress could be made in alleviating many of our most serious and complex social problems if nonprofits, governments, businesses, and the public were brought together around a common agenda to create collective impact. It doesn't happen often, not because it is impossible, but because it is so rarely attempted. Funders and nonprofits alike overlook the potential for collective impact because they are used to focusing on independent action as the primary vehicle for social change. (Kania and Kramer, p.38)

The mechanism for independent impact by organizations was the norm but there is little evidence that isolated approaches produce change (Kania and Kramer, 2011).

Sirgy (2014) posits that "an indicators project is typically motivated by a set of adverse conditions that bring committed leaders together to address the problem(s) and find the solution" (p.26). In this case, local governments and a nonprofit philanthropic funder decided to enable in meaningful collaboration to assess the local community, and make strategic decisions around scarce resources prior to distributing resources. Subsequent CGIC partners included two nonprofit hospitals (one national and one local); they joined the coalition seeking to comply with the needs assessment requirements pursuant to the Patient Protection and Affordable Care Act 2010 (Roberts, 2019). The Patient Protection and Affordable Care Act mandates that nonprofit (i.e. tax-exempt) hospitals conduct a community needs assessment every three years as well as develop and adopt a strategic plan to comply with its community benefit requirements to maintain tax-exempt status.[2]

The Savannah-Chatham nonprofit hospitals, like the city and county governments and the local United Way, the United Way of the Coastal Empire, were seeking meaningful community information for policy formation and decision-making. Needs assessment and strategic planning for nonprofit hospitals is not necessarily an adverse impact but it is one that was perceived as potentially daunting to fully capture overall community health needs. Intentional and process-oriented collective efforts will allow for more robust community development.

COMMUNITY DEVELOPMENT – BROADLY CONSTRUCTED

Community development according to Phillips and Pittman (2014) is a process. It is action and it is an outcome (Phillips and Pittman, 2014). It is also voluntary and participatory (Phillips and Pittman, 2014). Ploch (1976) suggests that development is "active voluntary involvement in a process to improve some identifiable aspect of community life; normally such action leads the strengthening of the community's pattern of human and institutional relationships" (cited in Phillips and Pittman, 2014, p.7). In this case, we see that two local governments and the local United Way, all resource providers for community organizations, recognized the need for systematic collaboration during the Great Recession. The Great Recession was a time of decreasing institutional resources to alleviate community needs and to be strategic with the allocation of resources for community improvement (Grusky et al., 2011). Individually, these organizations may have identified aspects of community life needing improvement through community service grantmaking and other initiatives on an ad-hoc basis. Declining resources during an economic downturn allowed the coalition's founding stakeholders to form an indicator coalition. This helped reshape both human and institutional relations focusing on community needs by committing to funding indicator studies to influence community change. In this case, the CGIC founding stakeholders and partners examined initial approaches to community development as a process, action and outcome.

Large-scale social change requires process. Process change formation requires evidence; the original partnership recognized the need to commission a feasibility study and, later, a study on the best governance practices for indicator coalitions. CGIC, from its founding, is a stand-alone organization with program partners seeking inclusivity across the community reflected in its commissioned indicator survey to other community initiatives.

From action and outcomes perspectives, CGIC recognized that a process for shared indicator measurement was essential for successful outcomes in assessing the community and for decision-making for allocating scarce resources. The voluntary nature of the coalition formation is a key factor for stakeholder coalition success around a common agenda. Kania and Kramer (2011) recognize that successful stakeholder initiatives are able to reach a common agenda when action is voluntary and inclusive of broad citizen dialogue and input. The founding coalition partners are high-powered stakeholders in the Savannah-Chatham community. It was important then for the organizations to recognize their power while being intentional about seeking public input for indicators projects' validity in a community.

Engaged citizen dialogue seeks to listen to the voices of a full range of community stakeholders not just "high-powered, high interest stakeholders" including government and philanthropic funders such as United Way organizations. Michael Quinn Patton (2011), a former community development practitioner, facilitator and proponent of utilization focused evaluation practice, suggests that there are high/low power and high/low interest stakeholders. High power/high interest are defined as "key players who are in a prime position to affect use" of the evaluation for themselves and other stakeholders" (p.73). High interest-low powered stakeholders are those who are affected by community development, evaluation and community initiatives who should be encouraged to participate to increase diversity and give community context (Patton, 2011, p.73).

CGIC's founding and subsequent intuitional stakeholders are "high-powered, high interest stakeholders" as defined by Patton (2011, p.73). These initial stakeholders sought to vet the selection of indicators through the community participation initially from the United Way of the Coastal Empire's Board of Directors as well as later seeking broader public participation through the indicator surveys (Toma, 2009). Intentional actions by the founding stakeholders and the university research partner sought public participation in the survey for indicator selection. Later, surveys recognized the need to incorporate the voice of the general public, essentially high interest, low powered stakeholders for the indicator project's validity (Toma, 2009). These intentional actions seeking public input and avenues for communication align with the goals of the collective impact model for a community seeking large-scale change. Kania and Kramer's (2011) collective impact model suggests that large-scale social change can be achieved when committed community actors abandon their separate agendas for social change. In order to develop these shared measures, "community actors need to engage in continuous communication with each other in order to develop a 'single-set' of shared goals, measured in the same way" (p.36). Kania and Kramer (2011) also suggest that collective impact will be achieved when there is a "coordinating organization to the process and infrastructure for the project. The collective impact model suggests that cross-sector collaboration moves beyond collaboration because the outcomes are shared and measured and can be used for both community learning as well as social change".

Process, actions and outcomes for community development should include participatory practices encouraging engaged "citizen dialogue around community change to develop a shared vision" in an "open and transparent process" (Vincent and Stephen, 2015, pp.105–7). CGIC is committed to a transparent process to allow for shared vision development by engaging experts with the university partner from the initial feasibility study to a governance best practices study and later indicators work inclusive of both high/low powered and high interest stakeholders. In order to fully assess the need for an indicators coalition and ultimately a neutral stand-alone community organization, it is important to understand the demographics and economic engine against a background of scarce resources from the 2008 Great Recession until today.

SAVANNAH-CHATHAM BACKGROUND INFORMATION

Savannah is the fifth largest city in the state of Georgia with a physical size of 108 square miles (Savannah, Georgia Population, 2019). In 2017, Savannah was called home by approximately 146,000 residents (US Census Bureau, 2017a). Physically, the City of Savannah and Chatham County are bounded by the State of South Carolina to the north, Florida to the south and the Atlantic Ocean to the east.

Savannah also is the oldest city in the state of Georgia dating back to the mid-1750s. The city was also the colonial capital in pre-Revolutionary America (Savannah, Georgia Population, 2019). By the mid-1800s, the City of Savannah in Chatham County had become a large commercial port in cotton and textiles as well as a commercial port for the slave trade until slavery was abolished.[3] Chatham County slightly post-dates the American Declaration of Independence in 1776; it was founded in 1777. It is the fifth largest county in the state of Georgia. Savannah-Chatham today continues to be a major center of commercial activity due to the Georgia Ports Authority (GPA), military bases, public employers including two universities, private employers including Gulfstream and the Savannah College of Art and Design as well as tourism (Chatham County, n.d.).

The Economic Engine and Employment

The GPA as of 2017 was the nation's largest container port (Georgia Ports Authority, 2017). "Georgia's deepwater ports industry consists of public marine terminals in Savannah and Brunswick owned by the Georgia Ports Authority as well as private marine terminals" (Humphreys, 2018, p 5.). In 2017, Georgia's deepwater ports "contributed two billion dollars to the state's economy, approximately 6 percent of the state's gross domestic product" (GDP). According to Humphreys (2018), the combined economic output in terms of impact was $106 billion. Sixty-three billon was from direct spending and the remaining 43 billion came from "indirect or induced" spending (p.7). The Savannah ports authority employs approximately 33, 692 people in its operations which is 87 percent of the combined Georgia deepwater ports (Humphreys, 2018).[4]

Tourism is the second largest economic driver due to the coastal beaches, wetlands and opportunities for outdoor recreation in addition to attractions in historic downtown Savannah. The Savannah-Chatham area "was visited by 14 million visitors, who spent $1.9 billion supporting 18,000 jobs in 2017" (Smith, 2019). In addition to the annual economic impacts reported by the Georgia Ports Authority (GPA) for Savannah including the city of Savannah's vibrant tourism industry, the United States military has two bases in the Savannah-Chatham area with "22 thousand soldiers, 3,500 civilian employees and a payroll exceeding a billion dollars" (Smith, 2019).

SAVANNAH-CHATHAM'S POPULATION DEMOGRAPHICS

The state of Georgia in 2017 had a population of just over 10 million people with just over 59 percent reporting as White/Caucasian and just over 31 percent reporting as African American (US Census Bureau, 2019). Savannah's African-American population is approximately 23 percent higher than the state of Georgia while Chatham County's is

Table 15.1 Savannah, Georgia and Chatham County, Georgia

Location and Year	Savannah 2017	Chatham County 2017	Savannah 2010	Chatham County 2010
Population	146,444	290,501	136,286	265,128
African-American	54.70%	46.70%	55.40%	40.13%
White/Non-Hispanic	36.20%	48.40%	38.29%	52.80%
Other	9.10%	4.90%	6.31%	7.07%
Over 65+	12.80%	14.80%	11.66%	12.40%
Population under 18	21.20%	28.70%	22.36%	22.64%
Civilian Workforce	61.80%	63.30%	57.50%	60.40%
Per Capita Income	22,497.00	28,765.00	16,921.00	21,152.00
Poverty Rate	24.00%	16.30%	14.20%	17.30%
Persons without Health Insurance	21.10%	14.70%	N/A	N/A
High School Graduate (includes equivalency)	26.10%	29.20%	32.70%	31.10%
Bachelor's Degree or Higher	17.30%	9.30%	15.30%	7.90%
Households with Disabilities under 65	10.10%	9.50%	N/A	N/A
Housing Units	N/A	125,090	61,883	123,555
Owner Occupied Housing	43.70%	53.90%	46.6%	57.7%

Source: US Census Bureau (2017b).

approximately 15 percent higher than the state of Georgia. By age, the greatest population demographic exists between ages over 18 to under 65 for Savannah-Chatham residents (US Census Bureau, 2019). See Table 15.1 for more information.

Savannah-Chatham's demographics differ minimally from 2010 to 2017 with the biggest shift in the racial demographics which trends differently than across the rest of the state of Georgia. From 2010 to 2017, the population grew in both Savannah city and Chatham County by approximately 1 percent. Chatham County is roughly double the population of Savannah city. In Chatham County, the African-American population grew by approximately 6 percent from 2010 through 2017 and just over 2 percent in the City of Savannah (US Census Bureau, 2010a, 2017a). See Table 15.1 for more demographic information.

Savannah-Chatham Poverty Rate, Per Capita Income, Education and Wages

The census data suggests that while the Savannah-Chatham area boasts recreational and historical attractions a robust economic engine driving employment, the poverty rate, per capita income and owner occupied housing is below the statewide average. Per capita income has increased in both Savannah and Chatham by approximately just under $5,000 and $7,000 respectively from 2010 to 2017 while the poverty rate has increased by approximately 10 percent in Savannah city and decreased in Chatham County during this same time frame (US Census Bureau, 2010a, 2017a). See Table 15.1 for more information. Per capita statewide income for Georgia in 2017 was approximately $44,000 compared to Savannah at just under $22,500 and Chatham at just over $28,500 (St Louis Federal Reserve, 2019).

Less than 30 percent of the population had a high school degree or the equivalent as of 2017, with only slight growth from 2010 to 2017 in that less than 18 percent had a bachelor's degree or higher in the City of Savannah (US Census Bureau, 2010b, 2017b). For Chatham County, that rate was lower than 10 percent with slight growth from 2010 to 2017 (US Census Bureau, 2010b, 2017b). See Table 15.1 for more educational attainment demographic information.

According to the Bureau of Labor Statistics, the 2018 average weekly wage for residents of Savannah city is $878 and for Chatham County is $887, which ranges from $157 to $168 below the national average at $1,055 (US Bureau of Labor Statistics, 2019). In 2018, the Savannah area total employment for nonfarm jobs was just over 179,000 (US Bureau of Labor Statistics, 2019). The largest percentage of jobs are in the trade, transportation and utilities industries with just under 43,000 jobs; education and health services jobs just over 267,000; leisure and transportation industries capturing approximately 25,500 jobs; civilian government capturing 24,500 jobs; professional and business services capture just under 19,000 jobs; and manufacturing at 17,000 jobs[5] (US Bureau of Labor Statistics, Current Employment Statistics, 2019). Given the data on the average weekly salary and the job classifications and coupled with the demographic data on education, the Savannah-Chatham area jobs are not on average white collar or professional jobs. See Table 15.1 for more information. The unemployment rates in 2017 trended slightly higher in Savannah city (4 percent) and Chatham County (4.1 percent) than across the United States (3.9 percent) (US Bureau of Labor Statistics, Local Area Unemployment Statistics, 2019). In 2018, these trends reversed with unemployment trending slightly higher than across the United States (93.9 percent) than in Savannah city (3.6 percent) and Chatham County (3.7 percent) (US Bureau of Labor Statistics, Local Area Unemployment Statistics, 2019).

Higher poverty rates in the Savannah-Chatham area than across Georgia and higher unemployment rates compared to the nationwide average combined with a low percentage of education attainment, lower per capita income and wages align with a lower rate of owner occupied housing. Statewide, Georgia owner occupied housing for 2017 is 63 percent while for Savannah, that rate is just under 20 percent at 43.7 percent and Chatham just under 10 percent at 53.9 percent (US Census Bureau, 2017b). In 2010, owner occupied housing was slightly higher by a few percentage points in Savannah-Chatham (US Census Bureau, 2010b). See Table 15.1 for more housing demographic information.

Savannah-Chatham's Population – Health Generally and Health Inequities

Health inequity is a great concern in the Savannah-Chatham area as it is across the United States and the globe. In the two decades plus since the 1990s, health coverage has become politicized in the United States. Access to care under federal programs is one in which conservative states coordinated an effort to resist federal programming in this area under the Patient Protection and Affordable Care Act during the Presidency of Barak Obama and after (Rigby, 2012).

Georgia is one of the states which did not expand its Medicaid program under the Patient Protection and Affordable Care Act (Malloy, 2012). Former Georgia Governor Nathan Deal stated in August of 2012 that the Medicaid expansion for the 650,000 low income residents of Georgia was too expensive (Malloy, 2012).[6] "No, I do not have any intentions of expanding Medicaid . . . I think that is something our state cannot afford" despite the promises of the federal government "to pay 100 percent of the expense for the first three years and 90 percent thereafter . . ." (Malloy, 2012).

While health inequities may be reported at a federal, state and county level, county level data allows for local decision-making and coordination of resources. The Centers for Disease Control (CDC) recognized that at least in the United States "diabetes belt" that at the local government level community leaders could utilize "community design encouraging healthy life style choices" (Georgia Healthy Cities, n.d.). While that is true, it is necessary to recognize that state and local government are intertwined. State level legislation impacts local places and has since the late 19th century to today (Burns and Gramm, 1997). One area where state political action has impacted local places is in health care and the resistance to Medicaid expansion under the Patient Protection and Affordable Care Act of 2010 (Belland et al., 2014; Rigby, 2012).

County level geographic data from the 2007 Behavioral Risk Factor Surveillance Survey has identified a diabetes belt in the United States (Barker et al., 2011). The diabetes belt stretches across 15 states in the southeastern portion of the country covering 644 counties (Centers for Disease Control, n.d.). Georgia is one of the identified states with a high proportion of counties situated in the diabetes belt and Chatham County is one of these 644 counties.

According to the CDC, persons living in the diabetes belt have diabetes diagnosed at an increase of 3.2 percent over those persons living outside of the diabetes belt. In the diabetes belt as of 2007, "11.7% of the people have diagnosed diabetes" (Centers for Disease Control, n.d.). Importantly, the CDC reported the data at a county level to allow for community leaders to be responsive in prevention efforts as opposed to only dealing with the issues from treatment (Centers for Disease Control, n.d.).

Across Savannah in the County of Chatham, 13 percent of persons are reported to have been diagnosed, which is a slightly higher average than across the diabetes belt in general (Centers for Disease Control, n.d.). In the poorest neighborhoods in Savannah, diabetes diagnosis is more than double the rate of the highest average across all 644 counties in the diabetes belt (Georgia Healthy Cities, n.d.). Data also shows that residents in these same tracts rank just under the 99th percentile of persons with "Chronic kidney disease among adults aged 18 and over" (Georgia Healthy Cities, n.d.). Care access for persons with diabetes is crucial for not only prevention but for maintenance preventing "complications, such as amputation and organ failure" (Georgia Healthy Cities, 2019). Diabetes is at epidemic proportions since the 20th century (Engleglau et al., 2004). Prevention efforts are greatly needed (Engleglau et al., 2004). "African Americans have a high risk for type 2 diabetes. . . ..African Americans have a high rate of diabetic complications, because of poor glycaemic control and racial disparities in health care in the USA" (Marshall, 2005). The Savannah-Chatham area according to the US Census data has a larger African-American population than in the state of Georgia. Demographics suggest then that community leaders need and should make diabetes detection, prevention and treatment a priority (US Census Bureau, 2010b, 2017b). See Table 15.1 for more demographic information.

Savannah was selected as part of Georgia Healthy Cities Project, a subset of the 500 Cities Project coordinated by the Center for Disease Controls and the Robert Wood Johnson Foundation (Georgia Healthy Cities, n.d.). The CGIC and member organizations were involved with the Georgia Healthy Cities Project. At the initial meeting in 2018, data showed that "[a]fter comparing Savannah's estimates with those of other cities

"... found that, relative to the country, many tracts in Savannah experience increased prevalence estimates of diabetes, stroke, COPD and low sleep", reported from Georgia Healthy Cities Hosts First Workshop (Bellows, 2018).

Additional data from the Georgia Healthy Cities Project established that "most of Savannah's neighborhood are in the 75th percentile or higher for stroke prevalence compared to the rest of the country's other largest 500 cities" (Georgia Healthy Cities, 2019). This rate is lower than diabetes and associated disease but it is in the 98th percentile compared to 500 of the largest cities in the United States (Georgia Healthy Cities, 2019).

Selecting health indicators around preventable diseases is important for communities to determine effective strategies to remedy health inequalities. Public–private partnership strategies will enhance efforts to reduce chronic disease (Butterfoss, 2009). Coordinated efforts to target and address health needs in the Savannah-Chatham area from the coordination of scarce resources through public and private grantmaking to compliance with the Patient Protection and Affordable Care Act's requirements were ripe for a public–private partnership through the CGIC.

SAVANNAH-CHATHAM'S PUBLIC AND PHILANTHROPIC RESOURCES

Collaboration has been seen as a priority for this coalition in order to both determine the needs but assess and coordinate efforts to distribute scarce resources. The Savannah-Chatham area's background demographics provide a snapshot of who lives in the area, jobs, income, poverty rate and identified health problems which we will see align with the desire to select meaningful indicator measures to assess and measure progress in the Savannah-Chatham, area. Quantifying the costs of addressing these identified needs are not explored; rather, a snapshot is given of the founding partner's community funding expenditures. Government expenditures are reported from 2006 through 2011.

Charitable expenditure are reported for the United Way of the Coastal Empire from 2007 through 2011 (United Way of the Coastal Empire, 2011–2017). These ranges were selected to capture what local resources were available through local government and the United Way for community services. These years are selected because they pre-date and post-date the years of the Great Recession in the United States from December of 2007 through 2009 (Rich, 2013).

Gordon (2012) suggests that during this time state and local government "undertook most of the direct services spending on public goods and services bearing primary responsibility for investments in education, social services . . ." (p.1). During this time period, losses at the state and local level were not offset by the federal government's stimulus programs. Between 2008 and 2011, 34 states reduced expenditures in K-12 education; 43 states cut university expenditures; 31 states lowered health care expenditures; 29 states cut services to the disabled and elderly; and 44 states reduced employee compensation (Gordon, 2012). Calls for partnership for philanthropic and volunteer programming were made and heard across the United States and in Georgia to help with the shortfall.

Table 15.2 reflects that the funding in Savannah city has an increase in funding for community services in 2008 and 2009. During this same time, Savannah municipal leaders recognized the need to address poverty in the area given the high poverty rate compared to

Table 15.2 City of Savannah general fund expenditures by service area (dollars reported in thousands)

Year	2006	2007	2008	2009 (Projected)	2010 (Projected)	2011
Public Development: Community Services	721,776	766,162	819,45.3	904,412	408,611	385,183
Department/Activity: Step Up	104,429	198,590	N/A	N/A	N/A	92,887
Public Development: Step Up Savannah	N/A	N/A	301,707	121,484	90,657	N/A

Source: City of Savannah Budget Allocations, 2009.

the state. The City of Savannah committed to funding poverty alleviation by funding the start of a nonprofit committed to reducing poverty in the city (City of Savannah Budget Allocations, 2006–2008 and see Table 15.2). The poverty reduction initiative also developed into a separate nonprofit organization ("Step Up Savannah") dedicated to increasing economic opportunity in the region. In the year 2008, the Savannah city government provided approximately the equivalent of 37 percent of the community services budget for poverty reduction to the poverty reduction initiative[7] (City of Savannah Budget Allocations, 2008 and see Table 15.2). In 2009, funding for Savannah city community services increased while funding to the poverty reduction organization decreased. In 2010, the city decreased its funding to both community services and to the poverty reduction initiative (City of Savannah Budget Allocations, 2009 and see Table 15.2).

Combined funding for community services and poverty reduction were highest during this time period at just over $1 million dollars in years 2008 and 2009 for the city. By 2010, this funding had decreased by approximately 50 percent and even to a greater extent by 2011 (City of Savannah Budget Allocations, 2009 and see Table 15.2).

Chatham County, larger than Savannah city, houses the larger economic engines in the area. Health and welfare budget expenditure ranged between 9 million to just over 11 million from 2006 through 2011 (Chatham County Budget Allocations, 2006–2011 and see Table 15.3). In 2010, these expenditures nearly tripled in 2011 and were nearly double the 2009 expenditures in 2011 (Chatham County Budget Allocations, 2006–2011 and see Table 15.3).

Reported budget data in Tables 15.2 and 15.3 do not specifically separate programming expenditures from administration and overhead costs. The tables reflect

Table 15.3 Chatham County actual general budget expenditures for health and welfare service area (dollars reported in millions)

Year	2006	2007	2008	2009	2010	2011
Amount of Funding	9,349,148	10,076,305	10,598,883	11,127,708	32,059,097	19,211,384

Source: Chatham County, Georgia.

that there was an uptake in local government funding during the period of the Great Recession in Savannah city with a decline just after while the opposite was true for Chatham County.

Calls for partnership with philanthropy, nonprofit and volunteer organizations were made by governments at the state and local levels to partner to alleviate the impacts of the Great Recession. However, charitable contributions declined during this same time period. Data on charitable contributions is collected from the Form 990, Return of Organization Exempt from Income Tax, with the Internal Revenue Service (IRS).[8] The IRS Form 990 was changed by the IRS which included revisions to the form and new schedules in addition for reporting on governance practices as well as reporting the compensation of officers, directors, trustees, key employees and highest compensated employees on an annual basis. Changes went into effect in 2008.

One change was that direct public support and government grants were not separately reported. Table 15.4 reflects these changes with year 2007 showing the separate lines and data from 2008–11 showing the combined reporting.[9]

Nationally, there was a drop in charitable giving by approximately 19 billion from 2008. While 2010 and 2011 saw increases in charitable giving, it was still below the 2008 giving rate[10] according to the Giving USA reporting (Giving USA Foundation, 2008–2012). Combined giving lines for health and human services were in the $20 million and $30 million range or 10 percent or less of total national personal giving/charitable contribu-

Table 15.4 IRS Form 990 data – United Way of the Coastal Empire (dollars reported in millions and thousands)

Year	2007	2008	2009	2010	2011
Direct Public Support (2007 only reported separately)	7,661,218	–	–	–	–
Government Grants (2007 only reported separately)	7,006	–	–	–	–
Contributions and Grants (2008–11)		10,161,228	8,361,204	8,676,665	9,173,935
Programming and Services	–	171,562	151,985	146,615	144,874
Investment Income	64,617	45,261	29,310	14,685	12,957
Special Events (2007 only reported separately)	37,203	–	–	–	–
Rental Revenue (2007 only reported separately)	−27,847	–	–	–	–
Other	9,356	33,087	79,449	60,879	82,818
Total Revenue	7,742,197	10,411,133	8,621,748	8,898,470	9,414,584
UWCE Grants Paid	5,697,509	8,383,405	6,321,592	6,538,470	6,889,856

Notes:
The dashes in the chart reflect a budget allocation of zero.
UWCE = United Way of the Coastal Empire.

Source: Pro Publica 990 Data.[11]

Table 15.5 Giving USA charitable contributions 2008–11 by contribution area (dollars reported in billions)

Year	2008	2009	2010	2011
Total Charitable Contributions	303.80	284.90	298.42	290.89
Health	21.1	19.9	23.8	31.1
Human Services	27.3	22.8	26.8	33.4
Public Society Benefit	24.3	25.6	23.8	37.8

Source: Giving USA Foundation.

tions. See Table 15.5 for trending of charitable contributions across the United States in discrete categories.

Charitable contributions to the United Way of the Coastal Empire area reflected the trends of the local government coffers rather than the declines in charitable giving across the United States.

Calendar year 2008 showed increases in giving of over $2 million from the year before and then a reduction of approximately $1.5 million in years 2009 and 2010 (Pro Publica data for United Way of the Coastal Empire, 2009, 2010). The years 2009 and 2010 saw giving decline from the 2008 national giving trends (Giving USA Foundation, 2009, 2010, 2011). Likewise, 2009 through 2011 saw decreased community grantmaking from the United Way of the Coastal Empire of more than $1 million from a high contribution intake and grantmaking output in 2008 (Pro Publica data for United Way of the Coastal Empire).

These tables reflect the trends during the Great Recession when funds were declining resources for increased community needs during that time. As noted, the demographics of the Savannah-Chatham area show steady or increased poverty rates, not much change in the education level of residents from 2010 to 2017, significant chronic health issues in the community and an economic engine that delivers low wages with the per capita income being significantly below the state level.

It is against this backdrop of economic, community and health needs, seeing shifts in public funding of community services at the city level downward and declining local charitable contributions during the Great Recession that the CGIC was formed. As we see the financial resources for community improvement from the city, county governments and philanthropic partners at their height of total combined funding allocations were approximately $17.5 million in 2009 and $38.9 million in 2010 (see Tables 15.2, 15.3 and 15.4). Twenty-three million dollars in the 2011 funding allocations were provided to community services with combined Chatham County and United Way funding only reported (see Tables 15.3 and 15.4). Combined financial resources then were not abundant in 2007 and 2008. Even with the uptake in funds from 2009 to 2011, we do not know which percentage went to the funding for the direct provisions of goods and services at the city and county level.

THE COASTAL GEORGIA INDICATORS COALITION

Regionalism is an important economic and community development priority in the Savannah-Chatham area. In the 1990s and through the formation of the CGIC, community leaders recognized that the coastal region of Savannah-Chatham could only grow through coordinated planning efforts seeking improvement in overall quality of life for its residents through participatory action. Community participatory action would not only help identify priorities for change, both community assets and needs would be identified (Beck et al., 2010, 2011; Toma, 2009, 2013; Toma et al., 2011; Toma et al., 2015).

From Visioning to Collective Action

Visioning is a community development process tool that allows for communities to imagine where they are and where they would like to go and what the community would like to become (Phillips and Pittman, 2014). Leaders in the Savannah-Chatham area came together in the early 1990s to engage in a visioning project. They came together again in the early 2000s to envision an indicators project (Vision 20/20 Commission, 1992). The Savannah-Chatham Indicators Project and later its organizational re-incarnation, the stand-alone CGIC seems to adopt the collective impact model. As discussed, the model for collective impact is one where organizational actors seek a common agenda for social change based on a system of shared measurement allowing for continuous communication through mutual reinforcing activities (Kania and Kramer, 2011).

The early visioning project of the 1990s is a first attempt to develop a common agenda for change but the apparatus was not in place for collaborative change. Collective impact calls for a backbone organization to coordinate these activities (Kania and Kramer, 2011). CGIC has become the type of coordinating backbone organization that Kania and Kramer (2011) originally saw as necessary for success for collective impact.

Envisioning the future

In 1990, Savannah city and Chatham County leaders came together through a community process to envision the future for 2020 in a multi-stakeholder initiative. Kania and Kramer (2011) define multi-stakeholder initiatives (MSI) as "voluntary activities by stakeholders from different sectors around a common theme" (p.41). In 1990, the MSI was when a committee, the Vision 20/20 Commission, was created to examine the state of the Savannah-Chatham area from an economic and community perspective. The mission of the Vision 20/20 Commission was to achieve consensus among a full range of community members through open dialogue and participatory practices, a "community that works together to achieve a quality of life for all citizens" (Vision 20/20 Commission, 1992). Kania and Kramer (2011) would call this process of Vision 20/20 as one where calling for "all participants to have a shared vision for change, one that includes a common understanding of the problem and a joint approach to solving it through agreed upon actions" (p.39).

In order to fully understand the development of the Savannah-Chatham Indicators project, later the Coastal Georgia Coalition, it is important to understand that the seeds for regional economic and community development were planted in the early 1990s in this voluntary multi-stakeholder initiative. That initial visioning planning process involved

Table 15.6   Vision 20/20: priorities

Priority Number	Priority Focus Area
Priority One	Neighborhood Livability and Housing
Priority Two	Economic Development and Job Creation
Priority Three	Youth and Education
Priority Four	Public Safety and Crime Prevention
Priority Five	Health and Human Services
Priority Six	Environmental Protection and Conservation
Priority Seven	Public Facilities and Infrastructure
Priority Eight	Community Preservation and Quality Enhancement

Source: Vision 20/20 Commission (1992).

public-private partners who facilitated a five-step process. The process for Vision 20/20 included surveying the community, hosting an electronic town hall, hosting a vision retreat, task force development and ultimately the preparation and dissemination of the Vision 20/20: A Blueprint for Community Action report (Vision 20/20 Commission, 1992).

Fourteen hundred people were surveyed in the community to determine the needs and priorities for change and future community and economic development (Vision 20/20 Commission, 1992). As part of the Vision 20/20, the Committee for Savannah's Vision 20/20 hosted a live televised town hall through a broadcast hosted by a local television station where viewers could watch the proceedings if unable to attend the actual town hall in person. This captured 90,000 viewers watching the town hall. Five hundred viewers also attended the meetings at 24 community sites across the county and city (Vision 20/20 Commission, 1992).[12] Eight priority areas came to light as a result of this planning process (Vision 20/20 Commission, 1992). See Table 15.6 for more information.

The Vision 20/20 Commission membership was an organized voluntary citizen effort. The 1992 report concludes with a recommendation that these efforts should continue on a voluntary citizen effort. Building social capital through coordinated community organizing is indeed a viable strategy (Gittell and Vidal, 1998). Essentially the efforts to produce Vision 20/20 constitute building social capital through participatory actions across the Savannah-Chatham area. The visioning committee was seeking community input around priorities and indicators. Kania and Kramer (2011) suggest that MSIs often "lack any shared measurement of impact and the supporting infrastructure to forge any true alignment of efforts or accountability for results" (p.4). The visioning plan did seek to shape common priority areas with indicators to assess and examine the community. The visioning process sought to be representative of the community in that the priority areas were elected from the electronic town hall and then vetted through a community survey (Vision 20/20 Commission, 1992).

Sustainable community development and planning requires more than voluntary action. It requires structure, committed finances and coordinated efforts to develop shared and measurable indicators. Kania and Kramer (2011) posit that solutions to complex problems through collaborations can only be achieved when through collective impact organizations. Collective impact organizations are "long-term commitments by a group of important actors from different sectors to a common agenda for solving a specific social problem. Their actions are supported by a shared measurement system, mutually reinforcing activities, and ongoing communication, and are staffed by an independent backbone organization" (Kania and Kramer, 2011, p.41).

While the initial multi-stakeholder initiative responsible for developing the voluntary Vision 20/20 Committee and producing Vision 20/20: A Blueprint for Community Action in 1992 may not meet the rigid criteria for collective impact, leaders wanted input from the general public to vet the vision 30 years out (Vision 20/20 Commission, 1992). The Vision 20/20 project is one that called for continuous communication among stakeholders based on a broad indicator scheme to shape a changed future 30 years out.

These early efforts in 1990 and the report in 1992 set the stage for government and community organizational leaders to be data focused for real change when the Savannah-Chatham Community Indicators project was formed. The purpose of the indicators project was to examine avenues for collaboration at the county and city levels to address community needs. It does not appear that this early work has any theoretical basis in community development indicators. That said, these common and cross-cutting priorities suggest that the city and community leaders seek to examine "developmental needs" or the "hierarchy of lower and higher-order needs such as health, safety, economic, social, esteem, knowledge, actualization and aesthetics" which allow for "assessment and monitoring (Sirgy, 2014, p.22). Priority areas which seek to benchmark community preservation and quality enhancement align with theories that focus on higher order actualization and aesthetics.

Developing the Indicator System in the Savannah-Chatham Area: Planning and Feasibility

Sirgy (2014) identifies a process for a community indicator committee to follow certain steps: (1) form a committee led by prominent leaders; (2) develop an initial set of indicators; (3) adopt applicable quality of life theories; (4) re-examine indicators and compare to comparable geographic location data; (5) incorporate residents perception through survey work for primary data; (6) examine relevant and credible secondary data; (7) write up a recommendation report; (8) disseminate the report to key stakeholders and decision-makers for action.

The Savannah-Chatham United Way initial partnership for the Savannah-Chatham Indicators Project appears to follow the best practices for the feasibility planning and coordinating project. In 2008, the three founding partners, Chatham County, Savannah city and the United Way of the Coastal Empire entered into a partnership with Armstrong Atlantic State University's Public Service Center to undertake an indicators project to help focus and coordinate collaborative efforts to distribute resources (Coastal Georgia Indicators Coalition, 2019).

Researchers at Armstrong Atlantic State University conducted a feasibility study from 2007 through 2009 in order to identify priority areas of funding (Toma, 2009). Community and business leaders were surveyed, households in the area were surveyed and focus groups were conducted (Toma, 2009). These surveys and focus groups identified four priority areas for which indicators availability needed to be assessed (Toma, 2009). Priority focus areas were then vetted for initial funding by the three founding stakeholders for the feasibility study, the City of Savannah, Chatham County and the United Way

Table 15.7   2009 Feasibility Study focus area priorities

Priority Number	Priority Focus Area
Priority One	Education and Youth Development
Priority Two	Economic Independence
Priority Three	Health and Wellness
Priority Four	Regionalism

Source: Toma (2009).

of the Coastal Empire (Toma, 2009). The feasibility study's purpose was to determine whether indicators are available or whether there were alternate suitable indicators to measure the four priority areas (Toma 2009). The four priority indicators have been benchmarked through 2015 (Beck et al., 2010, 2011; Toma, 2009, 2013; Toma et al., 2011; Toma et al., 2015).

In addition to availability of the indicators, the 2009 Feasibility Study in Table 15.7 above examines the problems with data collection relating to customized data that the United Way of the Coastal Empire had purchased from the US Census Bureau (Toma, 2009; Toma et al., 2011). Careful attention was paid to selecting meaningful indicator data existing and the need to survey.

The founding stakeholders were involved to the extent that they vetted the priority area indicators as relevant for the future surveys (Toma, 2009, 2013; Toma et al., 2011). These founding partners are legitimate and credible leaders with standing in the community. An initial set of priority areas was developed which reflect four of the eight priorities of the 1992 *Vision20/20: A Blueprint* (Vision 20/20 Commission, 1992).

Indicator Development

Indicators were selected by the three founding stakeholders, public and nonprofit organizations that fund community services. According to Sirgy (2014) "[m]ost communities use community quality of life indicators to assess the effectiveness of . . . planning efforts and programs" (p.9). Using "available secondary data" allows communities to be informed and "to make those assessments" (Sirgy, 2014, p.9). Top-down indicator development by experts allows for indicators to be developed but also to empirically test both theoretical and practical observations of the actual and perceived quality of life of the residents by residents and other key informants. A key objective when seeking representativeness in these types of study then is to have the indicators validated subjectively through surveys and other interview methods with a random sample of the population to determine the validity of the indicators as appropriate for the study (Sirgy, 2014).

Researchers conducting the feasibility study examined the priority areas and initial set of indicators with an eye to relevance, clarity, data availability and comparability. Further examination and refinement occurred because some of the data had potential bias or no data were available. Researchers at Armstrong Atlantic State University recommended two rounds of surveys to collect the data. Various community stakeholders were involved in the planning process from business and community leaders; surveys of households and focus group sessions were held in 2009 (Toma, 2009). Representativeness is a goal of the founding stakeholders and, through 2019, to reach a goal for the 1990 study, Vision 20/20 (Vision 20/20 Commission, 1992). The Board of Directors for the United Way of the Coastal Empire comprised community leaders from industry, small business, the professionals and community volunteers. Given the breadth of organizations and individuals represented on this board, the board was deemed appropriate for vetting the initial indicators selected (Toma, 2009).

Indicator Selection

The priority areas and indicators developed in the feasibility study created a framework for the indicators project for benchmarking "trends, opportunities and challenges as well as serving as a catalyst for conversation" (Toma, 2013, p.4). The ultimate goal in addition to informed decision-making about community resources by the founding stakeholders was to have an "informed citizenry on community matters that are important to them" (Toma, 2013, p.4). The priority areas and indicators in the feasibility study for the Savannah-Chatham Community Indicators Project were adopted as the four priority areas for the 2010, 2011 and subsequent indicator surveys. Selected indicators in the priority areas were selected on the basis of: "meaningfulness (measures a specific condition of interest to the public, government or agencies and spans community-wide interest); validity (consistently reliable sources, timely, readily available, accurate and measurable); understandable (easy to interpret and communicate to various constituencies); and, applicable (facilitates the establishment of priorities, development of policy and evaluation outcomes)" (Beck et al., 2010).

The indicators selected for the priority focus are of *education and youth development*: 1) children entering first grade ready to succeed; 2) fourth graders who are proficient in reading; 3) public high school students who graduate on time; 4) 18–24 year olds who are not working nor in school; 5) out of school suspensions for 9th and 12th graders; 6) teenage birthrate; and 7) economic independence; and 8) working families who are lower-income. Indicator data was available for all "except indicator four in this category" (Toma, 2009, p.4). A survey option would be proposed given the problems with data that the United Way of the Coastal Empire had access to through US Census data (Toma, 2009, 2013). The bias noted in the feasibility study for the priority focus of education and youth development indicator "youth 18–24 year olds who are not working nor in school" was eliminated due to the survey work (Toma, 2019a). The results from the first feasibility study and subsequent survey were reported in summary reports and disseminated to the public in written reports and presented to the founding stakeholders and in public meetings (Toma, 2019a).

The indicators selected for the priority focus area of *economic independence*:

1) working families who are lower-income (<250% of federal poverty income); 2) lower-income working families who have a checking or savings account with a minimum of $300 saved; 3) Homeownership rate for lower-income working families; and, 4) lower-income working families who spend more than 40% of their income on housing. (Toma 2009, p.4)

At the time of the feasibility study, no available indicator data was available for this priority area (Toma, 2009).

The indicators selected for the priority focus area of *health and wellness*:

1) healthcare insurance coverage among those under age 65; 2) non-emergency use of hospital emergency rooms; 3) students in grade 9 who are of healthy weight and 4) babies born with a healthy weight. (Toma, 2009, p.4)

Indicator data was available for these indicators (Toma, 2009).
The indicators selected for the priority focus area of *regionalism*:

1) coastal empire coincident economic index; 2) average commute time; 3) public transportation ridership per route mileage; 4) percent of commuters using alternative mode of transportation; 5) solid waste tonnage recycled per person and 6) air quality index, number of days exceeding "good" quality level. (Toma, 2009, p.4)

Indicator data was available for these indicators (Toma, 2009).

Community surveys were conducted in 2010 and 2011 on the four priority areas and the 20 indicators. Subsequent survey benchmarking the four priority areas has been completed every two years from 2013, 2015 and 2017.[13] The most recent data is presently to be completed and will be presented to the Chatham County Board of Commissioners in April 2019 (Jennings, 2018). In 2010, "we called 2400 persons. We sent a mailing of 3000. The response rate was roughly 16% for both procedures. The response was '*stunningly high*'" (Toma, 2019a; emphasis added). "In 2011, we mailed 30,400 surveys. The response rate was 6.3%" (Toma, 2019a).

MATURATION OF THE INDICATOR PROJECT STAKEHOLDER COALITION FROM A FUNDER COLLABORATIVE TO A COLLECTIVE IMPACT INITIATIVE

Kania and Kramer (2011) examine various types of collaborations for social and community change ranging from "funder collaborative, multi-stakeholder initiatives, public–private partnerships and social sector networks" to a more robust collaborative which they call "collective impact initiatives" (p.41). Kania and Kramer's (2011) frameworks for various collaborations will be used to assess the maturation of the indicator project stakeholder coalition.

In 1990, community actors came together to plan a vision for the future in 2020 in the Savannah-Chatham area; this visioning process is most like a stakeholder initiative rather than a social sector network. Social sector networks are defined as formal or informal networks which are often ad hoc and are focused on the short term (Kania and Kramer, 2011). The Vision 20/20 Commission was voluntary but it was a formal commission focused on longer term results, so the classification as a multi-stakeholder initiative is a better fit.

In 2008, when the two local governments and the United Way of the Coastal Empire formed a partnership for the Savannah-Chatham Indicators Project, a best description is as a funder collaborative rather than a public–private partnership. "Public–Private Partnerships are partnerships formed between government and private sector organizations to deliver specific services or benefits" (Kania and Kramer, 2011, p.41). The Savannah-Chatham Indicators Project was seeking information to make decisions about the allocation of scarce resources which is not specifically narrowed to particular services or benefits. Kania and Kramer (2011) define funder collaboratives as "groups of funders interested in supporting the same issue who pool their resources" (p.46). The founding stakeholders in the Savannah-Chatham Indicators Project were not seeking so much as to pool their resources in all cases; rather they were looking to make efficient and effective decisions for the distribution of these resources in the community during a time of great economic distress, initially. "Generally, participants do not adopt an overarching evidence-based plan of action or a shared measurement system, nor do they engage in differentiated activities beyond check writing or engage stakeholders from other sectors" (Kania and Kramer, 2011, p.41).

The Savannah-Chatham Indicator Project is evidence that a funder collaborative can and often does want to do more than write a check. Both government and the United Way affiliate funder recognized that priority funding areas, performance measurement and indicator development can shift community outcomes. Osborne et al. (1994) suggest that local governments can steer outcomes by focusing on community solutions adopting market practices (i.e. performance measurement). In the nonprofit sector, the United Way organizations are recognized early adapters and leaders in utilizing performance measurement techniques for community change for both strategy development and outcome reporting (Hatry, 2006; Kaplan, 2001).

While the Savannah-Chatham Indicators Project partners/founding stakeholders did not specifically articulate a community development theory or theories as a basis for the indicator selection they were seeking a shared performance measurement plan through an indicators project. The founding Savannah Chatham Indicators Project partners intended to deepen the coalition from just a funder/donor collaborative to a collaboration for community impact. This collaboration adopts several of Kania and Kramer's (2011) collective impact collaborative features, specifically around shared measurement, mutually reinforcing activities and ultimately with the establishment of a backbone organization.

The 2009 Savannah-Chatham Feasibility Study established a considered approach towards developing shared measurement and ultimately to allow for mutually reinforcing activities with a focus on vetting indicators through the community meetings, focus groups and surveys to a variety of community stakeholders (Toma, 2009, 2013). Additionally, the university research team sought evidence of best practices around the frequency of municipal indicator project benchmarking and output reporting and distribution (Toma, 2009, 2013). Researchers reviewed a sample of 25 municipal community indicator reports from across the country for substantive content, reporting intervals and distribution methods (Toma, 2009, 2013).

In 2012, the Savannah-Chatham Indicators Project began to mature as it moved from the founding three partners in 2008 by expanding to "15 partners including public, private and nonprofit organizations including institutions of K-12 public education and post-secondary education as well as hospitals in the area" (CGIC, n.d.).

Coalition Governance Initiatives

Early on the Savannah-Chatham Indicators Project founding stakeholders contracted with Armstrong Atlantic State University to examine and report on best practices for governing indicator coalitions (Beck et al., 2010, 2011; Toma, 2009, 2013; Toma et al., 2011). Researchers sent 48 surveys to "administrators of indicator projects and just under 42% or 20 were usable" (Toma et al., 2015, Executive Summary). The results revealed that "indicator projects required group decisions; sought and used advice from experts in the field; allowed community members to shape the direction of the project" (Toma et al., 2015, Executive Summary). The 2015 Chatham County Blueprint incorporates a project based approach for the CGIC and the establishment of priority area committees, recognizes the importance of seeking community input as well as the use of experts to inform survey work and conduct surveys (Jennings, 2015). In order to facilitate the transition to a stand-alone organization, the CGIC contracted with the Jacksonville Community Council, Inc. to assist with mission, vision development, priority areas and data collection techniques (CGIC, n.d.).

Coalition Partner Expansion and Committing to Developing a Recognized Community Brand to Increase Community Participation

The expansion allowed the members of the Savannah-Chatham Indicators Project to consider moving beyond the qualitative community assessment of the 1992 Vision 20/20: Blueprint for Action for more in-depth data collection to "develop a shared agenda to improve community well-being" in 2013 (Coalition for Coastal Georgia Indicators, n.d.b). In 2013, the Savannah-Chatham Indicators Project now referred to as the Coalition for Coastal Georgia Indicators ("CGIC") hosted a series of 16 neighborhood forums which were called community conversations to facilitate deep dialogue about the state of the community. Continuous communication is a cornerstone of the collective impact model because it allows for trust development (Kania and Kramer, 2011). Trust is essential for meaningful community development especially when there are both high and low powered stakeholders involved in the effort. The distribution of community findings to the public and other stakeholders allows for legitimacy and community buy-in or any government or other high-powered community stakeholder action changes from past responses in distributing resources.

A Community Summit was convened in 2014, where the data was presented from the forums and surveys ultimately resulting in Chatham County inviting the coalition to submit a plan to create a "community wide, long-range strategic plan" that was both data driven and based on community input (Coalition for Coastal Georgia Indicators, n.d.b). A plan for data collection continued through 2015 using community surveys (Toma et al., 2015). From the survey data, the Chatham County Blueprint (Blueprint) was created by the Coalition for Coastal Georgia Indicators (Jennings, 2015). Given the long history of the Coalition for Coastal Georgia Indicators, formerly known as the Savannah-Chatham Indicators Project, the coalition was seen as the natural project lead for community assessment (Jennings, 2015).

Despite the expansion of the coalition members, the community did not turn out in large numbers for these conversations. Brand awareness was identified as a primary reason for the lack of community participation. Committees of the Coastal Georgia Indicators Coalition recognized that there was limited brand awareness and there was clear focus by the coalition to establish its brand and establish CGIC and the legitimate and qualified project leader to oversee the implementation of the 2015 Chatham County Blueprint (Jennings, 2015).

Table 15.8 2015 Chatham Community Blueprint priorities

Priority Number	Priority Focus Area
Priority One	Economy
Priority Two	Education
Priority Three	Health
Priority Four	Quality of Life

Source: Jennings (2015).

"The Blueprint is a long-term plan for the community" (Jennings, 2015). The Blueprint benchmarks community progress in four areas, "the economy, education, health and quality of life" (Jennings, 2015, p.2). The priority areas track to the 1992 Vision 20/20: A Blueprint for Community Action and the 2009 Savannah-Chatham Indicators Project Feasibility Study. Regionalism was replaced with quality-of-life focus which incorporates aspects of regionalism from the coordination of the provision of goods and services, the reduction of government, crime reduction, neighborhood connectivity, the coordination of the provision of goods and services to the reduction of government (Jennings, 2015; Toma, 2009, 2013; Beck et al., 2010, 2011).

Vision statements were also established for each of the priority focus areas supported through goals and strategies and ultimately measured by indicators as shown in Table 15.8. In 2014, the Savannah-Chatham Indicators Project partnership transitioned into a stand-alone nonprofit organization, the CGIC (CGIC, n.d.). The CGIC assumed responsibility for the creation of the 2015 Chatham County Blueprint and the development of priority focus areas for change along with vision statements, strategies and goals (Coastal Georgia Indicators Coalition, n.d.b). These actions meet the requirements for collective impact in that the priorities for change were based on a shared vision for a common agenda around four priorities that would allow for continuous communication between the full ranges of stakeholders involved with the Coastal Georgia Indicators Coalition for community change as reported in Table 15.9 (Coastal Georgia Indicators Coalition, n.d.b).

A committee structure was instituted for the priority areas as well as a steering/leadership committee; a content committee, charged with managing logistics including the training and management of volunteers to facilitate community engagement; an outreach committee to assist with marketing and public relations; a data, evaluation and survey committee to assist coordinating the data in the four areas (Jennings, 2015). Today, the CGIC uses the committee structure for the substantive content areas and has a Leadership/Executive Committee (CGIC, n.d.). Separate committees meet bi-monthly to discuss strategies for action to address the identified needs of the community in the four areas.

> Collective impact initiatives depend on a diverse group of stakeholders working together, not by requiring that all participants do the same thing, but by encouraging each participant to undertake the specific set of activities at which it excels in a way that supports and is coordinated with the actions of others. (Kania and Kramer, 2011, p.40)

Table 15.9 2015 Chatham Community Blueprint priority vision statements

Priority Area	Vision Statements
Economy	Chatham County anchors a thriving, business-friendly, regional economy in which all workers are prepared for quality jobs and residents feel empowered to attain a high quality of life
Education	From early childhood education through post-secondary achievement, Chatham County's innovative and inclusive educational systems are a model of academic excellence that enables students to have the knowledge, skills and abilities to succeed as chosen jobs/career pathways
Health	Chatham County has a culture of health including equal access to quality and affordable health care, chronic disease prevention, health inclusive policies and environmental design
Quality of Life	Chatham County citizens achieve a superior quality of life within a safe and healthy environment inclusive of the area's history, natural resources, public mobility and efficient government

Source: Chatham Community Blueprint (2015).

THE COASTAL GEORGIA INDICATORS COALITION TODAY

The mission of the Coalition for Coastal Georgia Indicators Coalition is:

> [t]he coalition is a group of community members and advocates working together in a comprehensive, coordinated approach for planning and accountability; and serving as a resource for agencies addressing overall health and well-being while leveraging resources for community initiatives. The purpose of the coalition is to bring sponsors and partners together to: enhance awareness of issues; strengthen partnerships and connect efforts; facilitate development of collaborative strategies; leverage resources for community initiatives; and mobilize action by reporting on progress made to improve the community's well-being. (CGIC, n.d.; Internal Revenue Service, n.d.)

The Coastal Georgia Indicators Coalition as a stand-alone nonprofit organization continues to be the sponsor of the benchmarking of the bi-annual community surveys reporting on the four priority areas (CGIC, n.d.). Since the 2015 Chatham County Blueprint, the CGIC has created four project teams aligning with the priority areas. Project teams have co-chairs to champion the work of these teams (CGIC, n.d.). Project teams are actively involved in implementing the vision, goals and strategies identified in the 2015 Chatham County Blueprint specifically focusing on "building networks, leveraging resources and increasing partnerships" (CGIC, n.d.). In addition to the priority areas, the CGIC has created the Public–Private Stakeholder Community Improvement Council which will coordinate and promote the use of the 2015 Chatham County Blueprint to local government leaders "to work collectively for the adoption and implementation of strategic priorities that guide policy, programs and resource allocation" (CGIC, n.d.).

The coastal Georgia Indicators Coalition is also the Savannah-Chatham area for the Chatham County Family Connections Collaboration (CGIC, n.d.). The Family

Connections Program is a unique data driven coalition across Georgia's 159 communities "dedicated to the health and well-being of families and communities" (Georgia Family Connections Program, n.d. and Roberts, 2019). CGIC is a natural fit to be the host for Chatham County Family Connections Collaboration given the four priority areas from the 2105 Chatham County Blueprint (Roberts, 2019).

CGIC in addition to providing data on the four priority areas identified in the 2015 Chatham County Blueprint provides other community data sets such as the Centers for Disease Control-Robert Wood Johnson Foundation 500 Cities Data, Healthy Tracker 2020, and Disparity data sets among others on its website (Coastal Georgia Indicators Coalition, n.d.b). The Coastal Georgia Indicator's Coalition's website provides tools at no cost to users along with the ability to create custom data sets for unique use assessing over 120 indicators to help drive change in Chatham County, surrounding counties and the City of Savannah (Coastal Georgia Indicators Coalition, n.d.b).

Sustainability

The Coastal Georgia Indicators Coalition, an independent nonprofit, is in a better position to seek partnership revenues, apply for government and foundation grants and institute organizational and individual member fees. The Coastal Georgia Indicators Coalition continues to receive funding from one of its founding three stakeholders while continuing to leverage resources through various options. Nonprofits are encouraged to diversify funding streams. According to the Form 990, Return of Organization Exempt from Income Tax filed with the IRS for years 2015 and 2016, the Coastal Georgia Indicators Coalition's revenue originates from contributors which are predominantly organization sponsors and partners (Internal Revenue Service, 2015 and 2016). Table 15.10 shows the breakdown of governmental and nongovernmental agencies' partners and sponsors.

Currently, the Coastal Georgia Indicators Coalition has four sponsoring partners. One of these is a public university partner and three are nonprofit government partners and sponsors, as reflected in Table 15.10 (Coastal Georgia Indicators Coalition, n.d.a). Of the three original founding stakeholders, Savannah city, Chatham County and the United Way of the Coastal Empire only Chatham County appears to be an organizational

Table 15.10 The Coastal Georgia Indicators Coalition 2018–19 partners and sponsors

Government	Nonprofit
Chatham County	Georgia Family Connections
Coastal Health District of Chatham County	Healthcare Georgia Foundation
Gateway Regional Behavior Health	St Joseph's Candler Hospital System
Chatham-Savannah Metro Regional Planning Commission	
Georgia Southern University*	

Note: * Georgia Southern University consolidated with Atlantic Armstrong State University in 2018–19. The university is a public university, so it is listed as a government partner.

Source: Coastal Georgia Indicators Coalition (n.d.a).

sponsor today. Of the two nonprofit hospital sponsors to the Savannah-Chatham Indicators Coalition from 2012, only one, St Joseph's Candler remains a nonprofit hospital. Memorial Union Hospital was sold to the Hospital Corporation of America in 2018 after board approval in 2017. The sale produced $25 million in profit which will be administered through the Chatham County Hospital Authority to assist with health care needs for the indigent (Rawlins, 2018).

Sponsors and partners of the Coastal Georgia Indicators Coalition serve in decision-making roles for the organization. All roles are on a voluntary basis (Coastal Georgia Indicators Coalition, n.d.a). "The Coastal Georgia Indicators Coalition sponsors and partners guide the planning process by serving on committees, project teams, attending community functions in support of the coalition and advocating for a comprehensive community conscience plan", known as the Chatham Community Blueprint (Coastal Georgia Indicators Coalition, n.d.a). While each of the committees and project teams "provide suggestions and recommendations, it is the coalition's executive leadership that provide oversight to the director, review and approve materials and will be held responsible for deliverables related to CGIC contracts" (Coastal Georgia Indicators Coalition, n.d.a). See Table 15.11 for a breakdown of the executive leadership team and sponsoring organization. Coalition committee and project team members engage in task-oriented work to promote the goals of the Coastal Georgia Indicators Coalition and the four separate priority areas (Coastal Georgia Indicators Coalition, n.d.b). The organization has other partnerships including the public–private partnership committee which seeks to engage the business community into the work of the coalition at a high level by promoting the Chatham County Blueprint and its strategic initiatives (Coastal Georgia Indicators Coalition, n.d.b).

The CGIC's executive leadership has organizational oversight duties that align with the coalition carrying out its mission. It is notable, from the website, that three of the founding stakeholders, charter members of the original Savannah-Chatham Indicators Coalition are a part of the organizational oversight board while only one is currently a funding partner (Coastal Georgia Indicators Coalition, n.d.a). The Coastal Georgia Indicators Coalition has one paid staff member, an executive director who reports to the Executive Committee (Coastal Georgia Indicators Coalition, n.d.a). The coalition also will from time to time have both graduate and undergraduate students to assist with

Table 15.11   Coastal Georgia Indicators Coalition executive leadership team

Organization	Organization Category
St Joseph/Candler Hospital System	Coalition Sponsor/Chair
Memorial University Medical Center	Vice-Chair
Synovus Trust Company, NA	Past Chair/Business
Chatham County Commission	Charter Member Organization
City of Savannah	Charter Member Organization
United Way of the Coastal Empire	Charter Member Organization
Individual Community Member	Member at Large
CGIC Executive Director	Executive Director

Source: Coastal Georgia Indicators Coalition (n.d.a).

data collection and other tasks. Much of the work is carried out through the volunteers assigned to committees (Roberts, 2019). The Executive Committee is also responsible for convening the Community Advisory Council. The Community Advisory Council attracts members of the public as well as members of sponsoring and partner organizations and representatives of the three founding organizations to allow for discussion around the strategies and activities for the coalition (Coastal Georgia Indicators Coalition, n.d.b).

Data Center

Today, the Coastal Georgia Indicators Coalition offers on its website over 120 indicators of community change including the original indicators that continue to be reported on with the bi-annual Savannah-Chatham Indicators Project.[14] Data is open source and free and available to all visitors to the website. The CGIC presents a community dashboard with data from its four priority areas from the Savannah-Chatham project which can be compared against other data sets for a general or a custom report. Indicators can be tracked by location. Other dashboards include the Health Tracker and the 500 Cities report as well as a disparities dashboard. The disparities dashboard allows organizations, researchers and others to look at data for the area compared to Georgia and against the United States for comparable analysis. Additionally, a social-needs index produced by the Conduent Healthy Communities Institute tracking poor health outcomes is available as well as general demographic data. The website allows individuals to craft a location builder report for the City of Savannah, Chatham County and the smaller cities and towns in the area. The coalition's website also provides a similar city dashboard to track progress against cities of similar size (Coastal Georgia Indicators Coalition, n.d.b).

In 2018, the Coastal Georgia Indicators Coalition was selected to participate in the Georgia Healthy Cities Project. While this project did not provide direct funds to the organizations, the inclusion in the project expanded resources offered to coalition members and the general public. Participation in this joint Centers for Disease Control-Robert Wood Johnson Foundation project has also increased the organization's credibility as the repository of data for the Savannah-Chatham area (Georgia Healthy Cities, n.d.).

Priority area coalition teams are in place and meet on a regular basis as well as the full coalition, expanding networks and using the indicators for community projects which are both formal and informal grassroots projects. The data is available for decision-making regarding resource allocation that includes the original four priorities being tracked as well as access to other data. Data informed decision-making will help communities to make the best decisions possible to address complex community problems with scarce resources (National Research Council, 2002). Nonprofit and nongovernmental organizations that are underresourced are challenged to produce outcomes and community impacts. The lack of funding and other resources present a challenge for organizations to achieve mission, vision and goals.

A sustainable financial approach appears to be in place through organizational membership partnership sponsors and opportunities for public and foundation grants. Opportunities for financial viability also exist for community member attendance on a sliding scale to make sure all voices are not excluded.

Table 15.12 Collective impact indicators and CGIC

Collective Impact Indicator	Explanation	CGIC Application	Success Marker
Common Agenda	Shared vision for change and common understanding of the problem(s)	Blueprint 2020	Savannah-Chatham survey conducted by university partner to benchmark progress
Shared Measurement Systems	Shared measurement system is essential	Savannah-Chatham priority indicators and measurement tools in Blueprint 2020	Savannah-Chatham survey conducted by university partner to benchmark progress
Mutually Reinforcing Activities	Diverse stakeholder and participant involvement required	Committee structure	Clear roles and responsibilities
Continuous Communication	Trust through meetings over time	Committees meet bi-monthly and quarterly	Goals and strategies developed, alliances formed for project specific work and reported on during committee meetings
Backbone Support Organizations	Coordinating organization allows for successful implementation		

Sources: Kania and Kramer (2011) and Coastal Georgia Indicators Coalition (n.d., a and b).

Best Practices for Replication

Kania and Kramer (2011) offer the collective impact initiative as a model for replication. "Collective impact initiatives are long-term commitments by a group of important actors from different sectors to a common agenda for solving a specific social problem. Their actions are supported by a shared measurement system, mutually reinforcing activities, and ongoing communication, and are staffed by an independent backbone organization" (p.41). Table 15.12 shows the alignment by CGIC to Kania and Kramer's (2011) collective impact model.

CONCLUSION

The Coastal Georgia Indicators Coalition is a vanguard public–private partnership which is seeking change not only using measurement but allows for community to create a common agenda for the organization as well as its four priority areas. This coalition is an example of how Kania and Kramer's (2011) collective impact model is successful when there is a long-term commitment and intentionality for change.

NOTES

* Deep appreciation is due to Ms Lizann Roberts, Executive Director for the Coastal Georgia Indicators Coalition for making this case study possible. Special thanks are given to Dr Michael Toma, Parker College of Business, Department of Economics, Georgia Southern University for providing all indicator reports and surveys completed by him and his research team for this case study. Additional thanks are due to Ms Tara Jennings, Strategic Planning Administrator, Chatham County (formerly with the United Way of the Coastal Empire and the founding Executive Director of the Coastal Georgia Indicators Coalition) for sharing background information about the founding coalition.
1. "The University System of Georgia Board of Regents approved resolutions making the consolidation of Armstrong State University and Georgia Southern University official as of Jan. 1, 2018" (Georgia Southern and Armstrong State University, n.d.).
2. Nonprofit hospitals prior to the Patient Protection and Affordable Care Act (PPACA) (2010) generally categorized charitable or no/low cost service provision to low income individuals as meeting the community benefit (Leider et al., 2017). This chapter does not explore the community benefit reporting requirements of nonprofit hospitals. The PPACA requirement is mentioned only to the extent that the Savannah-Chatham based nonprofit hospitals had an incentive to voluntarily join the CGIC to meet these new federal requirements and shift how nonprofit hospitals allocate community benefits.
3. "The enslaved population of Chatham County grew from 8,201 in 1790 to 14,807 by 1860" according to aggregated US Census data from 1790 to 1970 (Lockley, 2000, p.6). This fact is mentioned because the slave trade was a robust form of the area's commercial strength in trade. It is also mentioned because there are generational poverty and health issues as well as other community issues which stem to the founding of the Savannah-Chatham area.
4. The economic impacts for the Brunswick, Georgia Ports authority are not reported in this chapter. Brunswick, Georgia is a separate city and located in a separate county than the area covered by the Coastal Georgia Indicators Coalition.
5. Bureau of Labor Statistics nonfarm job categories of under 17,000 jobs were omitted for this chapter (US Bureau of Labor Statistics, Current Employment Statistics, 2019).
6. Nathan Deal was the Governor of the State of Georgia in the United States from 2011 to 2019. Nathan Deal was a member of the Republican Party of the United States byname Grand Old Party (GOP) when he was the Governor.
7. The Step UP Savannah initiative became a nonprofit in 2008. Retrieved on 15 March 2019 from https://stepupsavannah.org/works.
8. Nonprofit organizations are tax-exempt organizations due to their charitable purposes as codified in the Internal Revenue Code under Title 26 of the United States Code.
9. Changes to the Form 990, Return of Organization Exempt from Income Tax. For more on the changes, readers can go to the following IRS link – https://www.irs.gov/pub/irs-tege/990_whatsnew_purpose.pdf (retrieved 3 March 2019).
10. The 2017 data from the USA Giving Report show that giving or charitable contributions were just over $410 billion. This amount was a 5 percent increase from 2016. Retrieved 3 March 2019 from https://givingusa.org/tag/giving-usa-2018/.
11. "ProPublica is an independent, non-profit newsroom that produces investigative journalism in the public interest", retrieved on 11 November 2019 from https://www.propublica.org/. Pro Publica is a free access online database.
12. There appears to no longer be available data on whether the survey used random sampling methodology or some other type of methodology for the Vision 20/20 project.
13. The findings from these surveys are not reported in this chapter. The chapter focuses on the background of the community and its demographic, economic, community and health needs and the process for developing an indicator project/coalition.
14. The 2019 report was in the process of being prepared in spring 2019. It will be delivered to the Chatham County Board of Commissions in a public meeting by Dr Michael Toma and his team at a public meeting in April of 2019.

REFERENCES

Barker, L.E., Kirtland, K.A., Gregg, E.W., Geiss, L.S. and Thompson, T.J. (2011). Geographic distribution of diagnosed diabetes in the United States: a diabetes belt. *American Journal of Preventative Medicine*, 40(4), 434–9. doi: 10.1016/j.amepre.2010.12.019.

Beck, J., McGrath, R., Sadaatmand, Y. and Toma, M. (2010). Community indicators for Savannah and Chatham County. Unpublished manuscript. Armstrong Atlantic State University, Savannah, GA.

Beck, J., McGrath, R., Sadaatmand, Y. and Toma, M. (2011). Community indicators for Savannah and Chatham County. Unpublished manuscript. Armstrong Atlantic State University, Savannah, GA.

Belland, D., Rocco, P. and Wadden, A. (2014). Implementing health care reform in the United States: intergovernmental politics and the dilemmas of institutional design. *Health Policy*, 11(1), 51–60.

Bellows, L. (2018). Georgia healthy cities hosts first workshop. Retrieved 15 March 2019 from http://smartcities.ipat.gatech.edu/news/georgia-healthy-cities-hosts-first-workshop.

Bureau of Labor Statistics (2019). *Savannah Area Economic Summary*. Retrieved 3 March 2019 from https://www.bls.gov/regions/southeast/summary/blssummary_savannah.pdf.

Burns, N. and Gramm, G. (1997). Creatures of the state: state politics and local government, 1871–1921. *Urban Affairs Review*, 33(1), 59–96.

Butterfoss, F.D. (2009). Evaluating partnerships to prevent and manage chronic disease. *Preventing Chronic Disease*, 6(2). Retrieved 10 March 2019 from https://www.ncbi.nlm.nih.gov/pmc/articles/PMC2687870/.

Centers for Disease Control (n.d.). *About Diabetes*. Retrieved 16 November 2019 from https://www.cdc.gov/diabetes/basics/diabetes.html.

Chatham County Budget Allocations 2006 (n.d.). Retrieved 3 March 2018 from https://www.chathamcountyga.gov/.

Chatham County Budget Allocations 2007 (n.d.). Retrieved 3 March 2018 from https://www.chathamcountyga.gov/.

Chatham County Budget Allocations 2008 (n.d.). Retrieved 3 March 2018 from https://www.chathamcountyga.gov/.

Chatham County Budget Allocations 2009 (n.d.). Retrieved 3 March 2018 from https://www.chathamcountyga.gov/.

Chatham County Budget Allocations 2010 (n.d.). Retrieved 3 March 2018 from https://www.chathamcountyga.gov/.

Chatham County Budget Allocations 2011 (n.d.). Retrieved 3 March 2018 from https://www.chathamcountyga.gov/.

Chatham County Actual General Budget Expenditures for Health and Welfare, 2006.
Chatham County Actual General Budget Expenditures for Health and Welfare, 2007.
Chatham County Actual General Budget Expenditures for Health and Welfare, 2008.
Chatham County Actual General Budget Expenditures for Health and Welfare, 2009.
Chatham County Actual General Budget Expenditures for Health and Welfare, 2010.
Chatham County Actual General Budget Expenditures for Health and Welfare, 2111.

City of Savannah Budget Allocations, 2006 (n.d.). Retrieved 3 March 2018 from https://www.savannahga.gov/DocumentCenter/Index/161.

City of Savannah Budget Allocations, 2007 (n.d.). Retrieved 3 March 2018 from https://www.savannahga.gov/DocumentCenter/Index/161.

City of Savannah Budget Allocations, 2008 (n.d.). Retrieved 3 March 2018 from https://www.savannahga.gov/DocumentCenter/Index/161.

City of Savannah Budget Allocations, 2009 (n.d.). Retrieved 3 March 2018 from https://www.savannahga.gov/DocumentCenter/Index/161.

Coastal Georgia Indicators Coalition (2019). *About Us*. Retrieved 18 December 2019 from https://www.coastalgaindicators.org/.

Coastal Georgia Indicators Coalition (n.d.a). *Where We've Been*. Retrieved 3 March 2018 from https://www.coastalgaindicators.org/index.php?module=Tiles&controller=index&action=display&id=9781999096554 9707.

Coastal Georgia Indicators Coalition (n.d.b). *Where We Are Headed*. Retrieved 3 March 2018 from https://www.coastalgaindicators.org/index.php?module=Tiles&controller=index&action=display&id=9781999115612 0518.

Engleglau, M., Geiss, L., Saddine, J., Boyle, J., Benjamin, S., Gregg, E., Tierney, E., Rios-Burrows, N., Mokdad, A., Ford, E. and Imperatore, G. (2004). The evolving diabetes burden in the United States. *Annals of Internal Medicine*, 140(11), 949–59.

Federal Reserve Bank of St. Louis (2019). *FRED: Federal Reserve Economic Data – Economic Research*. Retrieved 3 March 2019 from https://fred.stlouisfed.org/categories.

Georgia Family Connections Program (n.d.). Retrieved 3 March 2019 from http://gafcp.org/.

Georgia Healthy Cities (n.d.). Retrieved 15 March 2019 from https://georgiahealthycities.org/#.

Georgia Healthy Cities (2019). *Savannah Story Map and 5 High Health Risks*. Retrieved 15 March 2019 from https://georgiahealthycities.org/savannah-map-tools/.
Georgia Ports Authority (2017). *Fiscal Year 2017 Annual Report*. Retrieved 1 March 2019 from http://gaports.com/Portals/2/About/Annual%20Report/2017/GPA416-2017GPA-AR-Spread-Drft-16.pdf.
Georgia Southern and Armstrong State University (n.d.). *Consolidation*. Retrieved 15 March 2019 from Georgia Southern University, https://consolidation.georgiasouthern.edu/.
Gittell, R. and Vidal, A. (1998). *Community Organizing: Building Social Capital as a Development Strategy*. Los Angeles, CA: Sage Publications.
Giving USA Foundation. (2009). *Giving USA 2009: The Annual Report on Philanthropy for the Year 2008*. Chicago: Giving USA Foundation.
Giving USA Foundation. (2010). *Giving USA 2010: The Annual Report on Philanthropy for the Year 2009*. Chicago: Giving USA Foundation.
Giving USA Foundation. (2011). *Giving USA 2011: The Annual Report on Philanthropy for the Year 2010*. Chicago: Giving USA Foundation.
Giving USA Foundation. (2012). *Giving USA 2012: The Annual Report on Philanthropy for the Year 2011*. Chicago: Giving USA Foundation.
Gordon, T. (2012). State and local budgets and the great recession: a great recession brief. Russell Sage Foundation and the Stanford Center on Poverty and Inequality. Retrieved 3 March 2019 from https://web.stanford.edu/group/recessiontrends-dev/cgi-bin/web/sites/all/themes/barron/pdf/StateBudgets_fact_sheet.pdf.
Grusky, D.B., Western, B. and Wimer, C. (eds) (2011). *The Great Recession*. New York, NY: Russell Sage Foundation.
Hatry, H.P. (2006). *Performance Measurement: Getting Results*. Washington, DC: The Urban Institute.
Humphreys, J. (2018). The economic impact of Georgia's deepwater ports on Georgia's economy in FY 2017. Retrieved from https://www.terry.uga.edu/about/selig/GA%20Ports%202017.pdf.
IRS (n.d.). *About Form 990, Return of Organization Exempt from Income Tax*. Retrieved 3 March 2019 from https://www.irs.gov/forms-pubs/about-form-990.
Jennings, T. (2015). Chatham County Blueprint. Unpublished manuscript. Coastal Georgia Indicators Coalition, Savannah, GA.
Jennings, T. (2018). Personal Communication. 6 December.
Kania, J. and Kramer, M. (2011). Collective impact. *Stanford Social Innovation Review*, 9(1), 36–41.
Kaplan, R.S. (2001). Strategic performance measurement and management in nonprofit organizations. *Nonprofit Management and Leadership*, 11(3), 353–70.
Leider, J.P., Tung, G.J., Lindrooth, R.C., Johnson, E.K., Hardy, R. and Castrucci, B.C. (2017). Establishing a baseline: community benefit spending by not-for-profit hospitals prior to implementation of the affordable care act. *Journal of Public Health Management and Practice*, 23(6), e1–e9. doi: 10.1097/PHH.0000000000000493.
Lockley, T.J. (2000). Trading encounters between non-elite whites and African Americans in Savannah, 1790–1860. *The Journal of Southern History*, 66(1), 25–48.
Malloy, D. (2012). Deal rejects expansion of Medicaid. *The Atlanta Journal: Constitution*. 28 August. Retrieved 1 March 2019 from https://www.ajc.com/news/state--regional-govt--politics/deal-rejects-expansion-medicaid/3pNCCRcciYiNwKqpW4ZEnO.
Marshall, M.C. (2005). Diabetes in African Americans. *Postgraduate Medical Journal*, 81(962), 734–40.
National Research Council (2002). *Community and Quality of Life: Data Needs for Informed Decision Making*. Washington, DC: National Academies Press.
Osborne, D., Gaebler, T. and Holman, T.R. (1994). *Reinventing Government*. New York, NY: Plume Publications.
Patient Protection and Affordable Care Act, The (2010). Public law, 111(48), 759–62.
Patton, M.Q. (2011). *Utilization-Focused Evaluation*. Los Angeles, CA: Sage Publications.
Phillips, R. and Pittman, R. (2014). *An Introduction to Community Development* (2nd ed.). London: Routledge.
Ploch, L. (1976). Community development in action: a case study. *Journal of the Community Development Society*, 7(1), 5–16.
Rawlins, E. (2018). Sale of memorial health to hca final. *WTOC*. 1 February. Retrieved 13 March 2019 from http://www.wtoc.com/story/37402290/sale-of-memorial-health-to-hca-final/.
Rich, R. (2013). The great recession: December 2007–June 2009. Retrieved 11 November 2019 from https://www.federalreservehistory.org/essays/great_recession_of_200709.
Rigby, E. (2012). State resistance to "Obamacare". *The Forum*, 10(2), 1–16.
Roberts, L. (2019). Personal communication. 17 February.
Savannah, Georgia population (2019). Retrieved 1 March 2019 from http://worldpopulationreview.com/us-cities/savannah-population/.
Sirgy, M.J. (2014). The science of community indicators research: a certification manual. Unpublished manuscript. The International Society for Quality of Life Indicators, Blacksburg, Virginia.

Smith, L. (2019). Message from County Manager Lee Smith, viewed on 18 December, https://www.chathamcountyga.gov/.
Toma, M. (2009). Community indicators for Savannah and Chatham County. Unpublished manuscript. Armstrong Atlantic State University, Savannah, GA.
Toma, M. (2013). Savannah/Chatham community indicator database: feasibility study. Unpublished manuscript. Armstrong Atlantic State University, Savannah, GA.
Toma, M. (2019a). Personal communication. 17 January.
Toma, M. (2019b). Personal communication. 3 March.
Toma, M., Byrd, N. and Russell, B. (2011) Governing practices for community indicators projects: 2011 survey. Unpublished manuscript. Armstrong Atlantic State University, Savannah, GA.
Toma, M., Bennett, K., Smith, C. and Hoover, D. (2015). Chatham county speaks! The 2015 community-wide survey. Unpublished manuscript. Armstrong Atlantic State University, Savannah, GA.
United Way of the Coastal Empire (2007). *Return of Organization Exempt from Income Tax (IRS Form 990)*. Retrieved 3 March 2019 from Pro Publica Nonprofit Explorer https://projects.propublica.org/nonprofits/organizations/580623603.
United Way of the Coastal Empire (2008). *Return of Organization Exempt from Income Tax (IRS Form 990)*. Retrieved 3 March 2019 from Pro Publica Nonprofit Explorer https://projects.propublica.org/nonprofits/organizations/580623603.
United Way of the Coastal Empire (2009). *Return of Organization Exempt from Income Tax (IRS Form 990)*. Retrieved 3 March 2019 from Pro Publica Nonprofit Explorer https://projects.propublica.org/nonprofits/organizations/580623603.
United Way of the Coastal Empire (2010). *Return of Organization Exempt from Income Tax (IRS Form 990)*. Retrieved 3 March 2019 from Pro Publica Nonprofit Explorer https://projects.propublica.org/nonprofits/organizations/580623603.
United Way of the Coastal Empire (2011). *Return of Organization Exempt from Income Tax (IRS Form 990)*. Retrieved 3 March 2019 from Pro Publica Nonprofit Explorer https://projects.propublica.org/nonprofits/organizations/580623603.
US Census Bureau (2010a). *American Factfinder: Chatham County, Georgia*. Retrieved 3 March 2019 from https://factfinder.census.gov/faces/tableservices/jsf/pages/productview.xhtml?src=CF.
US Census Bureau (2010b). *American Factfinder: Savannah City, Georgia*. Retrieved 3 March 2019 from https://factfinder.census.gov/faces/nav/jsf/pages/community_facts.xhtml?src=bkmk.
US Census Bureau (2017a). *Quickfacts: Chatham County, Georgia*. Retrieved 3 March 2019 from https://www.census.gov/quickfacts/chathamcountygeorgia.
US Census Bureau (2017b). *Quickfacts: Savannah City, Georgia*. Retrieved 3 March 2019 from https://www.census.gov/quickfacts/savannahcitygeorgia.
US Census Bureau (2019). *Chatham, Georgia Population*. Retrieved 1 March 2019 from http://worldpopulationreview.com/us-counties/ga/chatham-county-population/.
Vincent, C. and Stephen, C. (2015). Local government capacity building and development: lessons, challenges and opportunities. *Journal of Political Sciences & Public Affairs*, 1–5.
Vision 20/20 Commission, The (1992). Vision 20/20: a blueprint for action. Unpublished manuscript. Savannah, GA. Retrieved 3 March 2019 from https://www.coastalgaindicators.org/content/sites/uwce/Vision_2020_Blue Print_for_Action.pdf.
Weffer, S.E., Mullooly, J.J., Sylvester, D.E., DeLugan, R.M. and Hernandez, M.D. (2014). Partnerships across campuses and throughout communities: community engaged research in California's central San Joaquin Valley. In Sirgy, M. Joseph, Phillips, Rhonda, and Rahtz, Don (eds), *Community Quality-of-Life Indicators: Best Cases VI* (pp. 119–41). Dordrecht: Springer.

16. Social indicator projects for rural communities: the case of the Northwoods Quality of Life Database
Brandon Hofstedt

INTRODUCTION

Quality of life research shares a common connection to a broader research agenda dating back over 50 years referred to as the social indicators movement.[1] Since the movement's inception in the 1960s (Bauer, 1966; Biderman, 1970), social indicator projects tend to bend toward applied research, or research for the sake of social change (Noll, 2018). Regardless of unit of analysis from the national to state to community, or neighborhood, the driving question remains the same: how are we doing? From the beginning, social indicator projects utilized tools, concepts and theories from a variety of academic disciplines as a way to organize, monitor and report important domains of human communities for the purpose of improvement of public policy and ultimately for the betterment of society (Bognar, 2005; Land et al., 2012). This initial vision of providing access to timely and relevant data about social well-being and quality of life has transformed over the decades into an effort to connect information and the tools to use this information to decision-makers in a way that both informs and empowers leaders and communities to gain control of their future (Warner, 2014, 55).

In this chapter, I provide a brief introduction to social indicators and quality of life projects describing underlying philosophical ideas and concepts. I then describe selected community and county level indicators projects in the United States that demonstrate the foundational and core elements of the social indicators movement. Specifically, I discuss the Jacksonville Quality of Life Project, National Neighborhood Indicators Partnership (NNIP), Minnesota Compass, County Health Rankings and Roadmaps, and Headwaters Economics. Finally, I turn to rural communities by highlighting the Northwoods Quality of Life Database – specifically, its background and geographic scope, conceptual framework and operational definitions, technical organization and data collection processes and ultimate uses and purposes.

SOCIAL INDICATORS AND QUALITY OF LIFE

Social indicators are defined as "statistics, statistical series, and all other forms of evidence – that enable us to assess where we stand and are going with respect to our values and goals, and to evaluate specific programs and determine their impact" (Bauer, 1966, 1). In the early 1970s, the US National Science Foundation, the Russell Sage Foundation and the Federal Government all funded efforts to develop social indicator measures, fine-tune methodologies for data collection and collect and report on key social indicators that would be useful to inform public policy and expand longitudinal data beyond strict measures of economic growth and development and simple demographic information (Land et al., 2012; Land and Michalos, 2018). From the beginning of the movement, social indicator projects were about democratizing information for monitoring public welfare with the hope of driving societal progress (Bognar, 2005). As a result, social indicator projects play both a descriptive and an evaluative role in society. Today, it is a commonly held belief that awareness, involvement and participation by community members are central to the long-term effectiveness and usefulness of social indicator projects (Sirgy et al., 2013).

One strand of the social indicators movement that has experienced tremendous growth and received a great deal of attention over the last 30 years is quality of life research (Land and Michalos, 2018). In fact, Sirgy et al. (2006) suggest that social indicators are statistics meant to capture quality of life, and therefore, the two terms are often used synonymously. Within the literature, quality of life is often described as something that individuals experience as well as communities, regions or societies try to improve. For the former, quality of life refers to the "degree to which a person's life is desirable versus undesirable" (Diener, 2005, 401) and "encompasses all fields of life . . . and the guarantee of natural conditions of life for present and future generations" (Noll and Zapf, 1994, 2). As for the latter, quality of life in a community or society refers to the quality of the community on such things as local amenities, natural beauty, social connections (Kolodinsky et al., 2013) and general livability of a place (Leitmann, 1999) as well as general proxies for overall social welfare (Bognar, 2005). In other words, quality of life can assess for an individual or a community.

Regardless of individual or community quality of life, practitioners and researchers view the concept as including both objective and subjective elements. The objective measures of quality of life refer to those elements of life that reflect the actual lived experiences, realities and behaviors of an individual or group or the places, services, natural resources and built capital of a location. Items such as meeting basic human needs for subsistence, security, health and social connections are often included as foundational domains of objective quality of life for individuals and communities. Subjective measures of quality of life are often referred to as subjective well-being, life satisfaction or happiness, and incorporate an individual's reactions and interpretations of their personal, lived experiences as well as a person's thoughts, feelings and values related to these things (Diener, 2005; Sirgy et al., 2006). Subjective evaluations of community amenities, services and livability can be aggregated to capture the subjective components of quality of life in communities (Kolodinsky et al., 2013). Overall, it is seen as best practice to develop an integrated approach to quality of life projects and studies that utilize both objective and subjective measures across a variety of domains to capture quality of life because objective measures related to human needs influence subjective interpretations in complex ways through culture, past experiences and context.

One of the best known and widely used instruments to capture individual quality of life is Cummins' (1997) Comprehensive Quality of Life Inventory, or ComQol. ComQol is a measurement instrument utilized by educational and psychological practitioners to capture quality of life among diverse populations. Operationally, ComQol measures quality of life as comprised of culturally relevant objective indicators (e.g., healthy days, material goods, and income) and weighted subjective components (satisfaction x importance) across seven domains – material well-being, health, productivity, intimacy, safety, community and emotional well-being (Cummins, 1997; Cummins, 2018). In a meta-analysis of 1,500 articles of domains of life satisfaction, Cummins (1996) found an overwhelming majority of quality of life articles identified domains that fit within one of the seven schema of ComQol. ComQol demonstrates one of the ways the concept of quality of life is measured to capture objective and subjective domains of quality of life in a way that is empirically reliable and valid.

In their influential report commissioned by the French President, Nicholas Sarkozy, Stiglitz et al. (2009) identified many of the same core quality of life domains captured by ComQol, with a slightly different classification schema, and added in education, political voice and governance, and environment. Whereas Cummins' work addresses individual quality of life, Stiglitz et al.'s (2009) domains of quality of life reflect examples of collective or communal domains of quality of life. More recently, Barrington-Leigh and Escande (2018) conducted a review of indicator projects since the 1970s looking for common themes related to conceptual framework, institutional lead, methodology and operationalized indicators, and found "considerable agreement with the general content of domains prevalent in many well-being and progress indicators . . ." and that fit within the categories of "material living standards; health; education; governance and civic participation; social connections, relationships, and community; environment; culture; accounts of time-use; and various forms of security" (918).

Quality of life indicators are collected, measured, reported, monitored and analysed on numerous levels from an individual level to an aggregated unit of analysis such as a community, region or nation. Furthermore, quality of life indicators can and are utilized as a sole indicator, or combined with other indicators to create indices. Often, individual indicators and indices are reported and monitored with other measures to provide an in-depth perspective into the multi-dimensional concept of quality of life.

Notwithstanding the proliferation of quality of life studies specifically and social indicators generally, the field still lacks a cohesive theoretical framework. Sirgy (2011) does suggest and describe six discrete areas with corresponding quality of life theories that underline and inform many quality of life and social indicator projects. The six areas are: (1) socioeconomic development; (2) personal utility; (3) just society; (4) human development; (5) sustainability; and (6) functioning. Each of these six areas share a common connection to the social indicators movement but are guided by distinct theoretical frameworks that point to different indicators, suggest different levels of analysis and ultimately draw unique conclusions. Although there is a clear delineation of quality of life projects that come from these six areas, Sirgy (2011) still strongly argues that researchers "invest much more time and effort to further develop the theoretical basis of their indicators projects" (18).

COMMUNITY INDICATOR PROJECTS ACROSS THE UNITED STATES[2]

A distinct form of social indicator projects meant to capture quality of life at a more localized level has been growing in popularity in the last 20 years (Kingsley and Pettit, 2014). Like other social indicator projects, one could classify community indicators as a form of social indicators meant to monitor and report quality of life for the purpose of guiding community leaders and local decision-makers. Additionally, community indicator projects typically emerge out of a need or concern from local stakeholders around questions of what is happening in and what ought to happen in a specific community (Phillips, 2003). The difference is simply the unit of analysis; community indicators are region, county, community or neighborhood specific. Most community indicator projects skew toward larger geographic regions (e.g., counties) and larger population centers (e.g., metropolitan areas, cities). This is most likely due in part to availability and reliability of secondary data.

Community indicator projects emerge from both a grassroots call from local residents and from leaders, policymakers or academics. More often than not, community indicator projects utilize a combination of both; either emerging under the guidance of one, but developing and changing by the incorporation of the other (Barrington-Leigh and Escande, 2018).This means many community indicator projects are developed, in part, through a top-down process (i.e., identification of desirable indicators and domains informed by a conceptual framework) as well as informed by a bottom-up approach where indicators, metrics and domains emerge from people and organizations on the ground.

To highlight the types and varieties of community indicator projects, I discuss five examples of community indicator projects from across the United States that differ in origin story (i.e., bottom-up versus top-down approach), unit of analysis (i.e., neighborhood to county), geographic scope (e.g., one community versus many), rural-urban continuum, and conceptual framework (i.e., few domains versus many).

Jacksonville Quality of Life Report

In 1974, the Jacksonville Community Council, Inc. (JCCI) convened numerous community leaders from government, business, nonprofits, workers and citizens to identify growth and development goals for Jacksonville, Florida. The purpose of the meeting was to fulfill JCCI's mission, "to bring people together to learn about [the] community, engage in problem solving and act to drive positive change" (Citizen Engagement PACT of Jacksonville, 2017). The result of the meeting was the identification of ten community domains and ten corresponding task force groups to examine each of these domains (Phillips, 2003). Ten years later, the JCCI released its first annual *Quality of Life Progress Report* examining ten domains comprised of 83 indicator measures (Warner, 2014). Now restructured and called the Citizen Engagement PACT of Jacksonville, the JCCI last put out its 31st edition of the *Quality of Life Progress Report* in 2016, making it the oldest and one of the longest running community indicator projects in the country (Brune Mathis, 2017). Like the initial report, the 2016 report covers ten domains including education, economy, environment, safety and security, arts and entertainment, distinctive neighborhoods and urban heart, health, government, transportation, and diversity and inclusion.

Today, as in 1985, this community indicators project follows a bottom-up approach where indicators are selected, monitored, reviewed and reported by citizen volunteers. For 31 consecutive years, JCCI's annual report provided meaningful and longitudinal data to the Jacksonville area allowing community leaders to track relevant information in a user-friendly manner for a diversity of community leaders and groups. The annual reports organize key indicators under domain names with previous and latest data points listed with one of three simple trend identifiers – "Better," "Same," or "Worse." This bottom-up, citizen driven and outcome-focused model is regularly identified as a gold standard for community level indicator projects (Warner, 2014).

National Neighborhood Indicators Partnership (NNIP)

Founded in 1996, the NNIP is a collaborative effort between the Urban Institute and local organizations or partners with the explicit mission "to improve low-income neighborhoods by empowering residents and local institutions to use data in their community building and policymaking" (Hendey et al., 2016, 3). NNIP is based on a model of three main functions: (1) compile and maintain neighborhood level data across various domains; (2) assist and show how to use data; and (3) stress the importance of using data to build and strengthen local capacity (NNIP, 2018). The idea behind the NNIP is to help local and regional community and neighborhood partners establish community indicator projects intended for use in community building efforts and policy decisions. These efforts vary in geographic scope including county, city and metropolitan area but all emphasize neighborhood level data.

Currently, NNIP is located in 31 cities throughout the United States, all of which are large metropolitan communities. Within each of these cities, one or more local organizations work with the Urban Institute to act as "data intermediaries" to develop, monitor and report key indicators using national secondary sources (e.g., American Community Survey, Decennial Census), state level secondary data (e.g., education or health) and locally supplied data (e.g., local government or nonprofit specific data). These intermediaries – which can be nonprofits, university research centers, community foundations, governmental agencies, among others – are the overseers of culling together, cleaning and maintaining these data. Intermediaries share information and apply it in ways that help the community see the impact of policy changes.

Intermediaries and the Urban Institute do not collect and choose data in a bubble; rather, they interact with local partners to identify which data are most relevant and needed while "strengthen[ing] civic capacity and governance" (Hendey et al., 2016, 6). Most commonly, NNIP partner communities focus on domain areas including child welfare, economy, education, health, land and properties, public assistance, and safety (NNIP, 2018). Similar to the underlying principles founded in the social indicators movement of the 1960s and 1970s, NNIP's neighborhood level indicators are guided by the belief that access to information for local communities creates better policy and engages community members (Kingsley and Pettit, 2014).

Minnesota Compass

Minnesota Compass is a social indicators project focused on counties, metropolitan regions and select communities in Minnesota, and is intended to monitor and track "progress" on a variety of domains including the following: demographics (aging, youth and immigration), arts and culture, civic engagement, economy, education, environment, health, housing, safety, transportation and workforce. Minnesota Compass is headed by Wilder Research, a part of the Amherst H. Wilder Foundation, and has numerous private, nonprofit, philanthropic and corporate sponsors. Guided by selection criteria (e.g., reliability and validity concerns, community relevant, policy-responsive and cost-effective), an advisory group made up of leaders and experts from academia and business identify broad categories and specific indicators. The project started as a result of a call from leaders in the business, government and nonprofit sectors within Minnesota to have reliable, relevant and accessible data for community leaders to guide and inform policy decisions (Minnesota Compass, 2018). Minnesota Compass provides access to publications and to simplified data visualizations for Minnesota communities and regions and works with interested clients on data compilation, analysis and interpretation; however, their primary function is to operate as a data warehouse where data gleaned from a variety of secondary sources is concisely labeled and cleaned and clearly organized and connected to domains of quality of life important to Minnesota communities.

County Health Rankings & Roadmaps

The County Health Rankings & Roadmaps (CHRR) is a program funded by the Robert Wood Johnson Foundation and maintained by the University of Wisconsin Population Health Institute (UWPHI). As suggested by its title, CHRR is a social indicators project focused on health-based outcomes for US counties and is intended to help local leaders and health officials understand the complex nature of what is needed for a healthy community so they can make informed decisions about what is going well, what are the areas of challenge and what should or could be appropriate interventions and policies to improve health for people across a county area. CHRR has its origins in the United Health Foundation's America's Health Rankings and interest among researchers at UWPHI as to why these state level health rankings changed from year to year and whether health outcomes were even more localized. This interest led to the first county level health ranking for all 72 Wisconsin counties in 2003. By 2010, UWPHI released the first national County Health Rankings. The following year, UWPHI and Robert Wood Johnson Foundation explored the Roadmaps to Health concept, where data and rankings could be utilized to move from knowledge to action (Catlin, 2014).

Unlike other social indicators that focus on many categories or domains of quality of life, CHRR is concerned with health. Despite this singular focus, the model is applied to understand, rank and guide communities on community health and is comprised of multiple, interrelated domains that affect and relate to health. For example, on an annual basis, CHRR ranks all US counties within each state. Rankings are developed using a model that weights a variety of secondary sources including health behaviors (e.g., smoking, diet and exercise, alcohol and drug use), social and economic characteristics (e.g., high school graduation rates, income, employment), natural-physical environment (e.g., air pollution, drinking water violations) and community healthcare infrastructure (e.g., access to health professionals). Each measure is standardized by calculating a z-score for each county within a given state. These standardized scores are then weighted and added to a composite measure (i.e., the sum of all standardized scores times the weight given to that indicator). Ranks are then assigned based on composite score – lowest score is highest ranked and considered to have the best health (County Health Rankings, 2018). Ultimately, the purpose of the ranking of each county within states is to allow local and regional leaders to "understand the many influences on health and vitality and inspire community-level change" (Catlin, 2014, 59).

Headwaters Economics

Located in Bozeman, Montana, Headwaters Economics is a nonprofit organization focused on research, community development and land management. Originally focused on the western US states, Headwaters Economics now covers the entire United States examining domains including economic development, energy, climate, public lands and wildfire. Like the other social indicator projects discussed in this section, Headwaters Economics strives to provide access to timely, relevant and clear data for local stakeholders and organizations to address local challenges and to develop appropriate, data-informed solutions (Headwaters Economics, 2018). Utilizing data from secondary sources such as the Bureau of Economic Analysis, US Census Bureau and the Bureau of Labor Statistics, Headwaters Economics do so through four different avenues: a profile system, data visualizations, original research and individual client projects.

Two of their more well-known profile systems are the "Economic Profile System" (EPS) and "Populations at Risk." Both profiling systems provide a user-friendly and accessible interface for local stakeholders to access comprehensive and customizable reports for select communities, counties, regions or states on a variety of topics. Each report also allows users to select benchmarks of interest such as other places or the country as a whole. For example, EPS allows users to create reports for 14 different topics including: overall summary, socioeconomic, tourism, non-labor income, timber, services, public land amenities, demographics, land use, agriculture, mining/oil/gas, government, federal land payments, and wildland urban interface. Reports are generated with selected benchmarks, can be downloaded in pdf or excel formats and have extensive yet user-friendly features throughout including guiding questions (e.g., what do we measure and why is it important) and data reliability (e.g., color coded data identifying issues with sampling error and level of reliability).

Headwaters Economics also creates and posts regular data visualizations and original research for public consumption and works with local clients across the country. For example, in March of 2015 and in partnership with the University of Michigan, Headwaters Economics produced an interactive map and data visualization of all US counties within the Great Lakes water basin examining how climate change may affect local and regional economies and infrastructure as well as vulnerabilities of local populations (i.e., resilience) to these changes (Gude, 2015). On their website, Headwaters Economics has an entire section devoted to "Data Viz" efforts. The organization works with individual clients on specific community and economic development and land management questions. Headwaters Economics' website has over 50 examples of working with communities and regions from across the United States on comprehensive reports and data visualization efforts under their "local studies."

NORTHWOODS QUALITY OF LIFE DATABASE

The rapid increase of social indicator projects over the last 50 years and the development of numerous community indicator projects notwithstanding, rural communities and regions often lack clear, consistent and identifiable efforts to develop community indicator projects (Phillips, 2003; Besser et al., 2012). Additionally, rural communities often conceive of quality of life in different ways (Arbuckle and Kast, 2012; Fernando and Cooley, 2016). Community indicator projects, as shown above, tend to skew toward metropolitan regions and to the county level unit of analysis. Both make sense. Larger populations have more resources making data collection, organization, monitoring, analysing and reporting much easier. When it comes to data collection and generalization of findings, larger populations with more resources often produce samples with less error that better reflect population parameters. Furthermore, drawing a sample of 1,000 people in a county of 100,000 is much easier to achieve than drawing a sample of 500 people in a community of 1,000 – yet, both produce roughly the same margin of error. Of the five projects examined in the previous section, only three include rural areas. Two of the three projects that cover rural areas do so at the county level only. Through their Populations at Risk profile system, Headwaters Economics is closest to providing community level data for rural communities by including select census designated places in their selection criterion setup. As anyone working in rural communities knows, aggregate county level data often hide important differences within communities, making problem and solution identification difficult. In this final section, I describe the Northwoods Quality of Life Database (NWQoL) including its origin story and geographic scope, its conceptual framework and operational definitions and its technical organization and data collection processes. I end the section by discussing its ultimate use and purpose for scholars, practitioners and community leaders.

Background

The NWQoL was founded in 2017 and is housed at the Center for Rural Communities (CRC) at Northland College, an applied research center focused on rural communities in the northwoods region of northern Minnesota, northern Wisconsin and the Upper Peninsula of Michigan. The idea behind the NWQoL emerged as researchers from the CRC worked with local stakeholders in the Chequamegon Bay region of northern Wisconsin; specifically, in the community of Ashland, Wisconsin. Like many rural, post-industrial communities, the narrative of Ashland was one of decline and loss – population decline, decline in industrial core (i.e., loss of ore docks, paper mill and logging), loss of downtown businesses, among others. A cursory examination of Ashland – for example, total population, driving the downtown and city neighborhoods and conversations with community members – would confirm this dominant narrative. Upon further investigation, the narrative was often ill-informed lacking actual data, and in cases where it did include data did so inappropriately – for example, using county level data for community level questions.

For example, total population loss trends provide an incomplete picture of what is occurring in Ashland and the surrounding area and require a more refined analysis that reveals population sorting based on education and income. Overall economic development has declined, but certain segments such as local food and outdoor recreation are growing and provide opportunities for further development. Downtown vacancies are commonplace, but in comparison to other similar-sized communities, Ashland is on par or better in terms of vacancy-to-store front ratio. Additionally, the footprint of the downtown is nearly double the length of the next community of similar size in the region. Additionally, diversity of business options skews heavily toward secondhand retail compared to other communities. The inaccurate perceptions of decline obfuscate viable economic and community development opportunities, while clear up-to-date information and analysis uncover unique insights and help generate practical and innovative solutions. This initial examination and exploration of what was happening in Ashland also revealed a lack of a meaningful yardstick to understand what was happening in this rural community and others like it in the northwoods (Hofstedt and Tochterman, 2015). Ashland, like so many other rural communities, is not alone in its narrative of decline, in its concerns about the future or in its difficulties accessing and interpreting information to guide decisions. It was out of this initial exploration and realization that the NWQoL was formed.

Today, the NWQoL is defined as a systematic and comprehensive cataloging of quality of life metrics for communities from across the northwoods region. Rural communities from the northwoods region share common experiences, opportunities, weaknesses and challenges. From climate and healthcare, to education and job creation, to attracting and retaining young families, to redefining community in light of post-industrial decline, rural communities struggle to address challenges with limited information at their disposal. Supported by funding from the Otto Bremer Trust, NWQoL is comprised of a collection of primary and secondary data organized into different domains of quality of life relevant to northern, Midwestern rural communities. Ultimately, the goal of NWQoL is to provide information to communities that allows them to understand what they have and to make decisions about what makes their rural community a good place to live, work and play.

Conceptual and Operational Framework

In the crucial report by the Commission on the Measurement of Economic Performance and Social Progress, Stiglitz et al. (2009) wrote: "What we measure affects what we do" (7) and "the time is ripe for our measurement system to *shift emphasis from measuring economic production to measuring people's well-being.* And measures of well-being should be put in a context of sustainability" (12). Following this line of thinking, NWQoL is intended to provide access to comprehensive data organized within a framework for understanding quality of life for rural communities, so they can sustain and thrive into the future. Simply, the NWQoL database is structured under three broad categories: (1) natural and built environment; (2) community characteristics and services; and (3) people. These three broad categories are made up of numerous quality of life domains populated with primary and secondary data or indicators (see Table 16.1).

For example, the category of natural and built environment is divided into two domains: natural environment and built infrastructure. Natural environment follows closely to Flora and Flora's (2008) definition of natural capital, "the landscape, climate, air, water, soil, and biodiversity of both plants and animals" (17–18). In NWQoL, natural environment specifically encompasses land use, food production, water and air quality, renewable energy potential and climate. Data for the aforementioned example indicators come from secondary sources such as National Oceanic and Atmospheric Administration (NOAA), United States Geological Survey (USGS) and USDA's Agriculture Census and includes both county and community level data. Similarly, built infrastructure aligns with Flora and Flora's definition of built capital, or "capital that . . . includes factories, schools, roads, habitat restoration, and community centers" (18). Built infrastructure in NWQoL includes a variety of indicators such as schools, hospitals, downtown buildings, housing, among many others. Data for these example indicators come from primary data

Table 16.1 Conceptual framework

Category	Domains and Example Indicators	Sources
Natural and Built Environment	Natural environment (e.g., land use, food production, water and air quality, renewable energy potential, climate)	National Oceanic and Atmospheric Administration (NOAA) United States Geological Survey (USGS) Agriculture Census (AgCensus)
	Built infrastructure (e.g., schools, hospitals, police stations, downtown buildings, housing, roads, trails, museums, library, music venues)	Center for Rural Communities (CRC) American Community Survey (ACS) Decennial Census
Community Characteristics and Services	Arts and culture (e.g., performing arts theaters, music venues, art galleries, museums, libraries)	Center for Rural Communities (CRC) National Center for Charitable Statistics (NCCS)
	Business and economy (e.g., jobs, employment, wage and salary, work, businesses, industries, big box retailers, downtown businesses, food businesses, economic development organizations)	Center for Rural Communities (CRC) American Community Survey (ACS) Bureau of Economic Analysis (BEA) Economic Census Decennial Census
	Education (e.g., youth and adult programs, school performance, daycare providers, college preparedness)	Center for Rural Communities (CRC) American Community Survey (ACS) Decennial Census Michigan Department of Education Minnesota Department of Education Wisconsin Department of Public Instruction
	Entertainment and leisure (e.g., activities, events, shopping)	Center for Rural Communities (CRC)
	Food (e.g., grocery stores, restaurants, specialty stores, farmers markets, emergency food organizations)	Center for Rural Communities (CRC) United States Department of Agriculture (USDA) Agriculture Census
	Health (e.g., physicians, programs, health rates, poor health days, smoking, suicide deaths)	Center for Rural Communities (CRC) Behavioral Risk Factor Surveillance System (BRFSS) Centers for Disease Control and Prevention (CDC)
	Nonprofit infrastructure (e.g., nonprofits organizations, anchor institutions)	Center for Rural Communities (CRC) National Center for Charitable Statistics (NCCS)
	Political governance (e.g., voter turnout, registered voters, political voice)	Michigan Secretary of State Minnesota Secretary of State Wisconsin Elections Commission
	Recreation (e.g., in- and out-door opportunities, youth and adult programs, recreational trails)	Center for Rural Communities (CRC)

Table 16.1 (continued)

Category	Domains and Example Indicators	Sources
People	Safety and security (e.g., crimes, public safety service, material well-being)	Center for Rural Communities (CRC) Uniform Crime Report
	Technology and communications (e.g., newspaper, social media, internet speed)	Center for Rural Communities (CRC) Federal Communications Commission (FCC) American Community Survey (ACS)
	Transportation (e.g., commuting, access to public transit, airports, four-lane highways)	Center for Rural Communities (CRC) American Community Survey (ACS) Decennial Census
	Demographics (e.g., age, education, income, race and ethnicity)	Center for Rural Communities (CRC) American Community Survey (ACS) Decennial Census
	Culture (e.g., values, attitudes, beliefs, norms, behaviors)	Center for Rural Communities (CRC)
	Social capital (e.g., interactions, involvement, connections)	Center for Rural Communities (CRC)
	Individual well-being (i.e., subjective and objective)	Center for Rural Communities (CRC)
	Personal evaluations of environment, built infrastructure, community characteristics, and services	Center for Rural Communities (CRC)

(e.g., annual comprehensive downtown inventories conducted through field visits) and secondary sources (e.g., US Census American Community Survey data on housing stock and characteristics).

Within each category, domains are made up of specific indicators gathered from existing data sources or collected through original data collection. Demographic data, which falls under the broader category entitled "People," is pulled from the Decennial Census and American Community Survey and includes specific indicators related to total population, age, income, education and race and ethnicity. Additionally, original data are collected for many domains under all three broad categories. For instance, under the broad categories of community services and characteristics and people, the CRC collects data on local food infrastructure, downtown vibrancy, recreational infrastructure, individual well-being and social capital. Overall, the NWQoL database is populated with information collected from original, proprietary data collection instruments and secondary data sources across 18 quality of life domains. Our proprietary data collection instruments and our original data fill gaps in existing data sources and bring relevant information to communities to help address region specific issues. The combination of bringing existing data together, supplementing it with CRC's original data found nowhere else, and providing meaningful comparisons creates a much-needed source of information for communities, organizations and leaders in the northwoods region of Minnesota, Wisconsin and Michigan.

Category and domain selection process

The indicators within each domain are identified through a combination of top-down and bottom-up processes. For example, categories of quality of life are gleaned from the community of science and verified on the ground in the northwoods through examination of important community documents (e.g., comprehensive plans) and conversations with local stakeholders. The CRC has also completed numerous community-based research projects over the last 3 years that provide insight into what is important to northwoods communities. The CRC has worked with nearly 50 local groups and organizations from the Chequamegon Bay and broader region on research projects, market studies, community forums, community-based projects and education events. These partnerships and collaborations have proven invaluable to the CRC's work overall and to the development of the database specifically. These relationships are essential in making our database accessible, useful and relevant to the communities and region we serve.

The final product and arrangement are far from perfect or complete and are looked at as a long-term effort that is mutable to meet needs and desires of northwoods rural communities in the future. The plan is to continue a circular process of broader domain identification (what we learn through research, community of science and from stakeholders), specific indicator inclusion (what is available, what is missing, what is attainable) and truth-testing relevance and value of indicators (what do local leaders and policymakers use). This information will allow us to fine-tune and revise the database over time.

Unit of analysis

The database is made up of both county and community level indicators – a region home to approximately 2.3 million people in small towns and rural communities across 78 counties (Figure 16.1). The objective of the NWQoL is to monitor, examine and report on rural communities in the northwoods region. As a result, NWQoL focuses on rural communities (n = 140) as defined by their relative location to metropolitan areas, overall population size and municipal boundaries. Only rural communities from across the northwoods region that are at or above 1,000 and below 30,000 in population are included. Additionally, rural communities are defined as those that are at least 10 miles from mid-size cities and their metro areas (i.e., <350,000 people) or 30 miles away from large cities and their metro areas (i.e., >350,000 population).[3]

County level data come primarily from secondary sources (e.g., US Census Bureau) but are occasionally supplemented with primary data (e.g., number of miles of recreation trails). These data are organized similarly to other county level social indicator projects; however, the county level data are intended to help situate smaller rural communities in a broader regional context. For instance, when examining tourism and recreation trends, the database tracks individual community trends but situates these trends in a broader county or regional context of natural resources, recreational infrastructure and demographic and economic activity.

Database structure, input and output

The NWQoL is comprised of disparate sources of data across community and county units of analysis. Data come from primary and secondary sources and are collected or coalesced at various points in time (i.e., during the year or over multiple years). Although the intention is to collect the same information over time, the database is set up as a

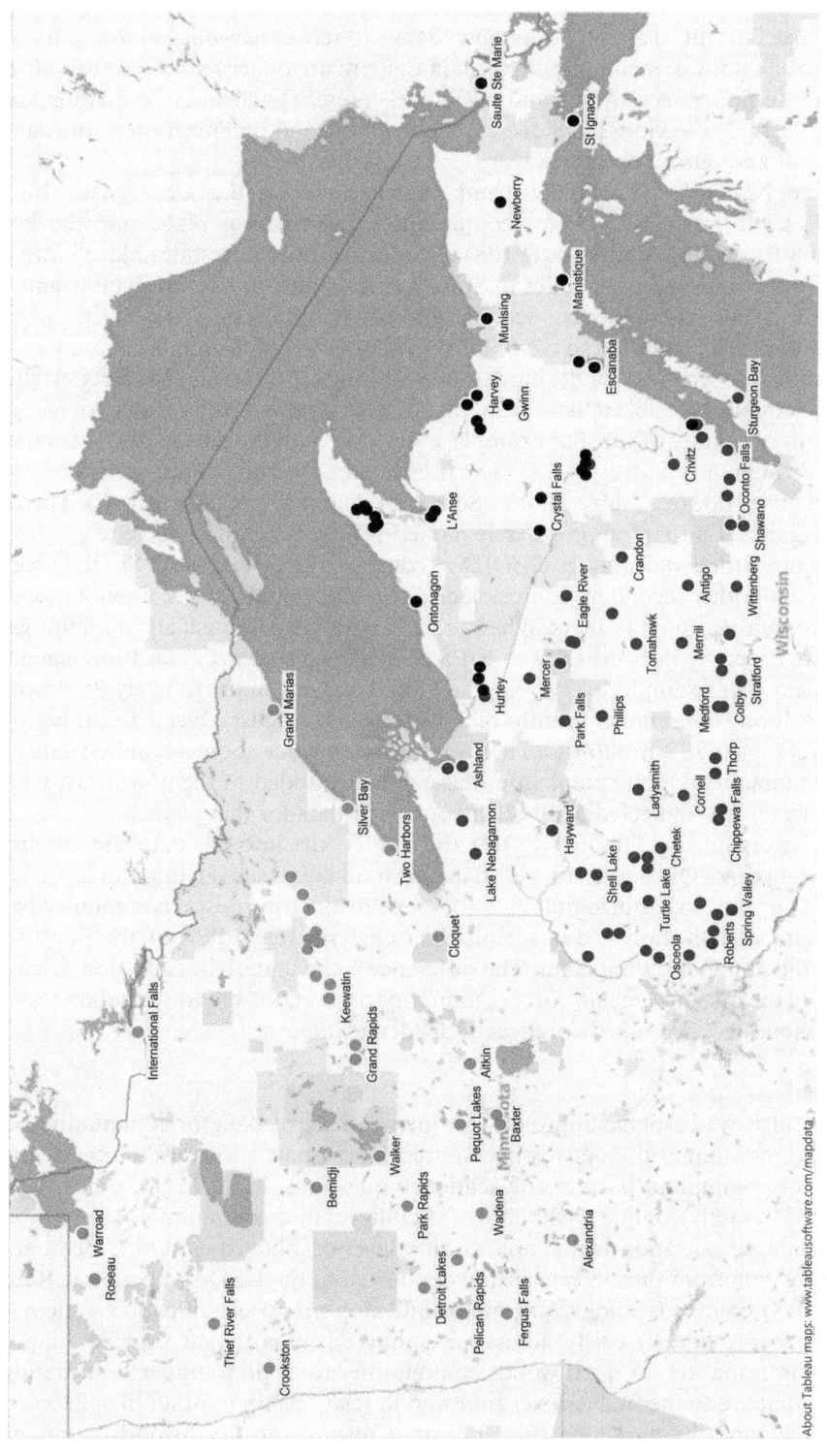

Figure 16.1 Northwoods region and population communities

"living" project meant to adapt, adopt and change to reflect new data sources, changes in indicators and removal or elimination of data that are no longer relevant or useful. These four things – unit, sources, timing and relevance – present challenges for maintenance of the NWQoL. In this section, I will discuss the structure and organization of the database and the input and output processes.

First, the NWQoL is structured and organized using the US Census Bureau's geographic identifiers (GEOID) for communities, identified as place, and the Federal Information Processing Standards (FIPS) code for counties and states placed into three individual but connected base tables in SQL Server – state table, county table and community table. These base tables connect place to county and state and act as the codes by which new data are connected to community and/or county locations.

This leads to the second part: the input process. Using GEOIDs and FIPS codes, all community and counties are able to have regularly added data points for a variety of categories and domains of quality of life. For example, every year with the release of the US Census' American Community Survey (ACS), 5-year estimates, we pull county and place data by designated table code (e.g., table DP03, "Selected Economic Characteristics"). These data are then organized into an appropriately named NWQoL database table (e.g., "housing characteristics") filed under either GEOID or county FIPS for that year. Similar processes are followed for other secondary sources such as the Decennial Census, Economic Census, Agriculture Census, among others. When original data are collected, all communities and counties are coded by their GEOID or FIPS code allowing us to connect original data to all other data. For example, every year the CRC conducts approximately 25 downtown inventories during the summer months of June, July and August. Over a 5-year period, we are able to complete 127 inventories. These inventories include about a hundred data points for each community. Upon completion, these data are added to the downtown vibrancy table and able to be connected to all other sources of data for that place.

Finally, by organizing data by GEOID and FIPS codes into one database, we are able to write syntax in SQL Server to pull data from all sources over time for a variety of purposes. Currently, we utilize pull data for three distinct purposes: community partner projects, data visualization and academic data analysis. Regardless of the purpose, the process is the same for pulling data. The difference is the underlying question driving the data pull. Is the question meant for a community partner, for broad scale data visualization and public access or for theoretical or academic interests?

Use and purpose

NWQoL utilizes a two-tiered approach[4] for intended use of data for communities, counties and regions: online, interactive open-source data visualizations and in-depth analysis or report for community partners or academic questions. First, over the next year, we will release interactive online dashboards for counties and communities about monthly. The first data visualization is a simple county level dashboard exploring demographic and housing data from the US census and connected to the USDA's Economic Research Service (ERS) county typology and metro and non-metro classification system.[5] The purpose is to provide an easily accessible online dashboard that provides important contextual information for northwoods communities to begin to understand trends and how they compare to themselves over time and in relationship to other like places.

Second, by actively marketing this important resource to northwoods communities, we plan to use these monthly dashboards as gateways to further community engagement, exploration and conversations. This can be completed through very simple and straightforward questions from community partners (e.g., what do our housing trends look like and how do people in our community feel about the topic?) or they can result in exploration of complex questions confronting a community (e.g., what are our most pressing housing needs, how are we doing with amenity development or how can we create a more vibrant downtown?). This can be used to help answer specific questions like a housing market study, to inform comprehensive community plans or to develop a strategic plan for the community.

CONCLUSION

Community indicators create access to and democratize information. They reflect a value system that provides guideposts and direction for communities of interest to follow as they strive for a desired outcome. Comprehensive community indicator projects integrate silos or domains of life that are often viewed in isolation into a framework that allows for practitioners, researchers, decision-makers and residents alike to examine the complex relationships among these domains (e.g., ecological, social and economic). The insights gained from examining and understanding how these domains interact with one another affords the opportunity for more technical, appropriate and effective policies as well as provide needed information and guidance for evaluating policies and other politically, socially or economically derived interventions.

The NWQoL database is different from other regional or national databases of community level information in a number of ways. First, starting with a wide array of secondary data sources, we bring together disparate sources from existing public and private sources into one location. By culling together existing data sources and understanding what is important to northwoods communities and the region, we are uniquely positioned to identify gaps in the data that allow us to develop original, innovative data collection instruments and will continue to allow us to adapt, change and add data as needed. The combination of bringing existing data together, supplementing it with original data found nowhere else and providing meaningful comparisons creates a much-needed source of information for communities, organizations and leaders in our region. Finally, we believe the true distinguishing feature of this work is found in the way the database framework is organized as critical categories of quality of life that interact and relate to one another in meaningful and exciting ways. Furthermore, this framework allows the people and communities we serve to imagine and examine how these various parts of community relate as well as put the CRC in a unique position to guide, understand and interpret data in hopes of establishing innovative and sustainable solutions to the challenges facing rural communities.

NOTES

1. Quality of life research goes back even farther having connections to Enlightenment thinkers and to the foundations of early Western philosophies in search of the good life (see Sirgy et al., 2006; Michalos and Robinson, 2012).

2. In this chapter, I focus on the community and county level indicator projects in the United States, but recognize the wealth, diversity and innovation of community and regional indicator projects across the globe – e.g., see Noll (2018) for rich history and current use of social indicators for monitoring and reporting across Europe.
3. We do include communities that are outside these parameters for some of our data collection efforts (e.g., Superior, Wisconsin and Hermantown, Minnesota). We also input every US county and US place identified by the US Census in our database.
4. We do have a third approach intended for academic questions. We see the database being utilized by CRC researchers, Northland College students and by other academics interested in questions that our data may be able to help answer.
5. See visualization at: https://www.northland.edu/sustainability/crc/north-woods-database/ (retrieved 1 March 2019).

REFERENCES

Arbuckle, J.G. and Kast, C. (2012). Quality of life on the agricultural treadmill: individual and community determinants of farm family well-being. *Journal of Rural Social Science*, 27 (1), 84–113.

Barrington-Leigh, C. and Escande, A. (2018). Measuring progress and well-being: a comparative review of indicators. *Social Indicators Research*, 135 (3), 893–925.

Bauer, R.A. (1966). *Social Indicators*. Cambridge, MA: The MIT Press.

Besser, T.L., Miller, N.J., and Malik, R. (2012). Community amenity measurement for the great fly-over zones. *Social Indicators Research*, 106 (2), 393–405.

Biderman, A.D. (1970). Information, intelligence, enlightened public policy: functions and organization of societal feedback. *Policy Sciences*, 1 (1), 217–30.

Bognar, G. (2005). The concept of quality of life. *Social Theory and Practice*, 41 (4), 561–80.

Brune Mathis, K. (2017). JCCI restructuring after over 40 years: Citizen Engagement Pact being announced Wednesday. *Jacksonville Daily Record*, 17 February. Retrieved 8 November 2018 from https://www.jaxdailyrecord.com.

Catlin, B. (2014). The county health rankings: 'a treasure trove of data'. In Federal Reserve Bank of San Francisco and the Urban Institute (eds), *What Counts: Harnessing Data for America's Communities* (pp. 58–73). San Francisco, CA: Federal Reserve Bank of San Francisco and Urban Institute.

Citizen Engagement PACT of Jacksonville (2017). *History of JCCI*. Retrieved 2 October 2018 from https://jaxpact.org/history/.

County Health Rankings (2018). *Calculating Scores and Ranks*. Retrieved 8 November 2018 from http://www.countyhealthrankings.org/.

Cummins, R.A. (1996). The domains of life satisfaction: an attempt to order chaos. *Social Indicators Research*, 38 (3), 303–28.

Cummins, R.A. (1997). *Comprehensive Quality of Life Scale: Intellectual/Cognitive Disability* (5th edition (ComQol-I5)). Melbourne, Australia: School of Psychology, Deakin University.

Cummins, R.A. (2018). Subjective wellbeing as a social indicator. *Social Indicators Research*, 135 (3), 879–91.

Diener, E. (2005). Guidelines for national indicators of subjective well-being and ill-being. *Journal of Happiness Studies*, 7 (2), 397–404.

Fernando, F.N. and Cooley, D.R. (2016). An oil boom's effect on quality of life (QoL): lessons from western North Dakota. *Applied Research in Quality of Life*, 11 (4), 1083–115.

Flora, C.B. and Flora, J.L. (2008). *Rural Communities: Legacy and Change* (3rd edition). Boulder, CO: Westview Press.

Gude, P. (2015). Socioeconomics and climate change in the Great Lakes region. Retrieved 8 November 2018 from https://headwaterseconomics.org/dataviz/great-lakes-atlas/.

Headwaters Economics (2018). *About Us*. Retrieved 8 November 2018 from https://headwaterseconomics.org/.

Hendey, L., Cowan, J., Kingsley, G.T., and Pettit, K.L.S. (2016). NNIP's guide to starting a local data intermediary. Retrieved 8 November 2018 from https://www.urban.org/sites/default/files/publication/80901/2000798-NNIP%27s-Guide-to-Starting-a-Local-Data-Intermediary.pdf.

Hofstedt, B. and Tochterman, B. (2015). *Assets & Amenities Comparison Report: Ashland, Wisconsin*. Ashland, WI: Center for Rural Communities at Northland College.

Kingsley, G.T. and Pettit, K.L.S. (2014). Data and community: foundation for an agenda. In Federal Reserve Bank of San Francisco and the Urban Institute (eds), *What Counts: Harnessing Data for America's Communities* (pp. 11–38). San Francisco, CA: Federal Reserve Bank of San Francisco and Urban Institute.

Kolodinsky, J., Roche, E., DeSisto, T., Sawyer, W., and Propen, D. (2013). Understanding quality of life in a northern, rural climate. *Community Development*, 44 (2), 161–72.

Land, K.C. and Michalos, A.C. (2018). Fifty years after the social indicators movement: has the promise been fulfilled? *Social Indicators Research*, *135* (3), 835–68.

Land, K.C., Michalos, A.C., and Sirgy, M.J. (2012). Prologue: the development and evolution of research on social indicators and quality of life (QOL). In Land et al. (eds), *Handbook of Social Indicators and Quality of Life Research* (pp. 1–22). New York: Springer Science.

Leitmann, J. (1999). Can city QOL indicators be objective and relevant? Towards a participatory tool for sustaining urban development. *Local Environment*, *4* (2), 169–80.

Michalos. A.C. and Robinson, S.R. (2012). The good life: eighth century to third century BCE. In Land et al. (eds), *Handbook of Social Indicators and Quality of Life Research* (pp. 23–61). New York: Springer Science.

Minnesota Compass (2018). *About the Project*. Retrieved 10 November 2018 from https://www.mncompass.org/.

NNIP (National Neighborhood Indicators Partnership) (2018). *NNIP Concept*. Retrieved 8 November 2018 from https://www.neighborhoodindicators.org/.

Noll, H.H. (2018). Social monitoring and reporting: a success story in applied research on social indicators and quality of life. *Social Indicators Research*, *135* (3), 951–64.

Noll, H.H. and Zapf, W. (1994). Social indicators research: societal monitoring and social reporting. In I. Borg and P.P. Mohler (eds), *Trends and Perspectives in Empirical Social Research* (pp. 168–206). New York: Walter de Gruyter.

Phillips, R. (2003). *Community Indicators* (Report No. 517). Chicago, IL: American Planning Association.

Singh-Peterson, L. and Underhill, S.J.R. (2017). A multi-scalar, mixed methods framework for assessing rural communities' capacity for resilience, adaption, and transformation. *Community Development*, *48* (1), 124–40.

Sirgy, M.J. (2011). Theoretical perspectives guiding QOL indicator projects. *Social Indicators Research*, *103* (1), 1–22.

Sirgy, M.J., Phillips, R., and Rahtz, D. (2013). *Community Quality-of-Life Indicators: Best Cases VI*. New York: Springer Science.

Sirgy, M.J., Michalos, A.C., Ferriss, A.L., Easterlin, R.A., Patrick, D., and Pavot, W. (2006). The quality-of-life (QOL) research movement: past, present, and future. *Social Indicators Research*, *76* (3), 343–466.

Stiglitz, J.E., Sen, A., and Fitoussi, J.P. (eds) (2009). *Report by the Commission on the Measurement of Economic Performance and Social Progress*. Paris: Commission.

Veenhoven, R. (2018). Co-development of happiness research: addition to 'fifty years after the social indicator movement'. *Social Indicators Research*, *135* (3), 1001–7.

Warner, J.B. (2014). The future of community indicator systems. In Federal Reserve Bank of San Francisco and the Urban Institute (eds), *What Counts: Harnessing Data for America's Communities* (pp. 43–58). San Francisco, CA: Federal Reserve Bank of San Francisco and Urban Institute.

17. An exploratory study of food deserts in Utica, Mississippi
Talya D. Thomas

INTRODUCTION

Cities and towns across the United States are experiencing an upsurge in the proliferation of a relatively new phenomenon referred to as a *food desert*. The US Department of Agriculture (USDA) defines a food desert as "any census district where at least 20 percent of the inhabitants are below the poverty line, and 33 percent live over a mile from the nearest supermarket" (USDA, 2015, n.d., p. 2). These are places where affordable and nutritious food is difficult to obtain, particularly for those with limited or no means of transportation. The increasing phenomenon of food deserts has not limited itself to large urban areas. Indeed, they appear more prevalent in small rural towns and cities, which have little to no access to resources that provide nutritious, healthy food choices.

In the deep Southern state of Mississippi, the small town of Utica historically and currently is predominantly rural and is one such locale, which has been identified as having all the common characteristics of a food desert. Characteristics of food deserts include "smaller populations, higher rates of abandoned or vacant homes, and residents who have lower levels of education, lower incomes, and higher unemployment" (Dutko et al., 2012, p. 1). For less dense urban areas, census tracts with higher concentrations of minority populations are more likely to be food deserts, while tracts with a substantial increase in minority populations between 1990 and 2000 may be more likely to be identified as food deserts (USDA, 2015, n.d.).

The purpose of my study is to investigate this community's status as a food desert, and explore potential approaches for improving and increasing its access to venues offering a wide variety of healthy and nutritious food products at lower prices, including the creation and opening of farmers' markets. It is the goal of this study to determine if there are other factors not previously indicated in typical reviews that can be identified and addressed. More specifically, an examination of the history of Utica is explored to provide possible indicators of subsequent trends and patterns of development.

The town of Utica, because it reflects so many of these identified factors, has been chosen to be profiled, in hopes that factors which have contributed to its current state of deprivation can be identified. The only grocery store (Sunflower) located in Utica closed its doors on 29 November 2014. It is also the goal of this study to determine if there are other factors, not previously indicated in typical reviews, that can be identified and addressed. I begin my study by examining the history and context of Utica, Mississippi, which is often an indicator of forthcoming trends and/or patterns of development.

BACKGROUND AND HISTORY OF UTICA, MISSISSIPPI

The history of Utica can also be said to be the history of Hinds County (home to the state's capital city), as the two are irrevocably related, sharing their roots of conception and settlement that can be traced back to the early 1700s. Towns and cities located in Hinds County include Bolton, Byram, Clinton, Edwards, Jackson, Learned, Raymond, Terry and Utica; it is also a part of the Jackson Metropolitan Statistical Area (MSA). Hinds County itself was originally part of the Choctaw Nation in 1809, and the town of Utica a small borough. Although documentation of the exact dates of settlement by others in the area is not available, the first records indicate that the town's development began when a few European families began settling in the Cane Ridge area of Hinds County just south of the Natchez Trace in the late 1700s. The original settlement was named Cane Ridge due to a large number of cane breaks (bamboo) in the area. Cane breaks landscape features were dominant in the Southeastern part of the United States during European settlement; to date they are endangered. The name was changed to Utica around 1835, after Utica, New York, home to an early settler who resided there, Ozias Osborn. During the Civil War, in May 1863, Union General U.S. Grant sent over 12,000 of his troops to Jackson with the intent to double back and isolate Vicksburg. These soldiers stopped in Utica for water, rest and to view the countryside, before moving on to the Weeks and Roach plantations on the Utica-Raymond road to camp, then on to the Battle of Raymond (Torp, 2017). Having little protection from the invading troops (as many of the town's men were away fighting on larger battlegrounds, such as in Vicksburg), the town of Utica was ravaged and many homes and churches in the area were burned.

The town did recover, and began a slow return to prosperity as more families began to settle in the area and rich natural resources were capitalized upon. The first mayor of the city was Harmon Jackson Sarrett, whose wife founded the first public school in Utica in 1852. In 1876, after the war and following reconstruction, rebuilding efforts included the construction by prominent timber landowners of a railroad extending from Jackson to Natchez. To ensure that the railway route would not bypass the town and hamper business recovery, Utica decided to incorporate as a town. A historian provides a view of the town at this time:

> In 1881, the "*Little J*" railroad was constructed east of the Civil War-era village (but west of the churches and cemetery), and business people moved their businesses and later their homes to the railroad track on what is now Depot Street and Main Street. After which, Utica became a railroad boom town, with numerous industries, almost 100 businesses, an opera house, and other signs of *progression*, and by the 1900 census had about 1,000 residents. Carpenter and Learned Mississippi owe their existence to the railroad, for Lebanon and Auburn residents moved to the railroad to what became Learned, Mississippi in 1882 and countryside citizens in north Copiah County moved to the railroad at what became Carpenter, Mississippi. (Landin, 1979, p. 3)

It is clear to see a pattern here – towns developed around the railroad access, the prominent and most effective means of transportation during that era. Mrs Dunbar Rowland, author of *History of Hinds County, Mississippi, 1821–1922*, further describes the town of Utica with the following description:

> [located] in the southwestern part of the county . . . an incorporated post-town. It is situated on the Y. & M.V. Railroad, 32 miles southwest of Jackson. It is hilly, well-drained, and surrounded

by a rich farming section. All kind of fruit and vegetables, especially watermelons, grow in abundance in the soil. The town is accessible to a large amount of hardwood timber. It ships annually about 10,000 bales of cotton. It has two banks with a combined capital of $90,000; two hotels; a public school; an industrial college for the education of Negroes; three churches, Methodist, Baptist, and Christian; and Democratic weekly newspaper, the Herald, established in 1897. Among its manufacturing enterprises are a brick plant, three steam cotton gins, and a sawmill. Many organizations that embrace intellectual as well as material progress are found in this thriving little city. (Rowland, 1922/2018, p. 54)

Though depicted in this 1922 chronicle as a thriving boom town (i.e., economically prosperous and growing), the town of Utica would in later decades fall victim to urbanization, population migration and economic decline. All these factors have contributed to its present circumstances and designation as a food desert.

Every city has its strengths and weaknesses, and cycles of growth and development. Utica is no different; once known as a boom town, it is now known as a food desert. One strength that the city of Utica can use to promote growth is the fact that the vast majority of the population is 19 years or younger. This should be seen as a resource or an asset, as it is known as an approach in community development such as asset-based community development where an emphasis on resources, rather than problems or deficits, is the focus (Haines, 2015). When you focus on success and small triumphs instead of the negativity of a place, then a positive community outlook and vision for the future can be clearly fostered (Haines, 2015). This also cultivates sustainability within the community. For Utica to become a boom town again, they will have to retain the energy and creativity from this younger generation. This age group needs to know the importance of keeping legacies alive and preserving their rich history, where they are encouraged to live, work and play. Another strength is the small size of Utica; this has enabled a strong sense of community, where everybody knows each other's name. Residents who have a shared sense of community, whether that is community interest or a geographic location typically will take responsibility for that community. In this description of community it implies to sharing; these shared experiences lead to bonding and or an attachment within the place where these experiences occur (Plunkett et al., 2018). Most residents want their community to do well, and in return, the community will give back to its residents.

Another strength of this town is the plethora of undeveloped land. These plots of land can be developed into parks, recreational fishing, and hunting facilities, for example. There is also strong potential for grocery stores, department stores and restaurants. In addition to strengths in Utica, there are weaknesses as well. One is the lack of diversity in the city, due to the city being predominately African American and Caucasian. This implies a deficit in the makeup of human capital for the town of Utica. The 20+ plus age groups in the community are the primary workforce and community leaders. There is a lack of businesses, there are no more grocery stores, limited eateries, department stores, and recreational venues. For one to live, work and play there must be a community where residents have these opportunities. Lastly, education is a fundamental weakness for the city of Utica. Research has proven that simply placing these things in a community will not sustain a community. To revive a community will take strong social community capacity and human capital, and a shared sense of community seems to play in building and developing this need for capacity and capital (Plunkett et al., 2018). For example, residents need to be educated on how to become business owners. Also, residents must learn how to improve social and human capital. These are critical aspects for community development to be effective – social capital is defined by the Organisation for Economic Co-operation and Development (OECD) as "networks together with shared norms, values and understandings that facilitate co-operation within or among groups" (Human Capital, 2009, p. 103). A better definition would emphasize that social capital refers to the resources that are made readily available through these networks, such as "information, ideas, leads, business opportunities, financial capital, power and influence, emotional support, even goodwill, trust, and cooperation" (Baker, 2010, p. 1). These are extremely important aspects of an individual's ability to have a chance at any form of success – through an expansive and diversified network. With such a high concentration of people with varying life experiences, knowledge, perspectives and values; it is sensible that people who live in cities tend to have completed a higher level of education and make more money than those who live in more rural areas. Research has repeatedly shown that when in a homogeneous environment, people are more likely to go along with the status quo and not rethink their point of view on various topics.

A BRIEF OVERVIEW OF FOOD DESERTS

The phenomena of food deserts has been increasingly drastic across the United States. This is becoming such a growing problem that new legislation has been drafted to alleviate the effects and the occurrences of food deserts.

What is a food desert and how does it affect the community? Food desert is a term that originated in Scotland in the 1990s to describe areas with poor access to affordable and healthy diets. Unfortunately, the majority of food deserts occur within low-income rural and urban neighborhoods which in turn have caused health care disparities. A food desert is not a complete absence of food; rather it is an imbalance in food choice, meaning a heavy concentration of nearby fringe foods, which are fast food restaurants and convenience stores, that are high in salt, fat and sugar (Hawkins, 2009).

However, there is a true lack of consensus in its meaning, which has sparked debates and questions about its actual existence. Often, this is because the definition has been modified by various researchers to reflect measures and variables of food deserts that align with their research focus. Therefore, food deserts have been identified with a focus ranging from accessibility and availability of food type, quantity and quality, as well as quantity type and size of food outlets for populations living in urban or rural areas. For instance, researchers Cummins and Macintyre (2002) define food deserts as "poor urban areas, where residents cannot buy affordable, healthy food" (p. 436). Hendrickson et al. (2006) defined food deserts as "urban areas with 10 or fewer stores and no stores that have 20 or more employees" (p. 372). The USDA has defined food deserts as areas of the country without fresh fruits, vegetables and other healthy whole foods, usually found in impoverished areas due to a lack of grocery stores, farmers' markets and healthy food providers (American Nutrition Association, 2015). Fringe locations are considered nearby locations that offer convenient low-quality foods, for example, fast food chains, convenience stores and dollar stores. These types of food services have been linked to hypertension, obesity and diabetes. Health disparities such as these are constantly plaguing the low-income families on a daily basis. Researchers have debated and theorized about the cause of a disease like obesity and other health disparities attributing them to poverty, genetics or psychological mindsets about food (NCEH, 2011). However, other researchers still argue that these adverse health outcomes and conditions are created by food choice and are a result of a more complex process than just the individual's choice to eat unhealthy foods. A consensus of the various literature on food choice and the built environment states that lack of availability and accessibility to healthier food options such as grocery stores plays a key role in people making unhealthy food choices (Horowitz et al., 2004). A number of recent studies have suggested that a person's food choice is primarily shaped by the dynamic interactions with their built environment, which includes the food environment as well.

In 2009, a study was conducted in four communities in Alberta, Canada where they took pictures of barriers and opportunities for healthy eating and shared stories during interviews. It was concluded that while availability and access to food outlets influence healthy eating practices, these particular factors may be eclipsed by other non-physical environmental considerations, such as food regulations and sociocultural preferences (Belon et al., 2016). The researchers also discovered that there were relationships between environmental attributes, people's perceptions and eating behaviors; by recognizing these relationships then effective community-based interventions can be developed to address how social, economic and political environments can and will interact with the world we live in today. According to Walker et al. (2010) they were able to investigate a systematic review of studies that looked at food access and food deserts in the United States. These researchers also focused on the role that local food environments play in residents' ability to purchase affordable, healthy and nutritious foods. Their findings gave insight into future research, policy development and program implementation in regards to food deserts.

In June of 2009, the USDA released a study that found approximately 23.5 million people live in low-income areas more than 1 mile from a supermarket or grocery store (Hawkins, 2009). USDA has also stated that more than 23 million Americans live in food deserts – areas where residents have no access to fresh, healthy, affordable food. Additionally, they report that urban core areas with limited access to food are characterized by higher levels of racial segregation and greater income inequality. In circumstances when low budget grocery chains are available, prices are often too high for quality food choices that are offered. Because of high prices and low-income budgets, food disparities remain, and the area is still considered a food desert.

USDA has been able to use census tracts as a way to see if a city or town qualifies as a food desert, by meeting the low-income and low-access thresholds, which are: (1) qualifying as a "low-income communities," is based on having a poverty rate of 20 percent or greater, or a median family income at or below 80 percent of the area median family income; and (2) qualifying as a "low-access communities," is based on the determination that at least 500 persons and at least 33 percent of the census tract's population live more than 1 mile from a supermarket or large grocery store (10 miles, in the case of non-metropolitan census tracts) (USDA, 2015, n.d.). According to USDA using census tracts to map the town, Utica meets both requirements to be considered a food desert.

EXPLORING APPROACHES FOR MITIGATING FOOD DESERTS

There exist several ideas and programs for alleviating difficulties caused by food deserts. These range in nature and type, from programs to grants, to community-based efforts. This section presents more information on these valuable approaches.

School and Community Garden Programs

The USDA Food and Nutrition Service (FNS) has announced the availability of $1 million for a People's Garden School Pilot Program. One grantee will be selected to enter into a cooperative agreement for developing and running community gardens at eligible high-poverty schools. This will teach students how to get involved in the gardens and about agriculture production practices, diet and nutrition. This program will also contribute to produce being provided as supplemental food to eligible schools, students' households, local food banks and senior center nutrition programs. Evaluations will be conducted on the funded project to learn more about the impact of school gardens.

This particular grant program is sponsored by the United States Environmental Protection Agency's Environmental Education Division (EED), Office of Children's Health Protection and Environmental Education, that supports environmental education projects to enhance the public's awareness, knowledge and skills to help people make informed decisions that truly affect environmental quality. The EPA Environmental Education Grants Program awards grants each year based on funding appropriated by Congress, annual funding for the program ranges between $2 million and $3 million (United States Congress, United States Senate, 2007).

Another program that could be useful to Utica is called the Green Thumb Challenge. It aims to connect children with nature and the healthy benefits of gardening as part of a nationwide movement to get kids interested in growing food. Whether they are sowing seeds during a class period, planting bulbs in another or planning an outdoor garden that comes back year after year, the teacher and students can be a part of the movement. There are gardening resources provided for the participants with helpful materials and strategies for gardeners of all experience levels. Every garden, no matter the size, will add beauty and life to the community. The Green Thumb Challenge has awarded over $10,000 worth of prizes to participants.

Food Enterprises

The Local Food Promotion Program (LFPP) offers grant funds with a 25 percent match to support the development and expansion of local and regional food business enterprises to increase domestic consumption of, and access to, locally and regionally produced agricultural products, and to develop new market opportunities for farm and ranch operations serving local markets. Two types of project applications are accepted under LFPP planning grants and implementation grants (Local Food Promotion Program, 2017, n.d.). Additional information about this program is provided as follows:

- LFPP planning grants are used in the planning stages of establishing or expanding a local and regional food business enterprise. Activities can include but are not

limited to market research, feasibility studies and business planning (Local Food Promotion Program, 2017, n.d.).
- LFPP implementation grants are used to establish a new local and regional food business enterprise or to improve or expand an existing local or regional food business enterprise. Activities can include but are not limited to training and technical assistance for the business enterprise and/or for producers working with the business enterprise; outreach and marketing to buyers and consumers; and non-construction infrastructure improvements to business enterprise facilities or information technology systems (Local Food Promotion Program, 2017, n.d.).

Eligible entities may apply if they support local and regional food business enterprises that process, distribute, aggregate or store locally or regionally produced food products. Such entities may include agricultural businesses and cooperatives; producer networks and associations; community supported agricultural networks and associations; other agricultural business entities (for-profit groups); nonprofit and public benefit corporations; economic development corporations; regional farmers' market authorities; and local and tribal governments (Local Food Promotion Program, 2017, n.d.).

Transportation

Government data shows that transportation to another grocery store is a problem for a lot of people in Utica (16 WAPT News, 2014). In some cases, the nearest store is up to 20 miles away from where people live. When I interviewed Mayor Kenneth Broome, he explained that, "It's mainly older people – the ones that are shut in and they don't have a vehicle or transportation." Most Utica residents will have to drive at least 15 miles to the Sunflower in Raymond, which is the next available grocery store. Some residents have stated that many of them do not have transportation to go that far for food and that is truly putting everyone in a bind. Other residents stated that they might have to travel as far as Clinton, Vicksburg or Jackson to shop for their groceries. Nearest statewide cities are as follows: Learned, MS (2.8 miles); Edwards, MS (4.0 miles); Raymond, MS (4.0 miles); Crystal Springs, MS (4.2 miles); Terry, MS (4.4 miles); Bolton, MS (4.4 miles); Hazlehurst, MS (4.6 miles); Byram, MS (4.6 miles) and the closest major city to Utica is Jackson, MS which is 28.7 miles away. Figure 17.1 shows nearest statewide cities.

Hinds County Human Resource Agency (HCHRA) provides rural transportation services for anyone living in Bolton, Byram, Clinton, Edwards, Raymond, Terry and Utica, Mississippi. The agency focuses assisting the elderly and disabled; however, our transportation services are available for anyone in rural Hinds County. The costs of the trips for seniors range from $1.00 to $2.50 per trip, non-seniors pay $2.50 one-way, $5.00 round trip. All trips must originate in rural Hinds County. This transportation service is available Monday through Friday from 6:00 am until 6:00 pm. Another transportation service that is even more affordable, convenient and accessible is available for seniors and disabled individuals in Hinds County through the Title XX Transportation Program. This particular program helps elderly residents maintain their independence by providing them with transportation to obtain goods and services, including medical and dental treatment, social services and other applicable services. Senior citizens and disabled are eligible for this service that is available Monday through Friday from 9:00 am until 1:00 pm. The cost

An exploratory study of food deserts in Utica, Mississippi 297

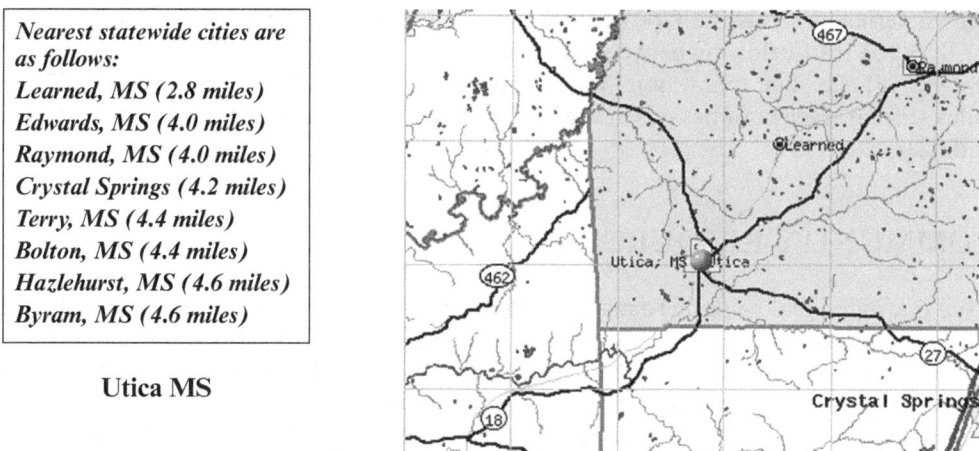

Nearest statewide cities are as follows:
Learned, MS (2.8 miles)
Edwards, MS (4.0 miles)
Raymond, MS (4.0 miles)
Crystal Springs (4.2 miles)
Terry, MS (4.4 miles)
Bolton, MS (4.4 miles)
Hazlehurst, MS (4.6 miles)
Byram, MS (4.6 miles)

Utica MS

Figure 17.1   Nearest statewide cities to Utica, MS

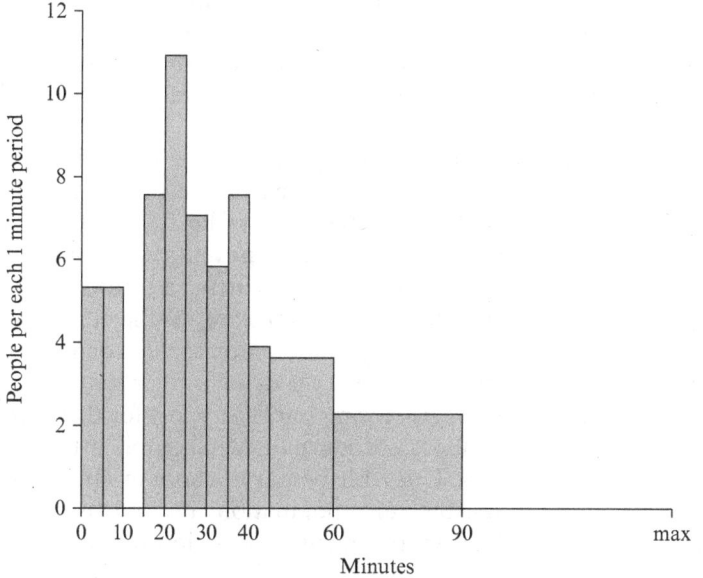

Figure 17.2   Transit time from Utica

of the program is $1.00 one-way, $2.00 round trip for seniors and $2.50 one-way, $5.00 round trip for non-seniors. Figure 17.2 shows transit time from Utica.

Mississippi Highway 27 (MS 27) is a state highway in Mississippi, which runs from south to north for 119.53 miles serving five counties, which are: Walthall, Lawrence, Copiah, Hinds and Warren. The segment between Vicksburg and Crystal Springs is known vernacularly as the Utica cutoff because it facilitates circumvention of Jackson for I-20/I-55 traffic flowing between Vicksburg and Crystal Springs. Mississippi Highway 18 (MS 18) is a state highway in Mississippi, which runs from east to west for 177.654 miles serving seven counties which are: Claiborne, Copiah, Hinds, Rankin, Smith, Jasper and Clarke.

COMMUNITY RESOURCES

Even one individual working to make the public aware of and to build support for health centers can have a major impact on the city. Many people working together can achieve exponentially greater gains for the particular community. The health center movement began with just two health centers in two communities over 45 years ago and has grown to encompass over 1,200 health centers providing quality, affordable health care to 20 million people in over 7,000 communities in every state and territory. Support for health centers has continued to grow and has grown through the economic and political shifts because there have always been people at the forefront to ensure that policy makers always knew what health centers meant back home (Community Health Center Association of Mississippi, n.d.). In the 1980s, the Federal Health Centers program was almost done away with; if it were not for a true grassroots uprising that hit Congress, we would not be talking about health centers today. That grassroots power was due to the fact that health centers are not just in, but they are part of the community too. Health centers are the grassroots and that is the power that sustains them (Community Health Center Association of Mississippi, 2019).

Consider how the benefits of past advocacy have helped the Utica health center and their community; it can help to inspire others when thinking about how much more still needs to be done. We can prepare to shape the future of health care in America by getting involved. Some examples of involvement are: creating events to celebrate National Health Center Week, writing letters to Congress, and lastly starting a petition with signatures that support significant legislation that deals with health care. As the old saying goes, "it's the squeaky wheel that gets the grease"; policy and opinion makers depend on advocates to help them understand and recognize issues that deserve their major support. Without grassroots advocacy, health centers risk losing public support for their needs in the community (Community Health Center Association of Mississippi, 2019). This community health services center is located at Utica Elementary School in Hinds County, which is community-based and is serving the rural population. Federal funds were granted to grantee Central Mississippi Civic Improvement Association to maintain the operation, but the center is open for limited hours with a part-time schedule and has been open since 2003. Central Mississippi Civic Improvement Association is operated through the federal funds that are received.

Other resources that are located within the city of Utica are Head Start Center, Non-emergency ambulance, fire and police services, Evelyn Taylor Majure Library, United States Post Office, and churches: two Baptist, one Methodist, one Church of Christ, one Pentecostal and one non-denominational. All of these resources serve as an outlet to help shape the community to an area that can be classified as learning, living and worshipping.

There are approximately 76 nonprofit organizations registered within the city of Utica, with reported assets of $2,473,363. There is reported income of around $710,919. Then the question comes to mind – why is there a lack of stores that sell nutritious food or farmers' markets that sell fresh fruits and vegetables when this community has some resources that can be of assistance?

CONCLUSIONS

From my study, it seemed that the most obvious factor determining Utica's food desert status was the closing of the Sunflower Food Market, formerly the area's primary source for fresh foods. Further exploration revealed that other factors existed, including Utica's geographic location, the socioeconomic status of the majority of its residents, educational attainment, land use patterns and infrastructure limitations. All these have contributed to the lack of resources needed to ensure healthy and affordable food choices. It takes more than the installation of a supermarket to eradicate the problem.

Recommendations include improved and increased access to venues that offer a wide variety of healthy, nutritious food products at lower prices, as well as the creation/opening of farmers' markets. The benefit of this enterprise would include:

- The provision of healthy, fresh fruits and vegetables to local residents, who would most likely not have to travel as far to reach them.
- Local vendors would benefit economically from the selling of their products. This would also encourage entrepreneurship for individuals who may have fewer skills and less educational attainment in more traditional enterprises.

Utica may also consider the creation of a community garden, which would strengthen the bonds of members in the community and may lead to further community development, such as the sharing of ideas for creative, low-cost, affordable yet still desirable menu options.

Improvements to the infrastructure would perhaps attract more settlement and diverse enterprise in the area.

This study explored why a food desert exists in a small Southern town. It uses a case study approach, relying on exploration of both the historical context and current conditions. From this, it is seen that food deserts have a variety of factors influencing their existence; at the same time, communities can help remedy the negative impacts of food deserts by a variety of approaches.

REFERENCES

American Fact Finder. (n.d.). Retrieved 15 July 2017 from http://factfinder.census.gov/faces/nav/jsf/pages/community_facts.xhtml.
American Nutrition Association. (2015). USDA defines food deserts. *Nutrition Digest*, *38*(1), 1–5.
Baker, W.E. (2010). *Achieving Success through Social Capital: Tapping the Hidden Resources in Your Personal and Business Networks*. San Francisco, CA: Jossey-Bass.
Belon, A.P., Nieuwendyk, L.M., Vallianatos, H., and Nykiforuk, C.I. (2016). Perceived community environmental

influences on eating behaviors: A Photovoice analysis. *Social Science & Medicine (1982)*, *171*, 18–29. DOI:10.1016/j.socscimed.2016.11.004.

Bureau of Labor and Statistics. (n.d.). Retrieved 25 April 2017 from Quarterly Census Employment Wages, http://www.bls.gov/cew.

Community Health Center Association of Mississippi. (n.d.). Retrieved 2 December 2019 from http://www.chcams.org/health-center-advocacy.

Cummins, S., and Macintyre, S. (2002). "Food deserts" evidence and assumption in health policy making. BMJ (Clinical Research Ed.), *325*(7361), 436–8.

Dutko, P., Ver Ploeg, Michele, and Farrigan, Tracey. (2012). Characteristics and influential factors of food deserts. ERR-140, US Department of Agriculture, Economic Research Service.

Haines, A. (2015). Asset-based community development. In R. Phillips and R. Pittman (eds), *Introduction to Community Development*, 2nd edition. London: Routledge, pp. 38–48.

Hardy, Stella P. (2015). *Colonial Families of the Southern States of America: A History and Genealogy of Colonial Families Who Settled in the Colonies Prior to the Revolution*. Lebanon, NJ: Franklin Classics.

Hawkins, S. (2009). Desert in the city: The effects of food deserts on healthcare disparities of low-income individuals. *Annals of Health Law Advance Directive*, *19*, 116–28.

Hendrickson, D., Smith, C., and Eikenberry, N. (2006). Fruit and vegetable access in four income food deserts communities in Minnesota. *Agriculture and Human Values*, *23*(3), 371–83.

Horowitz, C.R., Colson, K.A., Hebert, P.L., and Lancaster, K. (2004). Barriers to buying healthy foods for people with diabetes: Evidence of environmental disparities. *American Journal of Public Health*, *94*(9), 1549–54.

Human Capital (Summary in English). (2009). *Human Capital OECD Insights*. DOI:10.1787/9789264029095-sum-en.

Landin, M.C. (1979). *The History of Utica Baptist Church, Utica Mississippi: Written in Conjunction with the Celebration by the Utica Baptist Church, Utica, Mississippi's 150th Year since the Founding of the Church*. Austin, TX: Capitol Printing Company.

Local Food Promotion Program, 2017 (n.d.). Retrieved 15 July 2017 from http://www.ams.usda.gov/services/grants/lfpp.

Mindel, C.H., Habenstein, R.W., and Wright, R. (1998). *Ethnic Families in America: Patterns and Variations*. Upper Saddle River, NJ: Prentice Hall.

NCEH (National Center for Environmental Health). (2011). *Impact of the Built Environment on Health*. Retrieved 20 September 2017 from https://www.cdc.gov/nceh/publications/factsheets/impactofthebuiltenvironmentonhealth.pdf.

Plunkett, Dan, Phillips, Rhonda, and Ucar Kocaoglu, Belgin. (2018). Place attachment and community development. *Journal of Community Practice*, *26*(4), 471–82. DOI:10.1080/10705422.2018.1521352.

Rowland, Dunbar. (1922/2018). *History of Hinds County, Mississippi, 1821–1922: Published in Commemoration of the Centenary of the City of Jackson, the Capital of the State; 1821–22 1922*. Colchester, CT: Forgotten Books.

Teague, Michael, Mackenzie, Sara, and Rosenthal, David. (2019). *Your Health Today: Choices in a Changing Society*, 7th edition. New York: McGraw-Hill Education.

Torp, Kim. (2017). Mississippi genealogy trails. Retrieved 15 July 2017 from http://genealogytrails.com/miss/hinds/index.html.

United States Congress, United States Senate. (2007). *Federal Funding for the No Child Left Behind Act*. Washington, DC: Committee on Appropriations.

USDA, 2015. (n.d.). *Economic Research Service*. Retrieved 20 July 2015 from https://www.ers.usda.gov/data-products/food-access-research-atlas/documentation/.

Walker, R.E., Keane, C.R., and Burke, J.G. (2010). Disparities and access to healthy food in the United States: A review of food deserts literature. *Health Place*, *16*(5), 876–84. DOI:10.1016/j.healthplace.2010.04.013.

16 WAPT News. (2014). Utica's only grocery store shutting down. Retrieved 21 July 2017 from https://www.wapt.com/article/utica-s-only-grocery-store-shutting-down/2090791.

18. Impact of socioeconomic characteristics on neighborhood environment satisfaction in deteriorated areas
Mostafa Norouzi, Abolfazl Meshkini and Somayeh Khademi

One of the main problems being debated is the degradation of living conditions and its long-term consequences on the quality of residential environment in urban deteriorated and poor areas[1] (Amao, 2012; Salmani et al., 2012; Li and Wu, 2013; Asadi et al., 2014; Ray, 2017; Khaef and Zebardast, 2016). However, the concept of a place to the people who live in the deteriorated areas is not just dependent on physical characteristics that deteriorate over time, it is on the ratio of the relationship between residents and the place's characteristics. This mutual relationship is formed and becomes meaningful based on individual assessments. These values and assessments result in forming the feelings and attitudes of residents toward the built environment and creating satisfaction or dissatisfaction with it. In other words, people have the capacity for upgrading or deterioration of the environment quality through their behavior and attitudes (Lee, 2008; Haginegad et al., 2010; Islam et al., 2014; Bonaiuto et al., 2015). Thus, residents' mental image of the built environment in which they live should be considered. How does a resident understand the quality of the built environment and how do they accept it? In other words, residents' behaviors and objectives play a fundamental role in the definition of environmental quality and space, as far as the space can be considered as a personal experience (Rentfrow et al., 2008; Islam et al., 2014; Jokela et al., 2015). This mental image is shaped by filtering information through the experience and knowledge of the observable environment that is very influenced by the individual's beliefs and opinions gained during previous experiences in the environment, which can affect the perception of the environment (Amerigo and Aragonés, 1997; Cazzuffi and López-Moreno, 2018).

A very common approach to the built environment field is to examine it at a neighborhood level in urban areas. As an urban planner Kostas Mouratidis (2018) remarked that a neighborhood is a geographical area that lies between the micro level of a dwelling and the macro level of a city because neighborhoods can be selected for homogeneity in characteristics of interest. Most assessments of neighborhood environment quality consist of risk-taking and security (Baba and Austin, 1989; Okunola and Amole, 2012; Hur and Nasar, 2014; Chen and Chen, 2015), social environment (Forrest and Kearns, 2001; Adriaanse, 2007; Lenzi et al., 2012; Golabi et al., 2013; Hamdan et al., 2014), the sense of place and neighborhood identity (Oktay et al., 2009; Poortinga et al., 2017; Tucker and Abass, 2018), residential satisfaction and quality of housing (Galster and Hesser, 1981; Hashim, 2003; Cho and Lee, 2010; Mohit and Azim, 2012; Ibem and Aduwo, 2013; Tucker and Abass, 2018) and accessibility and public transportation (Abbaszadegan et al., 2011; Kheyroddin et al., 2014). However, so far few studies have been carried out in the literature based on the general pattern of satisfaction with the neighborhood and individual characteristics of residents of the deteriorated areas. The theoretical approach of environmental quality in deteriorated neighborhoods can be classified in two general categories: design-physical approaches and socio-spatial approaches. A distinction between these two forms of dealing with the issue of environmental quality can be recognized with the central role of indicators and physical variables in the first approach and regard for social phenomena and residents' mental and behavioral factors in addition to physical and environmental aspects in the second approach.[2]

According to the Statistical Center of Iran (2016) Mashhad city has 2247 hectares of deteriorated areas that comprises 6.4 percent of the city with 17 percent of population (Mashhad Municipality, 2018). The case study of this research, Ab-Kooh neighborhood in Mashhad city, has the history of a village format in the early eighteenth century; after the expansion of the city in the present time, it has combined with and merged into the city; consequently, it has become a deteriorated area due to its transformation. In terms of location, the neighborhood is among the upstate and high level neighborhoods, so that the imbalance among the neighborhood and its surrounded areas is clear to see. This fragmentation of socio-spatial structure has caused the loss of social interactions and gradually it lost its identity and regularity due to inaccurate mixing with the culture and the environment around it. The fundamental issue of perception of the neighborhood environment formed in the minds of residents is understanding the meaning of the quality of the neighborhood environment; because it refers to the nature of the settlement experience and may be expressed based on the residents' character, culture and lifestyle. Meaningfulness of the built environment and its impact on the residents' minds through the senses are the issues that are not only confirmed through logical reasoning, but also proved by experimental methods. For this purpose, this neighborhood was chosen as the study area with the aim of studying the built environment quality indicators in a deteriorated neighborhood in the central area of the city of Mashhad as well as studying the impact of socioeconomic characteristics (individual and household) on satisfaction with these components.

THEORETICAL CONSIDERATION

A Psychological–Geographical Approach

Environmental psychology is the most common approach to the relationship between space and human beings or the environment and society that investigates the impact of the environment on people's perceptions of and reactions to the environment (Barker, 1968; Ittelson, 1973; Gibson, 1979; Lang, 1987; Rapoport, 1990; Gifford, 1997; Wicker, 2002). The aim is to examine the issue of how a person looks at, understands, feels and reacts to the environment. In this regard, valuable theories are developed to examine the interaction between the environment and humans in the fields of psychology, behavioral sciences, behavioral geography and so on. The influence of psychology in geographical research has long been appreciated. In the 1960s, behavioral geography and environmental perception were widely considered as one of the most important topics of human geography (Amadeo and Golledge, 2003). In this field (Spencer and Blades, 1986) highlighted that geography has always adapted and developed idiomatic concepts, techniques and term ideas from the adjacent sciences. Regarding the fact that the subjectivism issue is the core of humanism research and the human geography system, this emphasis on subjectivity leads to behavioral geography and environmental psychology being close together. In behavioral geography, the place is the scene of economic, social and historical function as well as the psychological space based on preferences and perceptions (Golledge and Timmermans, 1990; Kitchin, 1996).

Space behavior in the social environment is a reaction of mental awareness and people's perceptions toward their living environment. Location, age, gender, race and cultural factors affect the size and shape of personal space. This space has its own identity, structure and definition and changes due to individual features in social and economic conditions (age, gender, time of residence, socioeconomic status of household, individual awareness and so on) and the living conditions (space perception, culture and consciousness). Thus, the internal and mental processes (subjectivity) ultimately lead to behavioral reflection in the individual and collective realm; therefore, it turns into objectivity.

Hence, in the pattern of behavior, obvious aspects and manifestations of human behavior in space are emphasized for investigation instead of inner emotions and reactions. In fact, mental images are the key to spatial behavior and decision making. They lead to improving and making better the places' planning. Knowing about the effective factors in people's decision making and their selection of geographical areas are necessary in geographical spaces planning.

Thus, behavioral geography considers the behavior of individuals rather than applying its own methods of analysing social objective and statistical data. It studies people's usage of their living spaces; moreover, it deals with the requirements of individual performance on the use of space, the scale of microgeography to individuals and space behaviors, satisfaction and quality of life (Cutter, 1985; Pacione, 1986; Pacione, 2003; Rentfrow et al., 2008).

Perceived Neighborhood Environment and Life Satisfaction

In the psychology literature, the concept of satisfaction is usually applied to all life experiences (Insch and Florek, 2008). Ibem et al. (2017) highlighted that satisfaction is a superior mental structure, which is influenced by a wide range of factors, including the individual experiences of the past, present realities and expectations. Common examples of this approach are in Campbell et al. (1976) in which life satisfaction is considered as satisfaction with various environmental aspects. This satisfaction is the result of the evaluation process, perception, prediction and action along with adjustment and has a hierarchical structure and makes a distinction between objective and subjective characteristics.

Ibem and Aduwo (2013) and Dong and Qin (2017) noted the influential factors of neighborhood satisfaction stem from three factors: individual characteristics, environmental factors and social environment of the neighborhood. In other words, these situations are derived from the form and content of several actions and reactions shaped between individuals and the physical and social environments of the neighborhood. It is logical that each level of data forms the findings of the other levels, and vice versa, considering the binary status between people on the one hand and the social and physical environment of the neighborhood on the other hand. This topic will be highlighted by considering the geographical differences in living environments. Many studies show that people in different geographic regions have different psychological perceptions. The research results of Jokela et al. (2015) regarding the role of individual-neighborhood character interaction in life satisfaction show that the level of life satisfaction and personality behaviors obey features of the neighborhood geographically. Their results show that personality behaviors of people in different parts of the metropolis of London are different in life satisfaction and these differences are associated with the specific characteristics of neighborhoods, such as population density and ethnic and cultural heterogeneity. Moreover, different characteristics of personality adopt life satisfaction from social and physical aspects of the environment.[3]

METHODS

This research is an applied and developmental research and in terms of research implementation is descriptive and analytic. Attention has been paid to the effect of individual and household characteristics due to analysing the residents' perception of environmental quality in the deteriorated and worn-out neighborhoods. Accordingly, this study seeks to answer the question that is which individual characteristics have the greatest influence on neighborhood environment satisfaction? To achieve the theoretical literature study, descriptive and documentation methods are used with a review of the literature on the concepts of perceptual quality of the environment and satisfaction based on theories of environmental psychology and behavioral geography. Furthermore, the analytical methods are applied in combination to measure the satisfaction of case study area as well as effectiveness and inspiration of individual variables (extended in Table 18.1) on residents' satisfaction. Satisfaction is considered as the dependent variable, and the independent variables include the six major indicators of housing quality, neighborhood security, environmental health, access to public facilities, transportation and mobility, and community status (extended in Table 18.2). The unit of analysis in this study is mainly a household. A 5-point Likert scale is used because we can provide more valid results in our research. The Likert scale is generally used to measure attitudes, feelings and opinions which are not observable, but can affect the respondents' behavior (Croasmun and Ostrom, 2011; Hallajow, 2018). Most studies on satisfaction and quality of life have also used this scale to measure the attitudes and opinions of individuals (Adriaanse, 2007; Cho and Lee, 2010; Haginegad et al., 2010; Abbaszadegan et al., 2011; Ibem and Aduwo, 2013; Bonaiuto et al., 2015; Huang and Du, 2015; Dong and Qin, 2017). It is a pleasingly simple way of gauging satisfaction. Each response is assigned a point value, and an individual's score is determined by adding the point values of all of the statements. Cochran's formula is used for calculating sample size because population size is finite and rather large (Hafeznia, 2010). A total of 370 completed questionnaires were prepared and filled in to test the hypotheses proposed according to the selection criteria of those who were selected as samples of statistical populations. The sampling method used in this study is the systematic method which means non-random selection of households. Thus, the study area was divided into several blocks. We therefore selected households within blocks of the study area in order to insure the sample households covered all the study area. Finally, data collected from the questionnaires was investigated by using statistical analysis methods in SPSS software. Satisfaction indicators were implemented according to the type of data through the single-sample T-test to evaluate the level of satisfaction. The meaningful relationship between the variables was significantly investigated in the 0.01 (99 percent) level of meaningful.

Moreover, the chi-square test was applied to determine the meaningfulness of the link between environmental quality indicators and individual variables and Kendall's tau-b tests and Cramer's V were utilized to determine the extent and direction of their correlation.

Study Area

The settlement characteristics in Mashhad city have undergone many changes throughout history, but from about the 1970s, it has become the second largest metropolis in Iran because of its rapid growth.[4] Consequently, it has created many deteriorated areas inside and in its surroundings due to its old age as well as its uncontrolled and unplanned spatial growth. Today, 8 percent of the total urban area in Mashhad is deteriorated, a proportion that is approximately equal to 2302 hectares. Ab-Kooh neighborhood is located in the central area of the city and in a region of Mashhad Municipality. This neighborhood measures 26 hectares (Tavanaei Marvi and Behzadfar, 2015) and the population is over 7299 people (Javdani Irani Nejad et al., 2014). Ab-Kooh's formation history as a village nearby Mashhad city dates back to the early eighteenth century (Saeedi Rezvani, 2006) and it had been located 6 kilometers to the west of Mashhad before the physical development of this city. The village had been on the margin of the city during the 1950s and then became a central part of it in the late 1960s due to the rapid physical development of the Mashhad metropolis (Figure 18.1).

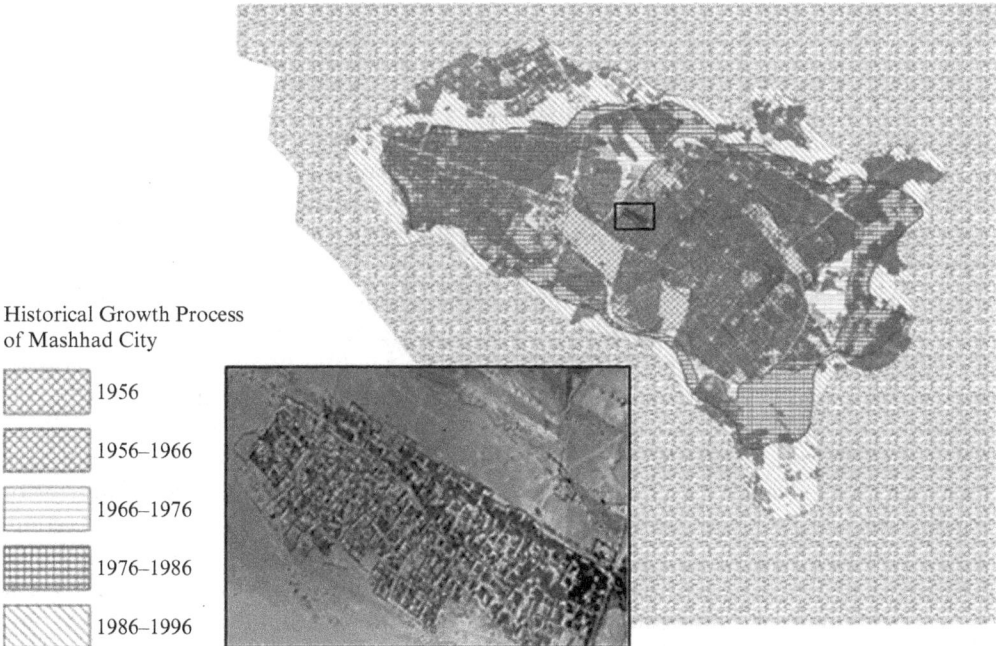

Source: Naghsh Azin Shargh Consulting Engineers (2006).

Figure 18.1 Location of Ab-Kooh neighborhood in process of historical growth of Mashhad since 1956

DESCRIPTIVE RESULTS

Respondents' Socioeconomic Characteristics (Individual Household Characteristics)

In many studies the individual characteristics such as age, gender, household size, income and education play a direct role in environment satisfaction (Baba and Austin, 1989; Diener et al., 1999; Huang and Du, 2015; Hannscott, 2016; Wang and Wang, 2016; Zhang and Lu, 2016). In most of these studies, neighborhood environment satisfaction is checked by two sets of factors: household characteristics (socioeconomic) and qualitative characteristics of the neighborhood (Lee et al., 2017). Thus, the relationship between the person and the neighborhood environment formed from personal and household characteristics consisting of socio-demographic factors (gender, age, education, marital status, family income, employment status) (Wang and Wang, 2016) and the environmental attributes (existent value in the environment or the capacity of its value, accountability and measurability of the special characteristics of the environment) (Van Poll, 1997), in addition to the above two factors, features resulting from the interaction between the individual and the environment, leads to giving value to residents from the neighborhoods and vice versa.

Regarding the socioeconomic status of individuals, Hannscott (2016) noted that individual and household characteristics take precedence over satisfaction. Additionally, while people tend to report high levels of satisfaction, still there is diversity in their responses which leads the researchers to ask the question, what factors may predict satisfaction? He replied that at the individual level, socioeconomic status appears because of its functions to the selection and implementation as a leading predictor of satisfaction.

Descriptive statistics results presented in this section offer the socioeconomic characteristics of the respondents shown in Table 18.1. According to this table, in terms of gender, more respondents are male (72.4 percent). In terms of age group, most of the sample (36.8 percent) belong to those between 30–45 years old, and the older group has the lowest number with 13.8 percent. In relation to marital status, 87 percent of respondents are married and the rest of them are single. In terms of education level, high school graduates with 38.1 percent have the highest rate of the five groups. The lowest is the illiterate, a total of seven people mostly belonging to the age group of 60 years and above. In terms of income levels, people in this neighborhood are living at a lower level, so that the average income of people living in nearby neighborhoods is estimated to be higher by at least twice. Among those, people with an average monthly income of less than 9300 thousand Rials (workers' minimum wage in Iran) account for the highest percentage. In terms of the household size, families of four are the largest number of respondents with 33.5 percent. Distribution of workers in different occupational groups shows that more respondents with 42.7 percent are self-employed. Also non-employee groups (pensioners, soldiers, students, homemakers and the unemployed) with a share of 24.1 percent are placed in the next category. Acquisitions within the scope of the study is as follows, 78.4 percent of the sample size are the owner and 21.6 percent are tenants. Most people over 15 years of residence in the neighborhood form the largest number of the sample size.

Table 18.1 Respondents' personal profiles

	n = 370	Percentage
Age group (years)[a]		
18–30	96	25.9
31–45	136	36.8
46–59	87	23.5
60 and above	51	13.8
Respondent's sex		
Male	268	72.4
Female	102	27.6
Marital status		
Married	322	87.0
Single	48	13.0
Education level[b]		
Illiterate	7	1.9
Primary education (elementary school)	124	33.5
Secondary education (guidance school)	62	16.8
Diploma (high school)	148	38.1
Graduate (university)	29	9.7
Average monthly income (IRR)[c]		
Below IRR 9,300,000 (low income)	109	29.5
IRR 9,300,000 – IRR 15,000,000	152	41.1
IRR 15,000,000 – IRR 30,000,000	33	8.9
IRR 30,000,000 and above	–	–
Household size (persons)		
2	49	13.3
3	113	30.5
4	124	33.5
5 and above	84	22.7
Employment status		
Unemployed (retirees, housewives and so on)	104	24.1
Public sector employee	72	23.5
Private sector employee	36	9.7
Self-employment	158	42.7
Ownership status		
Owner	290	78.4
Tenant	80	21.6
Length of residence		
Less than 5 years	17	4.6
5 years – 15 years	125	33.8
15 years and above	228	61.6

Notes: [a] 18 is the legal age for election and kinds of licenses in Iran. [b] Conventional academic classification in Iran. [c] 1US$ = IRR 4300 (January 2018).

Physical–Spatial Characteristics of Housing

What caused Ab-Kooh to be deteriorated includes: primary rural constructions, the formation of secondary development based on the basic pattern, dating to the formation of neighborhood, and the lack of ownership by the residents (most parts of the Ab-Kooh neighborhood lack property documentation and they are unable to obtain building permission because of Astan Quds[5] and the Endowments Organization,[6] since they are the official owners of land in this area and tend to transfer current residents to other places and apply this pleasant land for economic use) (Farnahad Consulting Engineers, 2009; Moghadam Ariaee et al., 2008). Looking at the shape of full and empty spaces in the Ab-Kooh neighborhood shows that the building mass is formed without regarding to the specific pattern or specified architectural criteria and is organic in form, only depending on the position and shape of the earth. Construction patterns are sometimes a central courtyard form, an L-shape or in some cases a simple rectangle. The number of building floors in the neighbourhood is a function of the passageways. This means that the height of buildings is more and more frequently three and four floors on the periphery of the main streets, nodes and major squares such as Dastghaib Street and Guidance Square. In the margin of narrow alleys, major buildings are formed with one and two floors due to the old and deteriorated texture (about 72 percent of buildings are one floor). Average occupancy levels of parcels is 88/66 percent and the average building density in the neighborhood is 134 percent. In terms of land use, the area's greatest portion is residential with a 57 percent share.

The neighborhood's old texture was disjointed by the construction of a communicational axis and divided into two parts in order to link this area as a new neighborhood in the central city of Mashhad with the implementation of detailed plans at the beginning of the 1980s. Subsequently, the sidelines of the axis were changed, while the residential texture lagged behind the current development with little change and became deteriorated with accelerated erosion and was increasingly in spatial and physical conflict with its surroundings (Figure 18.2).

In the following, several questions are raised about the dimensions and characteristics of the housing units from respondents. In terms of the housing age, among the participants, the homes of 277 people (74.9 percent) are more than 15 years old, and in terms of housing area, 276 people (74.6 percent) have less than 100 square meters area of home with less than two rooms.

INFERENTIAL RESULTS

Neighborhood Environmental Satisfaction (NES) Indicators Measurement

Local environment quality assessments are often done by understanding the preferences of residents of the physical, social and economic characteristics and quality of their neighborhood (Lee et al., 2013; Yin et al., 2016). This perception is derived from how both the conscious and the unconscious affect people's moods and emotions. On this basis, neighborhood environmental quality assessment is carried out not only on the objective environment but also on the basis of satisfaction with the environment in which

Figure 18.2   Ab-Kooh neighborhood morphology in comparison with surroundings

people live. This indicates that the issue of environmental quality is more a mental issue judged by the people and their experiences. These judgments are created due to positive or negative attitudes of people toward a particular aspect of the environment (Van Poll, 1997; Bonaiuto et al., 2003; Lee, 2008; Hur et al., 2010; Islam et al., 2014; Bonaiuto et al., 2015; Tucker and Abass, 2018).

Identification and classification of the satisfaction indicators is needed in order to develop measurement of the environmental quality at the neighborhood scale. Indicators are selected according to the literature, study goal and research area by reviewing the models and the other criteria (Table 18.2). In this study, due to residents' closeness to real conditions of neighborhood, feelings of satisfaction are used based on a subjective approach. This approach focuses on residents' perceptions of the neighborhood environment satisfaction and experimental procedures. In this method, the residents are questioned by questionnaire in order to evaluate the current state of the environment based on a set of satisfaction indicators.

In the questionnaire, the 5-point Likert scale was used for the questions relating to satisfaction and rankings of 1 to 5 were assigned to the items. Figure 18.3 shows the lowest level of residents' satisfaction of the relevant question and 5 represents the highest satisfaction rating. Thus, the number 3 was chosen for replies in the middle.

The normality of the distribution was studied in these groups to calculate the average satisfaction rating among the various indices. For this purpose, Kolmogorov-Smirnov

Table 18.2 Selected indicators to measure NES in current research

Housing Quality[a]	Internal facilities of the housing/dwelling unit strength against earthquakes/physical dimensions of the housing/housing health/external conditions of residence/dwelling unit costs
Neighborhood Security[b]	Night lighting of the public spaces/delinquent persons entering into the neighborhood/graffiti and posters on the walls/abandoned spaces and dilapidated parcels/haunted houses/bustle and crowds
Environmental Health[c]	The cleanliness of the neighborhood/how to collect garbage/who to gather surface water/having safe drinking water/odor/noise pollution/air pollution
Access to Public Facilities[d]	Shopping malls and stores/health centers/sports centers and recreational/cultural and educational centers/parks and green spaces/community centers and gatherings
Transportation and Mobility[e]	Public transport facilities/dimension access/parking space/quality of sidewalks/quality of streets and alleys cover/traffic and congestion of motor vehicles
Community Status[f]	A sense of belonging to the community/social interaction amongst neighbors/the charm of the neighborhood for residents/residents' happiness and vitality

Notes: [a] Van Poll, 1997; Bonaiuto et al., 2003; Adriaanse, 2007; Cho and Lee, 2010; Ibem and Aduwo, 2013; Asadi et al., 2014; [b] Amerigo and Aragonés, 1997; Van Poll, 1997; Haginegad et al., 2010; Asadi et al., 2014; Hur and Nasar, 2014; Bonaiuto et al., 2015; Wang and Wang, 2016; [c] Van Poll, 1997; Lee, 2008; Haginegad et al., 2010; Asadi et al., 2014; Bonaiuto et al., 2015; Ray, 2017; [d] Amerigo and Aragonés, 1997; Van Poll, 1997; Lee, 2008; Haginegad et al., 2010; Ibem and Aduwo, 2013; Asadi et al., 2014; Bonaiuto et al., 2015; Wang and Wang, 2016; Dong and Qin, 2017; [e] Van Poll, 1997; Haginegad et al., 2010; Asadi et al., 2014; Bonaiuto et al., 2015; Wang and Wang, 2016; Dong and Qin, 2017; [f] Amerigo and Aragonés, 1997; Van Poll, 1997; Adriaanse, 2007; Lee, 2008; Haginegad et al., 2010; Asadi et al., 2014; Bonaiuto et al., 2015; Wang and Wang, 2016.

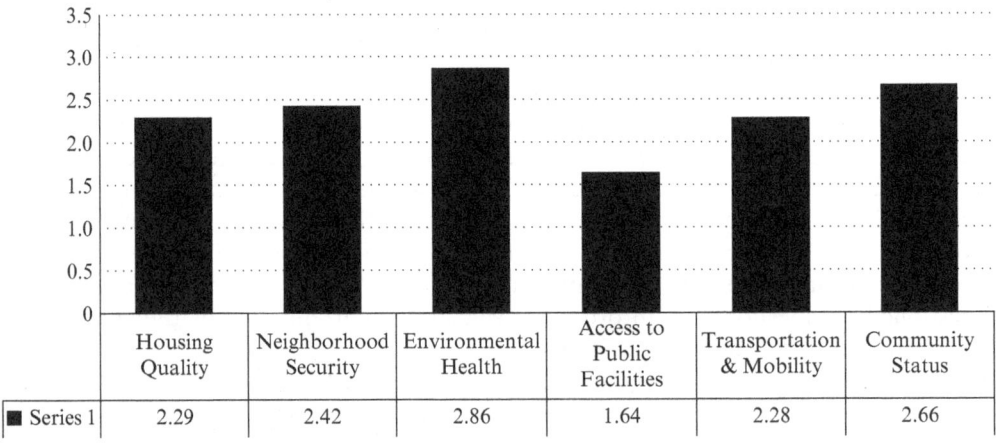

Figure 18.3 NES indicators mean

Table 18.3 T-test value results for NES indicators mean

Indexes	Standard Deviation	Mean	T-test Value	Sig.
Housing Quality	0.431	2.29	−31.602	0.000
Neighborhood Security	0.517	2.42	−21.212	0.000
Environmental Health	0.421	2.86	−6.146	0.000
Access to Public Facilities	0.280	1.64	−93.344	0.000
Transportation and Mobility	0.403	2.28	−34.070	0.000
Community Status	0.826	2.66	−7.822	0.000

one-sample test was used. According to the result, the distribution of data in the mentioned groups was normal. One sample T-test with test value equal to the number 3 (test value = 3) and the confidence interval of 95 percent (5 percent error) was used to achieve the satisfaction of residents. In the test, if the amount of P-value is greater than 0.05, the examined variables by the amount of test (i.e., number 3) will not have a significant difference; thus the amount of assessed quality indicators at the intermediate level will be evaluated; and if the amount of P-value is less than 0.05, the studied index by the test (number 3) has a significant difference. In this case, if average studied indicators are over 3, the quality of this index will be above the average, and if the index is lower than the number 3, the quality will be lower than the average. For this purpose, the relevant test is shown for each of the indicators in Table 18.3 and Figure 18.3.

As it can be observed, P-value is significant based on p<0.5; thus, there is significant difference between obtained average and the number 3. All in all, it can be concluded that the amount of overall satisfaction of the Ab-Kooh neighborhood residents is lower than the average.

NES from the Socioeconomic Characteristic Perspective

In the following, the effect of residents' individual characteristics on their perception of neighborhood environment satisfaction regarding answers to the main question of the research was evaluated after determining the level of resident satisfaction with the neighborhood environment quality. Accordingly, eight variables and economic and social characteristics of respondents addressed to describing the sample (age, gender, education level, monthly income, household size, employment status, marital status and length of residence in the neighborhood) and their relationship with satisfaction level were tested by satisfaction indicators of neighborhood in the study area. Kendall's tau-b correlation coefficient was used to determine the strength and direction of the relationship between age, family size, education level, income level and duration of residence in the neighborhood which have the rating scale, and Cramer's V correlation coefficient was applied for gender, marital status and employment status variables which have a nominal scale.

Each individual variable was examined with the index of satisfaction regarding to the intensity and direction of correlation. Table 18.4 shows the results of the correlation test between the six indexes and individual variable.

The highest correlation coefficient is related to the variable age which has positive correlation with indicators of satisfaction. The older the respondents are, satisfaction

Table 18.4 The correlation between socioeconomic characteristic and NES indicators

Indexes	Age	Household Size	Income	Education	Length of Residence	Gender	Marital Status	Occupation
Housing Quality	0.264	−0.080	0.069	−0.147	0.047	−0.147	0.194	0.193
Neighborhood Security	0.113	−0.011	0.006	−0.004	0.019	−0.004	0.176	0.188
Environmental Health	0.236	−0.064	−0.085	−0.102	0.085	−0.102	0.125	0.148
Access to Public Facilities	0.099	0.072	−0.043	−0.114	0.132	−0.114	0.152	0.171
Transportation and Mobility	0.080	0.080	−0.084	−0.006	0.019	−0.006	0.149	0.119
Community Status	0.044	0.109	0.132	0.377	0.391	0.377	0.096	0.256
Overall Satisfaction	0.340	0.029	−0.013	−0.205	0.182	−0.205	0.123	0.960

with the neighborhood environment increases as well. Among age groups, the age group of 60 years and above with the number of 51 (13.8 percent) has the highest satisfaction neighborhood environment index due to the sense of belonging to the neighborhood, a poor education and low expectations. Most previous studies have confirmed the positive effect of age in satisfaction of the place of residence (Baba and Austin, 1989; Van Poll, 1997; Mohan and Twigg, 2007; Wang and Wang, 2016). However, in some studies, the younger respondents show more satisfaction from the neighborhood especially if the neighborhood is modern and new (Zhang and Lu, 2016).

The household variable has a negative correlation with two indexes of satisfaction with housing and neighborhood security. In all the indexes, correlation is weak. Among the classified groups, the highest satisfaction of the indexes has been allocated to the respondents with a household size of four people with the number of 49 (13.3 percent), because of the deteriorated and small residential units, social and environmental insecurity for children and women, inadequate access to training, sport and recreation centers.

Income variable has a negative correlation with the three indicators of environmental health, access to public facilities and transportation and mobility, as well as overall satisfaction. Unlike most previous research (Baba and Austin, 1989; Hannscott, 2016; Zhang and Lu, 2016), among the classified groups in the part of descriptive statistics, those with the lowest income have the highest satisfaction rate index for the neighborhood environment. Inhabitants' economic failure of settling in the new urban context (surrounding neighborhoods) could be the important cause of their satisfaction with the neighborhood environment.

Test results for the education level variable show that with the exception of the community status index, which has a high positive correlation, the educational level variable has negative correlation along with other indexes and overall satisfaction. Among groups with different educational levels mentioned in the descriptive statistics, illiterate groups of seven people (1.9 percent) have the highest satisfaction on the neighborhood environment index, since increasing education leads to decreasing satisfaction. This index also confirms the findings of previous research (Cheung and Leung, 2008; Lee, 2008; Haginegad et al., 2010). Lack of knowledge of basic human needs and the institutionalization of traditional lifestyles with minimum facilities could be cited as a reason for higher satisfaction than the other accountable groups in terms of education level.

The variable of length of residence in the neighborhood has a positive correlation with the other indexes and overall satisfaction except the security index, because the perception of the neighborhood environment has always been tied to residents' emotional relationship with their neighborhood (Poortinga et al., 2017; Wang and Wang, 2016; Zhang and Lu, 2016). Residents with a duration of more than 15 years (228 people; 61.6 percent) have the highest satisfaction with their neighborhood environment indicators among groups with different durations of residence in the neighborhood mentioned in the descriptive statistics. The factors affecting individual satisfaction with long-term duration of residence in the neighborhood include a sense of belonging to the neighborhood, having good memories, acceptance of him/herself as a member of the neighborhood, close relationship with the community and so on.

Cramer's test results show that the gender variable is negatively correlated with other indexes and overall satisfaction, with the exception of the community status index. The correlation is relatively strong in the community status index, but it is weak on the security, environmental health, and transportation and mobility indexes. Between two groups of men and women mentioned in the descriptive statistics, women are more satisfied with Ab-Kooh neighborhood, with the number of 102 (27.6 percent), because women have better social interaction in neighborhoods and the local level as well as being present in public places like mosques. Zhang and Lu (2016) have also measured a higher level of satisfaction in their results for women and people who know their neighbors and interact with them. However, the results of Van Poll's (1997) research show that while gender is an important personal characteristic in the study of psychology, it does not seem to be an important characteristic for satisfaction evaluation.

Cramer's test results for the variable of marital status show that all indexes have shown a positive correlation. In terms of correlation, the marital status variable also has a higher correlation with the index of housing satisfaction which could be due to the high value of having a house among married people. Among two groups of married and single people mentioned in the descriptive statistics section, the married group, with 322 people (87 percent), is more satisfied with the quality of their neighborhood. The important role of marriage in satisfaction is due to being protected against hardship as well as having an economic and emotional sponsor that has positive effects on well-being and happiness (Diener et al., 1999) besides that of having better social interaction because of a spouse and children (Mohan and Twigg, 2007).

The present test results for the variable employment status show that all indexes are positively correlated. In terms of correlation, the jobs variable is more strongly correlated with the index of community status. Among occupational groups mentioned in the descriptive statistics section, more satisfied with the neighborhood belong to the group of non-workers (unemployed, pensioners, soldiers, students and housewives) with 104 people (24.1 percent).

Furthermore, the chi-square test was used to determine the existence of a significant relationship between personal variables and indicators of the quality of the neighborhood environment (Table 18.5).

As can be seen among six indexes of neighborhood environment, community status has been more affected by the individual characteristics by creating seven times the significant

Table 18.5 The relationship between socioeconomic characteristic and NES indicators

Indexes	Age	Household Size	Income	Education	Length of Residence	Gender	Marital Status	Occupation
Housing Quality	0.000	0.129	0.290	0.000	0.187	0.026	0.001	0.033
Neighborhood Security	0.008	0.319	0.532	0.001	0.826	0.000	0.003	0.042
Environmental Health	0.000	0.005	0.737	0.001	0.152	0.582	0.057	0.231
Access to Public Facilities	0.000	0.067	0.007	0.014	0.012	0.072	0.014	0.094
Transportation and Mobility	0.006	0.002	0.014	0.386	0.014	0.326	0.016	0.510
Community Status	0.000	0.003	0.001	0.000	0.000	0.000	0.183	0.000
Overall Satisfaction	0.000	0.079	0.248	0.000	0.010	0.303	0.061	0.760

relationship. Housing quality, neighborhood security, access to public facilities, transportation and mobility placed in the next rank, each by creating five times the significant relationship. Neighborhood environmental health index with three times the significant relationship has been less affected by the individual characteristics. Additionally, the variables of age, length of residence and level of education have had the most significant relationship with the satisfaction of the neighborhood residents.

CONCLUSION

The results in this research show that the overall satisfaction of the neighborhood environment in the study area is lower than the average. Environmental health indicators show the highest level of satisfaction with the neighborhood environment. Respectively, community status indicators, safety environment, environment security, housing satisfaction index, the index of transport and commuting have had the most residents' satisfaction. The lowest level of satisfaction index is related to benefiting from public facilities in the neighborhood. However, individual characteristics play an important role in the attitudes and perceptions of residents about the quality of residential environment. Age, duration of residence in the neighborhood and education level variables are highly correlated with people's perception of the quality of their neighborhood environment. So that, with boosting the education level, the satisfaction with the neighborhood comes down. Education is a vital factor for successful implementation of the study. In fact, education is applied as an instrument to overcome the psychological barriers such as ignorance or misinformation. Satisfaction also increases with increasing age and duration of residence in the neighborhood. There are differences between young people and the elderly in the process of environment perception and its results. Age differences arise from behaviors, actions and social attitudes over time. In relation to the duration of residence in the neighborhood, different opinions can be seen in the general population that determine the scope of their understanding of the quality of neighborhood environment. In general, elements of time and space have a significant impact in shaping the concept of life satisfaction. Although residents may live in different urban neighborhoods with different qualities, the impact of satisfaction or dissatisfaction with the quality characteristics of the individual's perception of the built environment are not negligible.

NOTES

1. There is no one single and agreed definition for deteriorated areas. Basically depressed, decayed, degeneration, erosion and blighted all refer to urban deteriorated areas where people deal with different issues. In fact, each of these concepts covers different dimensions of deterioration (Khaef and Zebardast, 2016).
2. The first approach deals with design of built environment; relies on master plan and capital rather than social aspects of people's lives. This approach insists heavily on the formal and aesthetic aspects of urban planning and it believes in physical determinism. It is a product-oriented approach and its final products would be something to look at. The second approach is empiricist in nature and insists on human sensory experiences of the built environments. It deals with users' needs and their aesthetic values and their everyday urban lives (Lawrence and Low, 1990; Besteliu and Doevendans, 2002; Gottdiener and Budd, 2005; Knox, 2005; Rennie Short, 2006).
3. In line with this topic also see (Rentfrow et al., 2008; Jokela et al., 2015; Wang and Wang, 2016).
4. The city of Mashhad, capital of Khorasan Razavi province, has an area of 35.234 hectares and consists of 13 districts and 158 neighborhoods (Mashhad Statistical Report, 2014, 2015: 44). The city's population had grown from 45,000 in 1897 to 176,000 in 1940, and then reached 241,000 in 1956. It had the highest population growth rates among the other cities of Iran from 1976 to 1986 (8.16 percent) (Akbari Motlaq and Abbaszadeh, 2012). It had a population of 2,766,258 in the last official census (Statistical Center of Iran, 2016) and 519,924 of them live in deteriorated areas (Mashhad Municipality, 2018).
5. Astan Quds is a public institution that is the custodian of the Holy Shrine and all its affiliated organizations and endowments in Mashhad.
6. The Endowment Organization is a public institution with the task of restoring, maintaining and managing country endowments for charity and welfare.

REFERENCES

Abbaszadegan, M., Rezazadeh, R., and Mohammadi, M. (2011). An investigation on the social and traffic effects of metro stations on neighborhoods. *International Journal of Architectural Engineering & Urban Planning*, 21(1), 45–51.

Adriaanse, C.C.M. (2007). Measuring residential satisfaction: A residential environmental satisfaction scale (RESS). *Housing and the Built Environment*, 22(3), 287–304.

Akbari Motlaq, M., and Abbaszadeh, G. (2012). The physical development of Mashhad city and its environmental impacts. *Environment and Urbanization ASIA*, 3(1), 79–91.

Amadeo, D.M., and Golledge, R.G. (2003). Environmental perception and behavioral geography. In Gary L. Gaile and Cort J. Willmott (eds), *Geography in America at the Dawn of the 21st Century* (pp. 133–48). Oxford, UK: Oxford University Press.

Amao, F.L. (2012). Housing quality in informal settlements and urban upgrading in Ibadan, Nigeria. *Developing Country Studies*, 2(10), 68–80.

Amerigo, M., and Aragonés, J.I. (1997). A theoretical and methodological approach to the study of residential satisfaction. *Environmental Psychology*, 17(1), 47–57.

Asadi, A., Movahed, A., Ahmadi, S., and Lorestani, A. (2014). Assessment and analysis on quality of life in old textures neighborhoods, Kuhdasht City. *Geography and Urban Planning Research*, 2(2), 195–218. (In Persian).

Baba, Y., and Austin, D.M. (1989). Neighborhood environmental satisfaction, victimization, and social participation as determinants of perceived neighborhood safety. *Environment and Behavior*, 21(6), 763–80.

Barker, R.G. (1968). *Ecological Psychology: Concept and Methods for Studying Environment of Human Behavior*. Stanford, CA: Stanford University Press.

Besteliu, I., and Doevendans, K. (2002). Planning, design and the post modernity of cities. *Journal of Design Studies*, 23(3), 233–44.

Bonaiuto, M., Fornara, F., and Bonnes, M. (2003). Indexes of perceived residential environment quality and neighborhood attachment in urban environments: A confirmation study on the city of Rome. *Landscape and Urban Planning*, 65(1–2), 41–52.

Bonaiuto, M., Fornara, F., Ariccio, S., Cancellieri, U.G., and Rahimi, L. (2015). Perceived residential environment quality indicators (PREQIs) relevance for UN-HABITAT city prosperity index (CPI). *Habitat International*, 45(1), 53–63.

Campbell, A., Converse, P., and Rodgers, W. (1976). *The Quality of American Life: Perceptions, Evaluations and Satisfactions*. New York: Publications of Russell Sage Foundation.

Cazzuffi, C., and López-Moreno, D. (2018). Psychosocial wellbeing and place characteristics in Mexico. *Health and Place*, 50, 52–64.

Chen, J., and Chen, S. (2015). Mental health effects of perceived living environment and neighborhood safety in urbanizing China. *Habitat International*, 46, 101–10.

Cheung, C.-K., and Leung, K.-K. (2008). Retrospective and prospective evaluations of environmental quality under urban renewal as determinants of residents' subjective quality of life. *Social Indicators Research*, 85(2), 223–41.

Cho, S.H., and Lee, T.K. (2010). Indoor environment quality related on residential satisfaction in old multi-family housing. SHB2010: 3rd International Symposium on Sustainable Healthy Buildings, 27 May, Seoul, Korea.

Croasmun, J.T., and Ostrom, L. (2011). Using Likert-type scales in the social sciences. *Journal of Adult Education*, 40(1), 19–22.

Cutter, S.L. (1985). *Rating Places: A Geographer's View on Quality of Life*. Washington, DC: The Association of American Geographers.

Diener, E., Suh, E.M., Lucas, R.E., and Smith, H.L. (1999). Subjective well-being: Three decades of progress. *Psychological Bulletin*, 125(2), 276–302.

Dong, H., and Qin, B. (2017). Exploring the link between neighborhood environment and mental wellbeing: A case study in Beijing, China. *Landscape and Urban Planning*, 164, 71–80.

Farnahad Consulting Engineers (2009). Excellence plan of well-being in Ab-kooh neighborhood, Mashhad.

Forrest, R., and Kearns, A. (2001). Social cohesion, social capital and the neighborhood. *Urban Studies*, 38(12), 2125–43.

Galster, G.C., and Hesser, G.W. (1981). Residential satisfaction composition and contextual correlates. *Environment and Behavior*, 6(13), 735–58.

Gibson, J. (1979). *An Ecological Approach to Visual Perception*. Boston: Houghton Miffin.

Gifford, R. (1997). *Environmental Psychology, Principles and Practice*. Boston: Allyne & Bacon.

Golabi, F., Aqhayari, T., and Ebrahimi, R. (2013). A study of the association between socioeconomic factors and satisfaction level of residential neighborhood. *Urban Sociological Studies*, 3(9), 125–54. (In Persian).

Golledge, R.G., and Timmermans, H.J.P. (1990). Application of behavioral research on spatial problems, I: Cognition. *Progress in Human Geography*, 14(1), 57–99.

Gottdiener, M., and Budd, L. (2005). *Key Concepts in Urban Studies*, Thousand Oaks, CA: SAGE Publications.

Hafeznia, M.R. (2010). *An Introduction to the Research Method in Humanities*. 17th edition, Tehran: SAMT.

Haginegad, A., Rafieian, M., and Zamani, H. (2010). Investigation of effective individual variables on citizens' satisfaction with environmental quality, case study comparison of old and new urban pattern of Shiraz. *Geography and Development*, 8(17), 63–72. (In Persian).

Hallajow, N. (2018). Identity and attitude: Eternal conflict or harmonious coexistence. *Journal of Social Sciences*, 14, 43–54.

Hamdan, H., Yusof, F., and Marzukhi, M.A. (2014). Social capital and quality of life in urban neighborhoods high density housing. *Social and Behavioral Sciences*, 153, 169–79.

Hannscott, L. (2016), Individual and contextual socioeconomic status and community satisfaction. *Urban Studies*, 53(8), 1727–44.

Hashim, A.H. (2003). Residential satisfaction and social integration in public low cost housing in Malaysia, Pertanika. *Journal of Social Science and Humanity*, 11(1), 1–10.

Huang, Z., and Du, X. (2015). Assessment and determinants of residential satisfaction with public housing in Hangzhou, China. *Habitat International*, 47, 218–30.

Hur, M., and Nasar, J.L. (2014). Physical upkeep, perceived upkeep, fear of crime and neighborhood satisfaction. *Journal of Environmental Psychology*, 38, 186–94.

Hur, M., Nasar, J.L., and Chun, B. (2010). Neighborhood satisfaction, physical and perceived naturalness and openness. *Journal of Environmental Psychology*, 30(1), 52–9.

Ibem, E.O., and Aduwo, E.B. (2013). Assessment of residential satisfaction in public housing in Ogun State, Nigeria. *Habitat International*, 40, 163–75.

Ibem, E.O., Opoko, P.A., and Aduwo, E.B. (2017), Satisfaction with neighborhood environments in public housing: Evidence from Ogun State, Nigeria. *Social Indicators Research*, 130(2), 733–57.

Insch, A., and Florek, M. (2008). A great place to live, work and play: Conceptualizing place satisfaction in the case of a city's residents. *Place Management and Development*, 1(2), 138–49.

Islam, M.S., and Rana, M.M.P., and Ahmed, R. (2014). Environmental perception during rapid population growth and urbanization: A case study of Dhaka city. *Environment Development and Sustainability*, 16(2), 443–53.

Ittelson, W.H. (1973). Environmental perception and contemporary perceptual theory. In W. H. Ittelson (ed.), *Environment and Cognition* (pp. 141–54), New York: Seminar Press.

Javdani Irani Nejad, M., Afzali, A., and Hatefi, F. (2014). Population information of Mashhad deterioration areas. Document report, Mashhad Municipality, Planning and Development Department, Information Analysis and Statistic Management.

Jokela, M., Bleidorn, W., Lamb, M., Gosling, S., and Rentfrow, P. (2015). Geographically varying associations between personality and life satisfaction in the London metropolitan area. *Proceedings of the National Academy of Sciences*, 112(3), 725–30.

Khaef, S., and Zebardast, E. (2016). Assessing quality of life dimensions in deteriorated inner areas: A case from Javadieh neighborhood in Tehran metropolis. *Social Indicator Research*, 127(2), 761–75.

Kheyroddin, R., Taghvaee, A., and Forouhar, A. (2014). The influence of metro station development on neighborhood quality (the case of Tehran metro rail system). *International Review for Spatial Planning and Sustainable Development*, 2(2), 64–75.

Kitchin, R. (1996). Increasing the integrity of cognitive mapping research: Appraising conceptual schemata of environment-behaviour interaction. *Progress in Human Geography*, 20(1), 56–84.

Knox, P.L. (2005). Creating ordinary places: Slow cities in a fast world. *Journal of Urban Design*, 10(1), 1–11.

Lang, J. (1987). *Creating Architectural Theory: The Role of the Behavioral Sciences in Environmental Design*. New York: Van Nostrand Reinhold Publisher.

Lawrence, D.L., and Low, S.M. (1990). The built environment and spatial form. *Annual Review of Anthropology*, 19, 453–505.

Lee, C.C., You, S.M., and Huang, L.Y. (2013). The influence of public facilities and environmental quality on residential satisfaction in Taiwan: Differences in neighborhood environment. *African Journal of Business Management*, 7(12), 915–25.

Lee, S.M., Conway, T.L., Frank, L.D., Saelens, B.E., Cain, K.L., and Sallis J.F. (2017). The relation of perceived and objective environment attributes to neighborhood satisfaction. *Environment and Behavior*, 49(2), 136–60.

Lee, Y.J. (2008). Subjective quality of life measurement in Taipei. *Building and Environment*, 43(7), 1205–15.

Lenzi, M., Vieno, A., Perkins, D.D., Pastore, M., Santinello, M., and Mazzardis, S. (2012). Perceived neighborhood social resources as determinants of prosocial behavior in early adolescence. *American Journal of Community Psychology*, 50(1–2), 37–49.

Li, Z., and Wu, F. (2013). Residential satisfaction in China's informal settlements: A case study of Beijing, Shanghai, and Guangzhou. *Urban Geography*, 34(7), 923–49.

Mashhad Municipality 2018, *Deputy of planning and human resources development*, viewed 31 December 2019, https://planning.mashhad.ir/.

Mashhad Statistical Report 2014 (2015). Development and planning department of Mashhad Municipality. Mashhad Municipality, Mashad.

Moghadam Ariaee, A., Izadi, S., and Tamiz, M. (2008). Feasibility study of land readjustment attainment in urban deteriorated areas (Mashhad, Ab-Kooh neighborhood). The First Conference on Regeneration and Revitalization of Urban Distressed Areas, Mashhad, Iran, 10–11 December.

Mohan, J., and Twigg, L. (2007). Sense of place, quality of life and local socioeconomic context: Evidence from the Survey of English Housing, 2002/03. *Urban Studies*, 44(10), 2029–45.

Mohit, M.A., and Azim, M. (2012). Assessment of residential satisfaction with public housing in Hulhumale Maldives. *Procedia Social and Behavioral Sciences*, 50, 756–70.

Mouratidis, K. (2018). Rethinking how built environments influence subjective well-being: A new conceptual framework. *Journal of Urbanism: International Research on Placemaking and Urban Sustainability*, 11(1), 24–40.

Naghsh Azin Shargh Consulting Engineers (2006). *Strategic Plan for Empowerment of Ab-Kooh Neighborhood, Mashhad*. Iran: Mashhad Municipality.

Oktay, D., Rustemli, A., and Marans, R.W. (2009). Neighborhood satisfaction, sense of community, and attachment: Initial findings from Famagusta quality of urban life study. ITU A|Z, 6(1), 6–20.

Okunola, S., and Amole, D. (2012). Perception of safety, social participation and vulnerability in an urban neighborhood, Lagos, Nigeria. *Social and Behavioral Sciences*, 35, 505–13.

Pacione, M. (1986), Quality of life in Glasgow: An applied geographical analysis. *Environment and Planning A: Economy and Space*, 18(11), 1499–520.

Pacione, M. (2003). Urban environmental quality and human well-being: A social geographical perspective. *Landscape and Urban Planning*, 65(1–2), 19–30.

Poortinga, W., Tatiana, C., Jones, N., Lannon, S., Rees, T., Rodgers, S.E., Lyons, R.A., and Johnson, R. (2017). Neighborhood quality and attachment: Validation of the revised residential environment assessment tool. *Environment and Behavior*, 49(3), 255–82.

Rapoport, A. (1990). *The Meaning of the Built Environment: A Nonverbal Communication Approach*, 2nd edition. Tucson: University of Arizona Press.

Ray, B. (2017). Quality of life in selected slums of Kolkata: A step forward in the era of pseudo-urbanisation. *Local Environment: The International Journal of Justice and Sustainability*, 22(3), 365–87.

Rennie Short, J. (2006). *Urban Theory: A Critical Assessment*. New York: Palgrave Macmillan.

Rentfrow, P.J., Gosling, S.D., and Potter, J. (2008). A theory of the emergence, persistence, and expression of geographic variation in psychological characteristics. *Perspectives on Psychological Science*, 3(5), 339–69.

Saeedi Rezvani, H. (2006). Evolution of old villages in today cities, a kind of emerging phenomenon of informal settlement (Mashhad, Ab-Kooh neighborhood). *Geographical Researches*, 21(3), 64–82. (In Persian).

Salmani, H., Taghvaiee, A.A., and Rafieeyan, M. (2012). Measurement of life quality in old habitat and its visualization, case study: The Hashemi district in 10th zone of Tehran metropol. *Geography and Territorial*, 2(4), 53–64. (In Persian).

Spencer, C., and Blades, M. (1986). Pattern and process: A review essay on the relationship between behavioral geography and environmental psychology. *Progress in Human Geography*, 10(2), 229–48.

Statistical Center of Iran (2016). 2016 Population and Housing Census.

Tavanaei Marvi, L., and Behzadfar, M. (2015). Local sustainability with emphasis on CPTED approach, the case of Ab-Kooh neighborhood in Mashhad. *Social and Behavioral Sciences*, 201, 409–17.

Tucker, R., and Abass, Z. (2018). Residential satisfaction in low-density Australian suburbs: The impact of social and physical context on neighborhood contentment. *Environmental Psychology*, 56, 36–45.

Van Poll, R. (1997). The perceived quality of the urban residential environment: A multi attribute evaluation (Doctoral dissertation). Center for Energy and Environmental Studies (IVEM), University of Groningen (RUG), The Netherlands.

Wang, D., and Wang, F. (2016). Contributions of the usage and affective experience of the residential environment to residential satisfaction. *Housing Studies*, 31(1), 42–60.

Wicker, A. (2002). Ecological psychology: Historical contexts, current conception, prospective direction. In Robert B. Bechtel and Arza Churchman (eds), *Handbook of Environmental Psychology* (pp. 114–25). New York: John Wiley and Sons.

Yin, J., Cao, X.(J)., Huang, X., and Cao, X. (2016). Applying the IPA-Kano model to examine environmental correlates of residential satisfaction: A case study of Xi'an. *Habitat International*, 53, 461–72.

Zhang, C., and Lu, B. (2016). Residential satisfaction in traditional and redeveloped inner city neighborhood: A tale of two neighborhoods in Beijing. *Travel Behavior and Society*, 5, 23–36.

19. Downtown revitalization, livability and quality of life in Tucson, Arizona*
Carlos J.L. Balsas

> Tucson is one of those American towns that has distinction.
> (Peterson, 1988, p. 451)

INTRODUCTION

Tucson is a great southwestern metropolis. Its modern settlement dates back to 1775. Its ubiquitous Hispanic influence is a result of having been part of Mexico until the Gadsden Purchase of 1856 (Lyons and Vasquez, 2010). Metropolitan Tucson is about one-quarter the size of its larger neighbor to the north, metro Phoenix, in both population and size of the economy. Phoenix's development has almost eclipsed Tucson's long history of becoming a great sunbelt city (Luckingham, 1982; Abbott, 1987). However, after having resided and studied Phoenix's development and planning for more than a decade it appears that Phoenix's dominance is more of the same decades-old growth machine, and less of what is important: Livable quality of life (Gober, 2006). Tucson's southwestern location, scale and built heritage may have worked to its advantage and current quality of urban life. This chapter answers the following research question: Are there any lessons to be learned from Tucson's recent urban revitalization efforts? The purpose of this chapter is to reflect on Tucson's urban and regional transformations and to extract a set of lessons learned that can be useful to cities similar to Tucson.

The guiding argument is that although Tucson may have had difficulty obtaining funding and attracting investment prior to the 2007–08 global financial crisis, from a quality of life perspective Tucson's residents appear to benefit from a whole array of small- and medium-size town amenities. These include such aspects as a compact and exciting downtown, affordable housing, alternative modes of transport, cultural institutions, a vibrant arts and culture scene, a land grant university and proximity to protected desert natural environments.

The methods comprised extensive literature reviews of specialized and professional materials; the participation in research and teaching engagements on funding redevelopment mechanisms (2007) and downtown revitalization (2010), respectively; and a post-studio reflections exercise within a context of macro-regional planning efforts to develop the Arizona Sun Corridor (IUPS, 2010; Balsas, 2017). Subsequent study visits were held in 2014 and 2016; and discussions with planners, residents, historians and scholars supplement the data, information and insights gathered.

In this study, I identify four lessons learned: First, the scale of urbanization makes Tucson relatively resilient to socio-economic shocks. Second, closer proximity to the border makes residents more understanding of the regions' strong Hispanic cultural heritage. Third, the urban renewal wounds of earlier decades (e.g., demolition of adobe buildings, the building of modern architectonic structures, forced relocation of residents) have been partially ameliorated (i.e., small-scale urban revitalization interventions such as the renovated train station depot, streetscape improvements, a new streetcar, a centrally located transit station, affordable housing, the renovation of the warehouse district, historic preservation efforts and so on). And fourth, home-grown advocacy informed by sensible planning research has strengthened the revitalization vision without curbing developmental ambitions.

This chapter is in four sections: Following this introduction, section one is the analytical mechanism with a literature review on good/great cities and initial considerations about the sunbelt. Section two is Tucson's case study in three sub-parts: An evolutionary regional perspective, downtown USA (urban renewal through the 1990s) and a 21st century city center (a "sewn" city – Rio Nuevo's great place). Section three is the case study's discussion according to Savitch's 4 Cs of Great Cities conceptual framework. Section four is the conclusion and lessons learned.

ANALYTICAL MECHANISM

This analytical mechanism identifies the concepts put forward to define and characterize memorable cities. Memorable signifies a combination of exceptional urban characteristics capable of creating and perpetuating the identity, character and happiness of a city to its residents' advantage (Hinshaw, 2007; Ballas, 2013). Various authors have written about what makes cities memorable. Some of those authors have conducted analyses at an international level, while others have discussed their ideas within more restricted national and regional contexts. A common feature of most studies appears to be the fact that they attempt to compare and, in certain cases, rank cities, according to pre-established criteria (Sperling and Sander, 2004). Table 19.1 is a synthesis of the Good/Great City and Quality of Life theories and their applications at various levels.

Contemporary work on good/great city and quality of life theories and their applications date back to the early 2000s with scholars of various genres attempting to conceptualize and test their ideas, theories and developments dealing with exceptionality of procedures and distinct evaluative outcomes (Norquist, 1998; Kirby, 2000). Collins's (2001) *Good to Great* bestseller examined how certain companies achieved superior performance and results in the business world. Markusen et al. (1999) edited a volume on *Second Tier Cities*. The various contributing authors were interested in understanding the reasons behind the rapid growth in this type of city. Based on case studies from throughout the world, they concluded that certain districts were leveraging those cities' growth. In the same vein, Ng and Hills (2003) examined the competitiveness of five large Asian cities at the beginning of the 21st century: Tokyo, Hong Kong, Singapore, Taipei and Shanghai. This study was followed by Savitch's viewpoint commentary (2010) which proposed the 4 Cs framework of great cities and applied it to four large cities in the United States.

The 4 Cs framework stands for *Currency* (or worth) – "the value of something and its ability to carry weight in crucial circumstances"; *Cosmopolitanism* (or pluralism) – "ability to embrace international, multicultural or polyethnic features"; *Concentration*

Table 19.1 *Synthesis of the Good/Great City and Quality of Life theories and their applications*

THE GOOD/GREAT CITY AND QUALITY OF LIFE		Scale of intervention and decision making		
		Metropolis/ city-region	City/ village	Inner-city/ downtown
Theoretical concepts and procedural approaches		Collins (2001) *Good to Great* Gladwell (2005) *Blink* Kahneman (2011) *Thinking, Fast and Slow*		
Scope	International	Ng and Hills (2003) examined the competitiveness of Tokyo, Hong Kong, Singapore, Taipei and Shanghai	Markusen et al. (1999) *Second Tier Cities*; Friedmann (2000, 2002) The good city; Amin (2006) theoretical discussion of the good city; Jacobs (2012) wrote *The Good City*	The British DoE (1994) articulated the 4 A's framework for reaching a vital and viable town center in contexts of urban regeneration
	National	Savitch (2010) examined the greatness of New York, Chicago, Los Angeles and San Francisco	Lynch (1981) *A Theory of Good City Form*; Sperling and Sander (2004) *Cities Ranked & Rated*; Bunnell (2002) *Making Places Special*; Benner and Pastor (2015) *Equity, Growth, and Community*	Bohl (2002) *Place Making: Developing Town Centers, Main Streets, and Urban Villages*; Speck (2012) *Walkable City: How Downtown Can Save America One Step at a Time*
	Regional	McAslan et al. (2013) studied quality of life issues in the US-Mexico border region	Balsas (1998) analysed quality of life and commercial activity in Coimbra, Portugal and Barcelona, Spain	Paumier (2004) *Creating a Vibrant City Center*; Balsas (2004) Measuring the livability of an urban centre

(or compactness) – "demographic density and productive mass": and *Charisma* (or emotion) – "magical appeal that generates enthusiasm, admiration or reverence" (Savitch, 2010, pp. 43–4). These analyses dealt with a large array of morphological and socio-economic characteristics. However, they did not fully engage with the theoretical aspects of the good city. Friedmann (2000, 2002) and Amin (2006) addressed what makes a good city from ideological, rights and equity perspectives. These studies were followed by detailed accounts of the good city and territorial development based on international consultancies, and equity, growth and community in the works of Jacobs (2012) and Benner and Pastor (2015), respectively.

Lynch (1981) in *A Theory of Good City Form* established the following characteristics as being important to a good city: Vitality, sense, fit access, control, efficiency and justice. Bunnell (2002) argued and exemplified that urban planning ought to be able to help make places special. Particularly relevant to quality of life at the regional level is the study of McAslan et al. (2013) who analysed quality of life issues in the US-Mexico border region. Balsas (1998) discussed how commercial urbanism and public markets in the Iberian Peninsula contributed to livable city centers. In fact, literature on city center livability has evolved considerably from the initial British DoE (1994) articulation of the 4 A's framework (see Table 19.1) to accomplishing a vital and viable town center in contexts of urban regeneration, to Bohl's (2002) *Place Making: Developing Town Centers, Main Streets, and Urban Villages*, Speck's (2012) *Walkable City: How Downtown Can Save America One Step at a Time*, Paumier's (2004) *Creating a Vibrant City Center*, to my own work attempting to measure the livability of downtown areas (see Balsas, 2004 for an example).

Common to many of these conceptualizations, interpretations and applications of special cities and places and their livability, one finds the emphasis on procedural decision making (Gladwell, 2005; Kahneman, 2011) and the need for holistic and integral planning leading to desirable results. Other medium-size cities in the sunbelt region known for their quality of life include: Santa Barbara and San Luis Obispo in California, Santa Fe in New Mexico and Fredericksburg in Texas. Although this selection is rather small and mostly based on individual observations, Sperling and Sander (2004) have been conducting advanced rankings and ratings of many cities in the United States since 2004. Common variables of analysis to the top performers include city size, climate, relative distance to large centers, the ocean and other bodies of water, affordability, jobs and regional amenities and overall quality of life, such as arts and culture and recreation.

The southwest of the United States is located within the sunbelt region (Bernard and Rice, 1983; Cunningham, 2017; Strom, 2017). The sunbelt has been variously conceptualized as a region possessing relatively warm weather, population migration, fast urban and suburban growth rates, job creation and real estate investment. This broad region ranging from the California coast to Florida's Atlantic Ocean front displays differences in terms of local and regional culture, lifestyles and susceptibility to global warming and sea level rise phenomenon. On the other hand, the US-Mexico border region is marked by the following issues: Cross-border migration, regional markets proximate distance to daily flows, border security, drug and crime traffic and a significant flow of goods (McAslan et al., 2013).

TUCSON'S CASE STUDY

This overview of the urban development history of Tucson is in four parts. It begins by addressing the origins of the city and its pueblo settlement in the Santa Cruz River valley. Then it discusses the growth of the city before and after WWII. This is followed by an examination of the downtown planning strategies at the dawn of the new millennium.

The origins of Tucson are centered on the old pueblo settlement on the northeast corner of the town site. Tucson's birthplace included "the Mission San Augustin and a portion of the walled Presidio – built in the 1770s by Mexican soldiers to repel the Apache attacks" (Stauffer, 2005, p. 20). Tucson is located in south-central Arizona at the foot of the Catalina Mountains, 60 miles north of the Mexican border. Tucson lies in a broad, flat valley with many dry riverbeds and washes. Tucson's comparative advantages are: high desert climate, arts and culture, attractive setting; and the pitfalls are: crime rate, cyclical economy, and long commutes. Sperling and Sander (2004, p. 677) have described Tucson in the following terms:

> Tucson is a large and growing Sun Belt city known for its attractive setting, pleasant climate, and cosmopolitan nature. It attracts retirees and a younger crowd. The area consists of a modern downtown, with a historic district, surrounded by suburbs laid out in a grid. The University of Arizona, about a mile north of the downtown, gives the entire area a college-town feel. Tall, forested mountains surround the city up to an elevation of 9,000 feet. The economy is mainly supported by the university, retirees and high tech-industry.

Tucson metropolitan area includes the surrounding land of eastern Pima County as well as the independent jurisdictions of Marana, South Tucson, Oro Valley and the Tohono O'odham Indian Reservation. Between 2000 and 2010, the population of Pima County grew by 16 percent from 843,746 to 980,263. Approximately 75 percent of population growth, and 70 percent of housing growth have occurred outside of the Tucson city limits in rapidly growing suburban areas. About one-third of metro Tucson is unincorporated territory in Pima County. Residents of unincorporated areas have high property taxes, relative substandard services and amenities, and little political leverage from an urban governance perspective. Although land is abundant, suburban growth has been partially curbed by a lack of readily available potable water (Le Tourneau and Dubertret, 2019).

Between 1990 and 2010 Pima County's growth boosted emissions of carbon dioxide and other greenhouse gases (GHG) faster than the national rate. However, Arizona and Tucson's worse than average economic recession in 2007 dropped emissions faster than national levels. Greenhouse gas emissions in the Tucson metropolitan area rose 41 percent in 1990–2010. Per person greenhouse gas emissions, though, dropped 3.7 percent in the Tucson metro area, while rising 0.7 percent in the city of Tucson. In 2010, transportation caused 32 percent of all metro Tucson greenhouse gas emissions, home energy use was responsible for 39 percent, industrial energy use contributed 20 percent and commercial energy use's share was 18 percent (PAG, 2010; Newman et al., 2016).

The construction of the Interstate highway I–10 through Tucson occurred later than in several other states. It created a major physical barrier that separates the western part of Tucson adjacent to the Tucson Mountain Park from the east valley, where most of the city of Tucson is located, in the proximity of the washes of the Santa Catalina Mountains, Rincon Mountains and the Saguaro National Park. Nonetheless, the construction of I–10 has enabled further dispersal throughout the metropolis and the bypassing of inner area locations. The I–10 freeway is a central transportation infrastructure to the CANAMEX Regional Corridor traversing the country from the Mexican border to the Canadian.

Like many other sunbelt cities, development before WWII was relatively slow and spatially contained. However, Tucson benefited from being connected to other regional cities through the railroad network. Over time, Tucson grew to encompass more peripheral neighborhoods, with downtown remaining as its main mercantile center for many decades. The advent of functionalist planning brought more orderly land parcellation. Land development occurred on a grid system interrupted only by the railroad, major arterial roads and the Santa Cruz River. The advent of the automobile and its generalized use enabled wider dispersal and easy movement between adjacent communities, and downtown Tucson and the city's various neighborhoods (Logan, 1995).

Post-WWII Tucson registered extensive suburbanization. Desert land, camp sites and mobile home parks slowly gave place to wildcat subdivisions and master planned communities. Suburban shopping malls, such as El Con Mall and Park Place Mall, were built to cater to the needs of suburban car-oriented residents. Major west-east and south-north roadways, such as Speedway and Broadway Boulevards, and Kino and Barazza-Aviation Parkways, respectively, emerged as important arterials linking the city core to distant suburban neighborhoods. Campbell Avenue between Grant Road and Fort Lowell is a major car-oriented shopping district located just outside of the inner city. The growth of the suburbs and of the peripheral jurisdictions slowly began to impact downtown and the core city.

The early attempts at remedying the impacts on the downtown area initially followed the national model of land clearance to build governmental facilities, a shopping plaza and a convention center (Bright, 2000). Particularly important in the case of Tucson was the modernist architecture utilized in the construction of the Pima County campus complex. Slum clearance on the southeastern corner of downtown constituted the favored plan to sanitize the city and to give place to the construction of the convention center.

The slow pace of urban revitalization in the neighborhoods untouched by the federal bulldozer and the urban renewal program has helped to maintain the urban morphology that we can still find in downtown today with its relatively narrow streets, cultural facilities and a just plain human and walkable scale-built environment marked with a distinct southwestern architectonic style (Whyte, 1988; Montgomery, 1998; Leinberger, 2008).

Historic preservation has occurred in many of the core areas. Important cultural anchors have remained downtown, and the warehouse district across from the train tracks is being revitalized. Some adobe style construction and Hispanic architectonic heritage, typical of the former territory under Mexican administration, have contributed toward the area's unique cultural vibrancy (Lara, 2018). In the last 15 years, urban conservation has not happened randomly. It has taken very committed planning advocacy and the search for consensus and compromise on reaching a balance between historic preservation and the construction of modern structures.

Downtown Tucson

Downtown Tucson comprises four sub-districts: Sentinel, Convention, Presidio and Congress with four other emblematic sub-districts in very close proximity: Lost Barrio District, Warehouse District, Fourth Avenue and Main Gate. Different intervention studies over the years have considered distinct boundaries for the downtown area. For instance, Lyons and Vasquez (2010, p. 22) stated that downtown Tucson was relatively small at about 1 square mile in size. The most recent study for the downtown area considered a much larger area at 4.5 square miles (IDA-TDP, 2018, p. 18). The downtown area now is roughly encircled by the El Presidio Trail.

Downtown Tucson has been subjected to many planning studies and interventions. In fact, Diaz et al. (2009, p. 8) have argued that "there is no other geographical area in the Tucson region that has been planned, prodded, poked, investigated, turned upside down, or examined with a microscope as much as downtown Tucson." A humorous joint venture of Worker, Inc., Pop Up Spaces and Design Co*op auto denominated the "Steering Committee of the Planning Division of the City Resources and Studies Unit of the Office of Urban Economic and Design Bureau of the City of Tucson" reviewed and synthesized a long list of about 100 planning documents for the downtown area (1932–2009). According to Diaz et al. (2009, p. 9), the Old Pueblo Redevelopment Plan (1967) and the Rio Nuevo (1999) were the only plans that had had some actual action in the downtown's 70-year history of planning because specific funding allocations were set aside for federal and state enabled interventions.

Urban renewal
Tucson was the only Arizona city to participate in the federal Model Cities Program. According to Williams (2011, p. 143), "Tucson received $49 million in federal funds, over the course of six years, to improve living conditions in blighted areas adjacent to its downtown." Although, various factors generated support for urban renewal in the city, perhaps the most important one was the fact that important retail stores moved to suburban shopping areas, irremediably affecting downtown's livability. In the words of Williams (2011, p. 144), "many business leaders fearing further decline of the city's commercial center, reversed their views on accepting federal dollars and actively campaigned to save the downtown business district." Gomez-Novy and Polyzoides (2003, p. 90) have argued that in Tucson "the promise of urban renewal collided with the historic urban and building fabric of the Barrio and Presidio neighborhoods."

The urban renewal in Tucson during the late 1960s involved the demolition of hundreds of homes and the forcible relocation of many residents (Otero, 2010). Gomez-Novy and Polyzoides (2003, p. 94) argued that:

> large roads for fast moving traffic were built in the place of intimate streets, (. . .) the many intimate scaled, typologically complex, multiuse buildings were replaced by a few large, single-type and single use ones, (. . .) [and] the new construction was of a generally undistinguishable form and [of] a banal, provincial, and internationalist modern style.

On the other hand, Stauffer (2005, p. 20) recognized that some 30 blocks of historic dwellings and businesses were razed for the construction of a mega block to be occupied with governmental buildings, the city's convention and visitor centers – this latter one within a *faux* Mexican shopping center called *La Placita*. The Tucson Government Center – comprised of City Hall, the Pima County Offices and the Federal Building – occupies the northern part of the urban renewal zone, while the convention center is located in the southern part. This "failed planning" story is similar to that of many other cities throughout the United States: Big ticket items built mostly with public funds, often wrapped in deceiving partnerships with the private sector, replaced traditional mixed-use and socially cohesive residential areas.

What is different is the partial neighborhood activism that developed in the city during the late 1980s? Marston and Saint-Germain (1991, p. 231) analysed the role of women in neighborhood activism in Tucson and arrived at three major findings:

(1) women's activism with respect to the negative externalities of urban restructuring emanated from their everyday experiences; (2) in a changing urban context, it had become the responsibility of neighborhoods to exert pressure on the city and county government; and (3) the importance of community and the physical (both natural and built) environment to a satisfactory quality of life.

Rio Nuevo

In 1999 voters in Tucson approved the creation of the Rio Nuevo District. Rio Nuevo stands for "new river." The Rio Nuevo District is a Tax Increment Financing (TIF) program utilized to help finance the revitalization of downtown Tucson. The Rio Nuevo scheme was "pitched to voters as a public private partnership project to revitalize downtown and recreate Tucson's birthplace, which includes the Mission San Augustin and a portion of the walled Presidio" (Stauffer, 2005, p. 20).

TIF is a special finance mechanism utilized by the state of Arizona to fund municipal improvement projects. However, contrarily to other TIF districts throughout the US, Tucson's is a sales tax rather than a property tax district. This provides a different tax-base to fund projects since the district's revenues are linked to the volume of sales instead of to the annual amount of property taxes paid by all property owners within the district (Balsas and Lathey, 2007). The TIF district is allowed to plan multi-faceted development projects such as cultural and recreational amenities and improvements, historic recreations, mixed-use developments and other projects that enhance and support the mission of the Tucson Convention Center. This is to be achieved by leveraging downtown Tucson's unique competitive advantage as the region's urban and cultural center (Johnson Consulting, 2016, pp. 3–4).

The TIF district is a 500-acre "zone shaped like a craggy iced-tea spoon with an extremely thin handle, where the bowl – the project's focus – starts just west of downtown at Tucson's birthplace near the base of 'A' Mountain and ends about a mile east of downtown" (Stauffer, 2005, p. 19). The western part of the district contains a large parcel of undeveloped land and new mixed-use developments with retail and housing. The eastern part runs down Broadway Boulevard traversing important attractors to the district and downtown Tucson, such as the Tucson Convention Center, the Congress Street Area and other historic buildings, hotels, museums and so on. The TIF funds are applied toward meeting the increased costs of construction and rehabilitation for existing projects like the Fox and Rialto theaters, the Tucson Museum of Art, Tucson Children's Museum and the Presidio Heritage Park, among others (Balsas and Lathey, 2007).

The initial Rio Nuevo ambitious projects of the early 2000s included the building of a long suspension Rainbow Bridge and a Science Center on the large tract of land west of highway I–10 and the Santa Cruz River. These projects have been abandoned partly due to the great recession and the underperforming of the TIF mechanism that was going to be used to pay for their construction. The projects were quite ambitious nonetheless, some of the attractions planned along or near the bridge included: "[A]n observatory, planetarium, giant screen theater, mineral museum, butterfly vivarium, educational resource center, space for touring science exhibits, and a rooftop café with unmatched views of Tucson's mountain scene" (Stauffer, 2005, p. 19).

Lyons and Vasquez (2010) have argued in the Tucson Downtown Partnership 2010 strategic plan entitled "Revitalizing Downtown Tucson: Building the New Pueblo" that the revitalization strategy had to be threefold: First, downtown Tucson had to be considered as part of a larger urban redevelopment area; Second, in order to enhance the uniqueness and authenticity of downtown, small buildings in the area had to also be included in the revitalization efforts. And third, the revitalization had to include private sector reinvestment in the downtown (see Robertson, 1999; Ford, 2003).

DISCUSSION

This discussion is structured according to three geographical scales: Downtown, city, and metropolis. Table 19.2 presents a synthesis of Savitch's (2010) 4 Cs framework applied to the Tucson case study. It demonstrates that urban revitalization is impacted by the planning processes, instruments and leadership at more encompassing levels of analysis and intervention than its rather circumscribed downtown location (Feehan and Feit, 2006).

Currency

Currency has been defined as "the value of something and its ability to carry weight in crucial circumstances" (Savitch, 2010, p. 43). Downtown's organic mix of functions, multiple uses, its walkable scale and various cultural heritage (e.g., St Augustine Cathedral, Fox and Rialto Theaters, Congress Hotel and so on) are great examples of the worth found in downtown Tucson. According to IDA-TDP (2018, p. 35), downtown has a higher Walk-Score (65 versus 42), Transit-Score (60 versus 34), and Bike-Score (94 versus 64) than the entire city. The newly launched Sun Link streetcar connects all six central downtown districts, has increased sustainable transportation options for residents and visitors and acts as a major catalyst for downtown revitalization.

Since 2008, downtown has drawn more than $1.2 billion in public and private investment (IDA-TDP, 2018, p. 21). At the city level, the place's currency is its urban quality of life when compared to for instance the city of Phoenix to the north. The city also benefits from having the University of Arizona – the land grant institution of the state – about 1 mile away from its downtown. Finally, at the metropolitan level Tucson's higher elevation and closer proximity to the nearby mountain ranges than Phoenix give it a slight comparative advantage in terms of quality of urban life.

The Mission San Xavier del Bac and the proximity to the border are two other important elements in the metropolis currency. However, perhaps the most important regional currency is the fact that Tucson is a part of the Sun Corridor megapolitan region that stretches from Nogales to Prescott (Balsas, 2017). The Sun Corridor has its economic core in Phoenix and, collectively, it defines the center of Arizona, while being the largest concentration of economic power in the eight states of the Intermountain West (Hunting et al., 2010; Ziegler, 2009). The same authors also argue that Tucson ought not to compete head-to-head with Phoenix; instead it ought to treat Phoenix and Maricopa as "assets that can promote prosperity in southern Arizona." Hunting et al. (2010) also claim that "development plans need to capitalize on metropolitan Tucson's strongest asset in relation to Phoenix: the incredible natural amenities of southern Arizona coupled with a lifestyle centered on enjoying that environment" (Hunting et al. 2010, p. 6). Thinking about Tucson within this megapolitan scale enables the whole metro area to compete with such large cities as Miami, Atlanta and Seattle (Hunting et al., 2010).

Table 19.2 *Synthesis of the 4 Cs framework applied to the Tucson case study*

	Downtown Tucson	City	Metropolis
CURRENCY (*or worth*)	Organic mix of functions, scale, various cultural heritage (e.g., St Augustine Cathedral, Fox and Rialto theaters, Congress Hotel and so on)	Arizona's second largest city, University of Arizona's land grant institution	Climate, proximity to nature preserves, desert environment; Mission San Xavier del Bac, proximity to the international border
COSMOPOLITANISM (*or pluralism*)	Hispanic and Anglo-Saxon architecture and culture	Hispanic and Anglo-Saxon architecture, language, culture and festivities; and knowledge and technology-oriented advancements	Simultaneous presence of Hispanic, Anglo-Saxon and Indigenous cultures (e.g., Tohono O'odham Nation)
CONCENTRATION (*or compactness*)	Presidio Historic District, historic barrios, Santa Cruz River Greenway Park, modern street car, downtown transit station, affordable housing, warehouse district	Sustainable urbanism, green infrastructure, environmental and social advocacy, Reid Park	Lincoln Regional Park, University of Arizona Science and Technology Park, Civano eco-master planned community, airport
CHARISMA (*or gusto*)	Rio Nuevo and Downtown Tucson Partnership's revitalization plans and leadership	Various city stakeholders such as City of Tucson, University of Arizona, Tucson Chamber of Commerce	Pima Association of Governments, Regional Transportation Authority, Tucson Regional Economic Opportunities, Pima County, Tohono O'odham Nation, Sonoran Institute

Finally, Hunting et al. (2010, pp. 15–19) identified the need for Tucson to "play to our competitive advantages" and those were enumerated as being: (1) protect aspects of the Tucson lifestyle that differentiate it from Phoenix; (2) build, maintain, and manage municipal and outdoor infrastructure; (3) education; (4) develop an industry based on climate-optimized architecture; (5) promote economic links to Mexico; (6) revise governance structures; (7) smart structures for smart growth; (8) separate management from politics; (9) research and quantify the links of the Sun Corridor; and (10) connect Tucson and Phoenix by Inter-City Rail.

Cosmopolitanism

Cosmopolitanism is explained by Savitch (2010, p. 43) as "the ability to embrace international, multicultural or polyethnic features." Tucson has a very strong Hispanic presence in the city and throughout the metropolitan area. The current downtown population is 15,779 people, which represents 3 percent of the city's share and a growth of 5 percent in 2010–18 (IDA-TDP, 2018, p. 18), when the city's population growth during the same time period was only 2 percent. Forty-six percent of residents are non-Hispanic White, 41 percent of residents are Hispanic, 4 percent are Black, 3 percent are American Indian and 3 percent Asian; the downtown workforce is slightly less diverse than the city's residential population (IDA-TDP, 2018, p. 25).

Peterson's (1988) study of Tucson's local symbols and place identity identified ten categories reflecting key physiography, socio-economic, historic and cultural place attributes such as Spanish-Mexican, desert, mountain, sun and cowboy-western. Hispanic culture is proudly displayed in the architecture, language, culture and festivities throughout the city (Cross, 2017). Furthermore, the presence of multinational firms and the university increase the number of knowledge and technology-oriented professionals in the city. Seventeen percent of downtown jobs – about 5,500 jobs – qualify as knowledge industry positions (IDA-TDP, 2018, p. 23). The indigenous presence in this cosmopolitan array of features is a result of the existence of one Native American reservation, the Tohono O'odham Nation (Jojola, 2008).

Concentration

Concentration in the words of Savitch (2010, p. 43) means the "demographic density and productive mass" of a particular place. Tucson's downtown possesses features that contribute to its concentration characteristics. Examples include the Presidio Historic District, the historic barrios, the Santa Cruz River Greenway Park, the modern street car and downtown transit station, affordable housing and the warehouse district (Bischoff, 1995). The daytime population in the downtown area was estimated to be 33,194 in 2015; this represented a 16 percent share of the total employment in the city (IDA-TDP, 2018, p. 21). Downtown Tucson is an economic generator for the entire city. According to the same source, in 2017 downtown had an assessed value of $1.8 billion, which represented 6 percent of the city's total assessed value (IDA-TDP, 2018, p. 21). As of 2015, downtown had 16 percent of the city's jobs or about 33,000 employees.

One less positive feature of concentration in Tucson is the fact that the city's median household income of $38,000 sits well below the US national median of $59,000. Also, according to IDA-TDP (2018, p. 27), downtown's household median is even lower, just above $29,000. While Tucson has a relatively low cost of living, downtown's low median income may be a sign of high poverty in the city. Although as the authors also recognize, the high presence of university students in the downtown area may have skewed these statistics. Furthermore, IDA-TDP (2018, p. 30) has also recognized that "while population across downtown, the city, and the region grew only modestly from 2010 to 2016, downtown's housing inventory boomed. New apartments catering to all income levels and to students have more than doubled since 2010."

Recently, downtown saw the opening of its first new hotel in four decades and several other new hotels are in the pipeline. These investments are being done together with the renovation of the Tucson Convention Center and reflect a slight change in direction of the Rio Nuevo's intervention in the revitalization of the downtown area. At the city and regional scale other measures of concentration include sustainable urbanism projects (Calthorpe, 2010; Meunier, 2012), with green infrastructure (e.g., Civano) (Balsas, 2018), environmental and social advocacy (Grantham, 2012), the existence of large parks such as Reid and Lincoln Regional Park, University of Arizona's Science and Technology Park and Tucson's airport.

Charisma

Charisma is the "magical appeal that generates enthusiasm, admiration or reverence" (Savitch, 2010, p. 44). Downtown Tucson provides many entertainment options and cultural amenities. The Congress Street District serves as the heart of downtown (see Figure 19.1). In fact, Tucson's Congress Street was recognized as a Great Street by the American Planning Association in 2017. Among the reasons for the designation was the fact that "Congress Street speaks to the value of connecting cultural resources to the larger city" (APA, 2017). Other important entertainment venues include Hotel Congress, the lively nightclub Club Congress and the historic Rialto and Fox theaters (Oldenburg, 1999). Tucson has also a strong bicycling culture. The annual El Tour de Tucson – departing and arriving downtown – is one of the largest cycling events in the nation attracting over 9,000 cyclists (IDA-TDP, 2018, p. 30).

Charisma is also frequently equated with leadership in governance. Rio Nuevo's and Downtown Tucson Partnership's leadership have been partially responsible for improving the city's downtown area. However, the current revitalization momentum runs the risk of getting dissipated in the city's and region's broader dynamics, partially under the purview of entities such as the Pima Association of Governments (PAG), the Regional Transportation Authority (RTA), Tucson Regional Economic Opportunities (TREO), Pima County, the Tohono O'odham Nation and the Sonoran Institute.

One important challenge for the metropolitan region and its governance leadership emerged in the late 1990s when "images of Tucson's downtown, once so central to the marketing pitch of boosters, were slowly overtaken by photos of undeveloped desert" (Prytherch, 2002, p. 782). This is framed by the same author as "the second contradiction of capitalism" meaning that:

> Tucson, in plain terms, may be killing the goose (the "natural" landscape it has worked so hard to market) that laid its golden egg (tourism and urban growth). Capital and capitalist local states have helped to establish the desert landscape as a condition for production, and they must then deal with the consequences as sprawl and haze cloud Tucson's scenery. (Prytherch, 2002, p. 787)

At the more localized downtown level, the fact that about 4,500 downtown residents lived in poverty in 2016 (representing a high 36 percent poverty rate), an increase from

Source: Courtesy of Ken Scoville and Carlos Balsas.

Figure 19.1 Congress Street, Tucson's historic street of commerce, circa 1909 (left) and 2007 (right)

3,700 in 2010, and that 47 percent of residents are rent burdened, meaning that they pay more than 36 percent of their income for rent (IDA-TDP, 2018, p. 30), raises additional leadership challenges. Grooms and Boamah (2018, pp. 2–3) have invoked Marcuse's (2009) call for "critical planning to expose the root causes of urban problems, propose programs, targets and strategies to achieve the desired solutions, and politicize by clarifying the political action implications of what was exposed and proposed."

In these authors' opinion this is essential toward exercising more political urban planning that learns from the growth machine – so prevalent in the Phoenix metropolitan area to the north (Vogel and Swanson, 1989) – and the social and environmental advocacy planning interventions in Tucson – for instance, in opposition to fringe suburban growth, water constraints, destruction of pristine desert, greenhouse gas emissions and climate change (Liverman et al., 2013) – and attempts to "*plannitize*" urban politics in effective governance institutions going forward (Acemoglu and Robinson, 2012; Nickerson and Dochuk, 2011).

CONCLUSION

The purpose of this chapter was to reflect on Tucson's urban and regional transformations at the turn of the millennium and to extract a set of lessons learned that can be useful to cities at the same latitude. About a year ago, the *Arizona Republic* newspaper published in Phoenix reported that Tucson was still far behind the Phoenix metropolitan area in recovering from the great recession (Wiles, 2018). Fink (1993, p. 328) has argued that "traditional models of urban form – such as those of the traditional city of the northeast and Europe – do not work for the Sunbelt."

Instead, Fink argued that there is need for a new sunbelt design model centered on the following five requirements: (1) the sunbelt's low densities and emphasis on the single family home; (2) the automobile and its passageways; (3) the sense of openness, which forms the central focus for sunbelt residents; (4) the historic and cultural context; and (5) urbanity (1993). Furthermore, Hunting et al. (2010) have defended that Tucson's unparalleled lifestyle of a livable city surrounded by preserved land fully surpasses that of Maricopa County and ought to be utilized to Tucson's advantage.

This chapter has argued that, although Tucson may have had difficulty obtaining funding and attracting investment prior to the 2007–08 global financial crisis, from a quality of life perspective Tucson's residents appear to benefit from a whole array of small- and medium-size town amenities such as a compact and exciting downtown, affordable housing, alternative modes of transport, cultural institutions, a vibrant arts and culture scene, a land grant university and proximity to protected desert natural environments. The *Arizona Republic* article also recognized that "over the past five years, $1.4 billion in state and local tax money and private investment funds has come into the area." A partial change of frameworks from the historic 4 Cs of state's prosperity: Copper, Cattle, Cotton and Climate to a more operationalized 4 Cs framework of Currency (or worth), Cosmopolitanism (or pluralism), Concentration (or compactness) and Charisma (or emotion) conceptualized by Savitch (2010) was fully discussed above.

This study has identified four lessons learned: First, the scale of urbanization makes Tucson relatively resilient to socio-economic shocks. Second, closer proximity to the border makes residents more understanding of the regions' strong Hispanic cultural heritage. Third, the urban renewal wounds of earlier decades (e.g., demolition of adobe buildings, the building of modern architectonic structures, forced relocation of residents) have been partially ameliorated (i.e., small-scale urban revitalization interventions such as the renovated train station depot, streetscape improvements, a new streetcar, a centrally located transit station, affordable housing, the renovation of the warehouse district, historic preservation efforts and so on). And fourth, home-grown advocacy informed by sensible planning research has strengthened the revitalization vision without curbing developmental ambitions.

NOTE

* I would like to thank the 21 undergraduate students in the Introduction to Urban Planning Studio in Fall 2010 who accepted the challenge of studying Tucson's downtown revitalization. I would also like to thank our Teaching Assistant Amanda Bosse, Tucson Downtown Partnership (TDP) Executive Director Mr Michael Keith and the local historian, Mr Ken Scoville, who readily accepted our invitation to share their knowledge of downtown planning with the group and guided us through the Presidio Historic District and through several other central neighborhoods.

REFERENCES

Abbott, C. (1987). *The New Urban America: Growth and Politics in Sunbelt Cities.* Chapel Hill: University of North Carolina Press.
Acemoglu, D., and Robinson, J.A. (2012). *Why Nations Fail: The Origins of Power, Prosperity, and Poverty.* New York: Crown Books.
Amin, A. (2006). The good city. *Urban Studies*, 43(5/6): 1009–23.
APA (2017). *An APA Great Street 2017: Congress Street Speaks to the Value of Connecting Cultural Resources to the Larger City.* https://www.planning.org/greatplaces/ (accessed 28 June 2019).
Ballas, D. (2013). What makes a "happy city"? *Cities*, 32(suppl. 1 July): S39–S50.
Balsas, C. (1998). Quality of life and city center commercial activity: A southern European study. *International Journal of Iberian Studies*, 11(2): 37–47.
Balsas, C. (2004). Measuring the livability of an urban centre: An exploratory study of key performance indicators. *Planning Practice and Research*, 19(1): 101–10.
Balsas, C. (2017). When markets reset, will we regain? Planning lessons from across the Atlantic Ocean. *Land Use Policy*, 65(June): 78–92.
Balsas, C. (2018). Sustainable urbanism in temperate-arid climates: Models, challenges and opportunities for the Anthropocene. *Journal of Public Affairs*, 18(4) DOI:10.1002/pa.1663.
Balsas, C., and Lathey, V. (2007). Tax Increment Financing and urban revitalization in Arizona (pp. 63–8). In: College of Design (ed.), *Unintended Consequences.* Tempe: College of Design at Arizona State University.
Benner, C., and Pastor, M. (2015). *Equity, Growth, and Community: What the Nation Can Learn from America's Metro Areas.* Oakland, CA: University of California Press.
Bernard, R.M., and Rice, B.R. (1983). *Sunbelt Cities: Politics and Growth since World War II.* Austin: Texas University Press.
Bischoff, A. (1995). Greenways as vehicles for expression. *Landscape and Urban Planning*, 33(1/3): 317–25.
Bohl, C.C. (2002). *Place Making: Developing Town Centers, Main Streets, and Urban Villages.* Washington DC: Urban Land Institute.
Bright, E. (2000). *Reviving America's Forgotten Neighborhoods: An Investigation of Inner-City Revitalization Efforts.* New York: Garland Pub.
Bunnell, G. (2002). *Making Places Special: Stories of Real Places Made Better by Planning.* Chicago: Planners Press.
Calthorpe, P. (2010). *Urbanism in the Age of Climate Change.* Washington: Island Press.
Collins, J. (2001). *Good to Great: Why Some Companies Make the Leap and Others Don't.* New York: Harpers Collins Publishers.

Cross, J.A. (2017). *Ethnic Landscapes of America*. Cham: Springer.
Cunningham, S.P. (2017). Sunbelt identities: The pursuit of place, process, and political sensibilities. *Middle West Review*, 4(1): 63–70.
Diaz, R., Eisele, K., Mackey, B., and Ray, J. (2009). *A Guide to Downtown Master Plans Tucson, Arizona 1932–2009*. Tucson: Worker, Inc., Pop Up Spaces and Design Co*op.
DoE (1994). *Vital and Viable Town Centres: Meeting the Challenge*. London: HMSO.
Feehan, D., and Feit, M. (eds) (2006). *Making Business Districts Work*. New York: The Haworth Press.
Fink, M. (1993). Toward a sunbelt urban design manifesto. *Journal of the American Planning Association*, 59(3): 320–33.
Ford, L.R. (2003). *America's New Downtowns: Revitalization or Reinvention?* Baltimore: Johns Hopkins Press.
Friedmann, J. (2000). The good city: In defense of utopian thinking. *International Journal of Urban and Regional Research*, 24(2): 460–72.
Friedmann, J. (2002). *The Prospect of Cities*. Minneapolis: University of Minnesota Press.
Gladwell, M. (2005). *Blink: The Power of Thinking without Thinking*. New York: Little, Brown and Co.
Gober, P. (2006). *Metropolitan Phoenix Place Making and Community Building in the Desert*. Philadelphia: University of Pennsylvania Press.
Gomez-Novy, J., and Polyzoides, S. (2003). A tale of two cities: The failed urban renewal of downtown Tucson in the twentieth century. *Journal of the Southwest*, 45(1/2): 87–119.
Grantham, D. (2012). Tucson works together, gets crisis center and care system right: Fast-growing city "raises the bar on quality of life" with new Crisis Response Center. *Behavioral Healthcare*, 32(4): 54–60.
Grooms, W., and Boamah, E. (2018). Toward a political urban planning: Learning from growth machine and advocacy planning to "plannitize" urban politics. *Planning Theory*, 17(2): 213–33.
Hinshaw, M. (2007). *True Urbanism: Living In and Near the Center*. Washington DC: Planners Press.
Hunting, D., Kalt, J., and Propst, L. (2010). *Tucson's New Prosperity: Capitalizing on the Sun Corridor*. Tucson: Sonoran Institute.
IDA-TDP (2018). *The Value of U.S. Downtown and Center Cities: Tucson Case Study*. Washington DC: International Downtown Association.
Introductory Urban Planning Studio (IUPS) (2010). *Revitalizing Downtown Tucson* (unpublished report). Tempe: School of Geographical Sciences and Urban Planning: Arizona State University.
Jacobs, A. (2012). *The Good City: Reflections and Imaginations*. New York: Routledge.
Johnson Consulting (2016). *Rio Nuevo Multipurpose Facilities District: Tucson, AZ*. Chicago: Johnson Consulting.
Jojola, T. (2008). Indigenous planning: An emerging context. *Canadian Journal of Urban Research*, 17(1): 37–47.
Kahneman, D. (2011). *Thinking, Fast and Slow*. New York: Farrar, Straus and Giroux.
Kirby, A. (2000). Moving from evaluation to regulation: Urban governance and the improvement of QOL (pp. 226–39). In: Foo, T. Lim, L., and Wong, G. (eds), *Planning for a Better Quality of Life in Cities*. Singapore: National University of Singapore.
Lara, J. (2018). *Latino Placemaking and Planning: Cultural Resilience and Strategies for Reurbanization*. Tucson: University of Arizona Press.
Le Tourneau, F.M., and Dubertret, F. (2019). L'espace et l'eau, variables clés de la croissance urbaine dans le Sud-Ouest des États-Unis: le cas de Tucson et du Pima County (Arizona). *L'Espace Géographique*, 48(1): 39–56.
Leinberger, C. (2008). *The Option of Urbanism: Investing in a New American Dream*. Washington DC: Island Press.
Liverman, D., Moser, S.C., Weiland, P.S., Dilling, L., Boykoff, M.T., Brown, H.E., Gordon, E.S., Greene, C., Holthaus, E., Niemeier, D.A., Pincetl, S., Steenburgh, W.J., and Tidwell, V.C. (2013). Climate choices for a sustainable southwest (pp. 405–35). In: Garfin, G., Jardine, A., Merideth, R., Black, M., and LeRoy, S. (eds), *Assessment of Climate Change in the Southwest United States: A Report Prepared for the National Climate Assessment*. Washington, DC: Island Press.
Logan, M.F. (1995). *Fighting Sprawl and City Hall: Resistance to Urban Growth in the Southwest*. Tucson: University of Arizona Press.
Luckingham, B. (1982). *The Urban Southwest: A Profile History of Albuquerque, El Paso, Phoenix, Tucson*. El Paso: The University of Texas at El Paso.
Lynch, K. (1981). *A Theory of Good City Form*. Cambridge: MIT Press.
Lyons, G., and Vasquez, T. (2010). *Revitalizing Downtown Tucson: Building the New Pueblo*. Tucson: Downtown Tucson Partnership.
Marcuse, P. (2009). From critical urban theory to the right to the city. *City*, 13(2–3): 185–97.
Markusen, A., Yong-Sook, L., and DiGiovanna, S. (eds) (1999). *Second Tier Cities: Rapid Growth beyond the Metropolis*. Minneapolis: University of Minnesota Press.
Marston, S., and Saint-Germain, M. (1991). Urban restructuring and the emergence of new political groupings: Women and neighborhood activism in Tucson, Arizona. *Geoforum*, 22(2): 233–36.

McAslan, D., Prakash, M., Pijawka, D., Guhathakurta, S., and Sadalla, E. (2013). Measuring quality of life in border cities: The border observatory project in the US-Mexico border region (pp. 143–69). In: Sirgy, M., Phillips, R., and Rahtz, D. (eds), *Community Quality-of-Life Indicators: Best Cases VI*. London: Springer.

Meunier, J. (2012). Making desert cities sustainable (pp. 107–21). In: Pijawka, D., and Gromulat, M. (eds), *Understanding Sustainable Cities: Concepts, Cases, and Solutions*. Dubuque: Kendall Hunt Publishing.

Montgomery, J. (1998). Making a city: Urbanity, vitality and urban design, *Journal of Urban Design*, 3(1): 93–116.

Newman, P., Beatley, T., and Boyer, H. (2016). *Resilient Cities: Overcoming Fossil Fuel Dependence*. Second edition. Washington: Island Press.

Ng, M.K., and Hills, P. (2003). World cities or great cities? A comparative study of five Asian metropolises. *Cities*, 20(3): 151–65.

Nickerson, M., and Dochuk, D. (eds) (2011). *Sunbelt Rising: The Politics of Space, Place, and Region*. Philadelphia: University of Pennsylvania Press.

Norquist, J.O. (1998). *The Wealth of Cities: Revitalizing the Centers of American Life*. Reading: Perseus Books.

Oldenburg, R. (1999). *The Great Good Place*. New York: Marlowe.

Otero, L.R. (2010). *La Calle: Spatial Conflicts and Urban Renewal in a Southwest City*. Tucson: University of Arizona Press.

PAG (Pima Association of Governments) (2010). *Regional Greenhouse Gas Inventory 1990–2010*. Tucson: Pima Association of Governments.

Paumier, C. (2004). *Creating a Vibrant City Center*. Washington: Urban Land Institute.

Peterson, G. (1988). Local symbols and place identity: Tucson and Albuquerque. *The Social Science Journal*, 25(4): 451–61.

Prytherch, D. (2002). Selling the eco-entrepreneurial city: Natural wonders and urban stratagems in Tucson, Arizona. *Urban Geography*, 23(8): 771–93.

Robertson, K.A. (1999). Can small-city downtowns remain viable? A national study of development issues and strategies. *Journal of the American Planning Association*, 65(3): 270–83.

Savitch, H.V. (2010). What makes a great city great? An American perspective. *Cities*, 27(1): 42–9.

Speck, J. (2012). *Walkable City: How Downtown Can Save America One Step at a Time*. New York: North Point Press.

Sperling, B., and Sander, P. (2004). *Cities Ranked & Rated*. Hoboken: John Wiley & Sons.

Stauffer, T. (2005). Downtown at the end of the rainbow. *Planning*, December: 18–21.

Strom, E. (2017). How place matters: A view from the Sunbelt. *Urban Affairs Review*, 53(1): 197–209.

Vogel, R., and Swanson, B. (1989). The growth machine versus the antigrowth coalition: The battle for our communities. *Urban Affairs Review*, 25(1): 63–85.

Whyte, W. (1988). *City: Rediscovering the Center*. New York: Doubleday Publishing Group, Inc.

Wiles, R. (2018). Tucson still far behind Phoenix metro area in recovery from great recession. *Arizona Republic*, 28 June.

Williams, J. (2011). To transform the inner city: Tucson's Model Cities Program, 1969–1975. *The Journal of Arizona History*, 52(2): 143–68.

Ziegler, E.H. (2009). The case for megapolitan growth management in the twenty-first century: Regional urban planning and sustainable development in the USA. *International Journal of Law in the Built Environment*, 1(2): 105–29.

PART III

EMERGING CONSTRUCTS AND THE FUTURE OF COMMUNITY DEVELOPMENT RESEARCH

20. Theories and concepts influencing sustainable community development
Maria Spiliotopoulou and Mark Roseland

INTRODUCTION

A sustainable future requires twinning improvements in quality of life with decreased consumption of materials and non-renewable resources. How can we chart our way conceptually and operationally in this unfamiliar territory? In the first section of this chapter, we briefly trace the history of sustainable development (SD) and sustainable community development (SCD) theory and practice, and demonstrate their interdisciplinary nature and influences by notable theories and concepts in fields such as ecology, economics and other social and natural sciences. In the second section, we present community productivity as an emerging theory of SCD and we argue that achieving SCD is possible through increased multi-factor productivity in the community. We explore how enhancing multiple forms of community capital can contribute to community productivity and well-being, and provide examples of community productivity metrics and current initiatives from around the world.

DEVELOPING A SUSTAINABLE COMMUNITY

SD emerged as a field of study after the 1987 Brundtland Commission report showed the interconnectedness between human activities and increasing environmental degradation: 26 percent of the world's population living in developed countries consumed 80–86 percent of non-renewable resources and 34–53 percent of food products (WCED, 1987). The Brundtland Report first popularized the term, stating that SD is "development that meets the needs of the present without compromising the ability of future generations to meet their own needs" (WCED, 1987, p. 41).

SD has been criticized as ambiguous and open to contradictory interpretations (Roseland and Spiliotopoulou, 2016), but most definitions present common characteristics: integration of environmental, economic and social aspects; systems thinking; and dynamic nature (Berke and Conroy, 2000). SD has represented a new way of thinking about economic development: "doing development differently" (Roseland, 2012, p. 3). This holistic view of sustainability is reflected in the new UN development agenda, aiming to tackle poverty, climate change and inequality in developed and developing countries (United Nations, 2015b), in the 2015 UN (Paris) Climate Change agreement (United Nations, 2015a), and in the UN New Urban Agenda (United Nations, 2017).

Importance of Communities and Urban Areas

Our research focus is sustainability at the community level, particularly in urban areas, where key components of both challenges and solutions are increasingly recognized (Roseland and Spiliotopoulou, 2016). For our research purposes, a community refers to "a group of people bound by geography and with a shared destiny, such as a municipality or a town" (Roseland, 2012, p. 12), and is considered as a complex, adaptive and interconnected system, requiring interdisciplinary study (Uphoff, 2014). An urban area is "a human settlement characterized – ecologically, economically, politically and culturally – by a significant infrastructural base; a high density of population, whether it be as denizens, working people, or transitory visitors; and what is perceived to be a large proportion of constructed surface area relative to the rest of the region"(James, 2015).

The UN Global Agenda for 2030 includes a goal for "inclusive, safe, resilient, and sustainable" cities, since they occupy 3–4 percent of the world's land surface, use 80 percent of resources, discharge most global waste (Girardet, 2015) and will be host to two-thirds of the world's population by 2050 (UN DESA, 2018). Urban areas are increasingly vulnerable to climate change and health challenges, and are linked to increased costs to the economy and the environment (Kanuri et al., 2016). The growing awareness that achieving sustainability requires societal change through collaborative decision-making and community engagement has brought SCD to the foreground (Clarke, 2012; Hermans et al., 2011). SCD is a holistic approach that integrates social, environmental and economic considerations into the dynamic processes and actions of communities on their path toward sustainability, while providing for current and future generations (Berke and Conroy, 2000; Roseland, 2012).

Theories and Concepts Influencing SCD

SCD, along with SD, has been influenced by a number of theories and has matured over the last few decades in academic, professional and popular discourse. Early work was carried out within a weak sustainability framework based on the assumptions that: humans should dominate over nature, natural resources are super-abundant and economic growth can continue indefinitely through resource efficiencies (Ayres, 2007; Solow, 1993; Williams and Millington, 2004).

Within this framework, SCD policies and initiatives have also been informed by ecological modernization and resource efficiency or eco-efficiency theories, aiming to create more efficient production processes and designs (Roseland and Spiliotopoulou, 2016). However, weak sustainability has been criticized for not incorporating important issues such as social equality, environmental justice, population trends and inter- and intra-generational equity (Agyeman et al., 2002; Bayulken and Huisingh, 2015).

In the recent years, SD and SCD have been gradually moving away from the uncertainties and debates involved in the weak sustainability viewpoint (Williams and Millington, 2004) toward a stronger sustainability model which acknowledges both the finite character of natural resources and the Earth's regenerative limits (Daly, 2005; Rockström et al., 2009), and also the need for socio-ecological and economic resilience "across temporal and spatial scales" (Meerow et al., 2016).

Community Economic Development (CED), Eco-localism, and Social Economy (SE) initiatives emerged as a community response to the negative impacts of weak sustainability and they evolved rapidly, from simple forms of local economic activity reflecting social or cultural values to ventures addressing broader social needs and environmental well-being (Gismondi et al., 2016; Hernandez, 2015). Under a strong sustainability model, social and ecological considerations are included in community analysis and policy-making through collaborative and systemic processes.

Weaknesses in Current Approaches on Sustainable Community Development

Despite the conceptual evolution of sustainability over the last decades, policies and initiatives have not always involved a balanced approach between environmental, economic and social concerns. The multitude of definitions and the lack of shared language or understanding of SD and SCD have contributed to limited and inconsistent application of sustainability principles through a variety of local agendas grounded in diverse theoretical backgrounds and frequently reflecting specific stakeholders' interests (Joss et al., 2015; Kristensen and Roseland, 2012; Roseland and Spiliotopoulou, 2017). Meanwhile, SE and CED initiatives have been criticized for operating inside the capitalist system without trying to change the system's rules (Roseland and Spiliotopoulou, 2016). At the same time, not all efficiencies translate to reduced resource extraction and consumption since other factors, such as population growth and industry interests, are also at play (Ang and Van Passel, 2012; Kopnina, 2015).

In the pursuit of sustainability or well-being, communities are challenged by the difficulties of addressing multiple objectives, thinking strategically and holistically about high-level goals, and meaningfully engaging their citizens, while also assessing projects and policies and tracking progress consistently (Caprotti et al., 2017; Connelly et al., 2013). The current abundance of SCD plans, assessment tools and community networks demonstrates acknowledgement of the need to take action and the desire to cooperate and exchange knowledge. Not all plans and agendas promote a whole-systems approach or are followed by implementation strategies, thus leading to lost opportunities, lack of credibility and increased public scepticism (Cairns et al., 2015; Roseland, 2012).

DEVELOPING A PRODUCTIVE COMMUNITY

Despite historical and theoretical debates as well as practical weaknesses, SCD should not be understood as a series of trade-offs between social, environmental and economic priorities; protecting ecosystems and promoting social inclusion at the local level need not mean job loss or economic downturn. Rather, SCD represents a new way of thinking about economic and other development over the long term: it is about "doing development differently" (Roseland, 2012). It requires fundamental changes to the status quo to stop "sustaining" an ill-functioning – and thus unsustainable – system and business-as-usual operations driven by quantitative increases, in favour of achieving meaningful improvements to community well-being, including the natural environment (Roseland and Spiliotopoulou, 2017). As Neuman explains, "to sustain an ecosystem or city over the long run assumes that it will be healthy" (Neuman, 2005).

Sustainability has in recent years expanded its scope to embrace advancements in resource and labour productivity (Jackson and Victor, 2011), collective action and SE (Connelly et al., 2013), local resilience, reorganization, self-reliance (Brugmann and Mohareb, 2012; Folke, 2006; Meerow et al., 2016) and resource regeneration (Robinson and Cole, 2015), as well as policies inspired by "just" sustainability (Agyeman, 2008) and a "shared ethical framework" (Earth Charter Initiative, 2010). Businesses have started to adopt green economy practices for efficiency in technology, design and management, and to promote green jobs (Kouri and Clarke, 2012), and communities are finding that they can actively pursue SD while improving their economic indices (Portney, 2013).

We suggest a way to overcome current problems in community sustainability planning and implementation, through a shift in mindset and subsequent action: from the current demanding, resource-extracting model of business-as-usual to a systemic, resource-regenerative model of a productive – and eventually sustainable – city. This transition, which has appeared in the SCD literature lately, involves shifting community development from a negative individualistic logic (reducing impact) to a positive systemic one (regeneration within a network of systems) (Brugmann, 2015; Girardet, 2015) and has the potential to contribute to achievement of sustainability goals so that the system we "sustain" thereafter is a well-functioning one.

During this shift, community, people and environment would be involved in a co-evolutionary process, engaging all related systems, sub-systems and stakeholders (Neuman, 2005). Traditional economic growth, based on weak sustainability principles, advises cities to maintain or increase their economic output by improving technology, accumulating capital and enhancing labour productivity. However, urban space that is planned using strong sustainability principles can lead to increases in human, resource and process productivity, improved urban assets performance and systemic interactions, ecological function regeneration and efficient use of resources (Brugmann, 2015; Girardet, 2015).

Community Productivity Conceptually

The concept of productivity is usually associated with economic and other resources: although economic, labour and resource productivity are quite developed concepts, there is not one widely accepted definition for ecological or social productivity within the urban setting. It is well recognized that economic and labour productivity is higher in cities that attract agglomeration economies and high-skilled employees – in developed and developing countries alike (Abel et al., 2012; Behrens et al., 2015; Glaeser and Xiong, 2017). We therefore posit that productivity has potential for great uptake by communities given its resonance with people as relevant to everyday life.

Conceptually, community productivity is multi-dimensional in the same way as sustainability and is grounded in strong sustainability principles. It seeks to move past the idea and practice of balance among community priorities toward the maximization and regeneration of the various forms of community capital. These forms represent the tangible and intangible assets and aspects of a community beyond the traditional triple-bottom line of SD. SCD incorporates natural, physical, economic, human, social and cultural dimensions of community development (Roseland, 2012).

Community productivity encompasses concepts and practices from various disciplines and theoretical backgrounds, including the above-mentioned notion of labour productivity. Another important concept, regenerative design is a decades-long concept rooted in ecology and living systems theory (Robinson and Cole, 2015) and has been applied to agricultural (Rodale Institute, 2014) and architectural practices (Thomson and Newman, 2018). This concept is also expressed through applications of net-zero and net-positive design, which are promising despite often being implemented with a mostly anthropocentric and technical focus (Mang and Reed, 2015).

Regenerative development principles constitute a more recent and broader way of thinking that emerged partly out of the necessity to develop or redevelop the built environment with a holistic socio-ecological systems worldview that is aligned and synergistic with the natural environment and resources (Mang et al., 2016; Mang and Reed, 2012; Robinson and Cole, 2015). Similarly, the concept of regenerative sustainability appeared in the literature recently to emphasize the need for processes of collaborative planning and participatory backcasting which ensure that all partners' perspectives are considered, including that of the natural environment and its intrinsic value (Robinson and Cole, 2015).

Economic and resource circularity is evolving rapidly, with various initiatives implemented around the world, mostly at local or sectoral scales. Most circular economy models urge resource regeneration, using waste as a resource, and closing technical cycles in production and consumption (Ellen MacArthur Foundation, 2017; World Economic Forum, 2018). By introducing the notion of productivity, circular economy approaches would not only imply resource extraction at a lower rate than that of resource regeneration, but they would also contribute to the recovery and restoration of the natural environment and improve social and human aspects (Geissdoerfer et al., 2017). Ecological productivity can be enhanced in an urban area and in its hinterland, thus adding value to the ecosystem and making up for the damage done during the Anthropocene, instead of extracting or simply maintaining the current balance (Mang and Reed, 2015).

Productive community development is not limited to economic and ecological concepts and practices. It also explicitly focuses on the social, human and cultural dimensions of community capital. Cultivating a sense of place is a fundamental component not only of an inclusive education but also of strong, healthy and regenerative communities. As David Orr explains, looking at the world through the lens of place promotes a sense of responsibility for and a sense of unity with the natural environment (Mang et al., 2016; Orr, 2013).

Likewise, reclaiming the urban commons and developing a sense of community within a city and its neighbourhoods can help build social capital and take it a step further than established notions of social equity and environmental equality toward newer notions of place-making and co-production of public space (Burden, 2014), of sharing common places in a city, and fulfilling the right to the city for every person and for nature and biodiversity (Agyeman and McLaren, 2017). By seeing the city as an ever-evolving organism and by making the city commons (built or natural) more open, creative and attractive, these spaces can be accessible for and inclusive of all – regardless of origin or status – and inviting to "produce" culture and social relationships (Landry, 2008; Smithsimon, 2008; Wahl, 2016).

Concepts that apply in productive community development are closely intertwined with systems thinking (Wahl, 2016). For instance, even though the UN Habitat report on

342 *Research handbook on community development*

The City We Need 2.0 (United Nations, 2016) includes a principle on creating a regenerative and resilient city, it only refers to resource regeneration and infrastructure/energy resilience. Community productivity can converge all the principles in this UN Habitat manifesto in the spirit of the concepts and theories mentioned above. For a complete application of systems theory in this approach, a community needs to carefully analyse the higher-level systems and networks to which it belongs and the sub-systems of which it is composed, and explore perspectives from many disciplines (Meadows, 2008; Uphoff, 2014).

Community Productivity Operationally

Operationally these concepts, combined under the umbrella of productivity, can inform and guide community visions and goals as well as the methods and metrics required to assess progress. Enhancing community productivity entails investment by many community actors to achieve improvements in all forms of community capital (Spiliotopoulou, 2019):

- Natural productivity, through ecological resource management, biodiversity and habitat restoration, local and regional food systems enhancement, promotion of urban and peri-urban regenerative agriculture
- Social productivity, through equity, connectedness, tolerance, inclusion, sharing of the commons, effective governance and justice, safety, individual and social resilience
- Cultural productivity, through heritage, arts, traditions, cultural continuity
- Human productivity, through lifelong learning, skills development and leverage, happiness and personal fulfillment, health
- Resource productivity, through infrastructure resilience, technological connectedness, resource efficient land-use and waste management, increased efficiencies, resource circularity and other regenerative practices
- Economic productivity, through economic diversity and resilience, inclusive economy, innovation fostering, living affordability.

Context matters when planning for and implementing community productivity strategies and actions. Political and other priorities and goals differ, and so do the issues and the decision-making processes. Best practices may not be transferable or easily implemented in every community and therefore contextual analysis and adaptable solutions are required (Roseland and Spiliotopoulou, 2017).

The concept of productivity can also contribute to measuring a community's progress toward achieving sustainability goals, by using objective (data-driven) and subjective (survey-based) information collected within a framework of productivity metrics. A productive community would seek to regenerate its resources of various types by being net-positive, that is, by producing more capital than it consumes. This can be measured using a combination of effective and widely accepted indicators of sustainability and several new indicators specifically geared toward the productive and regenerative aspects of the community. Here are some examples of community productivity indicators (Spiliotopoulou, 2019):

Theories and concepts influencing sustainable community development 343

- Growing space per dwelling unit
- Native plant preservation
- Energy from renewable resources (solar, wind, biomass, and so on), produced within the city or in adjacent cities
- Mix of land uses and compact development
- Regeneratively designed buildings (net-positive, energy label houses)
- Infill development
- Mix of housing options
- Modal split/share
- Industrial/commercial/construction solid waste reuse/recycle
- Local innovation (patents)
- Sustainable resource based jobs (in sustainable/organic agriculture, natural resource conservation or coastal restoration)
- Creative industry jobs
- Work opportunities for people with developmental disabilities
- Lifelong learning (vocational or other adult education opportunities)
- Positive individual health practices
- Life satisfaction and/or happiness perception
- Women and vulnerable populations' access to government
- Social service volunteering
- Confidence in local government
- Healthy and safe neighbourhood development initiatives
- Cultural access and participation
- Investment in public art and public art awareness
- Historic preservation initiatives

Community Productivity Examples

In our literature and practice review, we have identified numerous initiatives of community productivity. Examples of small-scale productivity cases exist in municipalities such as Adelaide, Australia (efficient use of local resources, dynamic public consultations, major organic waste composting schemes, impressive renewable energy development), Copenhagen, Denmark (energy efficiency initiatives, public transit and cycling uptake, extensive information campaigns and debates, exemplary waste management), The Hague (Central Innovation District, currently under development) and Amsterdam, the Netherlands (successful sharing and collaborative economy ventures), Bristol, UK (renewable energy initiatives, successful civil society partnerships, climate resilience actions), Medellin, Colombia (inclusive social practices, long-term participatory planning, efficient transportation system), Kigali, Rwanda (leader in knowledge-based sharing economy) and Guangzhou, China (cultural and social inclusion initiatives, large-scale urban development programs, efficient wastewater management) (Girardet, 2015; Razavi, 2017, 2018; Urban Innovation Community, 2015; WA Contents, 2018; Wahl, 2016).

Regenerative practices also exist around the world within specific sectors, such as energy and built environment (e.g. Beddington Zero Energy Development in the UK, Masdar in Abu Dhabi, Portland, Oregon's Eco-District initiative, the Arbed scheme in Wales, UK, or districts in Freiburg and Hamburg, Germany), social and human capital (personal and skills development research by the Theory U Lab at MIT, safety and well-being initiatives by the 8 80 Cities organization based in Toronto, Canada), natural environment protection and restoration (numerous regeneration projects mapped by Spherical Studio based in Oakland, California), and agriculture (e.g. urban farming programs in Havana, Cuba, community gardens in New York City and elsewhere, or energy efficient and hydroponic use of farmland in Shanghai and Beijing, China) (8 80 Cities, n.d.; Girardet, 2015; Hunt and de Laurentis, 2015; Roseland, 2012; Scharmer, 2018; Spherical Studio, n.d.).

CONCLUSION: THE PRODUCTIVE COMMUNITY POTENTIAL

As urban areas continue to grow and extract resources, they impose a disproportionate impact on the biosphere while suffering from economic and social challenges within their boundaries (Newman and Jennings, 2008). The significance of developing – as opposed to only growing – urban assets is increasingly being recognized by SCD researchers and practitioners. The traditional notion of urban economic growth is based on weak sustainability principles, but this century's realities and planetary constraints require that urban development is guided by strong sustainability values and whole-systems thinking.

Future community development research should emphasize performance enhancement, impact and user benefit increase, human productivity strengthening, effective and inclusive decision-making processes, co-production of community space and efficient use and regeneration of resources. These are values and outcomes that are not always recognized or successfully implemented or assessed using current SCD agendas (Clarke, 2012; du Plessis, 2012; Joss et al., 2015; Newman and Jennings, 2008; Roseland, 2012).

The application of community productivity principles has real potential to increase quality of life despite and perhaps in part because of lower levels of material expectations ("standard of living") by increasing the well-being of current and future generations. A productive community can be simultaneously livable, resilient, healthy, smart, regenerative, safe, creative and happy (Brugmann, 2015). These positive results can raise this dynamic paradigm to be the new normal for communities to sustain and achieve locally relevant long-term sustainability goals while also contributing to the achievement of the 2030 Sustainable Development Goals.

REFERENCES

8 80 Cities (n.d.). 8 80 Cities. Retrieved 27 December 2018 from https://www.880cities.org/.
Abel, J.R., Dey, I., and Gabe, T.M. (2012). Productivity and the density of human capital. *Journal of Regional Science*, *52*(4), 562–86. https://doi.org/10.1111/j.1467-9787.2011.00742.x.
Agyeman, J. (2008). Toward a "just" sustainability? *Continuum*, *22*(6), 751–6. https://doi.org/10.1080/1030431080 2452487.
Agyeman, J., and McLaren, D. (2017). Sharing cities. *Environment: Science and Policy for Sustainable Development*, *59*(3), 22–7. https://doi.org/10.1080/00139157.2017.1301168.
Agyeman, J., Bullard, R.D., and Evans, B. (2002). Exploring the nexus: bringing together sustainability, environmental justice and equity. *Space and Polity*, *6*(1), 77–90. https://doi.org/10.1080/13562570220137907.
Ang, F., and Van Passel, S. (2012). Beyond the environmentalist's paradox and the debate on weak versus strong sustainability. *BioScience*, *62*(3), 251–9. https://doi.org/10.1525/bio.2012.62.3.6.
Ayres, R.U. (2007). On the practical limits to substitution. *Ecological Economics*, *61*(1), 115–28. https://doi.org/10.1016/j.ecolecon.2006.02.011.

Bayulken, B., and Huisingh, D. (2015). A literature review of historical trends and emerging theoretical approaches for developing sustainable cities (part 1). *Journal of Cleaner Production*, *109*, 11–24. https://doi.org/10.1016/j.jclepro.2014.12.100.

Behrens, K., Duranton, G., and Robert-Nicoud, F. (2015). Productive cities: sorting, selection, and agglomeration. *Journal of Political Economy*, *122*(3), 507–53.

Berke, P.R., and Conroy, M.M. (2000). Are we planning for sustainable development? *Journal of the American Planning Association*, *66*(1), 21–33. https://doi.org/10.1080/01944360008976081.

Brugmann, J. (2015). The urban productivity imperative. In *The Productive City: Growth in a No-Growth World* (pre-publication draft). Toronto: The Next Practice Ltd.

Brugmann, J., and Mohareb, E. (2012). The productive city: defining the practices of urban ecological development. Paper presented at the 8th World Congress of ICLEI – Local Governments for Sustainability, Belo Horizonte, Brazil, 14–17 June.

Burden, A. (2014). How public spaces make cities work. *Ted2014*. TED Talk. Retrieved 4 January 2019 from http://www.ted.com/talks/amanda_burden_how_public_spaces_make_cities_work.

Cairns, S., Clarke, A., Zhou, Y., and Thivierge, V. (2015). *Sustainability Alignment Manual (SAM)*. Ottawa and Waterloo: Sustainable Prosperity and University of Waterloo.

Caprotti, F., Cowley, R., Datta, A., Broto, V.C., Gao, E., Georgeson, L., Herrick, C., Odendaal, N., and Joss, S. (2017). The New Urban Agenda: key opportunities and challenges for policy and practice. *Urban Research and Practice*, *10*(3), 367–78. https://doi.org/10.1080/17535069.2016.1275618.

Clarke, A. (2012). *Green Municipal Fund Passing Go: Moving Beyond the Plan*. Ottawa: Federation of Canadian Municipalities.

Connelly, S., Markey, S., and Roseland, M. (2013). We know enough: achieving action through the convergence of sustainable community development and the social economy. In *The Economy of Green Cities* (pp. 191–203). https://doi.org/10.1007/978-94-007-1969-9.

Daly, H.E. (2005). Economics in a full world. *Scientific American*, *293*(3), 100–107. https://doi.org/10.1038/scientificamerican0905-100.

du Plessis, C. (2012). Towards a regenerative paradigm for the built environment. *Building Research & Information*, *40*(1), 7–22. https://doi.org/10.1080/09613218.2012.628548.

Earth Charter Initiative (2010). The Earth Charter. Retrieved 9 September 2018 from earthcharter.org/discover/the-earth-charter/.

Ellen MacArthur Foundation (2017). *Cities in the Circular Economy: An Initial Exploration*. Cowes, UK: Ellen MacArthur Foundation.

Folke, C. (2006). Resilience: the emergence of a perspective for social–ecological systems analyses. *Global Environmental Change*, *16*(3), 253–67. https://doi.org/10.1016/j.gloenvcha.2006.04.002.

Geissdoerfer, M., Savaget, P., Bocken, N.M.P., and Hultink, E.J. (2017). The circular economy: a new sustainability paradigm? *Journal of Cleaner Production*, *143*(1), 757–68. https://doi.org/10.1016/j.jclepro.2016.12.048.

Girardet, H. (2015). *Creating Regenerative Cities*. Abingdon, Oxon, and New York: Routledge.

Gismondi, M., Connelly, S., Beckie, M., Markey, S., and Roseland, M. (2016). *Scaling Up*. Edmonton, Alberta: Athabasca University Press.

Glaeser, E.L., and Xiong, W. (2017). Urban productivity in the developing world. *Oxford Review of Economic Policy*, *33*(3), 373–404. https://doi.org/10.1093/oxrep/grx034.

Hermans, F.L.P., Haarmann, W.M.F., and Dagevos, J.F.L.M.M. (2011). Evaluation of stakeholder participation in monitoring regional sustainable development. *Regional Environmental Change*, *11*(4), 805–15. https://doi.org/10.1007/s10113-011-0216-y.

Hernandez, G. (2015). From spaces of marginalization to places of participation: indigenous articulations of the social economy in the Bolivian Highlands (Unpublished doctoral dissertation). Simon Fraser University, Canada.

Hunt, M., and de Laurentis, C. (2015). Sustainable regeneration: a guiding vision towards low-carbon transition? *Local Environment: The International Journal of Justice and Sustainability*, *20*(9), 1081–102. https://doi.org/10.1080/13549839.2014.894964.

Jackson, T., and Victor, P. (2011). Productivity and work in the "green economy." *Environmental Innovation and Societal Transitions*, *1*(1), 101–8. https://doi.org/10.1016/j.eist.2011.04.005.

James, P. (2015). *Urban Sustainability in Theory and Practice: Circles of Sustainability*. Abingdon, Oxon, and New York: Earthscan Publications Ltd.

Joss, S., Cowley, R., De Jong, M., Müller, B., Park, B.S., Rees, W.E., Roseland, M., Rydin, Y. (2015). *Tomorrow's City Today: Prospects for Standardising Sustainable Urban Development*. London: University of Westminster.

Kanuri, C., Revi, A., Espey, J., and Kuhle, H. (2016). *Getting Started with the SDGs in Cities: A Guide for Stakeholders (UN SDSN)*. Retrieved 29 September 2016 from http://unsdsn.org/resources/publications/getting-started-with-the-sdgs-in-cities/.

Kopnina, H. (2015). The victims of unsustainability: a challenge to sustainable development goals. *International*

Journal of Sustainable Development & World Ecology, *32*(2), 1–9. https://doi.org/10.1080/13504509.2015.111 1269.
Kouri, R., and Clarke, A. (2012). Framing "green jobs" discourse: analysis of popular usage. *Sustainable Development*, *22*(4). https://doi.org/10.1002/sd.1526.
Kristensen, F., and Roseland, M. (2012). Mobilising collaboration with Pando | Sustainable Communities. *Local Environment: The International Journal of Justice and Sustainability*, *17*(5), 517–23. https://doi.org/10.1080/13549839.2014.921388.
Landry, C. (2008). *The Creative City: A Toolkit for Urban Innovators*. London: Routledge.
Mang, P., and Reed, B. (2012). Designing from place: a regenerative framework and methodology. *Building Research and Information*, *40*(1), 23–38. https://doi.org/10.1080/09613218.2012.621341.
Mang, P., and Reed, B. (2015). The nature of positive. *Building Research and Information*, *43*(1), 7–10. https://doi.org/10.1080/09613218.2014.911565.
Mang, P., Haggard, B., and Regenesis (2016). *Regenerative Development and Design: A Framework for Evolving Sustainability*. Hoboken, New Jersey: John Wiley & Sons.
Meadows, D.H. (2008). *Thinking in Systems: A Primer*. London and Washington, DC: Earthscan Publications Ltd.
Meerow, S., Newell, J.P., and Stults, M. (2016). Defining urban resilience: a review. *Landscape and Urban Planning*, *147*, 38–49. https://doi.org/10.1016/j.landurbplan.2015.11.011.
Neuman, M. (2005). The compact city fallacy. *Journal of Planning Education and Research*, *25*(1), 11–26. https://doi.org/10.1177/0739456X04270466.
Newman, P., and Jennings, I. (2008). *Cities as Sustainable Ecosystems: Principles and Practices*. Washington, DC: Island Press.
Orr, D.W. (2013). Place and pedagogy. *The NAMTA Journal*, *38*(1), 183–8. Retrieved 4 January 2019 from https://files.eric.ed.gov/fulltext/EJ1078034.pdf.
Portney, K.E. (2013). Local sustainability policies and programs as economic development: is the new economic development sustainable development? *Cityscape: A Journal of Policy Development and Research*, *15*(1), 1–28.
Razavi, L. (2017). Building a city: regulating the sharing economy in Amsterdam. Retrieved 27 December 2018 from https://medium.com/@LaurenRazavi/building-a-city-regulating-the-sharing-economy-in-amsterdam-fa ecee8dfb0.
Razavi, L. (2018). How Rwanda's capital became an African tech leader. Retrieved 27 December 2018 from https://medium.com/s/story/how-rwandas-capital-became-an-african-tech-leader-66e00edc74d.
Robinson, J., and Cole, R.J. (2015). Theoretical underpinnings of regenerative sustainability. *Building Research and Information*, *43*(2), 133–43. https://doi.org/10.1080/09613218.2014.979082.
Rockström, J., Steffen, W., Noone, K., Persson, Å., Chapin, F.S.I., Lambin, E., Lenton, T.M., Scheffer, M., Folke, C., Schellnhuber, H.J., Nykvist, B., de Wit, C.A., Hughes, T., van der Leeuw, S., Rodhe, H., Sörlin, S., Snyder, P.K., Costanza, R., Svedin, U., Falkenmark, M., Karlberg, L., Corell, R.W., Fabry, V.J., Hansen, J., Walker, B., Liverman, D., Richardson, K., Crutzen, P., and Foley, J. (2009). Planetary boundaries: exploring the safe operating space for humanity. *Ecology and Society*, *14*(2), 32. https://doi.org/10.1007/s13398-014-01 73-7.2.
Rodale Institute (2014). *Regenerative Organic Agriculture and Climate Change (White Paper)*. Kutztown: Rodale Institute.
Roseland, M. (2012). *Toward Sustainable Communities: Solutions for Citizens and Their Governments* (4th ed.). Gabriola Island, British Columbia: New Society Publishers.
Roseland, M., and Spiliotopoulou, M. (2016). Converging urban agendas: toward healthy and sustainable communities. *Social Sciences*, *5*(3), 28. https://doi.org/10.3390/socsci5030028.
Roseland, M., and Spiliotopoulou, M. (2017). Sustainable community planning and development. In M.A. Abraham (ed.), *Encyclopedia of Sustainable Technologies* (pp. 53–61). Amsterdam: Elsevier.
Scharmer, C.O. (2018). *The Essentials of Theory U: Core Principles and Applications*. Oakland, California: Berrett-Koehler Publishers, Inc.
Smithsimon, G. (2008). Dispersing the crowd: bonus plazas and the creation of public space. *Urban Affairs Review*, *43*(3), 325–51. https://doi.org/10.1177/1078087407306325.
Solow, R.M. (1993). An almost practical step toward sustainability. *Resources Policy*, *19*(3), 162–72. https://doi.org/10.1016/0301-4207(93)90001-4.
Spherical Studio (n.d.). Regenerative projects around the world. Retrieved 19 December 2018 from https://www.google.com/maps/d/viewer?mid=1LZ8IVoeMYCplO7FkaueXTNZrdwY&ll=28.680773519821027%2C-118.5 233498&z=2.
Spiliotopoulou, M. (2019). Sustainable community development through the conceptual lens of productivity (Unpublished doctoral dissertation). Simon Fraser University, Canada.
Thomson, G., and Newman, P. (2018). Urban fabrics and urban metabolism: from sustainable to regenerative cities. *Resources, Conservation & Recycling*, *132*, 218–29. https://doi.org/10.1016/j.resconrec.2017.01.010.
UN DESA (2018). *United Nations. World Urbanization Prospects: The 2018 Revision [Key facts]*. Economic and Social Affairs. https://doi.org/10.1017/CBO9781107415324.004.

United Nations (2015a). *Adoption of the Paris Agreement (FCCC/CP/2015/L.9/Rev.1)*. Paris. Retrieved 30 October 2015 from http://unfccc.int/resource/docs/2015/cop21/eng/l09r01.pdf.
United Nations (2015b). *Resolution 70/1. Transforming Our World: The 2030 Agenda for Sustainable Development.* Retrieved 28 November 2015 from https://sustainabledevelopment.un.org/post2015/transformingourworld.
United Nations (2016). *THE CITY WE NEED 2.0: Towards a New Urban Paradigm*. Retrieved 2 December 2016 from http://worldurbancampaign.org/city-we-need.
United Nations (2017). *New Urban Agenda. Conference on Housing and Sustainable Urban Development (Habitat III)*. https://doi.org/ISBN: 978-92-1-132757-1.
Uphoff, N. (2014). Systems thinking on intensification and sustainability: systems boundaries, processes and dimensions. *Current Opinion in Environmental Sustainability, 8*, 89–100. https://doi.org/10.1016/j.cosust.2014.10.010.
Urban Innovation Community (2015). *Telling City Success Stories: Publication of Roundtable on SDGs and Urban Innovation*. Barcelona: UCLG World Council.
WA Contents (2018). *UNStudio Releases Design for Central Innovation District Test Site in The Hague*. Retrieved 27 December 2018 from https://worldarchitecture.org/article-links/epenf/unstudio_releases_design_for_central_innovation_district_test_site_in_the_hague.html.
Wahl, D.C. (2016). *Designing Regenerative Cultures*. Axminster, England: Triarchy Press.
WCED (1987). *Report of the World Commission on Environment and Development: Our Common Future (The Brundtland Report)*. https://doi.org/10.1080/07488008808408783.
Williams, C.C., and Millington, A.C. (2004). The diverse and contested meanings of sustainable development. *The Geographical Journal, 170*(2), 99–104. https://doi.org/10.2307/3451586.
World Economic Forum (2018). *Circular Economy in Cities: Evolving the Model for a Sustainable Urban Future (White Paper)*. Cologny/Geneva: World Economic Forum.

21. Re-imagining community development: the Cocoa360 model*
Shadrack Frimpong, Allison R. Russell and Femida Handy

INTRODUCTION

In an increasingly globalized and complex world, governments and nongovernmental organizations (NGOs) are seeking new and sustainable ways to address social challenges. Recently, the literature has examined these new ways and considered them to be "social innovations" if they both identify unique social needs and find novel ways of meeting these needs (Majumdar et al., 2015). These innovations must not only improve outcomes for individual service users but also seek to achieve a broader impact on the communities and experiences of vulnerable social groups. With this definition in mind, we will discuss how Cocoa360 has indeed offered a unique way of addressing a particular community development challenge in Ghana. We rely on the case study method to illustrate a social innovation, as case studies have been used widely in the examination of social innovations due to their heterogeneous nature (Bouchard et al., 2015).

Cocoa360 is an NGO that has pioneered a new model of community development that leverages existing community resources to improve access to education and healthcare for Ghana's cocoa farmers and their families without long-term reliance on foreign aid. This model, called the "farm-for-impact" model, intends to improve educational outcomes at Cocoa360's Tarkwa Breman Girls' School in rural Ghana. Cocoa360's model challenges traditional international development paradigms by placing decision-making in the hands of those whom the intervention is designed to impact (Sunog, 2018).

Taking into account the viewpoints of both community development scholars and practitioners, this chapter examines how social innovations can shape the future of community development. We discuss Cocoa360's model and its potential to be scaled up and applied in farming communities around the world. We also consider the potential challenges associated with its implementation and scaling, and outline potential implications of the model in other areas such as healthcare financing.

INTERNATIONAL COMMUNITY DEVELOPMENT IN THE AFRICAN CONTEXT

Historically, international development paradigms have adopted a top-down approach, in which wealthier nations sponsor funding and institutional strategies for overcoming social and economic challenges in developing nations (Williams, 2012). During the latter half of the twentieth century and in the present, these strategies have frequently been carried out through the day-to-day activities of NGOs, which lead the way in making decisions about how resources are best allocated and mobilized within communities, often with varying degrees of input from and interaction with community members (Brass et al., 2018; Lan, 2018; Elbers and Schulpen, 2015). However, development scholars have increasingly decried the negative impact of such approaches to aid financing (Easterly, 2006). They argue that these "hand-outs" not only stifle community participation and agency but may hinder the long-term sustainability of social interventions (Bridger and Luloff, 1999).

In the African context, neo-liberal policies advocated by powerful global actors favor a smaller role for the public sector, which is often regarded as inefficient or corrupt. This assumption has limited the influence of the state and resulted in a concomitant rise in the market and NGO sectors, especially in the development and economic arena (Mohan, 2002). Despite the phenomenal growth of the NGO sector in Africa, the intensity and complexity of social problems continue to increase, suggesting that current organizational approaches are falling short (Silverthorne, 2008; Brass, 2016). Furthermore, NGOs are largely dependent on donors, which opens up their programs to donor influence, often at the expense of local knowledge, thereby leaving aid beneficiaries as passive actors in the process. By contrast, community-based innovations have been hailed as sustainable as they often identify unique social needs and find new ways of addressing these challenges (Majumdar et al., 2015). For example, in a review of participatory processes in forestry projects in Ghana, Kenya, Nigeria, Senegal and Uganda, it was shown that the initiatives would have been more inclusive, sustainable and reflective of community needs had the organizations and agencies engaged local leaders instead of creating and working through parallel institutions (Ece et al., 2017).

While "social innovation" remains a somewhat ambiguous concept, especially in the community development arena, two central components are an overarching goal of social change and a collaborative approach to achieving this goal (e.g., van der Have and Rubalcaba, 2016; Phillips et al., 2015; Cajaiba-Santana, 2014). Furthermore, Neumeier (2012) suggests that, in the case of rural development, the community or collective element of social innovations should be foremost, stating that social innovations "refer to the effort, method, result or change initiated by *collaborative actions*" (pp. 49–50, emphasis added). He offers the following definition of social innovations: "changes of attitudes, behavior or perceptions of a group of people joined in a network of aligned interests that in relation to the group's horizon of experiences lead to new and improved ways of collaborative action within the group and beyond" (Neumeier, 2012, p. 55).

In the African context, Patel and colleagues (2012) argue that long-established, "indigenous" approaches to social welfare provision represent one possible and promising strategy for improving regional community development. Drawing on examples from South Africa and Burundi, the authors suggest that because of the historically top-down, western-centric approach to international development, any effort to integrate local knowledge and established social structures into this realm emerges as an innovation in and of itself. According to Howaldt, Kopp, and Schwarz (2015), "The question is not about how to introduce solutions into society, but rather how to transform existing solutions to better arrangements" (p. 44). In this way, organizations that seek to incorporate existing community practice into improved models of development can be seen to be undertaking social innovations.

Social innovations improve outcomes for individual service users and achieve a broader impact on both the communities and experiences of vulnerable social groups. Thus, social innovations offer a unique way of looking at and addressing community development challenges. Indeed, they may represent a critical component of sustainable development strategies, especially in rural contexts (Neumeier, 2012).

In Africa, there are several NGOs that have developed and implemented social innovations to maximize their impact and put their constituents at the forefront of social change. One such NGO is Indego Africa, which provides market access, business education and vocational training for women artisans in Ghana and Rwanda. They pay artisans approximately 50 percent of the wholesale price of each item up front, and then sell the products in high-end stores such as Nordstrom and NomadTribe in the United States and Europe (Indego Africa, 2018). One hundred percent of product sales are then reinvested in business and entrepreneurship programs for their partner artisans. The results have been stunning: they have engaged over 1,200 artisan partners, who have collectively impacted over 4,800 lives in their immediate circles (Indego Africa, 2018). One Acre Fund is another NGO with a similar approach. To reduce hunger and poverty, One Acre Fund supplies smallholder farmers in East Africa with asset-based financing and agriculture training services. Farmers who worked for the fund in 2017 had a 249 percent return on their investment and significantly increased farm income on every planted acre (One Acre Fund, 2017).

By embracing business models to successfully effect social change, these NGOs and other similar NGOs function as social enterprises. Yet, the same cannot be said of many community-based NGOs, particularly ones that seek to provide educational and healthcare services. For example, there exist NGOs such as Lwala Community Alliance (LCA) and Village Health Works (VHW) in Kenya and Burundi, respectively, that provide education and healthcare while utilizing agriculture to improve economic conditions. However, neither of these organizations has developed business models for using agricultural proceeds to financially support their work in healthcare and education; instead, the programs are conducted separately (LCA, 2017; VHW, 2017). Given that both VHW and LCA operate in communities that rely on subsistence farming of small-scale, low profit-yielding crops such as sugar cane and corn, it is understandable why relying on these proceeds may not provide sufficient financial sustenance to support their other programs. In the context of cash-crop farming communities, however, the potential for merging such programs together, with the goal of fully or partially supporting education and health-related programs through revenue generated locally by agricultural industries, has emerged as a new possibility for sustainable and innovative community development practice. In the next section, we turn to the case study of Cocoa360 to illustrate how this innovative concept has become a reality for agricultural communities in rural Western Ghana.

CASE STUDY: COCOA360

Farming is the primary occupation of rural Ghanaians. As the world's second leading exporter of cocoa, Ghana generates approximately $2 billion in cocoa revenue annually. Yet, the average cocoa farmer earns less than 50 cents a day (Simoes, 2016). Cocoa farmers and their children are trapped in cycles of generational poverty. The problem is widespread: Ghana is home to 1.6 million farmers in over 1,300 cocoa-growing communities who are facing dire health inequities due to lack of access to education and medical care (Wolter, 2007).

In 1995, the government of Ghana implemented the Free Compulsory Universal Basic Education (FCUBE) as a policy to achieve universal education in the country by 2005 (Akyeampong, 2009). Under this scheme, salaries of teachers are paid by the government and education is tuition-free for all students in public primary schools. Despite these accommodations, the FCUBE could not fully achieve its 2005 goal. FCUBE's failure lies in its inability to offset the cost of schooling for the poorest households. Costs that students must pay include certain fees, the cost of textbooks, uniforms and transportation, costs which disproportionately impact households in rural and remote communities (Nudzor, 2015).

In response to these challenges, Cocoa360 was established in 2015 and operates in Tarkwa Breman, which is in rural Western Ghana. Cocoa360 pioneered a new model of development that leverages existing community resources to improve health and education outcomes for Ghana's cocoa farmers and their families. The "farm-for-impact" model is simple yet impactful: in exchange for tuition-free education and subsidized healthcare, community members must work on a community-run cocoa farm. With Cocoa360's guidance, community members apply the farm's revenues toward improving health or educational outcomes in a tiered manner. In the tiered funding allocation approach, farm revenues are spent in a gradual approach based on community prioritization of needs such as books, uniforms, medications, among others. By addressing access to both education and medical services, two major social determinants of health, Cocoa360's model seeks to increase community members' overall health and well-being in their daily lives.

BACKGROUND AND HISTORY OF COCOA360

Cocoa360's inception was guided by the deep-seated frustration that accompanied the grave poverty that its founder experienced first-hand as a child growing up in Tarkwa Breman, a cocoa farming village in rural Western Ghana without electricity or running water. At age nine, he contracted an infection that nearly resulted in the amputation of his legs. Several months passed before his parents could find enough money to send him to the nearest hospital, five hours away. Years later, he would become the first person from his village to attend college in the United States at the University of Pennsylvania. In 2015, he graduated with a BA in Biology after being awarded the $150,000 President's Engagement Prize (PEP), one of Penn's highest competitive honors for graduates (Ozio, 2015).

With seed funding from the PEP and another $15,000 from the Samuel Huntington Public Service Award, he founded Tarkwa Breman Community Alliance (later renamed Cocoa360). A crucial step after winning the Samuel Huntington and PEP awards was forming a team and establishing an organization for operations. It was easy to recruit people to pursue the vision; peers from Penn, including some young visionary Ghanaians, were inspired to join him. The founder led the team as the Executive Director; his colleagues took over roles in financial management, partnerships and fundraising and creative design. Together, they went to Tarkwa Breman where they were welcomed with support in guidance and land donation. Led by the community leaders, including the Chief and the Council of Elders, the Tarkwa Breman community provided 50 acres of land for Cocoa360's work. They also worked with the community members to create a

Village Committee (VC), a decision-making body comprised of the most respected citizens in the community, to ensure that the organization's work reflects the interests of the community and to provide operational counsel for Cocoa360's leadership (Larbi, 2015).

CONCEPTUALIZING COCOA360'S "FARM-FOR-IMPACT" MODEL

In line with the conceptualization of social innovation offered by Neumeier (2012), the social innovation in this case is Cocoa360's "farm-for-impact" model, a concept that effected changes in community attitudes and approaches to education and healthcare delivery as well as in organizational approaches to development practices. The concept, now fledged as the core of Cocoa360's mission, has real and material effects on the lives of community residents.

Community members work on a community-run cocoa farm that is managed by Cocoa360. Led by Cocoa360's farm managers and supervised by local leaders, community members take care of the cocoa crop through weeding, pruning, fertilizer application and spraying of pesticides. During harvesting, they use sickles and other tools to gather the ripe cocoa pods, and then break the pods to gather the beans for drying and eventual compounding for sale. The profits from the harvest of the cocoa crop are then used to subsidize costs associated with education and healthcare services for community members who work on the farm.

Cognizant of the importance of taking the time to master the model's effectiveness and success, Cocoa360 is currently carrying out extensive monitoring and evaluation of its application in education only and not yet in healthcare. This approach is a vital step in their future plans to implement the model toward their work in saving lives through medical care. For instance, a partnership between Cocoa360 and Ghana's Ministry of Health would enable Cocoa360 to manage the Ghana Health Service (GHS) Community-Based Health Planning Services (CHPS) compounds or clinics in rural areas, many of which are not being effectively used due to high user fees and exorbitant costs of medications. Through the "farm-for-impact" model, Cocoa360 could synergistically work with rural community members to apply revenues from community-run farms to offset healthcare expenses such as user fees, and also to subsidize the cost of medications. It is well-documented that user fees significantly impact health outcomes and negatively affect healthcare access in rural areas (Johnson et al., 2012). Using Cocoa360's model to eliminate user fees would therefore improve infant and maternal care and increase earlier detection of cancer, malaria and HIV/AIDS in rural areas.

Cocoa360's model is hinged on three (3) key pillars: farm labor, financial reporting and decision-making. Figure 21.1 illustrates these pillars.

Farm Labor: Use of the "Farm-for-Impact" Model Card

Each farmer has a communal labor voucher card used to log their work hours (Figure 21.2). Each checkbox on the card is equal to one hour, and Cocoa360 pegs it against a monetary value.[1] Currently, the card system is only used for Cocoa360's educational component, its tuition-free girls' school at Tarkwa Breman in the Western Region

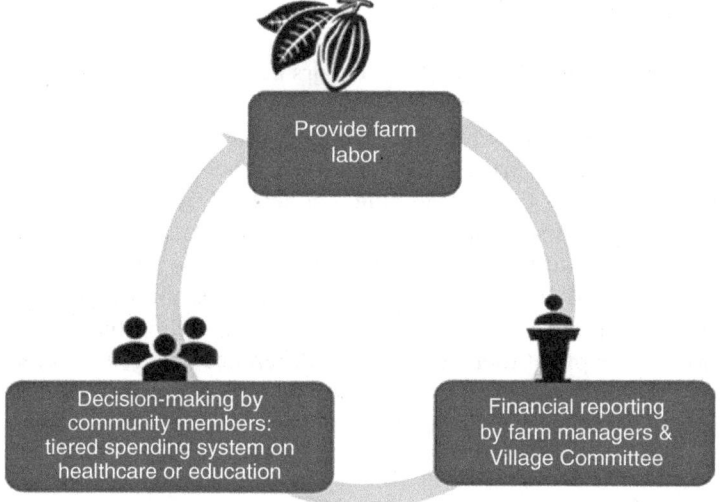

Figure 21.1 Cocoa360's "farm-for-impact" model

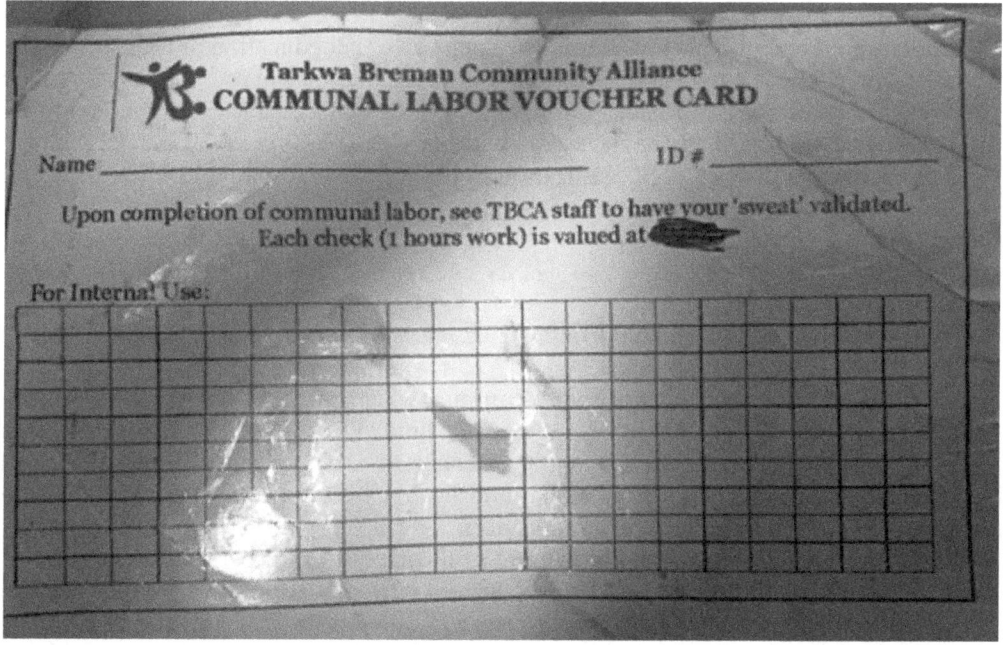

Figure 21.2 Cocoa360's communal labor voucher card

of Ghana. All parents are expected to log an equal number of hours at the end of the academic year, since they work on the farm at the same time(s) on the same day(s). When a parent misses an hour, an appropriate financial penalty, commensurate with the cost associated with that labor hour, is applied.

Financial Reporting and Debriefing

The main crop harvesting season for cocoa occurs between September and March, while the mid-crop harvesting season takes place between May and August of every year. An acre of a well-nourished and well-tended cocoa farm can produce eight to ten bags of cocoa. Ghana has a producer price for cocoa at GH₵7,600 per ton or GH₵475 per bag of 64 kilograms (Ayitey, 2018). Crop harvesting is collectively done by community members as part of farm labor. At the end of the harvest cycle, the VC, Parent-Teacher Association (PTA), Cocoa360's staff, and community members come together to debrief on the crop year. Farm managers update the number of bags harvested, funds gained from their sale (sale price) and the funds invested in fertilizer application, herbicides and other farm inputs (cost price). The profit then becomes the difference between the cost price and the sale price. Next, the farm managers, the VC and the PTA answer any and all questions that the audience may have. The goal of this financial reporting is to ensure transparency and boost parental engagement.

Decision-Making

Farm managers and the VC present options for investing farm revenues to parents and community members. Typically, community members vote on a tiered system for spending the farm revenues exclusively toward improvements in either education or healthcare. The tiered system allows them to prioritize funds allocation since farm revenue may fluctuate from year to year. This approach to decision-making helps to ensure that community members remain active participants in each pillar of Cocoa360's work – a crucial asset for long-term sustainability of the organization's interventions (Esman and Uphoff, 1982). In this way, the model cultivates not only cocoa but also direct community involvement and collective ownership.

TARKWA BREMAN CENTER OF EXCELLENCE

Currently, Cocoa360 maintains an agro-campus, which includes the Tarkwa Breman Girls' School, the Tarkwa Breman Community Clinic (TBCC) and the Tarkwa Breman Community Farm. For the past 3 years, Cocoa360 has cultivated community buy-in by continually evaluating the educational and healthcare impact of its model through dialogue with community members and leaders.

During this time, Cocoa360 has grown to 37 full-time staff members and cared for over 3,000 patients. The Tarkwa Breman Girls' School currently educates 126 young female students. By utilizing community-owned cocoa resources to raise funds that go directly to improvements in education infrastructure, coupled with comprehensive training and curriculum development provided by Cocoa360's experienced education team and university partners, this model seeks to transform educational outcomes for students in Tarkwa Breman. It also recognizes the central role of education and community development in both helping to stem the tide of young people abandoning the cocoa industry and transforming it from an engine of wealth for the elite to a resource that builds futures for farmers and their families. Cocoa360's agro-campus in Tarkwa Breman serves as its

Center of Excellence and "living learning laboratory," where the organization researches how best it can scale its efforts in education and healthcare to other communities.

KEY STAKEHOLDERS OF COCOA360'S MODEL

Community Leaders and Community Members

Almost every rural community in Ghana has a chieftaincy council of elders, kings and queen-mothers. These local decision-makers are crucial to the success of any initiative in their communities. By lending their voice to the initiative, they encourage community members to participate in the project's successful implementation. At the grassroot levels of rural and remote communities, local leaders coordinate communal labor, contribute land, provide informal technical advice and work hand-in-hand with NGO leaders to ensure the long-term sustenance of any social innovations in their communities (Ozor and Nwankwo, 2008).

Village Committee

In each community, Cocoa360 collaborates with community members to create a VC, a decision-making body comprising the most respected citizens in the community. The VC has a diverse membership of young women and men (25–39 years) and elderly men and women (40 years and older), who come from varied religious, ethnic and occupational backgrounds. Additionally, one member of the Village's Chieftaincy Council sits on the VC to represent the chief and his elders. With the exception of the chieftaincy council representative, VC members chosen by the community are not members of the chieftaincy council. The VC serves as a liaison between Cocoa360 and the community to ensure the organization's operations are always in line with the needs of the community and are reflective of the voice of community members. For instance, due to the relatively long gestation period of cocoa (about 5 years), the VC advised Cocoa360's leaders to acquire a 10-acre already-matured cocoa farm to serve as a "demo farm" for the model and a site for the organization to learn best farming practices while it seeds its 40-acre farm. Colloquially, Cocoa360 refers to the VC as the "local board," since it is the ultimate decision-making entity representing the community's needs (Sotunde, 2019).

Parent-Teacher Association

The PTA is the decision-making body established exclusively to deliberate on Cocoa360's school. The PTA executives meet regularly with parents to decide on issues pertaining to school uniforms, textbooks, disciplinary actions for students, organization of communal labor for Cocoa360's farm as well as all matters related to the successful operation of the school. The PTA Committee operates under the mandate of the VC (Baillie Unger, 2016).

Ghanaian Government

The key arms of the government of Ghana supporting Cocoa360's work include the Ministry of Health (MOH), the Ministry of Education (MOE) and the Ghana Cocoa

Board (Ministry of Health, 2019; Ministry of Education, 2019). The MOH promotes and ensures quality health of the nation's citizenry. It provides public services, manages and supervises the country's healthcare industry and builds Ghana's hospitals and medical education system. Ghana's MOE, on the other hand, provides educational infrastructure as well as curriculum development. Currently, Cocoa360 partners with the MOE and MOH to ensure efficient delivery of educational and healthcare services at their Tarkwa Breman Center of Excellence. In the future, the organization plans to partner with these government entities to scale its model further across rural cocoa-growing communities in the country. The Ghana Cocoa Board, the sole government entity responsible for quality cocoa production and sale to international markets, will assist Cocoa360 with agricultural extension services as well as fertilizer and seedling support in future expansion. Together, these stakeholders assist Cocoa360 to successfully implement its pioneering "farm-for-impact" model in rural cocoa-growing communities in Ghana.

Funders

As a 501c(3) tax-exempt non-profit organization that is registered in both Ghana and the United States, Cocoa360 raises funds from a variety of sources, including foundations, individual donors and fundraising events. Family foundations, both small and large, have so far provided the organization with grants, both restricted and unrestricted, that range from $5,000 to $50,000. Individual donors also provide funds toward initiatives such as student tuition coverage, maternal care and community health outreach. Further, the organization enjoys corporate philanthropy from several private entities such as the private equity firm Warburg Pincus. Partnerships with companies such as Vodacom and Google have allowed Cocoa360 to secure in-kind services, such as free internet on their campus and pro-bono financial consulting, respectively. Finally, the organization also generates revenue from its sale of cocoa crops and fees from patients receiving medical care provided at their community clinic. Both revenue sources are directly applied toward the sustenance of Cocoa360's programs in education and healthcare. These funders have supported Cocoa360's vision for the "farm-for-impact" model, allowing village leaders and community members to remain at the heart of organizational decision-making and operations (Kahn, 2017).

IMPACT OF COCOA360'S WORK TO DATE

A defining factor that led Cocoa360 to establish the tuition-free Tarkwa Breman Girls' School was the prevailing view that household heads (typically males) held about the importance of educating girls. When asked which gender (male or female) they would choose to educate given limited funding, 44.5 percent said they would choose the male child, 27.7 percent opted for the female child, 26.5 percent stated that it would depend on the specific circumstances and 1.3 percent refused to answer (Stack, 2017). While Cocoa360 has yet to conduct a follow-up study to ascertain whether these perceptions have changed, there are clear indications that point to the fact that household heads (mainly men) now recognize the role of educating the girl-child, especially when money is not a barrier. Many of these family heads willingly and consistently show up every

Labor Day to work on the organization's farm and are also actively engaged in Tarkwa Breman Girls' School's PTA. Furthermore, prior to the establishment of the girls' school, many of the students attended schools in the village that had dilapidated structures, lacked classroom furniture and were woefully understaffed. Likewise, students from the neighboring villages, who now commute to school via the school bus, previously had to walk several miles before getting to school, often showing up late.

Recognizing the impact of educational challenges on the largely rural and remote communities that it serves, Cocoa360 decided to work with the community to use cocoa farm revenues to offset non-salaried educational expenses at its Tarkwa Breman Girls' School. Their results so far have been astonishing: Their student attendance rate is 98 percent, compared to the national rural attendance rate of about 70 percent for similar schools under the FCUBE program (EPDC, n.d.). The 2 percent difference in attendance rate at the Tarkwa Breman Girls' School is largely attributed to student absenteeism due to illnesses. This preliminary finding now serves as a guiding light for the organization's expansion into similar communities in their quest to marshal local community efforts to improve educational outcomes.

Moreover, the organization's work has significantly impacted healthcare accessibility within the community. Prior to the opening of Cocoa360's TBCC, the villagers had only an understaffed government-built CHPS compound, which was not equipped to handle childbirth nor any serious medical treatment beyond First Aid (Stack, 2017). Consequently, in instances of life-threatening illness, a patient had to be sent to the nearest comprehensive medical facilities in the cities, in Tarkwa or Prestea, about three hours away via deplorable roads. Staffed with a physician's assistant, a midwife, a nurse, a pharmacist's assistant and a records clerk and equipped with an ultrasound machine, TBCC has been able to bridge the majority of the healthcare gaps that existed in the community, particularly those in infant and maternity care. In less than a year of operating, TBCC has seen over 3,000 patients, treated over 1,200 malaria cases and delivered over 40 babies. Additionally, the clinic has two motorbikes that enable its staff members to proactively reach out to patients in neighboring communities who are not able to quickly commute to Tarkwa Breman during medical emergencies in the middle of the night.

Given the organization's focus on targeting social determinants of health such as education and healthcare access, its success is heavily tied to improved health outcomes among community members. The benefits of education on health outcomes for young girls are well-documented and include reduced child and maternal deaths, improved child health and lower fertility. Women with at least some formal education are more likely than their uneducated peers to use contraception, marry later, have fewer children and be more informed on the nutritional needs of children (PRB, 2011).

By using the "farm-for-impact" model to improve education outcomes for young girls coupled with medical care provided for the community, Cocoa360 aims to work with community members to attain health equity. The model's potential for success has been evaluated and confirmed by researchers from the Wharton School of Business at the University of Pennsylvania. Particularly, the study confirmed that proceeds from the cocoa farm could successfully support educational improvements (Stack, 2017).

SCALING AND REPLICABILITY

As a social innovation, the Cocoa360 "farm-for-impact" model represents an opportunity to not only scale up current operations in rural Western Ghana but also develop effective ways of implementing the model in other communities and geographies. In the Ghanaian context, Cocoa360 plans to extend its scope by applying the lessons learned at the Tarkwa Breman Center of Excellence in partnership with the Ghanaian government to strengthen educational and healthcare systems in rural areas. Currently, the organization services eight surrounding communities through its operations at the Tarkwa Breman Center of Excellence. Plans are far advanced to scale the "farm-for-impact" model to other rural cocoa-growing communities in Ghana. In each new community, Cocoa360's success will be heavily hinged on community involvement and the community's recognition of the benefits of cocoa revenue being invested to improve education for their children.

LIMITATIONS AND CHALLENGES

While Cocoa360 has experienced great success to date, several key challenges arise with this approach, including gaining (and maintaining) community support, recruiting operational staff and losing significant farm revenue due to a bad crop season. However, Cocoa360 has learned to manage these challenges over the years. The organization attracts highly qualified employees – including teachers, nurses, a midwife and a physician – to staff the Center of Excellence by incentivizing them with competitive salaries that take into account national averages, qualifications and the "hardship" factor of working in a rural environment. Cocoa360 must continue to get community buy-in by involving leaders and members of the community in the VC, to ensure transparency and accountability.

In an effort to avoid the "one-crop" failure, the organization is currently diversifying its cash-crop fold to include rubber and palm. Together, these experiences and lessons that the organization has amassed over the past 3 years uniquely position Cocoa360 to succeed in a partnership effort with the government of Ghana to scale their model to other rural Ghanaian farming communities.

CONCLUSION AND FUTURE DIRECTIONS

As a social innovation, the "farm-for-impact" model offers a paradigm shift for development practitioners around the world with implications far beyond rural Ghana. However, there is still some more room for growth. To ensure efficiency in data collection, Cocoa360 is exploring the digitization of the labor voucher card. In addition, initiatives such as awards for the most hard-working community members as well as activities like intra-community games should be organized to ensure continual participation of community members in farm work. Finally, they should also work to set fixed monetary values for the hour's worth of work that community members do on the farm. These steps, if taken, will drastically help to ensure transparency, sustain community engagement and secure the model's long-term success.

Through its innovative financial model, which leverages rural communities' strengths, Cocoa360 has pioneered a program that challenges traditional development interventions by putting communities at the forefront of decision-making and reducing dependence on foreign aid. Though currently operational in rural Ghana, the model's applicability goes beyond Ghana into any rural community around the world with similar cash-crop growing dynamics. Whether in rural cocoa-growing communities in Indonesia or Ivory Coast or in rural coffee-growing communities in Kenya, the message remains the same: community members have the power to leverage their resources to improve services such as education and healthcare.

NOTES

* The authors would like to thank Susi Neher, Courtney Kramer and members of the Cocoa360 Board of Directors for their feedback on an earlier draft of this chapter. Requests for data associated with Cocoa360's work should be sent to Cocoa360 at info@cocoa360.org.
1. These monetary values are set by Cocoa360 and are not fixed since crop harvest revenue fluctuates from year to year.

REFERENCES

Akyeampong, K. (2009). Revisiting Free Compulsory Universal Basic Education (FCUBE) in Ghana. *Comparative Education*, *45*(2), 175–95. DOI: 10.1080/03050060902920534.

Ayitey, C. (2018). Ghana maintains cocoa producer price at GH₵7,600 per tonne. *JoyBusiness*, 1 October 2018. Retrieved 16 December 2018 from https://www.myjoyonline.com/business/2018/October-1st/ghana-maintains-cocoa-producer-price-at-gh7600-per-tonne.php.

Baillie Unger, K. (2016). Penn President's Engagement Prize launches effort to empower Ghanaian girls. *PennToday*, 7 January 2016. Retrieved 16 December 2018 from https://penntoday.upenn.edu/news/penn-president-s-engagement-prize-launches-effort-empower-ghanaian-girls.

Bouchard, M.J., Trudelle, C., Briand, L., Klein, J., Levesque, B., Longtin, D., and Pelletier, M. (2015). A relational database to understand social innovation and its impact on social transformation. In Nicholls, A., Simon, J., Gabriel, M., Whelan, C. (eds), *New Frontiers in Social Innovation Research* (pp. 69–85). London, UK: Palgrave Macmillan.

Brass, J.N. (2016). *Allies or Adversaries: NGOs and the State in Africa*. Cambridge, UK: Cambridge University Press.

Brass, J.N., Longhofer, W., Robinson, R.S., and Schnable, A. (2018). NGOs and international development: A review of thirty-five years of scholarship. *World Development*, *112*, 136–49.

Bridger, C.J., and Luloff, A.E. (1999). Toward an interactional approach to sustainable community development. *Journal of Rural Studies*, *15*(4), 377–87.

Cajaiba-Santana, G. (2014). Social innovation: Moving the field forward – A conceptual framework. *Technological Forecasting & Social Change*, *82*(C), 42–51.

Easterly, W. (2006). *The White Man's Burden: Why the West's Efforts to Aid the Rest Have Done So Much Ill and So Little Good*. New York: The Penguin Press.

Ece, M., Murombedzi, J., and Ribot, J. (2017). Disempowering democracy: Local representation in community and carbon forestry in Africa. *Conservation and Society*, *15*(4), 357–70.

Elbers, W., and Schulpen, L. (2015). Reinventing international development NGOs: The case of ICCO. *European Journal of Development Research*, *27*(1), 1–18.

EPDC (n.d.). *Primary School Net and Gross Attendance Rates, Ghana: Education Policy and Data Center*. Retrieved 16 December 2018 from https://www.epdc.org/sites/default/files/documents/Ghana_coreusaid.pdf.

Esman, M.J., and Uphoff, N. (1982). *Local Organizations: Intermediaries in Rural Development*. Ithaca, NY: Cornell University Press.

Howaldt, J., Kopp, R., and Schwarz, M. (2015). Social innovations as drivers of social change: Exploring Tarde's contribution to social innovation theory building. In Nicholls, A., Simon, J., Gabriel, M., Whelan, C. (eds), *New Frontiers in Social Innovation Research* (pp. 29–51). London, UK: Palgrave Macmillan.

Indego Africa (2018). *Social Impact Report*. Retrieved 16 December 2018 from https://cdn.shopify.com/s/files/1/0031/6622/t/7/assets/2018_ImpactReport.pdf?15120703900175138242.

Johnson, A., Goss, A., Beckerman, J., and Castro, A. (2012). Hidden costs: The direct and indirect impact of user fees on access to malaria treatment and primary care in Mali. *Social Science and Medicine*, 75(10), 1786–92. Retrieved 16 December 2018 from https://www.ncbi.nlm.nih.gov/pubmed/22883255.

Kahn, N. (2017). With the help of cocoa and Amy Gutmann, this 2015 Penn graduate is promoting gender equality in Ghana. *The Daily Pennsylvanian*, 12 January 2017. Retrieved 16 December 2018 from https://www.thedp.com/article/2017/01/penn-alum-ghana-school-hospital.

Lan, Y. (2018). History and paradigm shift: NGOs in international development aid. *The China Nonprofit Review*, 10(1), 108–33.

Larbi, K. (2015). Giving it back, Tarkwa-Breman. *Design233*, 8 December 2015. Retrieved 16 December 2018 from https://www.design233.com/giving-it-back-tarkwa-breman/.

Lwala Community Alliance (LCA) (2017). *Annual Report.* Retrieved 17 December 2018 from http://lwalacommunityalliance.org/wp-content/uploads/Lwala-Community-Alliance_2017-Annual-Report.pdf.

Majumdar, S., Guha, S., and Marakkath, N. (eds) (2015). *Technology and Innovation for Social Change.* New Delhi, India: Springer India. DOI: 10.1007/978-81-322-2071-8.

Ministry of Education, Ghana (2019). About Us. Retrieved 15 March 2019 from http://moe.gov.gh/index.php/about-us/.

Ministry of Health, Ghana (2019). Home. Retrieved 15 March 2019 from http://www.moh.gov.gh/#.

Mohan, G. (2002). The disappointments of civil society: The politics of NGO intervention in northern Ghana. *Political Geography*, 21(1), 125–54.

Neumeier, S. (2012). Why do social innovations in rural development matter and should they be considered more seriously in rural development research? Proposal for a stronger focus on social innovations in rural development research. *Sociologia Ruralis*, 52(1), 48–69.

Nudzor, H.P. (2015). Taking education for all goals in sub-Saharan Africa to task: What's the story so far and what is needed now? *Management in Education*, 29(3), 105–11.

One Acre Fund (2017). *Annual Report.* Retrieved 17 December 2018 from https://oneacrefund.org/about-us/reports/.

Ozio, R. (2015) President Gutmann announces 2015 President's Engagement Prize winners at Penn. *Penn Today*, 25 March 2015. Philadelphia, PA. Retrieved 17 December 2018 from https://penntoday.upenn.edu/news/president-gutmann-announces-2015-president-s-engagement-prize-winners-penn.

Ozor, N., and Nwankwo, N. (2008). The role of local leaders in community development programmes in Ideato local government area of Imo State: Implication for extension policy. *Journal of Agricultural Extension*, 12(2), 63–75.

Patel, L., Kaseke, E., and Midgley, J. (2012). Indigenous welfare and community-based social development: Lessons from African innovations. *Journal of Community Practice*, 20(1–2), 12–31.

Phillips, W., Lee, H., Ghobadian, A., O'Regan, N., and James, P. (2015). Social innovation and social entrepreneurship: A systematic review. *Group & Organization Management*, 40(3), 428–61.

PRB (Population Reference Bureau) (2011). *The Effect of Girls' Education on Health Outcomes: Fact Sheet.* Washington, DC. Retrieved 17 December 2018 from https://www.prb.org/girls-education-fact-sheet/.

Silverthorne, S. (2008). Putting entrepreneurship in the social sector. *Harvard Business School: Working Knowledge.* Retrieved 17 December 2018 from https://hbswk.hbs.edu/item/putting-entrepreneurship-in-the-social-sector.

Simoes, A. (2016). *Ghana: MIT Observatory of Economic Complexity (OEC).* Retrieved 17 December 2018 from https://atlas.media.mit.edu/en/profile/country/gha/.

Sotunde, O. (2019). Farming to impact with Cocoa360. *HowWeMadeitInAfrica*, 4 February 2019. Retrieved 17 December 2019 from https://www.howwemadeitinafrica.com/farming-to-impact-with-cocoa360/62823/.

Stack, R. (2017). The effects of an NGO development project on the rural community of Tarkwa Breman in Western Ghana. *Wharton Scholarly Commons.* Retrieved 17 December 2018 from https://repository.upenn.edu/sire/47/.

Sunog, E. (2018). The innovation point: Shadrack Frimpong on education and healthcare in Ghana. *The Center for Social Impact Strategy*. Retrieved 17 December 2018 from https://socialimpactstrategy.org/the-innovation-point-shadrack-frimpong-on-education-and-healthcare-in-ghana/.

Van der Have, R.P., and Rubalcaba, L. (2016). Social innovation research: An emerging area of innovation studies? *Research Policy*, 45(9), 1923–35.

Village Health Works (VHW) (2017). *Annual Report.* Retrieved 17 December 2018 from https://static1.squarespace.com/static/56b50316b654f9372fd3a4e7/t/5bb395c5eef1a163dcb1c1c2/1538496029764/2017+VHW+Annual+Report.pdf.

Williams, D. (2012). *International Development and Global Politics: History, Theory and Practice.* New York: Taylor & Francis Group.

Wolter, D. (2007). Ghana: Seizing new agribusiness opportunities. *OECD Report.* Retrieved 17 December 2018 from http://www.oecd.org/dataoecd/35/30/41302232.pdf.

22. Community development and place attachment using an inductive social media approach*
Justin B. Hollander and Max Page

INTRODUCTION

For this chapter, we begin by defining community development in broad terms: the practices of improving places for people through community engagement (Hollander, 2018). While community development can have a range of social, economic and political dimensions, we employ a physical place lens in this research.

When it comes to physical places, we propose that few material objects matter more than historic buildings. However, in places across the US and the world, these structures that shape the places we live, work and play in, are under threat. Despite 50 years of the National Historic Preservation Act much of the historic fabric of the country remains at risk of demolition and development (Page, 2016). While the State and National Registers of Historic Places have been key policy tools for identifying and protecting those resources, the listing process is only a small percentage of all historic places recognized, and the legal powers of protection are weak, leaving the vast majority of historic sites vulnerable. Conventional historic preservation processes are weak and rely on the activism of community leaders. Given the ways that these historic structures contribute to broader community well-being, it becomes a critical task of community development professionals to understand and respond to this crisis. However, few avenues exist for the ideas of ordinary residents to be incorporated into formal processes of identifying places that matter, with the exception of the rare public meeting or survey. So, what does it even mean for a community development professional to take steps to address the physical artifacts of a community, the places that matter to people?

With the advent of low-cost, widely available social media data we argue that opportunities exist for community development practice to improve. These social media data can provide unobtrusive insight into ordinary people's online conversations and public photo posts to shape the ways that historic preservation professionals identify places in need of preservation. For this chapter, we used microblogging data in conjunction with a mix of quantitative and qualitative methods, including content analysis and advanced multivariate statistics. The research used a novel data collection and analysis framework to collect, organize and analyse social media data for the purposes of identifying places that matter to people. The first section of the chapter provides an overview of relevant literature in well-being, sentiment analysis and historic preservation, the next section offers an introduction to the tools we employed in this analysis. This is followed by the methods and results from our study. We end the chapter with a conclusion, suggestions for future research, and a discussion about the implications of this study for community development.

REVIEW OF THE LITERATURE

Assessing Well-Being

Subjective well-being is a commonly used concept in psychology and is comprised of two dimensions: life satisfaction and affect (along a positive or negative continuum) (Boniwell, 2008; Diener et al., 1999). Both dimensions can be measured at the individual scale, where life satisfaction is how satisfied a person is with their progress toward some kind of ideal state and affect is about the kinds of emotions a person experiences on a daily basis (Diener et al., 1999; Boniwell, 2008). A third, lesser agreed-upon dimension is homeostatically protected well-being, where the individual feels overall contentment, as well as a lesser sense of excitement and happiness (Cummins, 2009).

While it is important to distinguish these dimensions of subjective well-being, the three are all clearly related. How content a person is will bear on their affect, which, in turn, relates to their overall life satisfaction (Davern et al., 2007).

Subjective well-being may be a far more important way to assess the status of a community or society than the economic indicators that are usually employed. Diener and Seligman (2004, p.1) define well-being as "peoples' positive evaluations of their lives, includes positive emotion, engagement, satisfaction, and meaning." They point out that while economic factors do influence well-being, social relationships and physical health have a greater impact, arguing that well-being is much more predictive of worker productivity, mental and physical health, and social and community relationships than is economic status. Diener et al. (2013) show that self-reported well-being is also influenced by the politics of the country in which one lives. However, it is a necessarily imprecise task to try to measure or even define something as subjective as well-being. Van Kamp et al. (2003) describe a similar concept, "quality of life," as the overlap of human community, natural environment, economics and geography (Wang, 2015). Diversity within the natural environment also contributes to well-being (Panagopoulos et al., 2016). On an individual scale, self-reported "well-being" can be measured with a survey or questionnaire, but this would be cost-prohibitive to administer on a large scale (Quercia et al., 2012). Fortunately, studies have found that results of sentiment analysis of social media content correlate strongly with self-reported life satisfaction for individuals (Kramer, 2010). This can also be applied to communities or even whole countries. For example, Kramer (2010) uses a sentiment analysis of Facebook posts to measure Gross National Happiness over time, though Wang (2015) has critiqued this approach due to weaknesses in statistical correlation of key variables. Quercia et al. (2012) did find a strong relationship between overall sentiment detected from Twitter data and economic status at the community and neighborhood level. Researchers at the Computational Story Lab of the University of Vermont and their colleagues provide a website called the "Hedonometer"[1] that continually monitors Twitter posts and assigns each an average happiness score (Dodds et al., 2011). Among its many applications, the Hedonometer project tracks sentiment over both time and space, and provides happiness rankings of cities and states.[2]

Microblog sentiment analysis

As an emerging field, the analysis of microblog data[3] as a means of gathering information about social issues has both strengths and weaknesses. It is a relatively fast and low-cost method of collecting freely volunteered opinions in real time from a wide range of the public on a wide range of topics. This is much simpler, cheaper and faster than conducting surveys or interviews, for example. However, there are also limitations to consider. Use of social media to express opinions and sentiment is much more pervasive among certain age groups and among those who have more access to smartphones and computers than it is among other groups. A Pew Research Center report from 2012 found that 15 percent of adults in the US used Twitter, and these individuals were most likely to be between the ages of 18 and 29 and live in urban areas (Smith and Brenner, 2012). Mislove et al. (2011) additionally found that Twitter users were significantly more likely to be male and live in densely populated areas. There are socioeconomic, linguistic and cultural factors that may also impact use of social media. Thus, any social media or microblog data collected in this way cannot be said to be a random sample of the population to be studied, and it is important to be aware that key demographic groups may be underrepresented.

Uses of sentiment analysis
Sentiment analysis is a quasi-qualitative analytical method, a form of content analysis that can be applied to large data sets including social media data sets. Unlike traditional content analysis, in which a researcher reads through a document and codes certain words and phrases, sentiment analysis is a more automated process, using a sentiment dictionary and a computer program to analyse large data sets.

Sentiment analysis of microblogging data has been used to consider social issues in a variety of studies. For example, sentiment analysis can be used to assess the public mood in response to events. Bollen et al. (2011) conducted a Twitter sentiment analysis in which they considered nationwide sentiment over a six-month period in 2008. They calculated a daily mood for their entire pool of data and correlated that with external events such as elections and holidays. Several other studies have compared microblogging sentiment analysis with the results of elections (Ceron et al., 2014; Gordon, 2013; O'Connor et al., 2010). Of special interest are those studies in which sentiment analysis has been used to compare different geographic areas. For example, Quercia et al. (2012) compared sentiment analysis of tweets geotagged to different areas of London, and found a strong correlation between expressed positive sentiment and higher socioeconomic variables for each area. A number of other studies have also used geotagged tweets to look at differences between different geographic areas (a. Antonelli et al., 2014; Balduini et al., 2013; Bertrand et al., 2013; Lovelace et al., 2014; Mearns et al., 2014; Mitchell et al., 2013).

Planning applications of microblog sentiment analysis
Our research falls within the category of urban and regional planning applications of sentiment analysis. While the field is still new, Twitter sentiment analysis has been applied successfully to urban studies and design topics. (b) Antonelli et al. (2014) and (b) Balduini et al. (2013) look at Twitter as a way to assess reactions to city-scale events, while MacEachren et al. (2011) and Beigi et al. (2016) apply similar methods to crisis management. (b) Lovelace et al. (2014) consider a very small scale, comparing how many visitors frequent different museums in Yorkshire, England based on tweets about the museums or tweets sent from the geographic locations of the museums. Geotagged tweets have also been used to track movement of people over time (Fujisaka et al., 2010), to determine

land use in urban environments (Frias-Martinez et al., 2013) and to map the location of self-identified hipsters, bankers and artists (Poorthuis and Zook, 2014).

The majority of these studies use only the quantitative data available from Twitter, rather than qualitatively analysing the content of specific tweets. Two key exceptions are (b) Mitchell et al. (2013) and (a) Hollander and Renski (2017), which both use sentiment as a proxy for happiness in comparing conventional indicators of well-being with Twitter posts. Both papers correlated happiness from sentiment analysis of Twitter with census data, where (c) Mitchell et al. (2013) found a strong correlation between cities with a higher percentage of white, married, higher-income residents and cities with higher happiness scores, (b) Hollander and Renski (2017) identified that declining cities had, on average, equal or higher happiness scores than growing cities (d. Mitchell et al., 2013). (b) Bertrand et al. (2013) conduct a sentiment analysis to tweets at a much finer level of geographic detail, to explore how sentiment varies across different areas of New York City and changes over time.

A related stream of research evaluates Twitter and other social media venues as a potential tool of public engagement. Schweitzer (2014) finds evidence that transit agencies engage more actively with other Twitter users, as opposed to simply blasting out information without the potential for a dynamic dialogue, and experienced a significantly improved level of Twitter discourse surrounding public transit. Evans-Cowley and Griffin (2012) evaluate tweets and other online microblog forums in the development of the Austin, Texas (USA) Strategic Mobility Plan. The authors conclude that microblogging can be used to stimulate engagement from a more expansive public than typical of conventional forums and effectively measure sentiment and public opinion. However, public officials struggled to turn the reams of information into "stories that could resonate with decision makers" (p.97). López and Zaragoza (2015) examine a similar planning process in Mexico City, reaching similar conclusions regarding its potential value. Lastly, Hollander et al. (2016) analysed the potential and pitfalls of using Twitter data in a variety of local urban contexts and found promise for the methodology.

Identifying historically significant properties
This section begins with some context on the challenges of identifying historically significant properties and then reviews key findings and methodological contributions of research in historic preservation on these questions.

The legal and regulatory approaches to historic preservation in the US are, as in many areas of policy, shaped heavily by state and local and non-governmental actors. While the National Historic Preservation Act of 1966 established the National Register of Historic Places, it is in fact a national list produced not centrally, but through applications coming from states, cities and towns. Even furthermore, most nominations to be on the National Register are driven by individuals and nonprofit organizations. In other words, there is no systematic effort to assess the sites of greatest importance and acknowledge them and protect them (Page, 2016). Further exacerbating the situation, has been the bias toward centering historic preservation efforts – listing and protecting – on buildings because of their architectural importance. Bluestone (2016) notes that the "curatorial" approach to historic preservation has meant that places of enormous community significance but little architectural value (in the eyes of architectural historians) too often fall by the wayside.

Efforts to expand federal, state and local lists of historically significant sites have started to have their effect. Increasingly sites that are of importance to ethnic and racial groups, or to groups subject to ostracization, are finding advocates for recognition. The elevation of the Stonewall Inn in New York City – which became the catalyst for the modern gay rights movement – to National Landmark status (a subcategory of listing on the National Register) is a case in point.

But perhaps most exciting has been a broader movement to identify "places that matter." Moving well beyond the focus on architectural significance, some individuals and organizations have developed a new approach to preservation which, first, looks for any places – building, landscape, even absent structures – that hold significance to a community (a town or a particular group) and, second, comes about these places through intentional community engagement. Organizations such as Place Matters in New York City do not use their own expertise to identify the "most important" places but rather cultivate conversation among community members so that the places that matter are developed organically and democratically.

The roots of this direction in historic preservation lie in the effort to identify "sites of conscience," or more broadly, simply "difficult places" – sites of controversy and violence that have often been covered over or erased. In seeking to tell the stories of lynching or confront the sites of state terrorism, or places related to the civil rights movement, these groups see historic places as crucial to community dialogue in the hopes of achieving healing and reparation. A growing ecosystem of human rights organizations and local advocacy groups have become intent on uncovering these "difficult places." And they are doing so equally committed to the process by which they identify and narrate these places, as to the identification of the sites alone (Page, 2016; Sevcenko, 2010; Rutherford and Rutherford, 2013).

As the US becomes increasingly diverse, and the varied ethnic and racial groups gain growing political stature, the pressure to broaden the definition of historic sites, and to engage a wider range of communities in the discussion of what is significant, will grow.

RESEARCH DESIGN, DATA AND METHODS

To execute this project, we needed first to develop a framework, to identify significant places in a new way. The process employs an urban social listening approach (Hollander et al., 2016) whereby we "listen to" social media activity for several months in the location of our study, the city of Holyoke in Hampden County, Massachusetts. Holyoke was selected because of our research team's pre-existing relationships with community leaders there and the city's rich architectural history. Most social media apps are not publicly available or not at a fine level of geography. The exception is both Twitter and Flickr, which make their application programming interfaces (APIs) open to the public and provide precise latitude and longitude of those posts for which users make their location available. We collected geotagged posts from both services over several months, mapping their locations (see Figure 22.1).

366 *Research handbook on community development*

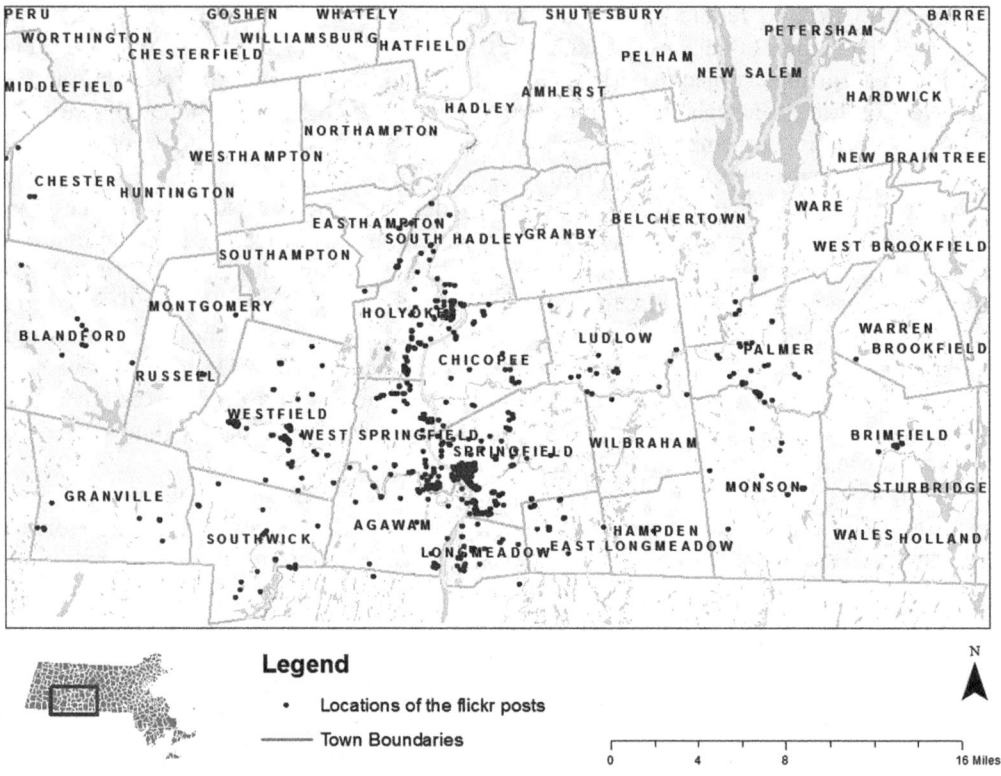

Figure 22.1 Flickr data collection for Hampden County, including Holyoke (Massachusetts)

Monitoring Sentiment through Twitter and Flickr Posts

For studying Twitter and Flickr posts, the first step is to identify bounding coordinates for Holyoke, and to capture the Tweets and Flickr posts posted within these bounds using our Urban Attitudes program. The Urban Attitudes program tracks a random sample of geographically identified (or "geotagged") tweets that fall within a specified rectangle (see Figure 22.1).[4] Because we are limited to rectangles, the sampled area does not perfectly match the municipal jurisdictional boundaries. We purposely set our bounding rectangles at the outmost edge of the broader Hampden County region and then clipped the posts to the boundaries of Holyoke.

We continuously collected tweets over a roughly four-month period from January to April 2017. Flickr was collected from January 2016 to January 2017. We determined that this was an acceptable timespan to distinguish durable and consistent differences in community sentiment, from those induced by temporary and/or one-time events.

We collected over 13 000 tweets and over 6000 Flickr posts across Hampden County and, in Holyoke, 668 usable tweets and 184 Flickr posts during this period. Many of the initially captured tweets were commercial solicitations, which do not truly reflect resident attitudes or opinions, which left only the usable ones. Most commonly, these are job announcements that are easily identified by a hash-tag (#) including keywords such as #Jobs, #Job, #Hiring and the like. These tweets contain little sentiment, and their inclusion may make resident sentiment appear to be more neutral than it really is. We thus exclude job announcements from our analysis. It was difficult to identify other forms of commercial solicitation, as they lack a set of consistent and common hash-tag keywords to aid in their identification.

As part of this project, we conducted a basic sentiment analysis of captured tweets, using a sentiment dictionary. The AFINN dictionary was developed by Finn Årup Nielsen, and has been used in multiple research studies, including an identification of anti-vaccine sentiments from tweets (Brooks, 2014), evaluation of more than 5000 advertisements in business magazines (Abrahams et al., 2013) and as part of a model predicting fluctuations in global currency markets (Jin et al., 2013). The version of AFINN used in this study includes 2477 sentiment words, all of which are ranked on an ordinal scale ranging from five to negative five depending upon the overall tone (positive or negative) and perceived degree of intensity. For example, the word "abusive" is given a score of negative three, while "satisfied" is given a score of two. The score of each sentiment word is summed up for every tweet, which are then aggregated to develop overall sentiment scores.

Developing a Framework for Identifying Significant Properties

To understand how these posts might be related to the underlying attitudes of people living in Holyoke, we also mapped all buildings listed on the National Register of Historic Places in the city. For each of the listings (12 in total), we established a 500-foot buffer and clipped all Tweets and Flickr posts within that buffer, for each of the listings. Next, we manually looked at each tweet (and if there was a link to a photo, typically Instagram photos) and each Flickr post (and its photo link) to determine whether the posts had any relationship with the National Register property. Then, we ran a sentiment analysis for all of those tweets (Flickr posts are not as well suited to sentiment analysis; they tend to be low on text).

Details on the results of this testing of the framework are presented below in the Results section. Because it yielded useful results, we then proceeded to conduct an open reading of all Tweets and Flickr posts within the city to identify significant properties (not already on the National Register). We began this by looking at the map of all posts (Figure 22.3) and attempted to determine if any obvious clustering was occurring that might suggest an underlying important place. We began with the Flickr posts. Fourteen locations (Potential Places that Matter) where some level of clustering appeared and we then followed links to photos and were able to identify architectural and landscape elements that might deem each as significant – in some cases we found clear identifying features for the location, in other cases we simply labeled the location with descriptive terms, such as "industrial canal area."

We then repeated the same process we ran for the National Register analysis, creating 500-foot buffers around each of the Potential Places that Matter clipping all Twitter and Flickr posts for each. Then, we read through those posts, following links whenever they existed, searching for social media post references to those actual Potential Places that Matter and running sentiment analysis of those posts. Details on what we found are presented below.

Finally, we took those maps and tables with Potential Places that Matter and presented them to Holyoke residents at two public meetings held in June 2017. Thirty-three residents attended the meetings, where a local community organizer assisted the research team with recruitment and facilitation. The meetings each lasted two hours and involved a discussion around the inductively derived places and whether they matter. It was a chance to determine if a handful of Flickr and Twitter posters' attention can be generalized to a broader community, whether the sentiment of the posts matters, and what other factors should be considered in refining this framework.

RESULTS

Across the data collection period, Twitter activity remained remarkably reliable and consistent, giving us confidence in using the data for additional analysis (see the Appendix for additional details). To develop a framework for using social media data to detect places that matter, we looked at the nature of social media activity in and around sites in Holyoke listed on the National Register of Historic Places. Nine sites were examined (excluding historic districts), see Table 22.1 and Figure 22.2.

For each site, we examined all Twitter and Flickr posts within 500 feet, searched each for a reference to that place, then performed sentiment analysis for each. See the results in Table 22.2.

Each location had very few tweets, as low as three for Robert Clovis Block and as high as 38 for Holyoke City Hall. The Percent Positive ranged from 33 percent to 100 percent as well, but the low number of tweets makes this not very useful information. Most important is that for Wistariahurst, the historic mansion of the Skinner family and the city's historical museum, there were two Twitter mentions within 500 feet. This was a place that appeared to matter more than other named locations among Twitter users.

Next, we conducted the same analysis for the entire city, using an open reading approach

Table 22.1 National Register of Historic Places in Holyoke, MA

Ref. no.	State	County	City	Resource name	Address
80000473	MA	Hampden	Holyoke	Holyoke Canal System	Front and South St, and CT River
79000346	MA	Hampden	Holyoke	Caledonia Building	185–193 High St
75000259	MA	Hampden	Holyoke	Holyoke City Hall	536 Dwight St
83003980	MA	Hampden	Holyoke	Maplewood Hotel	328–330 Maple St
73000295	MA	Hampden	Holyoke	Wistariahurst	238 Cabot St
86000122	MA	Hampden	Holyoke	US Post Office–Holyoke Main	650 Dwight St
02001472	MA	Hampden	Holyoke	Robert Clovis Block	338–348 Main St
02001473	MA	Hampden	Holyoke	Friedrich Block	449–461 Main St
04000931	MA	Hampden	Holyoke	Prospect Park	Maple St, Arbor Way, Connecticut River

Source: www.nationalregisterofhistoricplaces.com.

Community development and place attachment 369

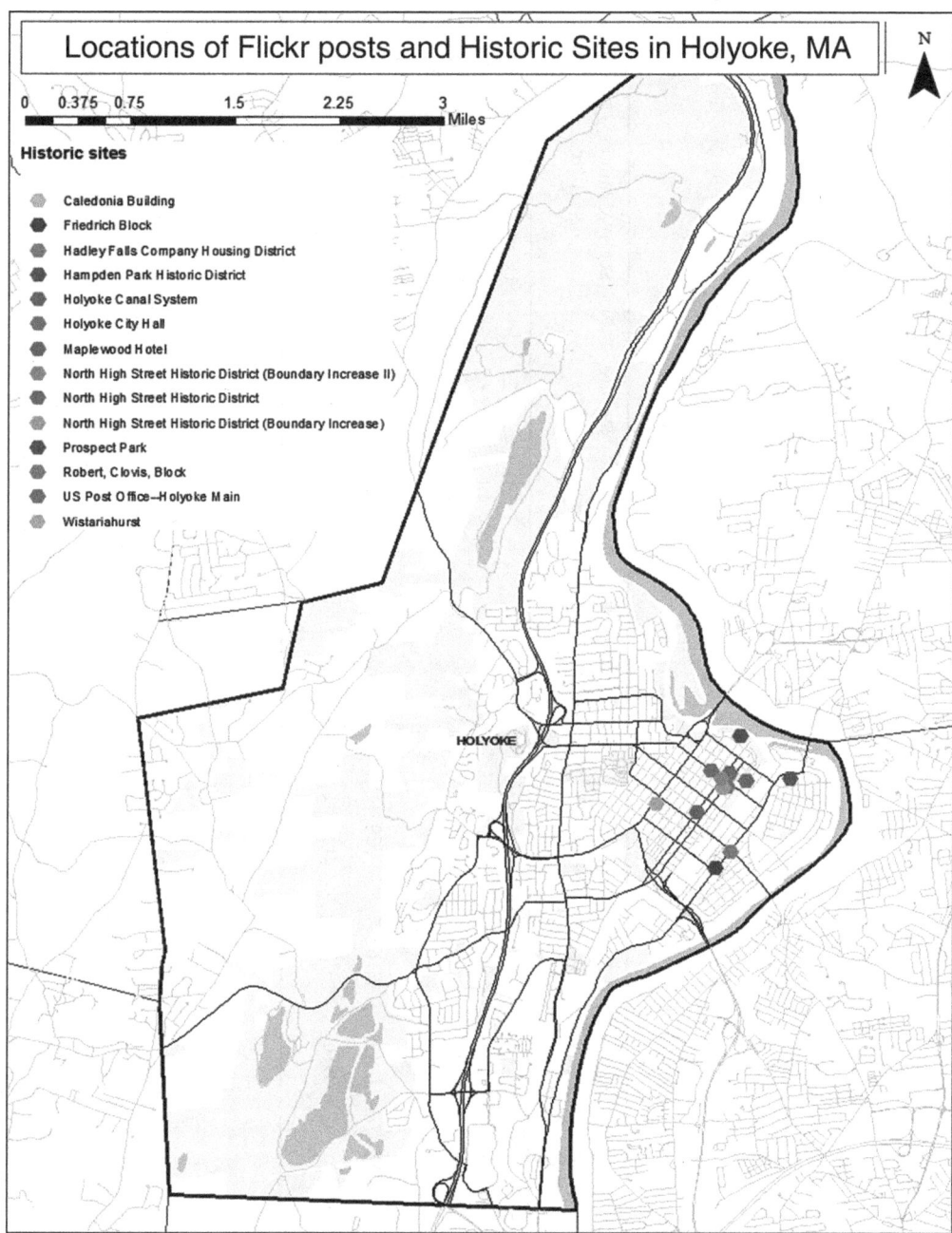

Figure 22.2 Locations of National Register sites in Holyoke, MA

Table 22.2 Tweets within 500 feet of National Register sites with sentiment scores

Site_name	Sent_Contain_Twt	Num_Positive	Pct_Positive (%)	Num_Negative	Pct_Negative (%)	Ratio_Pos-Neg	Positive_Scrs	Negative_Scrs	Net_Scr	Mentions_Twtr	Mentions_Flckr
Caledonia	17	14	82	7	41	2.0	53	−22	31	0	0
Friedrich	2	2	100	2	100	1.0	10	−2	8	0	0
Holyoke City Hall	38	36	95	5	13	7.2	126	−10	116	0	0
Robert Clovis	3	1	33	3	100	0.3	2	−5	−3	0	0
Wistariahurst	72	49	68	29	40	1.7	125	−64	61	2	0
Total for Hampden	5365	3986	73	2156	41	1.8	11631	−4278	7353		

to the Flickr data first. All the Twitter and Flickr posts are presented in proximity to the National Register historic sites, in Figure 22.3. We found important clustering of Flickr images, suggesting important places, accompanied then by Twitter activity. From this we identified 11 places, Potential Places that Matter (see Figure 22.4).

We searched for mentions on both Twitter and Flickr for each place; the results are presented in Table 22.3. Like the National Register sites, very few of the proximate posts concerned the particular place of community importance, but some did. For Gateway City Arts, three Twitter messages referred specifically to the location. We also ran sentiment analysis for each location where there was Twitter activity, but like with the historic analyses, the numbers are low so drawing conclusions here would not be prudent.

Community Engagement Meetings

The Twitter and Flickr data proved to be an enormously fruitful way to launch the two public meetings we held. By describing the almost unconscious "listing" that was conducted by residents and visitors, through tweeting and taking photographs, community members were jumpstarted into rethinking what constitutes an historic site. At the public meetings, community members offered feedback on the Twitter- and Flickr-identified locations and made only minor adjustments to the places we selected, in general confirming the validity of the framework we employed to select places that matter. The community feedback was distilled into edits we made to the proposed Places that Matter list and is presented below as an underlined document (see Figure 22.5).

Through a facilitated series of exercises and group discussions, we also solicited community members' ideas about 63 other places in Holyoke that matter to them, including individual streets, parks, private homes, and museums in the city. The community feedback validated the social media generated list, but also provided an even broader list of places that people care about.

CONCLUSION

Through an innovative merger of quantitative and qualitative methods – "listening" both to social media and to real people in discussion – this project sought to validate a strategy for identifying places that matter to people using social media data. Community meetings helped refine the social media results, as well as inspire reflection on locations not picked up by the Twitter and Flickr analysis. The general support for the 11 social media generated locations validated this approach. Community feedback suggested that, along with in-person public outreach, using social media posts can serve as a useful starting point for cataloging places that matter to people.

The short period of Twitter data collection is a major limitation of this research. Small numbers of posts make any sentiment analysis difficult to interpret, so we recommend any future project allows for at least a full year of social media data collection prior to analysis.

Future research ought to produce easily adaptable tools – a robust, intuitive website, mobile app, a system for obtaining Twitter and Flickr data and training manuals for conducting "places that matter" community meetings – to build on this research and advance the community development field. Expanded to other cities in a region or state,

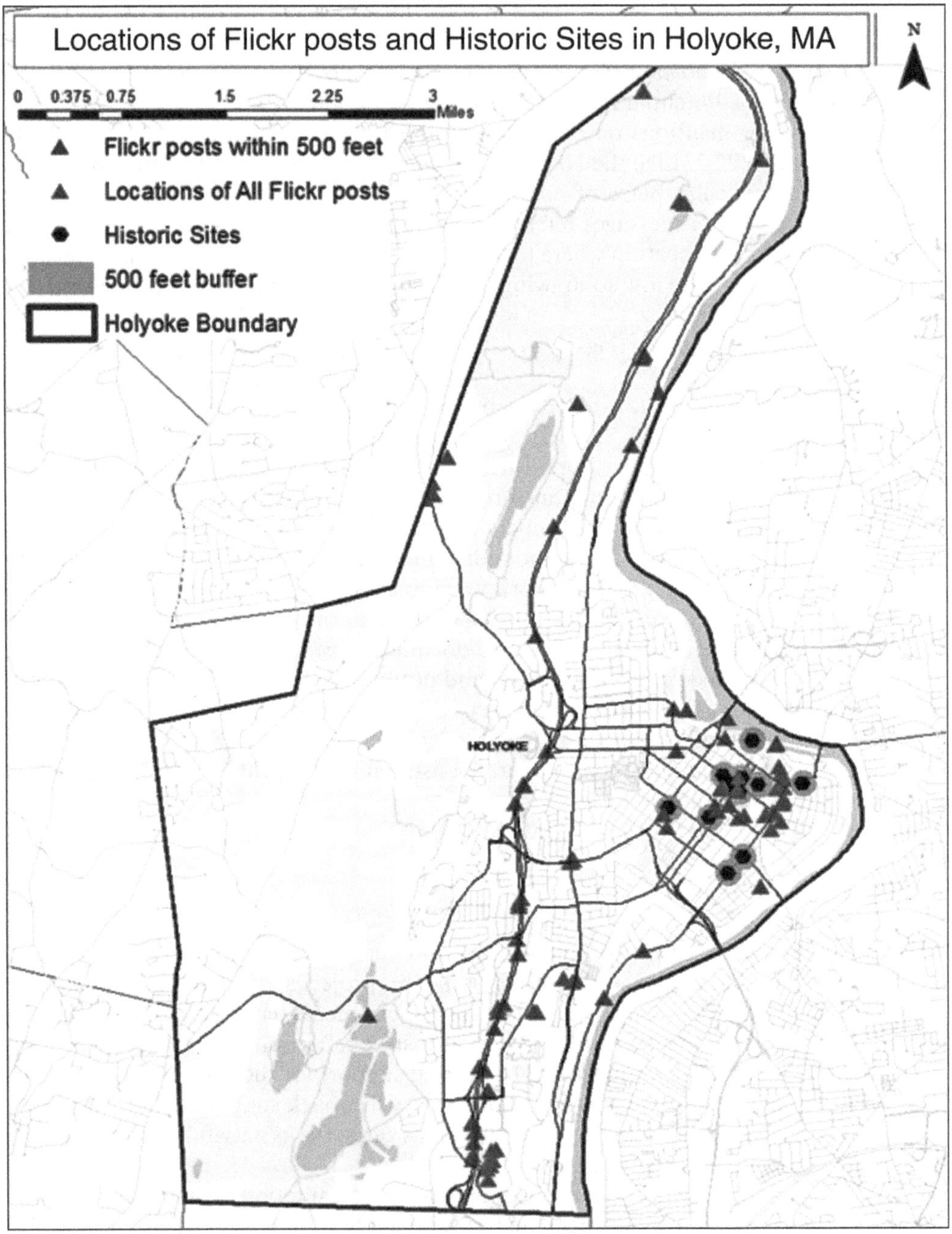

Figure 22.3 Twitter and Flickr posts in relation to National Register sites within Holyoke

Community development and place attachment 373

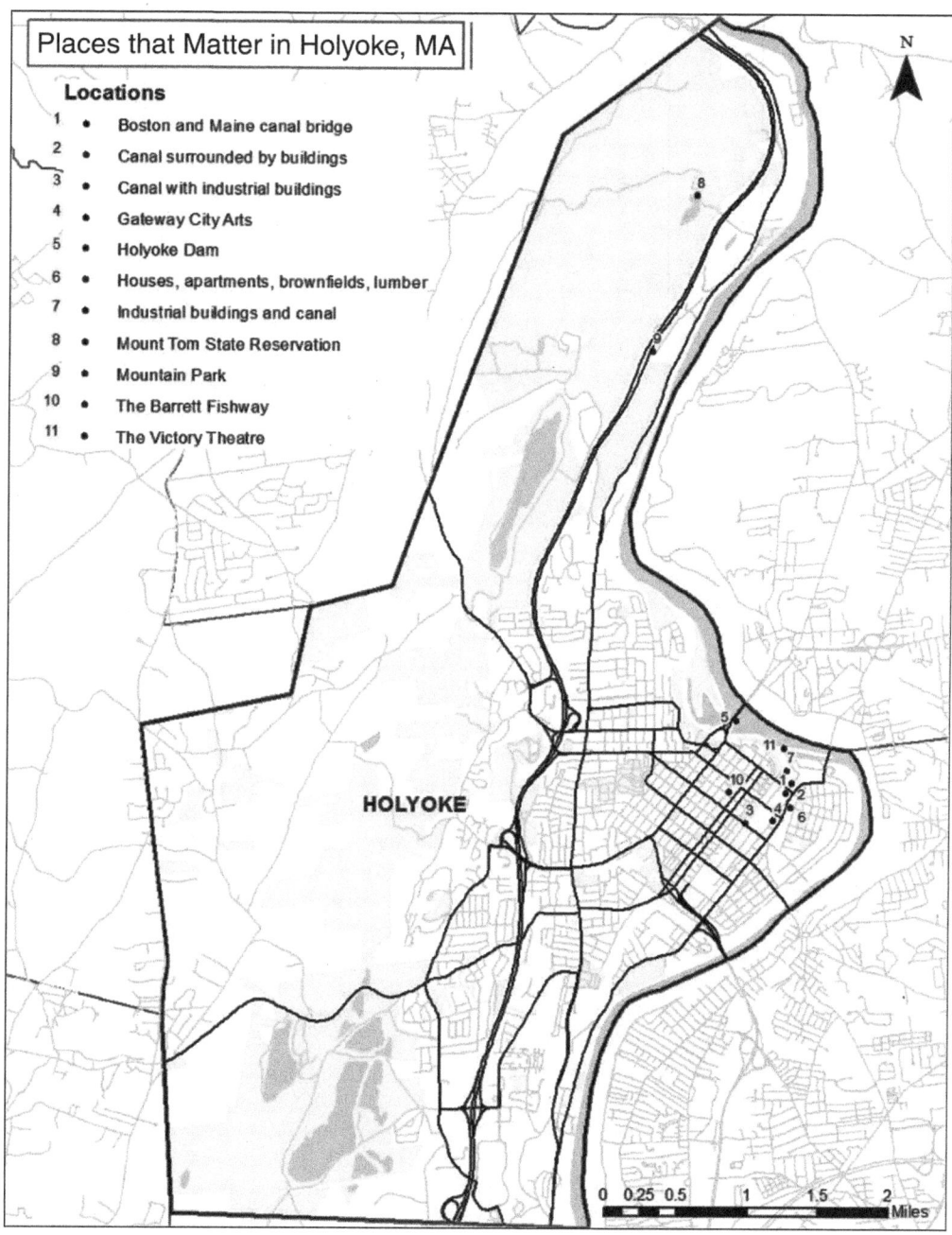

Figure 22.4 Potential Places that Matter in Holyoke, MA

Table 22.3 Tweets about Potential Places that Matter with sentiment scores

	Sent_Contain_Twt	Num_Positive	Pct_Positive (%)	Num_Negative	Pct_Negative (%)	Ratio_Pos-Neg	Positive_Scrs	Negative_Scrs	Net_Scr	Mentions_Twtr	Mentions_Flckr
Places that matter											
The Victory Theatre	34	35	88	5	13	7.0	125	−10	115	0	6
Gateway City Art	2	1	50	1	50	1	2	−4	−2	3	1
Canal surrounded by building	2	1	100	0	0	0	6	0	6	0	2
Canal with industrial building	1	1	100	0	0	0	3	—	3	0	1
Boston and Maine canal bridge	—	—	—	—	—	—	—	—	—	0	2
Holyoke Dam	—	—	—	—	—	—	—	—	—	0	1
Houses, apartments, brown.	—	—	—	—	—	—	—	—	—	0	3
Industrial buildings and canal	—	—	—	—	—	—	—	—	—	0	1
Mount Tom State Reservation	—	—	—	—	—	—	—	—	—	0	31
Mountain Park	—	—	—	—	—	—	—	—	—	0	5
The Barrett Fishway	—	—	—	—	—	—	—	—	—	0	2
Total for Hampden	5365	3986	73	2156	41	1.8	11631	−4278	7353		
Total for Holyoke	171	140	82	52	30	2.7	452	−117	335		

10. The Victory Theatre
4. Gateway City Arts
2. Boat Lock Station overlooking Conklin Office [aka Hampden Mill] (Canal surrounded by building)
3. Heritage State Park (Canal with industrial building)
1. Open Square (Boston and Maine canal bridge)
5. Holyoke Dam (combined with #9 The Barrett Fishway)
6. Houses, apartment, brown,(remove this, it is NOT a place that matters)
7. The Where House [Jimmy Kern's Banquet Hall] (Industrial buildings and canal)
8. Mount Tom State Reservation
9. Mountain Park
~~The Barrett Fishway~~

Figure 22.5   Places that matter edited based on community feedback

this approach represents an important new way for professionals to identify and mark places that matter most to a community and could lead to an entirely different type of historic "register."

NOTES

* We gratefully acknowledge the financial support of the 1772 Foundation and Mary Anthony's leadership for making this research possible. Research assistance was provided by Judy Fung of Tufts University.
1. See http://hedonometer.org/index.html (accessed 4 May 2019).
2. See http://hedonometer.org/maps.html for state rankings and http://hedonometer.org/cities.html for city rankings (both accessed 4 May 2019).
3. Microblogs are short (in the case of Twitter, generally less than 140 characters) voluntary posts that people make online about their "lifeworld".
4. The Twitter API allows access to a (roughly) 1 percent random sample of the all tweets. Geotagged tweets are available from any Twitter user who makes their location available, which is currently not the default option for new users. Scholars estimate that nearly 20 percent of all Twitter users have made their location available (Weidemann and Swift, 2013).

REFERENCES

Abrahams, A.S., Coupey, E., Zhong, E.X., Barkhi, R., and Manasantivongs, P.S. (2013). Audience targeting by B-to-B advertisement classification: A neural network approach. *Expert Systems with Applications*, *40*(8), 2777–91. https://doi.org/10.1016/j.eswa.2012.10.068.
Antonelli, F., Azzi, M., Balduini, M., Ciuccarelli, P., Della Valle, E.D., and Larcher, R. (2014). City sensing: Visualising mobile and social data about a city scale event. Conference paper. In *AVI'14 Proceedings of the 2014 International Working Conference on Advanced Visual Interfaces* (pp. 337–8). Como, Italy, 27–29 May. https://doi.org/10.1145/2598153.2600032.
Balduini, M., Della Valle, E., Dell'Aglio, D., Tsytsarau, M., Palpanas, T., and Confalonieri, C. (2013). Social

listening of city scale events using the Streaming Linked Data Framework. In H. Alani, principal editor (eds), *The Semantic Web: ISWC 2013* (pp. 1–16). Berlin Heidelberg: Springer.

Beigi, G., Hu, X., Maciejewski, R., and Liu, H. (2016). An overview of sentiment analysis in social media and its applications in disaster relief. In W. Pedrycz and S.-M. Chen (eds), *Sentiment Analysis and Ontology Engineering: An Environment of Computational Intelligence* (pp. 313–40). Cham: Springer International Publishing. https://doi.org/10.1007/978-3-319-30319-2_13.

Bertrand, K.Z., Bialik, M., Virdee, K., Gros, A., and Bar-Yam, Y. (2013). Sentiment in New York city: A high resolution spatial and temporal view. *arXiv preprint arXiv:1308.5010*. [physics.soc-ph]. https://arxiv.org/abs/1308.5010v1. Accessed 12 May 2019.

Bluestone, D. (2016). Dislodging the curatorial. In M. Page (ed.), *Bending the Future: Fifty Ideas for the Next Fifty Years of Historic Preservation in the United States*. Amherst: University of Massachusetts Press.

Bollen, J., Mao, H., and Pepe, A. (2011). Modeling public mood and emotion: Twitter sentiment and socioeconomic phenomena. Conference paper. In *Fifth International AAAI Conference on Weblogs and Social Media*. July. https://www.aaai.org/Conferences/ICWSM/icwsm11.php. Accessed 7 January 2020.

Boniwell, I. (2008). *Positive Psychology in a Nutshell: A Balanced Introduction to the Science of Optimal Functioning*. New York: McGraw-Hill.

Brooks, B. (2014). Using Twitter data to identify geographic clustering of anti-vaccination sentiments. PhD dissertation. University of Washington.

Ceron, A., Curini, L., and Iacus, S.M. (2016). *Politics and Big Data: Nowcasting and Forecasting Elections with Social Media*. Abingdon: Routledge.

Ceron, A., Curini, L., Iacus, S.M., and Porro, G. (2014). Every tweet counts? How sentiment analysis of social media can improve our knowledge of citizens' political preferences with an application to Italy and France. *New Media & Society*, 16(2), 340–58. https://doi.org/10.1177/1461444813480466.

Culotta, A. (2014). Estimating county health statistics with twitter. Conference paper. In *Proceedings of the 32nd Annual ACM Conference on Human Factors in Computing Systems* (pp. 1335–44). ACM. Toronto, Ontario, Canada, 26 April–1 May.

Cummins, R.A. (2009). Measuring population happiness to inform public policy. Paper. In *The 3rd OECD World Forum on "Statistics, Knowledge and Policy" Charting Progress, Building Visions, Improving Life* (pp. 27–30). Busan, Korea, 27–30 October 2009.

Davern, M.T., Cummins, R.A., and Stokes, M.A. (2007). Subjective wellbeing as an affective-cognitive construct. *Journal of Happiness Studies*, 8(4), 429–49.

Diener, E., and Seligman, M.E. (2004). Beyond money: Toward an economy of well-being. *Psychological Science in the Public Interest*, 5(1), 1–31.

Diener, E., Inglehart, R., and Tay, L. (2013). Theory and validity of life satisfaction scales. *Social Indicators Research*, 112(3), 497–527.

Diener, E., Suh, E.M., Lucas, R.E., and Smith, H.L. (1999). Subjective well-being: Three decades of progress. *Psychological Bulletin*, 125(2), 276–302.

Dodds, P.S., Harris, K.D., Kloumann, I.M., Bliss, C.A., and Danforth, C.M. (2011). Temporal patterns of happiness and information in a global social network: Hedonometrics and Twitter. *PloS One*, 6(12), e26752. https://journals.plos.org/plosone/article?id=10.1371/journal.pone.0026752. Accessed 12 May 2019.

Evans-Cowley, J.S., and Griffin, G. (2012). Microparticipation with social media for community engagement in transportation planning. *Transportation Research Record*, 2307(1), 90–98. https://doi.org/10.3141/2307-10.

Frias-Martinez, V., Soto, V., Hohwald, H., and Frias-Martinez, E. (2013). Sensing urban land use with twitter activity. Telefonica Research, Madrid, Spain.

Fujisaka, T., Lee, R., and Sumiya, K. (2010). Exploring urban characteristics using movement history of mass mobile microbloggers. Conference paper. In *Proceedings of the Eleventh Workshop on Mobile Computing Systems & Applications* (pp. 13–18). ACM. Annapolis, Maryland, 22–23 February.

Gordon, J. (2013). Comparative geospatial analysis of Twitter sentiment data during the 2008 and 2012 US Presidential elections. Master's thesis. University of Oregon.

Hollander, J.B. (2011). Can a city successfully shrink? Evidence from survey data on neighborhood quality. *Urban Affairs Review*, 47(1), 129–41.

Hollander, J.B. (2018). *A Research Agenda for Shrinking Cities*. Cheltenham, UK and Northampton, MA, USA: Edward Elgar Publishing.

Hollander, J.B., and Renski, H. (2017). Measuring urban attitudes embedded in microblogging data: Shrinking versus growing cities. *Town Planning Review*, 88(4), 465–90.

Hollander, J.B., Graves, E., Renski, H., Foster-Karim, C., Wiley, A., and Das, D. (2016). *Urban Social Listening: Potential and Pitfalls for Using Microblogging Data in Studying Cities*. New York: Springer.

Jin, F., Self, N., Saraf, P., Butler, P., Wang, W., and Ramakrishnan, N. (2013). Forex-foreteller: Currency trend modeling using news articles. Conference paper. In *Proceedings of the 19th ACM SIGKDD International Conference on Knowledge Discovery and Data Mining* (pp. 1470–73). ACM. Chicago, Illinois, 13 August 2013.

Kramer, A.D. (2010). An unobtrusive behavioral model of gross national happiness. Conference paper. In

Proceedings of the SIGCHI Conference on Human Factors in Computing Systems (pp. 287–90). ACM. Atlanta, Georgia, USA, 10–15 April.

López-Ornelas, E., and Morales Zaragoza, N. (2015). Social media participation: A narrative way to help urban planners. In G. Meiselwitz (ed.), *Social Computing and Social Media* (pp. 48–54). New York: Springer International Publishing.

Lovelace, R., Malleson, N., Harland, K., and Birkin, M. (2014). Geotagged tweets to inform a spatial interaction model: A case study of museums. *arXiv preprint arXiv:1403.5118*. https://arxiv.org/abs/1403.5118. Accessed 12 May 2019.

MacEachren, A.M., Robinson, A.C., Jaiswal, A., Pezanowski, S., Savelyev, A., Blanford, J., and Mitra, P. (2011). Geo-twitter analytics: Applications in crisis management. Conference paper. In *25th International Cartographic Conference* (pp. 3–8). Paris, France, July.

Mearns, G., Simmonds, R., Richardson, R., Turner, M., Watson, P., and Missier, P. (2014). Tweet my street: A cross-disciplinary collaboration for the analysis of local twitter data. *Future Internet*, 6(2), 378–96.

Mislove, A., Lehmann, S., Ahn, Y.Y., Onnela, J.P., and Rosenquist, J.N. (2011). Understanding the demographics of twitter users. Conference paper. In *Fifth International AAAI Conference on Weblogs and Social Media*. July. https://www.aaai.org/ocs/index.php/ICWSM/ICWSM11/paper/view/2816/3234. Accessed 12 May 2019.

Mitchell, L., Frank, M.R., Harris, K.D., Dodds, P.S., and Danforth, C.M. (2013). The geography of happiness: Connecting twitter sentiment and expression, demographics, and objective characteristics of place. *PloS One*, 8(5), e64417. https://journals.plos.org/plosone/article?id=10.1371/journal.pone.0064417. Accessed 12 May 2019.

O'Connor, B., Balasubramanyan, R., Routledge, B.R., and Smith, N.A. (2010). From tweets to polls: Linking text sentiment to public opinion time series. Conference paper. In *Fourth International AAAI Conference on Weblogs and Social Media*. May. https://www.aaai.org/ocs/index.php/ICWSM/ICWSM10/paper/viewFile/1536/1842. Accessed 12 May 2019.

Page, M. (2016). *Why Preservation Matters*. New Haven: Yale University Press.

Pallagst, K., Wiechmann, T., and Martinez-Fernandez, C. (eds). (2013). *Shrinking Cities: International Perspectives and Policy Implications*. New York: Routledge.

Panagopoulos, T., Duque, J.A.G., and Dan, M.B. (2016). Urban planning with respect to environmental quality and human well-being. *Environmental Pollution*, 208(A), 137–44.

Poorthuis, A., and Zook, M. (2014). Artists and bankers and hipsters, oh my! Mapping tweets in the New York Metropolitan Region. *Cityscape*, 16(2), 169–72.

Quercia, D., Ellis, J., Capra, L., and Crowcroft, J. (2012). Tracking gross community happiness from tweets. Conference paper. In *Proceedings of the ACM 2012 Conference on Computer Supported Cooperative Work* (pp. 965–8). ACM. Seattle, Washington, USA, 11–15 February.

Rutherford, S., and Rutherford, P. (2013). Geography and biopolitics. *Geography Compass*, 7(6), 423–34.

Schweitzer, L. (2014). Planning and social media: A case study of public transit and stigma on Twitter. *Journal of the American Planning Association*, 80(3), 218–38.

Sevcenko, L. (2010). Sites of conscience: New approaches to conflicted memory. *Museum International*, 62(1/2), 20–25.

Smith, A., and Brenner, J. (2012). Twitter use 2012. *Pew Internet & American Life Project*, 4. https://www.pewinternet.org/wp-content/uploads/sites/9/media/Files/Reports/2012/PIP_Twitter_Use_2012.pdf. Accessed 12 May 2019.

Tumasjan, A., Sprenger, T.O., Sandner, P.G., and Welpe, I.M. (2010). Predicting elections with twitter: What 140 characters reveal about political sentiment. Conference paper. In *Fourth International AAAI Conference on Weblogs and Social Media*. May. https://www.aaai.org/ocs/index.php/ICWSM/ICWSM10/paper/viewPaper/1441. Accessed 12 May 2019.

Van Kamp, I., Leidelmeijer, K., Marsman, G., and De Hollander, A. (2003). Urban environmental quality and human well-being: Towards a conceptual framework and demarcation of concepts – a literature study. *Landscape and Urban Planning*, 65(1/2), 5–18.

Wang, Jiun-Hao. (2015). Happiness and social exclusion of Indigenous peoples in Taiwan: A social sustainability perspective. *PloS One*, 10(2), e0118305. https://www.ncbi.nlm.nih.gov/pmc/articles/PMC4335048/.

Weidemann, C., and Swift, J. (2013). Social media location intelligence: The next privacy battle – An ArcGIS add-in and analysis of geospatial data collected from Twitter.com. *International Journal of Geoinformatics*, 9(2), 21–7.

APPENDIX

Before examining the social media data, it was important to first assess the reliability of the data across time. In Appendix Table 22A.1 we show all of the tweets across the time period of data collection, with sentiment scores indicated.

Table 22A.1 *Number of tweets collected with sentiment scores*

Week	Week_Num	Num_Tweets	Sent_Contain_Twt	Num_Positive	Pct_Positive	Num_Negative	Pct_Negative	Ratio_Pos-Neg	Positive_Scrs	Negative_Scrs	Net_Scr
1/2/2017	1	264	107	77	0.72	46	0.43	1.67	194	−79	115
1/9/2017	2	251	124	82	0.66	63	0.51	1.30	205	−129	76
1/23/2017	3	820	342	249	0.73	154	0.45	1.62	804	−305	499
1/30/2017	4	864	366	263	0.72	147	0.40	1.79	723	−276	447
2/6/2017	5	210	82	55	0.67	34	0.41	1.62	180	−65	115
2/13/2017	6	778	334	242	0.72	133	0.40	1.82	661	−260	401
2/20/2017	7	931	377	284	0.75	152	0.40	1.87	853	−282	571
2/27/2017	8	466	194	141	0.73	83	0.43	1.70	402	−154	248
3/6/2017	9	990	412	318	0.77	161	0.39	1.98	901	−325	576
3/13/017	10	1210	429	319	0.74	173	0.40	1.84	881	−342	539
3/20/2017	11	990	397	290	0.73	188	0.47	1.54	806	−495	311
3/27/2017	12	1077	398	268	0.67	181	0.45	1.48	769	−339	430
4/3/2017	13	956	359	269	0.75	130	0.36	2.07	830	−248	582
4/10/2017	14	1063	430	344	0.80	168	0.39	2.05	1016	−328	688
4/17/2017	15	1047	455	368	0.81	140	0.31	2.63	1143	−258	885
4/24/2017	16	1081	412	319	0.77	137	0.33	2.33	952	−254	698
5/1/2017	17	374	147	98	0.67	66	0.45	1.48	311	−139	172
Total		13372	5365	3986	0.73	2156	0.41	1.81	11631	−4278	7353

This table shows how fairly consistent the Twitter activity was, across Hampden County, over the 17 weeks of data collection. We ran additional statistics and determined that the indicator "Ratio of Positive-Negative" was the most valid and reliable measure across time (see Figures 22A.1–22A.3).

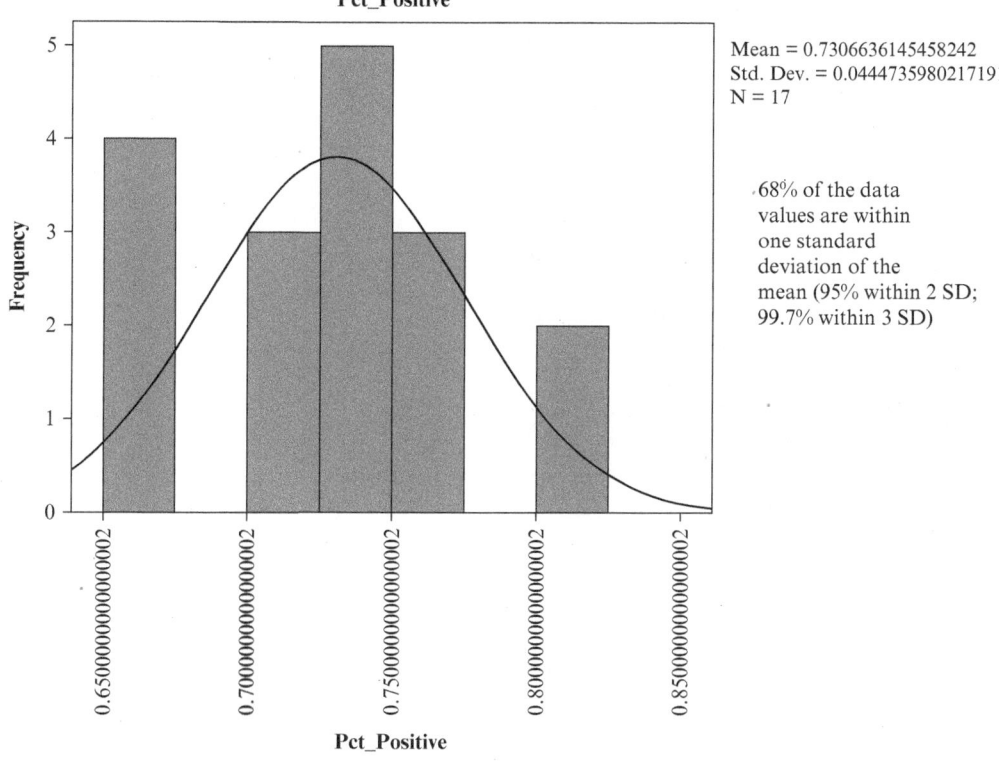

Figure 22A.1   Frequency of tweets based on positive sentiment scores

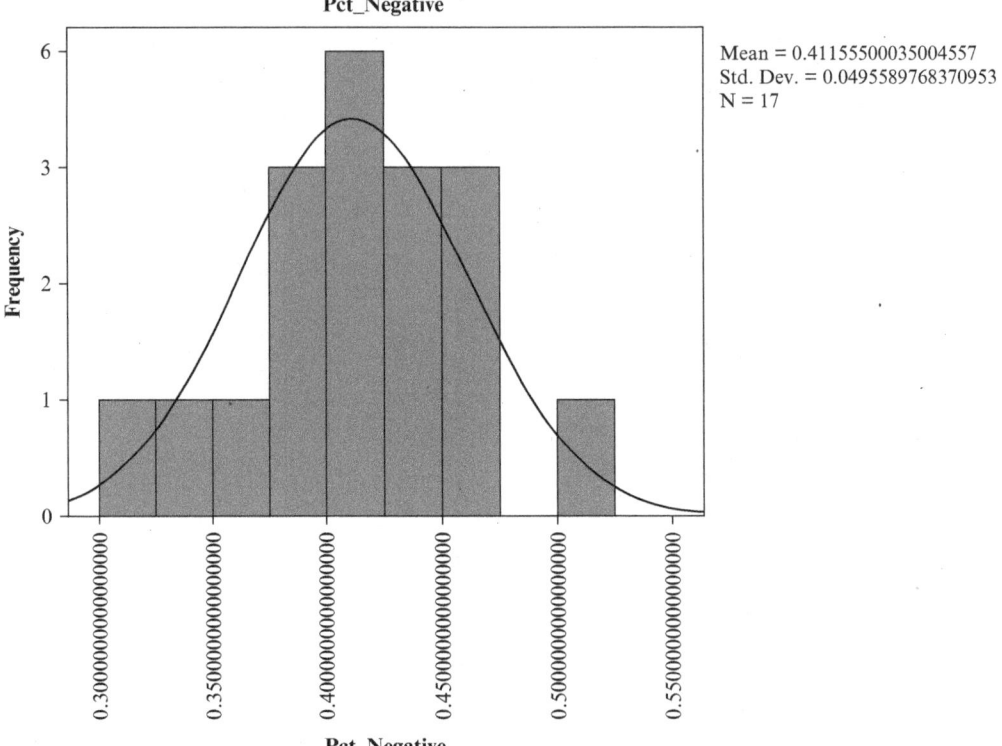

Figure 22A.2   Frequency of tweets based on negative sentiment scores

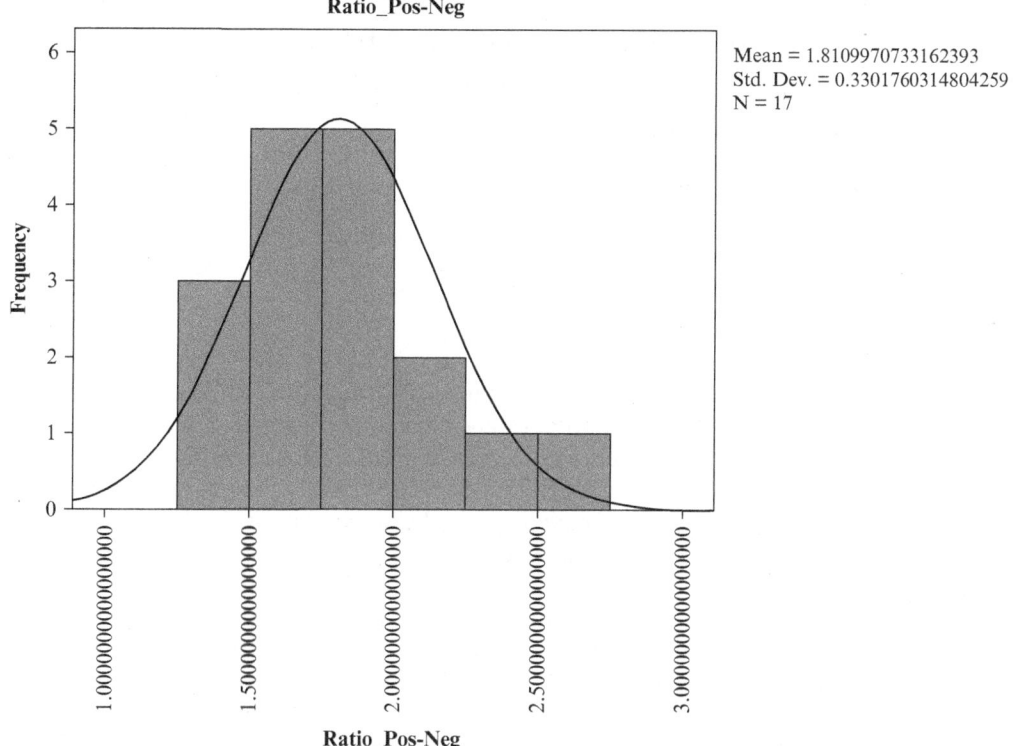

Figure 22A.3   Frequency of tweets based on ratio of positive to negative sentiment score

23. Re-imagining democratic research processes in community-based development: a case for photovoice
Camille Sutton-Brown

INTRODUCTION

Photovoice is a qualitative research method that asks participants to document, reflect on and represent their using a specific photographic technique (Wang et al., 2000). As part of the research process, participants engage in critical group dialogue to identify their individual and collective concerns. They then present these concerns about their community to policy makers for the purpose of enacting social change. Photovoice has been used to conduct community health needs assessments (Wang et al., 1998), and this chapter demonstrates that photovoice can also be used to conduct needs assessments, as well as formative and outcome evaluations, of development initiatives. Specifically, photovoice can be used for these purposes in the context of community-based development (CBD). CBD is a capacity-building model that international development practitioners use in their attempt to enhance communities. It asks the local beneficiaries to identify their community's needs and to help design initiatives that can potentially benefit them and their community. There are commonalities between the methodology of photovoice and the CBD model, including their valuing of experiential knowledge, the adherence to a participatory model and their orientation toward community enhancement. The similarity of their processes permits photovoice to be an appropriate research methodology to use in CBD.

This chapter uses the following structure. First, a brief overview of CBD and participatory research is presented followed by a more detailed description of photovoice before discussing the tenets that characterize both CBD and photovoice to present the commonalities between the two models. The chapter then concludes with a critical examination of photovoice and CBD and strategies to minimize potential limitations.

International development is "the process of expanding human freedoms" (Sen, 1999, p. 36) to improve the quality of life for all of humanity. Such freedoms relate to the Universal Declaration of Human Rights put forth by the United Nations, some of which include access to health care and education as well as access to enough food to prevent starvation and malnourishment (Escobar, 1995; Lindenberg and Bryant, 2001; Sen, 1999). To address the 736 million people who live in extreme poverty (The World Bank Group, 2019), who likely do not experience such freedoms, international development initiatives are typically directed toward poverty-alleviation strategies.

Historically, development practitioners have used a top-down development model, which contemporary international development theorists critique for its inherent unequal power relations. The top-down development model refers to a centralized, hierarchal model (Teshome, 2007) that fails to consult the local stakeholders in decision-making processes. Typically, it is Western-based program managers who top this hierarchy, a standpoint from which they apply the current best practices to make key decisions for the intended beneficiaries. Decades of time and billions of dollars have been spent on this unsuccessful development model (Easterly, 2007) that is criticized by development theorists for its paternalistic approach (Escobar, 1995). The desire for a more effective poverty-reduction approach led to a reform in international development, resulting in a shift toward a bottom-up model.

The application of the bottom-up model, hereinafter referred to as participatory models, to international development discourse and practice has increased in recent years. Participatory models position the intended beneficiaries at the center of decision-making processes to direct the development initiatives. Due to its heavy reliance on local input, theorists and practitioners consider participatory models to be an effective means to promote and encourage sustainable development. One such participatory model is CBD. CBD is a generic term used to describe international development models which have a primary objective of including beneficiaries in project design and management (Mansuri and Rao, 2003). Practitioners adhering to a CBD model focus their efforts on community capacity-building by involving the beneficiaries in all aspects of program design, implementation and evaluation activities. This encourages the beneficiaries to assume more responsibility for their community than was permitted in the top-down development models, and the resulting initiatives are direct reflections of the community's self-identified needs and desires. As Dickmann et al. (2018) contend, the goal for CBD initiatives should be not only to engage, but also to empower communities. Doing so, they argue, will better equip communities to manage short-term crises and to build long-term resilience.

The problem, however, is that though many international development practitioners have shifted to participatory models for program delivery, the research methods that they use within these practices continue to reflect a top-down model. This is highly problematic and provides a strong rationale for the appropriateness of using a participatory research model with CBD rather than researcher-directed research methods. This will allow for better alignment between the program delivery and the research model.

Participatory research is a relatively new model within the social sciences, though it is now increasing in popularity and being used in a variety of ways (Moletsane et al., 2007). Similar to CBD, participatory research models were developed in response to critiques of the power dynamics inherent in traditional research methods. In traditional research models, the researcher usually makes the key decisions that pertain to the research process. These typically include choosing the topic of study, designing the research process, collecting and analysing data and disseminating findings. Participants' roles are usually limited to data generating activities with little, if any, control over the research process. In participatory research models the participants assume greater decision-making agency than in traditional researcher-directed approaches, and the research process itself is considered to be beneficial to the participants. Photovoice is one example of participatory research methods.

PHOTOVOICE

Photovoice is a participatory research methodology wherein participants document their community's strengths and concerns using a specific photographic technique (Wang,

1999). Participant-produced images and testimonies defy traditional research norms by asking participants to lead the research process and through prioritizing their experiential knowledge to advocate for change. Borne out of the three theoretical perspectives of feminism, documentary photography and Freire's education for critical consciousness, photovoice is advocacy-driven and typically used with marginalized populations who represent their community from their perspective (Foster-Fishman et al., 2005). Freire's work centers on democracy and critical pedagogy, particularly his seminal text, Pedagogy of the Oppressed (1972). Each of these concepts are embedded in the philosophical underpinnings of photovoice as, above all else, the outcomes for photovoice projects are aimed at benefiting the actual participants involved through advocating for social change.

As photovoice is participant-directed the preliminary questions that guide a study are malleable. This openness allows for increased responsiveness to the participants' identified needs, as well as for the exploration of new areas of discovery as they emerge (Patton, 2002). The flexibility of the research design makes it highly adaptable to best fit the participants' specific needs as well as the socio-political context of the community in which the project is situated (Burris and Wang, 1997). Though this methodology lacks a regimented uniform structure for implementation, there is a general framework and certain procedures to characterize it as a photovoice study (Sutton-Brown, 2014). Wang and Burris outline the following steps for conducting a photovoice study. These are:

1. Select and recruit a target audience of policy makers or community leaders
2. Recruit a group of photovoice participants
3. Introduce the photovoice methodology to participants and facilitate a group discussion
4. Obtain informed consent
5. Pose an initial theme for taking pictures
6. Distribute cameras to participants and review how to use them
7. Provide time for participants to take pictures
8. Meet to discuss photographs
9. Plan with participants a format to share photographs and stories with policy makers or community leaders (Wang, 1999).

Photovoice has been applied to many disciplines including education, disability studies, health care and mental health, among others. In relation to CBD, there are photovoice projects that are specifically considered to be community-based participatory research. Examples are a project conducted in Mali focused on women's development in the context of microfinance (Sutton-Brown, 2014), a project to change water, sanitation and hygiene habits in Kenya (Bisung et al., 2015), a project to examine barriers to maternal health in Uganda (Musoke et al., 2015), a project to identify health needs in Kenya (Kingery et al., 2016) and a project to inform a diabetes intervention in Boston (Florian et al., 2016). In addition, Teti et al. (2012) examined the methodology itself in terms of it being a community-based participatory research strategy specifically in the context of women living with HIV/AIDS. The findings reveal that it is a useful methodology to empower participants as they define their own health priorities, while there were ethical challenges related to the photographs disclosing their HIV status and/or potentially illicit activities. The aforementioned examples illustrate the appropriateness of using photovoice as a research tool in CBD in both local and international contexts to inform stakeholders using a bottom-up approach.

COMMON CHARACTERISTICS OF COMMUNITY-BASED DEVELOPMENT AND PHOTOVOICE

The processes used in CBD and photovoice are similar, sharing three main characteristics: valuing experiential knowledge, participant-driven and oriented toward social change. It is for these reasons that I view them as being complimentary models.

Experiential Knowledge

The underlying philosophies of photovoice and CBD are securely rooted in an intermediary position on a continuum that stretches between absolute relativism and absolute truth. From this standpoint, it is assumed that lived experiences are the ontological root of knowledge (Ramazanoglu and Holland, 2002). Photovoice and CBD are most often used with marginalized and/or disadvantaged populations who are typically spoken for by others. It is often the case that foreign development practitioners and local policy makers are far removed from the experiences of the people for whom they make the policies (Wang, 1999). Recognizing this, photovoice and CBD draw upon the participants/beneficiaries' experiential knowledge to mobilize communities toward positive change. As such, they are both conceptualized as a medium of knowledge transfer, whereby the developers, researchers and policy makers are on the receiving end of instruction (McIntyre, 2003; Wang, 1999).

Both models value an insider perspective, stemming from the participants and beneficiaries in photovoice and CBD, respectively. CBD practitioners value the beneficiaries' expertise, seeking their input and involvement in all stages of development initiatives. Since photovoice participants are assumed to be experts of their own situation (Wang et al., 1996), the participants are encouraged to draw upon their experiences to make demands for policy changes. Photovoice encourages each participant to discuss their personal histories and experiences, emphasizing the importance of the participants telling *their* story, rather than trying to tell *the* story. This indicates that photovoice recognizes that there are many 'truths' of experience. However, the assumption is also that a geopolitical reality exists; one that influences and partially shapes their everyday experiences. It is the connections between the participants' 'truths' and this contextual reality that serve as the catalysts for political advocacy.

In addition to being experiential, photovoice and CBD both assume that knowledge is also localized as a function of time and place. Paralleling transnational feminist theorists' rejection of the notion that there is a universal women's experience (Mohanty, 2003), photovoice acknowledges that experience, perspective and identity are all mitigated by demographic, historical and social factors. It is for this reason that photovoice researchers ask participants to document their everyday experiences (Burris and Wang, 1997) and situate them within their own specific geographic, socioeconomic and political context (Moss and Al-Hindi, 2007). It is for similar reasons that CBD initiatives in one community are not automatically assumed to be appropriate for other communities.

Development theorists assert that the activities that promote sustainability are contextually specific. Therefore, rather than being mass-replicated, CBD projects are implemented on a case-by-case basis to account for the particular needs of each community.

The high regard for experiential knowledge is the reason that photovoice and CBD use the local participants' experience as a knowledge base from which to initiate social change. It also explains why they encourage the participants/beneficiaries to be actively involved in the processes.

Participant-directed agenda
Contrary to the top-down model, participatory models are collaborative approaches in which the developing agencies/researchers act as facilitators rather than as authoritarian directors. They engage the participant/beneficiaries in the decision-making activities, making them the key informants of outcome initiatives. CBD asks the beneficiaries to identify their community needs, to design, implement and monitor the outcome projects. Similarly, in photovoice the participants determine the issue that they want to address and they decide how to represent it using photographs. They generate possible solutions to their identified problems and decide on the dissemination procedures they would like to use to share their images and narratives to a broader audience. In CBD, this collaborative model effectively combines the talents of the private, public and community sectors (Squazzoni, 2008) to use resources in the most efficient manner (Yen and Luong, 2008). In photovoice this participant-led approach leads to an increased sense of participant ownership (Castleden and Garvin, 2008) and community building (Killion and Wang, 2000).

Proponents of participatory models associate its participant-directed approach with empowerment. Photovoice is described as empowering (Booth and Booth, 2003), while Buccus et al. (2008) assert that CBD is an empowering process for the beneficiaries, because it allows poor citizens to be change agents by giving them more control over the development initiatives that directly impact their community. It has the ultimate objective of mobilizing the community through locally initiated development activity, and it encourages the beneficiaries to assume more responsibility for their community. It promotes capacity-building through its recognition and valuing the roles of community members. Proponents of photovoice and CBD recognize the importance and empowering potential associated with having members of disenfranchised populations take control of the decisions that directly impact their lives. Having control over initiatives that are geared toward social change has the potential to result in individual and community self-efficacy.

Political advocacy
CBD and photovoice share a common objective to promote and initiate positive social change in communities, thus the outcomes of both models are directed toward community improvement. The difference between the two approaches is that photovoice typically involves more advocacy, and is more politically charged than CBD.

CBD is, by definition, oriented toward social change. The purpose of international development initiatives is to enhance the quality of life for citizens who have difficulty meeting their basic needs. Thus, any development initiative, regardless of whether or not it is community based, has social improvement as its primary objective. Most often the social changes center on issues that are either a cause or consequence of poverty, such as health services, education, infrastructure, water and sanitation, and housing. The distinguishing feature between CBD and other development models is that the former attempts to achieve its social goals via a capacity-building model.

Photovoice, on the other hand, is grassroots activist research. It was developed as a research method to increase awareness about specific issues and to enact social change. It is rooted in problem-based inquiry (Burris and Wang, 1997) that is intended to address participant-identified concerns and community needs through exploration of a specific problem. Through the photovoice process, participants propose possible solutions to their identified problem(s), and steps are taken to initiate social change. This often includes a public forum within which the participants tell their story in their own words to a selected audience. Photovoice methodology includes an explicitly political agenda, and it is this commitment to social action that distinguishes it from the photo-elicitation method.

Recognizing the parallels between CBD and photovoice for their participatory approach, their valuing of experiential knowledge and their orientation toward social change, it is clear that photovoice is an applicable research method to use in conjunction with CBD. In summary, the photovoice approach to inquiry aligns with the participatory philosophy that also underlies the CBD model. Placing the participants at the center of the study and having them dictate the research process defines the researcher's role as a facilitator. Both photovoice and CBD recognize the empowerment potential that lies in the relinquishment of decision-making agency by the researcher/"developer", in favor of designing the study according to the locals' collective self-identified needs and preferences. The preliminary questions that guide a photovoice study are malleable, accounting for the likelihood of qualitative research designs to be altered during the course of a study (Pink, 2001). This flexible research design allows for increased responsiveness to the participants' identified needs, as well as for the exploration of new areas of discovery as they emerge (Patton, 2002).

As a cautionary note, all participatory models are not created equal. Jurkowski (2008) notes that photovoice was developed to be a participatory means in which to engage the participants in all phases of the project; however, it is not always applied as such. Photovoice can also be implemented in controlled, researcher-directed settings, creating an authoritarian, rather than authoritative, research model. Participatory research methods cite participant empowerment as one of its main benefits; however, Foster-Fishman et al. (2005) raise an important question about the possibility of photovoice to be disempowering. In their use of photovoice with a community building initiative, they pose the question "to what extent could photovoice be a vehicle for disempowerment if a new awareness of the need for change were not coupled with opportunities for improvement?" (p. 289). Photovoice is oriented toward social action, and this is communicated to the participants throughout the course of the project. Thus, the participants invest their time and energy in the process with the hopes that it will lead to individual and/or community improvement. If the project is constrained in its ability to initiate and implement the changes that the participants propose, then there may be the potential for the participants to feel disempowered. This may be particularly relevant to cases in instances where the photovoice project raises new awareness in participants of community needs and problems that they had not previously noticed.

To address this possible backlash, Foster-Fishman et al. suggest that photovoice researchers should be acutely aware of contextual and resource constraints, and to not over promise the extent of social change. In the critical discussion it may be best to discuss short-term goals that are realistic given the parameters of the project as well as long-term goals that they can continue to work toward after the official project ends. To maintain fidelity to the original concept, the participants should be involved in all phases of the process to the largest extent possible.

In international CBD and cross-cultural research, practitioners and researchers should be even more aware of the contextual basis of ethics. Denzin and Lincoln (2008) draw upon the works of key scholars on Indigenous research to demonstrate that in Indigenous communities the researcher has a moral obligation to the research participants and communities to which they belong. Linda Smith (2000) asserts that the procedures that Western Institutional Review Boards require, such as consent forms and assurance of accountability, are not enough for researchers wishing to study Maori culture, which has its own set of moral codes. Photovoice typically includes a researcher who is considered an outsider to the group they are studying. With culturally relevant pedagogy and respect and a willingness to learn from the group, they can attempt to conduct the study in a respectful manner.

CONCLUSION

Despite its limitations and ethical considerations, photovoice is a useful research method to use in the social sciences. Following in the inductive tradition of qualitative research (Bogdan and Biklen, 2007), it is an exploratory method with the potential to yield rich data. Photovoice is particularly useful to use with CBD, as they have a similar philosophical stance. This is particularly true with regard to their endorsement of a participatory approach. Like CBD, photovoice has high regard for experiential knowledge and highly values the insider perspective, which justifies their participant-directed process. Both photovoice and CBD aim to have a positive social impact on the communities in which they are implemented; however, they achieve this via different means. Photovoice uses grassroots activism to influence policy, while CBD is less politically charged. Photovoice's acute political orientation is useful for CBD practitioners who want to conduct needs-based assessments for funding purposes and/or to garner support for a particular initiative. Even in recognizing that inequitable power relations mediate participatory model processes, photovoice and CBD have the potential to be empowering processes for the people and communities involved. By putting the participant/beneficiaries in decision-making positions, photovoice and CBD often lead to increases in individual and community enhancement (Killion and Wang, 2000), and individual self-confidence (Foster-Fishman et al., 2005). When a major goal of CBD is empowerment, it is only appropriate that the research method also has the potential to be empowering.

REFERENCES

Bisung, E., Elliott, S.J., Abudho, B., Schuster-Wallace, C.J., and Karanja, D.M. (2015). Dreaming of toilets: Using photovoice to explore knowledge, attitudes, and practices around water-health linkages in rural Kenya. *Health & Place*, *31*, 208–15.

Bogdan, R.C., and Biklen, S.K. (2007). *Qualitative Research for Education: An Introduction to Theories and Methods*. Boston, MA: Pearson Education Inc.

Booth, T., and Booth, W. (2003). In the frame: Photovoice and mothers with learning difficulties. *Disability & Society*, *18*(4), 431–42.

Buccus, I., Hemson, D., Hicks, J., and Piper, L. (2008). Community development and engagement with local governance in South Africa. *Community Development Journal*, *43*, 297–311.

Burris, M., and Wang, C. (1997). Application of photovoice to participatory needs assessment. *Health Education & Behavior*, *24*, 369–87.

Castleden, H., and Garvin, T. (2008). Modifying photovoice for community-based participatory indigenous research. *Social Science & Medicine*, *66*, 1393–405.

Denzin, N.K., and Lincoln, Y.S. (2008). *Strategies of Qualitative Inquiry*. Thousand Oaks, CA: Sage Publications, Inc.

Dickmann, P., Katua, A., Apfel, F., and Lightfoot, N. (2018). Kampala manifesto: Building community-based One Health approaches to disease surveillance and response – The Ebola legacy: Lessons from a peer-led capacity-building initiative. *PLOS Neglected Tropical Diseases*, *12*(4). Retrieved 5 December 2019 from https://www.ncbi.nlm.nih.gov/pmc/articles/PMC5880333/.

Easterly, W. (2007). *The White Man's Burden*. New York: Penguin Books.

Escobar, A. (1995). *Encountering Development: The Making and Unmaking of the Third World*. Princeton, NJ: Princeton University Press.

Florian, J., St. Omer Roy, N.M., Quintiliani, L.M., Truong, V., Feng, Y., Bloch, P.P., Russinova, Z.L., and Lasser, K.E. (2016). Using photovoice and asset mapping to inform a community-based diabetes intervention, Boston, Massachusetts, 2015. *Preventing Chronic Disease*, *13*, 1–11.

Foster-Fishman, P., Nowell, B., Deacon, Z., Nievar, M.A., and McCann, P. (2005). Using methods that matter: The impact of reflection, dialogue, and voice. *American Journal of Community Psychology*, *36*(3/4), 275–91.

Freire, P. (1972). *Pedagogy of the Oppressed*. New York: Herder and Herder.

Jurkowski, J.M. (2008). Photovoice as a participatory action research tool for engaging people with intellectual disabilities in research and program development. *Intellectual and Developmental Disabilities*, *46*(1), 1–11.

Killion, C., and Wang, C.C. (2000). Linking African American mothers across life stage and station through photovoice. *Journal of Health Care for the Poor and Underserved*, *11*(3), 310–25.

Kingery, F.P., Naanyu, V., Allen, W., and Patel, P. (2016). Photovoice in Kenya: Using a community-based participatory method to identify health needs. *Qualitative Health Research*, *26*(1), 92–104.

Lindenberg, M., and Bryant, C. (2001). *Going Global: Transforming Relief and Development NGOs*. Bloomfield, CT: Kumarian Press.

Mansuri, G., and Rao, V. (2003). Evaluating community-based and community-driven development: A critical review of the evidence. Retrieved 25 December 2019 from http://siteresources.worldbank.org/INTECAREGTOPCOMDRIDEV/Resources/DECstudy.pdf.

McIntyre, A. (2003). Through the eyes of women: Photovoice and participatory research as tools for reimagining place. *Gender, Place and Culture*, *10*(1), 47–66.

Mohanty, C.T. (2003). *Feminism without Borders: Decolonizing Theory, Practicing Solidarity*. Durham, NC: Duke University Press.

Moletsane, R., De Lange, N., Mitchell, C., Stuart, J., Buthelezi, T., and Taylor, M. (2007). Photo-voice as a tool for analysis and activism in response to HIV and AIDS stigmatization in a rural KwaZulu-Natal school. *Journal of Child and Adolescent Mental Health*, *19*(1), 19–28.

Moss, P., and Al-Hindi, K.F. (2007). *Feminisms in Geography: Rethinking Space, Place, and Knowledges*. Rowman and Littlefield: Lanham, MD.

Musoke, D., Ekirapa-Kiracho, E., Ndejjo, R., and George, A. (2015). Using photovoice to examine community level barriers affecting mental health in rural Wakiso district, Uganda. *Reproductive Health Matters*, *23*(45), 136–47.

Patton, M.Q. (2002). *Qualitative Research and Evaluation Methods*. Thousand Oaks, CA: Sage Publications, Inc.

Pink, S. (2001). *Doing Visual Ethnography*. Thousand Oaks, CA: Sage Publications, Inc.

Ramazanoglu, C., and Holland, J. (2002). *Feminist Methodology: Challenges and Choices*. Thousand Oaks, CA: Sage Publications, Inc.

Sen, A. (1999). *Development as Freedom*. New York: Anchor Books.

Smith, L.T. (2000). Kaupapa Maori research. In M. Battiste (ed.), *Reclaiming Indigenous Voice and Vision* (pp. 225–47). Vancouver: University of British Columbia Press.

Squazzoni, F. (2008). Local economic development initiatives from the bottom-up: The role of community development corporations. *Community Development Journal*. Retrieved 5 December 2019 from https://www.future-agricultures.org/wp-content/uploads/pdf-archive/Bottom_Up_Policy_Process_Ethiopian_Herald.pdf.

Sutton-Brown, C.A. (2014). Photovoice: A methodological guide. *Photography and Culture*, *7*(2), 169–85.

Teshome, A. (2007). Bottom up policy process: An agenda for future agricultures in Ethiopia. *Economy and Development* (*Ethiopian Herald*, 23 June).

Teti, M., Murray, C., Johnson, L., and Binson, D. (2012). Photovoice as a community-based participatory research method among women living with HIV/AIDS: Ethical opportunities and challenges. *Journal of Empirical Research on Human Research Ethics*, *7*(4), 34–43.

The World Bank Group. (2019). *Poverty and Shared Prosperity: Piecing Together the Poverty Puzzle*. Retrieved 5 December 2019 from http://www.worldbank.org/en/publication/poverty-and-shared-prosperity.

Wang, C. (1999). Photovoice: A participatory action research strategy applied to women's health. *Journal of Women's Health*, *8*(2), 185–92.

Wang, C., Burris, M.A., and Ping, X.Y. (1996). Chinese village women as visual anthropologists: A participatory approach to reaching policymakers. *Social Science & Medicine*, *42*(10), 1391–400.

Wang, C.C., Cash, J.L., and Powers, L.S. (2000). Who knows the streets as well as the homeless? Promoting personal and community action through photovoice. *Health Promotion Practice*, *1*, 81–9.

Wang, C., Yi, W., Tao, Z., and Carovano, K. (1998). Photovoice as a participatory health promotion strategy. *Health Promotion International*, *13*(1), 75–86.

Yen, N.T.K., and Luong, P.V. (2008). Participatory village and commune development planning (VDP/CDP) and its contribution to local community development in Vietnam. *Community Development Journal*, *43*(3), 329–40.

24. Centering aesthetics in community development: approaches from the Banff Centre for Arts and Creativity
Jerrold McGrath

INTRODUCTION

The challenges faced by community development professionals and their stakeholders are complex, evolving and too often resistant to status quo approaches. There is an ongoing need for creative solutions, good ideas from other places and capacity for ideas to transform existing systems and structures. This chapter focuses on models for understanding and organizing spaces and activities necessary to support grassroots creativity. Specifically, it discusses approaches to support the generation of novel and useful ideas, the transfer of ideas from and to other communities, and the transformation of sociocultural systems and structures.

In this chapter, community is understood as a "place-oriented process of interrelated actions through which members of a local population express a shared sense of identity while engaging in the common concerns of life" (Theodori, 2005, p. 661). The cases discussed here involve communities that are "place-oriented" though much of the work has not been local. Cases are principally centered in conversations of resource extraction and use in the Alberta context. Furthermore, the orientation to community development in this chapter is focused on structure building, often delocalized with interventions being place-based and grounded in interactional theory (Kaufman, 1959).

First, a summary of the research on creativity is outlined. Second, an approach to understanding the organization of spaces is offered to show how power dynamics and cultural assumptions become embedded in physical and relational spaces and how this can impact the interactions available to community members in a development process. Third, an approach to designing interactions is described that reduces barriers to generating, evaluating and instantiating novel and useful ideas. The approach also serves to facilitate the exchange of creative ideas which increases the efficacy of development activities and encourages the internalization of novel approaches within the community, with adjacent community actors and with existing power structures. Finally, two case studies are summarized that demonstrate how the approaches are applied and the opportunities and challenges that result.

CREATIVITY RESEARCH AND PRACTICE

Creativity research and practice have informed the development of the design methodologies outlined in this chapter. Creativity, for our purposes, is understood as a communicatively constituted process and this section outlines the evolution of the literature on creativity toward that understanding (Amabile, 1996; Csíkszentmihályi, 2014; Sawyer, 2012). By showing the progression within the field of creativity research and practice toward a collaborative and communicative understanding, the design methodologies can be situated in the broader conversation around the importance of a nuanced understanding of creativity as a phenomenon of social communication (Bartels, 2010; Cooren, 2015; Cooren et al., 2016). Creativity is too often understood as the reserve of the lone genius and the mythologies of creativity often hinder the ability of practitioners to support spaces where new ideas can emerge (Csíkszentmihályi, 2014; Paulus and Nijstad, 2010; Sawyer, 2012).

Creativity has become a central feature of approaches to addressing intransigent environmental, economic, social and other issues. In parallel, creative work is increasingly defined by ad hoc arrangements and concatenations of diverse talents around finite projects (Henry and de Bruin, 2011). How creative work happens and the outcomes toward which it is aimed become important considerations when seeking novel and useful approaches to addressing multi-faceted social, environmental and cultural challenges. Furthermore, engaging those most affected in generating and implementing solutions is critical. The creative sector, both non-profit and for-profit, can serve an important role in informing how acts of co-creation are organized, how symbols can be arranged and distributed to frame communications around key issues, and how attention is drawn to system dynamics where interventions are best suited for success. While this chapter looks to Japanese architecture and to the world of video game design, numerous approaches have emerged to harness the creative energy of diverse groups of people.

Recently, there has been increased attention on the social dimensions of creativity and the perspective that creativity can best be understood as a social phenomenon has come to dominate the field. However, questions of how creativity is enacted or communicated within these social dimensions and approaches to applying a social frame of creativity to design methodologies in educational, professional and community settings are less explored.

A review of the history of creativity research (Isaksen et al., 1993) suggests that singular or unifying definitions or models are unlikely (Sawyer, 2012). Currently, creativity is broadly understood as the capacity to generate ideas, processes, products or solutions that are novel and appropriate (Amabile, 1996; Ford and Gioia, 1996). Calls for creativity have become ubiquitous across countries, sectors, communities and disciplines. Issues such as climate change, inequality, racism and reconciliation with Indigenous peoples are proving intractable to status quo approaches and the need for new and appropriate responses has become clear.

The call for creative solutions is hardly recent. In 1992, the Preamble of the United Nation's "Agenda 21" called for innovation and problem solving to address worsening outcomes related to poverty, health, literacy and the natural systems on which we collectively depend (UNCED, 1992). More recently, the UNESCO 2030 Sustainable Development Agenda acknowledges the critical importance of creativity and culture in reaching the 2030 Sustainable Development Goals (UNESCO, 2015). The environmental and social challenges described by international programs such as Agenda 21 demand increased local competency to respond and capacity to implement solutions in an interconnected and interdependent way (Boutellier et al., 2008). The nature of the challenges being addressed demands an interdisciplinary perspective and broad engagement in the generation and implementation of solutions. Creativity has psychological, historical, cultural, social and other aspects. Contexts that recognize and support the various aspects that influence creative behaviour are necessary tools for community development.

Furthermore, creativity is a characteristic of social relations and communications. When these relations and communications reflect a narrow framing of a challenge or opportunity, the capacity for creative action is restricted to those fluent in the habitual frame. Engaging broader expertise requires different approaches. Community development professionals know this all too well. Without greater diversity in creativity, innovation and innovation systems cannot realize their full potential (Dubina et al., 2012).

Creativity can be understood as comprising three activities or dimensions. *Generative activities* are the means by which new ideas emerge and are directly tied to human imagination (Funk and Woodward, 2004). Much of the mystification attached to creativity relates to this dimension. Everyone has the capacity to respond to situations, although the opportunity to apply that capacity is often limited to those within existing power structures. Asking a group of people to come up with solutions with no intention of evaluating them for appropriateness and no means of embedding them in existing mental models or real-world systems drives many of the critiques of "creative action" in community contexts.

Evaluative activities introduce criteria of value, recognition, relevance, usefulness and tradition (Csíkszentmihályi, 1996). An idea emerges and the social context assesses both the newness of the idea and its appropriateness (Cooren et al., 2006). Generative and evaluative processes are non-linear as ideas are shared, transformed, reviewed, revised and rejected. Creativity often involves the introduction of established ideas into unfamiliar contexts and the combination of existing ideas in novel ways.

Most creative interactions occur in complexity. Accomplishments in community development are rarely solitary events, but rather rely on humans working together. Creativity research and practice has increasingly recognized this over time. Theresa M. Amabile's Component Theory considers the influence of potential cognitive, personal, motivational and social attributes on the specific phases of the creative process (Amabile, 1996) and on how individuals can contribute to collective processes. Amabile also considers the social appropriateness and novelty of new ideas and the role of expertise in judging whether a specific solution should be considered within a domain (Amabile, 1996). Narrow participation in creative action limits both the space for generating useful ideas and the evaluative process of situating it within real contexts or domains.

Mihály Csíkszentmihályi elaborates the sociocultural approach by relating the process to awareness, adoption and evaluation of new ideas by certain social groups (Csíkszentmihályi, 1996). Creativity, through this framing, starts to become understood as a sociological, historical and cultural phenomenon as well as a psychological one.

Sociocultural activities describe how creative phenomena embed in human activities. Creative solutions can be deemed novel and appropriate, but they may also generate a transformational impact on the knowledge, norms, modes of action, attitudes or perspectives of a system. Margaret Boden describes conceptual spaces of possible solutions (Boden, 1994) and how creativity can transform the space of what is understood as possible. Foucault points to new discourses (Foucault, 2017) and their intimate relationships with power. Power shapes the criteria for legitimacy of knowledge or truth. Creative ideas can transform or disturb the rules and categories of a discursive order and in the process the arrangement and relationships of power.

To both Boden and Foucault, creativity can become a space of opportunity wherein a creative subject can act. More than a new way of thinking about solutions, the frameworks by which the solutions are generated is validated by the social system. In a community development context, ideas must not only be generated and assessed for fit but must also alter perceptual and/or material systems and thereby allow subsequent creative interactions to occur. Essentially, does an idea shift what is conceived of as possible within a given context? How does the idea perturb how other aspects of the system are understood?

For creativity in a community development context, the translation of newness and uniqueness into existing conceptual and concrete structures is essential. A creative pattern that fails to consolidate itself into a social context will struggle to persist beyond its articulation. Therefore, creativity is necessarily co-creative. *Persistence of a creative idea is impossible without reciprocity and co-involvement of a creator or creators with other people, cultures and societies.*

Communication is critical for a social conception of creativity. Relations with others are essential to generate, evaluate and contextualize a creative pattern. While recognizing the importance of communication in the sociocultural approach, few researchers conduct detailed study of the element of interaction. Paulus and Nijstad (Paulus and Nijstad, 2010) attempt to address the unknowns around collaborative creativity and place central importance on the need for heterogeneity in groups. They also point to the need for an open culture of exchange. How to do this remains an open question.

Collaborative creativity should not be reduced to situational incidences of working together toward a common goal. Focusing only on the situation and the social factors that contribute to understanding that situation ignores the internal dynamics that define how a group constructs and evaluates meaning. Creativity should also refer to a contextual capability for meaningful novelty or novel ideas which emerge from interaction (Bartels, 2010). An idea may arise during conversation or within a small team. The development of that idea, however, relies on it being expressed within various contexts and with different fluencies. Communication becomes the means by which solutions evolve, are evaluated and eventually come to transform systems of activity. Solutions must survive interaction with social regulation such as norms, expectations, rules, anticipation and requirements.

This theoretical framework for creativity has informed the application of design methodologies in community development activities. Practitioners not only attend to the barriers to participating in community action and the context for the action but also the spaces and structures within which interactions occur.

The subsequent section illustrates how creativity understood predominantly as a collaborative process is animated in design methodologies applied in community contexts constituted by and in social communication.

CONCEPTIONS OF SPACE IN JAPANESE ARCHITECTURE AND DESIGN

This methodology and the one that follows prioritize the subjective experience of people in a process while simultaneously evaluating the dynamic behaviour of the system being constructed. The core assumption is that positive interactions with other participants and learning objects, such as generated hypotheses, encourage better outcomes. The design approaches are not concerned with manipulating affective responses but rather about designing spaces and experiences that are more likely to inspire the application of perspectives and heuristics required to address the adaptive challenges of the system. The approaches are situated in broader conversations about human-centered design and place primacy on the experience of participants in the systems communities co-create. Design methodologies allow for diverse expertise and experiences to jointly participate in an act of shared creation. Individuals bring their own cultural and disciplinary assumptions into co-creative processes, and a shared orientation to the work allows for those involved to apply themselves to the situation with a minimum of distraction or uncertainty.

The first design methodology to be described draws on different conceptions of space between Japanese and Western cultures and languages. Western ideas of space are heavily influenced by Euclidean geometry and mathematics generally (Nitschke, 1966; Oosterling, 2000, 2005). Spaces are conceptually empty until something happens within them. Japanese conceptions of space are far less historically influenced by advances in mathematics in Europe and understand and describe spaces somewhat differently (Kodama, 2017; Nitschke, 1966; Oosterling, 2005). The core idea is that ideas of Japanese space are full – interactions are inherent and embedded in spaces and the people come later. Space in Japanese culture is understood primarily by how it shapes relationships (Oosterling, 2000). A room at a community center in Tokyo might appear empty to a Western perspective but would appear full of symbols and instructions about how interactions can and should occur to an observer fluent in Japanese ways of seeing the world. From this perspective, spaces are full of invisible structures regardless of their occupants.

The Japanese language has at least four different words that correlate to the English word of "space" and most are quite different from the English equivalent. How the Japanese create and talk about spaces has long been of interest to Western designers and architects (Nitschke, 1966; Oosterling, 2005). However, the application to community development efforts has also generated positive and meaningful outcomes.

Japanese words and ideas about space center the interactions and relationships among people. A focus on interactions and uncertainty is an interesting complement to a Western approach that tends to focus first on the physical arrangement of people and objects like walls, doors and other inanimate objects and then later the relationships that are enabled by them (Oosterling, 2000).

A simple way to see this is in how houses are set up. In the West, walls tend to be fixed and rooms segregated from each other. Each room also has a fixed purpose – living room, dining room, bedroom – which may vary from home to home but tends to vary less over the course of a day. In the West, a bedroom in the morning remains a bedroom throughout the day. In Japan, houses are divided up by sliding doors, often made of paper, that allow for a quick reconfiguration of space to meet the needs of the occasion (Kodama, 2017; Nitschke, 1966). Families may pull out their futons and sleep in the same room that they earlier used to socialize or greet guests. Boundaries are ambiguous and contingent.

Mitsuru Kodama, professor at Nihon University, argues that Japanese concepts of space derive from two foundational traditions: Shinto, an Indigenous spiritual tradition in Japan, and Buddhism, which was imported from mainland Asia (Kodama, 2017). From Shinto came the high value placed on harmony in relationships and a focus on the connections, spoken and unspoken, that tie people together. From Buddhism came the ideas of emptiness and selflessness. Even the word for person in Japanese, 人間, reflects differences in how interactions and identity are understood. The first character, 人, represents a human being and the second, 間, stands for space, or in-between. So, the understanding of a person is not distinct and atomistic, but rather is the connections and relationships that people form as they interact with each other. Similarly, Japanese spaces tend to focus on structuring interactions, contingency and connections to other people and to society. For example, traditional tea houses have doors that are narrow and low. This forces guests to lower their head and, historically, for samurai to leave their swords outside by the door. The doors serve to remind entrants of their relationship to the host by requiring them to lower their heads, and to the broader culture by banning weapons. The doors shape interactions physically and symbolically.

Western Ideas of Space

The West draws on different traditions and this has inevitably generated different ideas about how to understand space. Architect Gunter Nitschke argues that, in the West, "as a box viewed from the outside is an object, so the inside is space" (Nitschke, 1966, p. 117). He goes on to describe how space, in the West, is related to how the culture understands science and mathematics at any given time. Edward T. Hall argues that Westerners think and talk about space as the distance between objects (Hall, 1990). In the West, we are taught to perceive and react to the arrangements of objects and to think of space as empty. As Oosterling offers, in the West, "a room is empty until someone enters this space" (Oosterling, 2005, p. 1). The Western focus leads to spaces that are appropriate for the function they are intended to serve, rather than the experience they are meant to generate. Spaces are laid out to accommodate the people and objects that will make use of them. Generally, the designer takes a less active role in prioritizing how users interact.

The Four Kinds of Japanese Space

Relational space (私), or WA, is often translated as harmony but can also be understood as an awareness of interpersonal connection. Every space has qualities and symbolic referents that suggest the types of relationships that are possible. Being fluent in reading the relational qualities of an environment is a skill admired in Japanese culture. In fact, teens use an expression, "KY," in electronic communication that is an acronym for 空気読めない or "*kuuki yomenai*" which literally translates as an inability to read the air and refers to people who are unable to assess the relational qualities and assumptions defining a context.

From a community development perspective, how much time is committed to ensuring that intimacy and shared purpose are surfaced and understood? Activities and spaces that explicitly center the interpersonal relationships among community members can too easily be removed or truncated in the name of task completion. However, a lack of focus on interpersonal connections can curtail the fluency with which new ideas are generated, evaluated and contextualized.

Knowledge mobilizing space (場), or BA, plays an important role in models in the literature of business innovation (Nonaka and Konno, 1998). While 私 focuses on relationships, 場 is concerned with the spaces within which knowledge is formed and shared. While 私 is about contexts of social and interpersonal harmony, 場 is about spaces to engage jointly in the creation of meaning and value (Nonaka and Konno, 1998). The Japanese prioritize interdisciplinary teams because they believe that the concentration of different ways of seeing the world will lead to breakthroughs (Kodama, 2017; Nonaka and Konno, 1998). There is often a lack of efficiency when bringing together different specializations, but 場 requires shared space for different relationships and experiences to be brought forward. 場 shows up in all aspects of society. Both open office concepts and the culture of group dating in Japan are reflections of 場 as a design principle.

For community development practitioners, attention to 場 asks that spaces and activities be designed to actively mobilize the experiences of individuals in each process. As the designer has imperfect knowledge about the nature and range of knowledge available, ambiguity becomes a useful frame to surface and explore unexpected sources and directions of knowledge.

場 is about the arrangement of elements to create connections that are more likely to produce new knowledge or experiences (Nonaka and Konno, 1998). Simple spaces that prioritize 場 might involve participation by outside communities in local issues; events or conferences outside of the described area of focus; and meeting and interacting with people that might not normally engage in the discourse. The challenge is to be intentional about designing more open boundaries so that new ideas can enter and exit. The key to 場 is getting involved and transcending one's current perspectives.

Location (所), or TOKORO, is used to describe the location or site of something. It is also used to describe a state of being. In many ways, it resembles the Western idea of place. The idea of 所 also implies the idea of context as the place is inevitably connected with all the activities happening in and around it. In the West, the idea of place may or may not capture the relationships concomitant with that location. In Japan, the idea of place is indistinguishable from the historical, cultural, social and other contextual phenomena related to it (Lopes, 2007).

As spaces configure relationships, 所 situates activity within larger contexts. The same is true in the West, but the differences are important. Western concepts of space have an inside and outside and a boundary between the two (Kodama, 2017). It is often easier to think about things as being contained within larger things and containing smaller things. A community center is in a community which is in New York City, which is in the United States. The team is inside the office and the practitioner is inside the team. Japanese concepts of space are ambivalent about boundaries, so being a part of a place means being in a dynamic relationship with it. In Japan, a building cannot be in Tokyo without Tokyo being in the building.

Practically, we might borrow this idea and think about location as both an opportunity and an obligation. It is tempting to limit participation in spaces and activities in order to focus in on specific needs or opportunities. However, associations with the broader context within which the community development activity is occurring are necessary to ensure that generative, evaluative and sociocultural phenomena are actively shaping the interactions that occur.

Negative space (間), or MA, is often translated as negative space but is better understood as a free zone that allows for dissimilar things to co-exist (Oosterling, 2005). When communicating it is often assumed that the recipient will understand the communication in the way in which it was intended. This is often not the case. If a person tells us that they are hungry, it may be interpreted as information, as a request, as an indictment of the hosting or something else entirely. 間 is the space where the distance between what is communicated and what is received can be processed and synthesized.

The character for 間 shows the light of the sun (originally the moon) shining through a window or gate. The idea here is time, openings and experience. The Japanese idea of 間 is that we need to create interruptions or absences that allow for difference to be reconciled. Shrines are often built at the end of long uphill hikes. The long and tiring walk prepares the mind to enter the shrine and leave behind other distractions and worries. Japanese cities are scattered with small parks that appear suddenly and offer winding trails for quiet reflection. Even conversation in Japanese is marked by long pauses that would be unsettling for Western ears (Panalaks, 2001). Being intentional about creating spaces that allow for reflection and integration might allow practitioners to better address some of the contradictions and tensions of community development work.

The need to focus on personal relationships, to mobilize knowledge effectively, to contextualize work appropriately and to support pauses to address tensions and contradictions is hardly new. However, design and delivery of community development processes can too often be done uncritically. Habits developed over time emphasize some qualities of space and de-emphasize others. The practitioner's subjective preferences should not become the formula by which the work is organized. Japanese conceptions of space can therefore serve as a shared language and a useful reminder of the different elements that contribute to intersubjective co-creation. The unintentional omission or minimization of any of the four dimensions can limit the efficacy of any process or activity.

MECHANICS-DYNAMICS-AESTHETICS

The second methodology presented in this chapter emerged from the desire to convene diverse stakeholders at the Banff Centre for Arts and Creativity in Alberta, Canada in acts of co-creation in the period from 2012 to 2016. The Banff Centre is a non-parchment post-secondary that had committed in 2012 to being "a generative space to explore how the application of creativity could give rise to innovative ideas and solutions that were shaped by collaborative approaches, diverse perspectives and forward-thinking ideas" (Jeff Melanson, personal communication, 12 September 2012).The Banff Centre's core community consists of artists, Indigenous leaders and creators, corporate partners, conference attendees, researchers, government at all levels and the local community (*The Banff Centre Mandate*, n.d.).

The design approach is known as MDA. MDA as a design framework comes from the world of video game design and research. MDA is an acronym for Mechanics-Dynamics-Aesthetics and was described in 2004 (Hunicke et al., 2004). The primary benefit of MDA is that it supports coordinated experiences in designed systems. By centering the subjective experience of participants (aesthetics) in a designed process, collaborative and creative approaches to generating and acting on solutions to complex challenges emerge. The design methodology is organized around co-creative projects that ask participants to interpret the sense- and meaning-making of their co-creators in order to make progress on targeted issues.

The MDA research argues that "thinking about games as designed artifacts helps frame them as systems that build behavior via interaction" (Hunicke et al., 2004, p. 2). A set of rules or expectations (mechanics) describe the range of decisions available to players. As players make decisions and thereby act, emergent properties of the game are revealed (dynamics). Examples of dynamics include competition, sharing, evaluation, critiquing and so on. These dynamics are usually intentional but iterative processes are necessary to assess the relationship between selected mechanics and desired dynamics. How the player experiences the dynamic is described as the aesthetic and is highly subjective but meaningful as it informs what the player does next. If a set of rules or conditions create interactions that are experienced as unpleasant, a player will be unlikely to further commit to the game or will commit in ways to reduce their negative experience of the interactions. For example, a game that uses mechanics such as a prominent scoreboard, an opponent and winning or losing is likely to create dynamics of competition. One individual may experience competition as motivating and will increase the energy and attention she/he assigns to the game. Another may experience competition as inappropriate or unpleasant and decrease or alter her/his engagement.

By focusing on the aesthetic, or how the designer wants participants to experience a development process, interactions can be selected that are most likely to generate that subjective experience. Then rules and expectations can be defined that are most likely to result in the desired dynamics. Practitioners recognize the need to support different spaces in order to activate community members. MDA provides a formalized structure to engage others in the design and embodiment of these spaces.

MDA offers a different way of approaching the design of development spaces. If a social environment, like a game or a community development process, only exists through its interactions, then the design focus needs to be on the nature and subjective interpretations of those interactions (aesthetic). Rather than prioritizing what the designer wants the participants to do, the focus is on how the participants need to subjectively experience the design space to encourage non-habitual interaction. It is tempting to focus on the mechanics of the space such as actions, behaviours and controls, because this is where the designer has greatest influence. However, mechanics can have unpredictable impacts on system behaviour and the experience of members. For example, discursive cues that advance a process may pre-suppose next steps and discourage participants from contributing potentially novel directions. The physical layout of a room that elevates an individual or group may communicate that certain knowledge is privileged and shape participation from those drawing on knowledge outside of that sphere. Something as simple as asking that people raise their hand to speak may generate associations with institutional models of education that serve to quiet subsets of the community.

The MDA model focuses attention on the experience of the learner and the dynamic behaviour of the system. Iterative loops and testing focused on enhancements in the experience of the learners. Ad hoc and diverse community processes require a fluency in working across disciplinary and experiential boundaries and in collectively navigating uncertainty. An approach that focuses on the experience of the participant allows for a greater likelihood of new desirable behaviours to emerge that are intrinsically derived and centered on positive interactions with other community members and symbolic constructs such as ideas, representations or models.

Designers need to select an aesthetic that is likely to encourage participants to invest in persisting in the space (Han, 2017; Sanders and Stappers, 2012). The aesthetics present in the subsequent cases include "curiosity," "celebration" and "generosity." Measurement becomes simplified as testing and iteration focus on the suitability of the aesthetic in soliciting investment; the appropriateness of the aesthetic in generating adaptive approaches to the issues at hand; and the efficacy of selected mechanics in generating dynamics that inspire the selected aesthetic.

The MDA model is an iterative methodology that simultaneously considers the learner's experience of system dynamics and the controls that enable those system dynamics. As the learning spaces are targeting transformational adaptations in perspective and behaviour, focus on controls is eschewed in favour of a focus on aesthetics (Sanders and Stappers, 2012). Structured reflection on aesthetic experience and collective processes of meaning-making allow groups to interrogate existing patterns of behaviour and apply new behaviours as demanded by the designed context (Woodward and Funk, 2010). Through collective inquiry into a shared aesthetic experience a group can begin a process of adaptation to the emerging context. The groups transition from "what does this mean to me" to "what does this mean to us" to "how might we move forward."

The following cases show how these models can be applied and where there is potential for further exploration as well as what challenges need to be overcome.

CASE STUDY: PETER LOUGHEED LEADERSHIP INITIATIVE

The author has applied the described methodologies to the design and delivery of over 100 programs over the past seven years. Programs have supported capacity building, collective action and co-creation around issues such as economic inequality, hope and hopelessness, precarity, creative placemaking, cultural leadership, navigation of the Canadian intellectual property regime in racialized communities, interdisciplinary youth leadership and others. Instantiations of the approaches center on co-creative processes and activities that bring together individuals from communities that face structural barriers to participation, outside communities facing similar or related challenges, and individuals representing the various interests of dominant systems and structures. The approach has been effective in opening and supporting alternative views, patterns of thinking, perceptions and interactions; synthesizing different disciplines and approaches to meaning-making; dismantling invisible or stubborn behaviour patterns; and recognizing and reconciling to paradox and contradiction.

The first case study was designed from the outset as a collaborative and processual event. The case, a partnership between the University of Alberta and the Banff Centre for Arts and Creativity, exemplifies how pedagogical interventions can support cross-disciplinary collaboration around issues of broader community or societal concern within educational settings. The second case describes a multi-sectorial, national exploration of hope and hopelessness by individuals and participating organizations facing issues of persistent hopelessness within their own practices and within their stakeholder communities. Co-constructed processes were developed and deployed in four organizations addressing homelessness, Indigenous families in crisis, LGBTQ+ theatre and youth-run environmental activism.

The Peter Lougheed Leadership Initiative is an institutional partnership between the University of Alberta and the Banff Centre. The author of this chapter served as director of innovation and program partnerships for the Banff Centre's leadership programming from 2012 until 2016. In 2015 and 2016, the Banff Centre was able to apply the previously described design methodologies to the orientation of a cross-disciplinary cohort of undergraduates from the University of Alberta. The program was assessed as effective in promoting improved communication and collaboration among the participants, particularly across disciplinary boundaries, and suggests a potential way toward innovation in capacity-building programs.

In August 2015, the first cohort from the Peter Lougheed Leadership College came to the Banff Centre for three days of orientation prior to the start of the school year. The College "was developed as an inclusive community of undergraduate students to learn, grow and connect with other change makers through an experiential, interactive, interdisciplinary leadership program" (Peter Lougheed Leadership College, n.d.).

The aesthetics of "curiosity" and "confusion" were selected as most likely to engender the interactions that would facilitate cross-disciplinary collaboration and inter-subject dialogue. In year two, based on feedback following the orientation in year one, an aesthetic of "paradox" was selected.

One of the benefits of centering aesthetics is that designers have a criterion to evaluate collective assumptions about the appropriateness of routine design elements in constructing a program or experience. Habitual activities such as room setup or meals come into question as assessments are made about their suitability as contributors to senses of "curiosity" or "paradox." The selection of an aesthetic also contributes to coherence as disparate elements are connected thematically and emotionally throughout the process. Reflections from the 2015 cohort indicate the efficacy of the embedded aesthetic:

> I had not imagined that our forum of such different people from different faculties would mesh so well in creating discussions and collaborating together.

> This weekend has shown me how little I know.

> It's extremely refreshing coming from a very competitive faculty to enter a community where everyone is so eager to learn from one another and hear different perspectives.

Two design elements highlight the execution of this aesthetic.

Elder Participation

Program design was centered on a local Stoney Nakoda Elder. Mechanically, this meant that each day began with a smudge and a ceremony; protocol was followed in introductions and process; and the Elder controlled the speaking spaces and was invited to offer response at any point during the program, including during keynotes or presentations by invited speakers. The design fostered confusion and curiosity by privileging different forms of knowledge. The Elder was not constrained by the agenda, so punctuality and agenda-making were also disrupted. By asking the Elder to contextualize presented content, further confusion and curiosity could emerge as the learners struggled to synthesize their impressions with the second-order observations of the Elder. An unexpected consequence of this design choice was that the physical organization of the work shifted. The Elder preferred to work in circles.

The implications of this design choice show up in student reflections:

After taking part in a traditional smudge ceremony and seeing the respect Elder Powderface bestows on her sacred land, I began to respect the land of Banff more.

What surprised me most about orientation was the emphasis to Aboriginal traditions. I was able to attend two sessions with Elder Powderface. I could never have prepared for the experience I had of laying on a buffalo skin while a medicine woman sang to us in her native language.

I was especially surprised when I got to participate in the session with Elder Corleigh. I have never experienced smudging or participating in a sharing circle environment. Also, being able to be a part of the leadership which she is familiar with. It really opened my eyes to a different style and perspective of leadership. One that I will never forget and take with me everywhere.

Fragmentation

Another design choice involved a dynamic of fragmentation. The multi-year program requires that the cohort be divided into forum groups that work together. Fragmentation encouraged participants to reflect on the incompleteness of personal knowledge and experience and to integrate more effectively the experiences of others. The opening evening brought the cohort into a room, furnished only with a boardroom table that they were forbidden to use. A group of actors entered and took their positions in tableau at the boardroom table. The cohort was told that they would be watching a play and that they would need to make choices. The play began, and the plot offered a fictionalized account of Ray Anderson's decision to make Interface Corporation a sustainable enterprise. Twelve actors began in the room with five of the actors soon departing to take their positions around the campus. As the first scene concluded, the actors began to move, though not together. The audience was forced to choose who to follow and the first fragmentation of experience occurred. Throughout the play actors met and separated. Audience members were each individually accountable for deciding who and what to follow. It was impossible to see more than a small part of the play and the assembly of the narrative could only occur afterward. Structured reflection drew attention to how choices were made, how individuals experienced the fragmentation and how they experienced the partial nature of their experience. Fragmentation as an organizing dynamic was clear in the qualitative assessments that followed the program.

> I realized as this weekend progressed that the image that I held of what leadership "is" is but one of many different conceptions of the topic.

> These three days showed me practically how one can excel through diversity.

> In my degree, so far, I've mostly been taught technical concepts that generally allude to a single correct answer. One of my main motivations to join the PLLC was to be exposed to education that would open my mind, help me to grow as a person, and challenge my ways of thinking. This weekend's programming surpassed all of my expectations and made me extremely excited for the upcoming year.

The application of Japanese design principles and the MDA model to cohorts of 70–150 cross-departmental undergraduates from the University of Alberta suggests the efficacy of the approach for interdisciplinary collaboration. The approach bypassed disciplinary strategies and centered interactivity and direct experience in order to encourage personal and collective reflection and adaptation.

CASE STUDY: HOPE DECODED

Hope Decoded was an intensive five-day residency hosted at the Banff Centre that looked at fostering hope around four complex societal issues: income inequality, Indigenous rights, arts and culture and the environment. Rather than having a discussion rooted in theory, the conversation was framed around the work of four "Cause Champions" – organizations at the frontline of combating hopelessness. Thirty-five creative thinkers and leaders representing a diversity of knowledge, experience and background worked with these case studies over five days to develop new strategies.

Hope Decoded was organized around the aesthetic of celebration. By centering the sharing of successes in adjacent communities, participants were able to step away from preconceived notions to generate new strategies to foster hope. As with the previous case, program design revolved around an Indigenous Elder. Mechanics encouraged sharing, suspending judgment, play, co-creation and empathy.

Open Mic

The opening evening was hosted in the Banff Centre's Club, a space primarily intended for jazz performance. Each participant was asked to share a success with the broader community on stage and with the microphone. By structuring the sharing as a performance, successes were reframed as artifacts of creative action rather than advice on how to proceed. Collaborative design of new processes brought communities together that might otherwise not share best practices in service to live challenges in communities.

The design aimed at disrupting potential hopelessness in community-based work by supporting spaces where new ideas could be introduced, proposed and demonstrated. On the final day, Hope Decoded opened the doors to 150 influencers from across North America to hear the ideas and approaches generated during the residency and continue the discussion about application across sectors of society. This larger group included business executives, First Nations leaders, not-for-profit managers, elected officials, activist athletes, authors and speakers. The larger event was hosted by improvisational comedians and supported by local musicians. The mood was celebratory, and the group was charged with helping to carry the conversation into the future by leading long-term social change initiatives that contribute to the well-being and resiliency of communities. From this event came direct and indirect investment in multi-year residencies to support market-based approaches to economic inequality and persistent collaborations across and within communities.

Feedback from the Four "Cause Champions"

> The Hope Decoded residency gave us access to a high calibre constellation of thinkers who are also grappling with this challenge.
>
> I have new ideas, new language, and an expanded network to further our organization's mission and increase the effectiveness of our efforts at arts advocacy.

We came away with some very tangible products: a better question that will guide our work moving forward as well as two games that we will further develop as tools for our work.

There are at least three individuals that I met who I intend to contact and engage directly.

I know I am part of a community of change makers that will have impact well beyond our week together.

There were profound discoveries that came out of the experience, many important connections that were made and new ideas that have taken root as a result of the week.

By intentionally curating diverse perspectives and experiences, leaders could work on meaningful projects with collaborators with very different ways of seeing the world and the work. The group shared a strong commitment to supporting the four partner organizations and needed to work across significant boundaries to do so. Every interaction became source material for self-directed interpretation.

CASE STUDY LIMITATIONS

The implementation of the approach demonstrates successes, but points of tension and limitations of the process were also evident. Integrating the preferences of participants and the design team was not always possible and this led to design elements that did not support systematic coherence and behaviour that acted against the intended aesthetic. Often this occurred when attempting to accommodate tools that stakeholders wished to include. Expectations around privilege and position were often unspoken and needed to be addressed during delivery.

Another limitation relates to different interest and comfort in unfamiliar spaces. The design process often results in spaces and activities that are outside the standard frame for development experiences. Some participants expressed a desire for more traditional approaches and felt overwhelmed by the ongoing expectations of interactivity and collective meaning-making. Some individuals were more suited to these environments by temperament or by experience. Cultural factors can play a role, as disruptions in patterns of deference and etiquette were more directly felt by some.

Incentives and contexts that extended beyond the process continued to impact behaviour. Participants were very conscious of structures of power and frequently inquired into the relationship of different actors to the process. In Hope Decoded, the group created a space to discuss the role of the primary funder of the event, a foundation created by a large energy company. Awareness of structures of power led to moments of deference or resistance. These were often expected but complicated the consistent application of a desired aesthetic.

Finally, as the programs were delivered in Banff, a UNESCO World Heritage Site, the designers were concerned that negative feedback was diminished due to a participant desire to be invited into future opportunities. Greater anonymization of feedback could help to offset this limitation.

CONCLUSION

This chapter bridges current theory and practice in creativity to applications in community development. By understanding creativity as a socially and communicatively constituted phenomena, design methodologies can be applied that enhance the relational field within which development processes occur. Community development activities are constructed through interactions. By focusing on aesthetic experiences that encourage contribution, participation and interest, systems can be enacted that persist and that increase the frequency of new and appropriate ideas, the transfer of ideas within and across communities, and the transformation of existing material and symbolic social structures.

REFERENCES

Amabile, T.M. (1996). Creativity in context: Update to *The Social Psychology of Creativity*. Boulder, Colorado: Westview Press. https://doi.org/10.1146/annurev.psych.093008.100416.

Bartels, G. (2010). The creative small group: towards a framework of collaborative creativity within the creative sphere. Master's Thesis. Mount Saint Vincent University, Nova Scotia.

Boden, M.A. (1994). Precis of "*The Creative Mind: Myths and Mechanisms*," London: Weidenfeld and Nicolson 1990 (Expanded edn., London: Abacus, 1991). *Behavioral and Brain Sciences*, 17(3), 519–31.

Boutellier, R., Gassmann, O., and Zedtwitz, M. Von (2008). Managing global innovation. *Industrial Research*. https://doi.org/10.1007/978-3-540-68952-2.

Cooren, F. (2015). *Organizational Discourses*. Cambridge, UK: Polity Press.

Cooren, F., Bartels, G. and Martine, T. (2016). Organizational communication as process. In A. Langley and H. Tsoukas (eds), *The SAGE Handbook of Process Organization Studies* (pp. 513–28). London: SAGE Publications Ltd. doi: 10.4135/9781473957954.n32.

Csíkszentmihályi, M. (1996). *Flow and the Psychology of Discovery and Invention*. New York: Harper Collins. https://doi.org/10.1037/e586602011-001.

Csíkszentmihályi, M. (2014). Society, culture, and person: A systems view of creativity. In Csíkszentmihályi, M., *The Systems Model of Creativity: The Collected Works of Mihaly Csíkszentmihály* (pp. 47–62) New York: Springer. https://doi.org/10.1007/978-94-017-9085-7_4.

Dubina, I.N., Carayannis, E.G., and Campbell, D.F.J. (2012). Creativity economy and a crisis of the economy? Coevolution of knowledge, innovation, and creativity, and of the knowledge economy and knowledge society. *Journal of the Knowledge Economy*. https://doi.org/10.1007/s13132-011-0042-y.

Ford, C.M., and Gioia, D.A. (1996). Multiple visions and multiple voices: Academic and practitioner conceptions of creativity in organizations. In Ford, C.M., and Gioia, D.A., *Creative Action in Organizations: Ivory Tower Visions and Real World Voices* (pp. 3–11). Thousand Oaks, California: SAGE Publications Ltd.

Foucault, M. (2017). What is an author? In Goldblatt, D., Brown, L.B., and Partridge, S. (eds), *Aesthetics: A Reader in Philosophy of the Arts: Fourth Edition*. New York: Routledge. https://doi.org/10.4324/9781315303673.

Funk, C., and Woodward, J.B. (2004). The aesthetics of leader development: A pedagogical model for developing leaders. In *The Second Art of Management and Organization Conference* (pp. 1–17). Paris, France: ESCP-EAP European School of Management.

Hall, E.T. (1990). *The Hidden Dimension*. New York: Random House Inc.

Han, B.-C. (2017). *Psycho-Politics: Neoliberalism and New Technologies of Power*. Brooklyn: Verso.

Henry, C., and de Bruin, A. (2011). Introduction. In C. Henry and A. de Bruin (eds), *Entrepreneurship and the Creative Economy: Process, Practice and Policy* (pp. 1–6). Cheltenham, UK and Northampton, MA, USA: Edward Elgar Publishing.

Hunicke, R., LeBlanc, M., and Zubek, R. (2004). MDA: A formal approach to game design and game research. https://doi.org/10.1.1.79.4561.

Isaksen, S.G., Murdock, M.C., and Firestien, R.L. (1993). *Understanding and Recognizing Creativity: The Emergence of a Discipline*. Norwood, NJ: Ablex.

Kaufman, H.F. (1959). Toward an interactional conception of community. *Social Forces*. https://doi.org/10.2307/2574010.

Kodama, M. (2017). Knowledge convergence through "Ma thinking." *Knowledge and Process Management*. https://doi.org/10.1002/kpm.1541.

Lopes, D.M. (2007). Shikinen Sengu and the ontology of architecture in Japan. *Journal of Aesthetics and Art Criticism*. https://doi.org/10.1111/j.1540-594X.2007.00239.x.

Nitschke, G. (1966). "MA": The Japanese sense of place in old and new architecture and planning. *Architectural Digest*, (March), 113–56.

Nonaka, I., and Konno, N. (1998). The concept of "Ba": Building a foundation for knowledge creation. *California Management Review*. https://doi.org/10.1016/j.otsr.2010.03.008.

Oosterling, H. (2000). A culture of the "Inter." In Kimmerle, H., and Oosterling, H., *Sensus communis in Multi- and Intercultural Perspective: On the Possibility of Common Judgements in Arts and Politics* (pp. 61–90). Würzburg: Königshausen & Neumann.

Oosterling, H. (2005). MA or Sensing Time-Space. Lecture "Japanese Inter-Esse: 'MA' as In-Between" voor Transmediale.05 BASICS Berlin, 5 February 2005, Haus der Kulturen der Welt.

Panalaks, M.S. (2001). *Ma of Taiko*. Master's Thesis. Dalhousie University.

Paulus, P.B., and Nijstad, B.A. (2010). *Group Creativity: Innovation through Collaboration*. Group Creativity: Innovation through Collaboration. Toronto: Oxford. https://doi.org/10.1093/acprof:oso/9780195147308.001.0001.

Peter Lougheed Leadership College (n.d.). Retrieved 14 January 2019 from https://www.ualberta.ca/lougheed-leadership-college.

Sanders, L., and Stappers, P.J. (2012). *Convivial Design Toolbox: Generative Research for the Front End of Design*. Amsterdam: BIS.

Sawyer, K. (2012). *Explaining Creativity: The Science of Human Innovation* (2nd ed.). Toronto: Oxford University Press.

The Banff Centre Mandate (n.d.). Retrieved 21 February 2019 from https://www.banffcentre.ca/sites/default/files/Banff Centre Publications/banff_centre_mandate_jan_09.pdf.

Theodori, G.L. (2005). Community and community development in resource-based areas: Operational definitions rooted in an interactional perspective. *Society and Natural Resources*. https://doi.org/10.1080/08941920590959640.

UNCED (1992). (Agenda 21) *Earth Summit '92. The UN Conference on Environment and Development. Reproduction*. https://doi.org/10.1007/s11671-008-9208-3.

UNESCO (2015). *Re-shaping Cultural Policies: A Decade Promoting the Diversity of Cultural Expressions for Development – 2005 Convention Global Report*. Paris: UNESCO. https://doi.org/978-92-3-100136-9.

Woodward, B., and Funk, C. (2010). Developing the artist leader. *Leadership*. https://doi.org/10.1177/1742715010368768.

25. The new role of the university in community development
Graciela Tonon

INTRODUCTION

The role of the university has changed during the last decades from mere production and reproduction of scientific knowledge to the construction of a social role for people and communities. If we further regard the university as an organization which forms part of the community and interacts as one of the social actors in development processes, its activities should be carried out, not only within the university buildings but also in the community. The aim of this chapter is to present the social role of the university in relation with community development. To do that we analyse the concepts: community, neighborhood, community development, participation and human agency. Finally we will present a study that shows the opinion of Social Work students about university-community relations.

THE SOCIAL ROLE OF THE UNIVERSITY IN THE COMMUNITY

The twenty-first century presents us with a university which has broadened its traditional role of producer and reproducer of scientific knowledge and has, furthermore, acquired relevance as a space of social interaction for the people (Tonon, 2012, p. 515) who require the university to be committed to community life and with the needs of the citizens who form part of the community – on the basis of respect for the diversities of each human group.

In principle, it ought to be made clear that, in Latin America, universities have played a different and unique role as compared to the rest of the universities in the world, and that they have been acknowledged as the major educational instrument of the new local political elites (Brunner, 2007). For in Latin America universities have not only been devoted to higher education and research but also to the development of political leaders, thus giving way to more innovative ideological debates aimed at the promotion of social change (Mollis, 2006).

According to Mendoza Álvarez (2006, p. 100) the university is, simultaneously, an institution and an organization; it is an institution because its very nature has been established, defined and related to society's global plan, and it further possesses a collective recognition of its legitimacy and scope of autonomy; it is an organization because it possesses academic instrumental knowledge and operational effectiveness. From the point of view of Murcia Peña (2012, p. 36) the university is a social, cultural and political scenario in which ideas, feelings and projects are confronted, and where it is possible to share experiences and theories that contribute to maintain, construct and develop the individuals, societies and cultures.

There is a traditional development of a university-society scheme of work, pre-eminently based on knowledge transfer as a form of interaction with the environment. If we further regard the university as an organization which forms part of the community and interacts as one of the social actors in development processes, its activities should be actively carried out, not only within the university buildings but also in the community scenario. In this manner, it would be possible to identify the university's active role in the construction of situational diagnoses of the communities' needs – regarding the conception of needs as defined by Max-Neef, Elizalde and Hopenhayn (1993) in the sense that needs are social. This conception is described as:

> The present situation calls for a radically different conception of the social context of human needs as compared to the, so far, general conception sustained by social planners and designers of development policies. It is no longer a question of merely relating needs to goods and services – which presumably satisfy these needs – but also to relate them to social practices, forms of organization, political models, and values which exert an influence over the manners in which the aforementioned needs are expressed. (1993, pp. 51–2)

Concrete decisions are thus required in order to establish an accurate link between university and community, on the basis of a development of joint action such as the construction of community situational diagnoses, focused on the protagonists' point of view, and which might be taken into account by state organs in charge of generating public policies.

Therefore contributing with innovative information to enhance the traditional methodologies, namely: the systematization of a permanently updated community resource guide, accessible to all its members; the organization of a variety of training courses on topics that may be of interest to the members of the community, as well as lectures held in different community organizations related to topics of their concern. At the same time the generation of a community space for training in research and development aimed at graduate and post-graduate university students, and the edition of virtual and printed periodical publications which may reflect the productions resulting from the work that has been carried out.

We embarked upon this task since the creation of UNICOM, Institute of Social Studies devoted to the university-community relationship, of the Faculty of Social Sciences of Universidad Nacional de Lomas de Zamora, Argentina.

COMMUNITY AND NEIGHBORHOOD

When we make allusion to community, we are not making reference to uniformity, since community implies inclusion of diversity, that is, to succeed in sharing in a community – for community is a synonym of voluntary social interaction and sharing in a context of diversity; thus, we conceive community as unity in diversity (Tonon, 2009). Community is further defined as:

> The different definitions of community which emerge from the Latin American scenario are consistent with the idea that a community is a group of people who cohabit and share various

elements; namely, a territory, a geographical space and/or place, life styles, everyday situations, customs, identity, language, culture, activities, norms, history, social bonds; people who construct a network of relationships, disagree and accept each other's peculiarities, people who are committed and help each other when in need, who seek/share a common aim/the greater good. (Tonon, 2017, p. 12)

Yet the conception of community has undergone changes in the last decades, going beyond the traditional and exclusive idea that it is restricted to merely belonging to a geographical space inhabited by a group of persons – it has been conceptualized in terms of the relationships among the persons sharing a certain scenario. "'Communities' can be regarded as groups of people which are interest-based, or formed on the basis of participation in a common enterprise, or some shared spiritual experience based on belief which may involve no physical propinquity" (Jenks and Dempsey, 2007, p. 158).

It ought to be added, though, that the notion of being part of a community not only involves a sense of belonging but also feelings of loyalty and responsibility; thus, the notion of citizenship is articulated on the basis of the recognition that the members of a community share features which identify them and distinguish them from those who are alien to it (Tonon, 2012, p. 3). When we make concrete reference to spatial locality we are not only making allusion to a certain geographical space regarded as a synonym of neighborhood. It is at this point that the concept of neighborhood ought to be reconsidered in order to identify its differences and/or similarities to the concept of community – though there is current literature on the subject whose authors use both concepts as if they were synonyms.

Jenks and Dempsey (2007, p. 154) sustained that neighborhood can be considered either as a "spatial/functional construct" or as a "social construct". Hallman (1984, p. 45) points out that "the territory becomes a neighborhood only through occupancy and use by its residents". Barton, Davis and Guise (1995) defined neighborhood as a functional entity which provides residents with services and facilities. "The neighborhood is conceptualized by some theorists as a functional entity, namely a provider of services, a physical construct that supports the needs of the people living there." (Hallman, 1984; Barton (2000) quoted by Jenks and Dempsey, 2007, p. 156).

Galster (2001, p. 2112) states that "neighborhood" consists of: structural characteristics of the residential and non-residential buildings; infrastructural characteristics; demographic characteristics of the resident population; class status characteristics of the resident population; tax/public service package characteristics; environmental characteristics; proximity characteristics; political characteristics; social-interactive characteristics; and sentimental characteristics. Further, Galster (2001, p. 2114, quoting Suttles, 1972) identified four neighborhood scales: the first level of the scale is the block or area in which children may be allowed to play without supervision; the second level is the smallest area possessing a corporate identity as defined by mutual opposition or contrast to another area; the third level consists of some local governmental body districts in which individuals' social participation is selective and voluntary; and the fourth level is the highest geographical neighborhood scale which comprises an entire sector of the city. The author completes the idea by expressing that a person's externality space in the neighborhood "is the area over which changes in one or more spatially based attributes, initiated by others, are perceived as altering the well-being of the individuals from that particular location" (Galster, 2001, p. 2114).

Kallus and Law-Yone (2000, p. 819) believe in the development of a stratified process of transformation in the meaning of the concept of neighborhood. They identified three types of approaches: humanistic, instrumental and phenomenological. The humanistic idea highlights the concept of neighborhood as a means for social change, for the improvement of personal life, and the betterment of the city (p. 821). The instrumental approach deals with practical and substantial topics which attempt to resolve the making of the neighborhood as an agent in a methodical and systematic design process (p. 822). The phenomenological approach regards the neighborhood as a cultural phenomenon, with a deeper significance than a social organization, generated by physical proximity; it focuses on the neighborhood as a cultural entity rather than a social construction. "Accordingly, the neighborhood is regarded as a unique urban entity in which the knowledge and awareness of a place are embedded. It is a spatial pattern whose meaning originates from profound and continuing bonds between place and people" (Kallus and Law-Yone (2000, p. 823).

The concept of neighborhood has traditionally been based on the spatial aspect, and on a specific location. Yet the various abovementioned definitions make it clear that the concept of neighborhood may be applied by different authors, either as a mere geographical space of residence or as a synonym of community, when they focus on the persons who inhabit that space (Briggs, 1997, p. 208; Galster, 2001). "The different uses of the term are repeated in urban design, planning, and urban sociological literature, where an interchange of physical and social terminology are not uncommon" (Barton (2000) quoted by Jenks and Dempsey, 2007, p. 155).

CONCEPT OF COMMUNITY DEVELOPMENT

The concept of community development has been long debated and, in certain cases, even interchanged with the concept of local development (Rothman (1968) quoted by Bhattacharyya, 2004, p. 6). In this text, though, we shall consider the definition given by Bhattacharyya (2004), who regards it as "different from other endeavors in that it aims at building solidarity and agency by adhering to three practice principles, namely, self-help, felt needs, and participation" (p. 5). In explaining the three practical principles, Bhattacharyya (2004, pp. 21–3) points out that self-help builds and mobilizes people's cultural and material assets and avoids dependency, felt needs (or demands), it further affirms human variation – thus resisting developmental impositions from above – and concludes that participation means inclusion in the definition and problem solving processes, as well as in the ways of solving them. Finally, and in methodological terms, the author expresses that these three principles allow decision-making – methodologically speaking – which will elicit the techniques to be applied in a community development process (p. 21). In the light of this definition, we coincide with the author in the sense that community development is a distinctive field (Bhattacharyya, 2004, p. 28) and part of the Democracy Project (Bhattacharyya, 2004, p. 14).

THE IMPORTANCE OF PARTICIPATION

The concept of participation requires a special consideration, for it plays an important role in community development, particularly in Latin American countries. When making reference to local development, Crocker (2014, pp. 100–102) distinguishes different types of participation in group decision-making, and identifies seven types of participation, namely: *Nominal Participation* which implies that a person is member of a group, though not attending the meetings; *Passive Participation* in which case the people who form part of the group attend the meetings in which decisions are made, yet passively listen to the reports on the decisions taken by others; *Consultative Participation* in which the members offer information and air their views, but it is the leaders who decide; *Petitionary Participation*, when the members demand certain decisions or action from the authorities, in most cases, to allay injustice; the so-called *Participant Implementation*, that is, the case in which the leaders decide and the members merely put the decisions into practice; *Negotiation* in which case all the parties involved attempt to achieve consensus formation; and *Deliberative Participation* in which case both members and leaders participate in deliberation to examine propositions and motives in order to reach acceptable agreements that may obtain a majority rule. The author concludes that, as the list progresses, "each type of participation becomes narrower or deeper, in the sense that individual or collective agency is expressed more completely" (Crocker, 2014, p. 102). Bhattacharyya (2004) conceives participation as a way of:

> taking part in the production of collective meanings. Thus the principle of participation means inclusion, not merely taking part in the electoral process, or endorsing decisions, but in developing an agenda for debate and decision; it means inclusion in the processes of problem solving and defining the way those problems ought to be solved. (p. 23)

The concept of participation plays a major role in Latin America. It emerged in the 1970s in the context of new social and political practices, with the aim of giving way to a process of social development which might put an end to injustice and inequality. In the particular case of Argentina, the work carried out by Sirvent in the 1980s and 1990s, is a clear illustration of the progress made in this field. Sirvent (1999) acknowledges the existence of two types of social participation: real participation and symbolic participation. The real type takes place when the members of a group or institution exert an influence over all the institutional processes and upon the nature of their decisions. Symbolic participation refers to actions which have little or no bearing on institutional policies and management, thus creating – both in individuals and groups – the illusion that power is non-existent (Sirvent, 1999, p. 129). From our point of view, the essential concept of participation lies in joint decision-making, which implies distribution of power, rather than simply acting together.

Human Agency

Regarding human agency, Sen (2000, p. 35) makes it clear that the term "agent" has been traditionally applied in literature related to economics with reference to a person who acts on behalf of another – and whose achievements must be assessed in the light of someone else's objectives. The author proposes the use of the term "agent" in the original sense of the word, which considers that persons act and provoke changes whose achievements may be assessed on the basis of their own values and objectives, independent from any form of external assessment criteria. According to this point of view, a person ceases to be a passive receptor of the actions produced by organs of state and professional bodies, to become an actor and protagonist of development processes.

METHODOLOGY

Our research work is a descriptive study, developed from a qualitative perspective, which has allowed a more accurate understanding of the participants' personal perceptions thus giving way to new categories, in other words, to a new signification of the same categories which may have been so far observed in previous research studies. Research studies conducted through the qualitative method are characterized by a search of the significance actors attribute to concrete actions, placing special emphasis on the study of the phenomena in their own particular context.

The sampling was composed of 350 students of different courses of study in the Faculty of Social Sciences of Universidad Nacional de Lomas de Zamora. We used the simple sampling method for qualitative studies, which requires the researcher to make a list of the essential attributes of each selected unit – in this case, we prioritized the selection of students of both sexes and different ages, taking different courses of study.

The technique used was the individual text, which each of the students was required to hand in, in the context of the classroom space. This text contains information related to basic identification data: age, sex, neighborhood in which they live, course of studies they are enrolled in, year of access to university, number of subjects passed, and open questions organized according to thematic axes of reflection – one of which dealt with the university-community relationship, so as to identify the type of activities the university might conduct in order to fortify the community development process.

The methodological strategy, applied for the analysis of the collected data, is the thematic analysis proposed by Braun and Clarke (2012) to identify, analyse and provide information on topics and structures, thus revealing the subjects' experiences as well as their meanings and realities, and moreover examine the ways in which the events, realities, meanings and experiences may be the effects of society as a discourse marker. In this work we have used the research modality known as theoretical categorization, defined by the researchers' specific theoretical interests. Thus, the latter identified the topics on the basis of what each subject verbally expressed, progressively working their way from an organized description to the interpretation of the data retrieved. Taking into account the abovementioned theory, we have attempted to understand the significance the subjects give to this data. Braun and Clarke (2012) consider that thematic analysis is a methodological strategy in qualitative research, through which it is possible to systematize results.

ANALYSIS OF RESEARCH RESULTS: OPINIONS OF SOCIAL WORK STUDENTS

The Social Work students who took part in the survey are at present taking and/or have completed the course Theory and Practice of Social Work I (Community), whose main object is to become an academic space that may achieve a real theoretical-practical academic integration for Social Work – the community constituting its field of action. The theoretical topics developed in the abovementioned course originate in Social Work as well as in other academic fields, and are particularly characterized by their application in the specific field of Social Work. Moreover, they are characterized by constituting thematic axes in the social workers' daily professional activity, and are considered to be of the utmost importance in the context they operate in.

The axis of the aforementioned subject is methodological, and its aim is to encourage the students into developing a critical attitude and commitment with the present social reality, so that they may discover and construct possible structures in the community context. The sample student survey responses show different proposals for action that may facilitate the relationship between the university and the community, which can be organized into two research blocks:

- Activities to be developed in the community
- Policies for action in order to develop the aforementioned activities in the community (i.e., philosophical and technical aspects that lead to the shaping and development of those policies).

We hereby present a table and a box which introduce a summary of questionnaire responses: Table 25.1 refers to the activities surveyed, while Box 25.1 includes the policies for action related to the aforementioned activities.

The classification of students' responses shows their concern about the type of activities that the university might develop in order to enhance its relationship with the community – in particular, those activities which might typically apply to other organs of state. The proposals which make reference to organizing: legal service and counseling, community health programs, assistance for access to social services programs, sports and recreation activities in deprived areas, festivals in squares open to the community, community solidarity through different types of charity donations, are considered activities which ought to be included in the public policy agenda of every specific governmental area in interaction with the university (in the role of collaborator) – yet the fact that students should propose that the activities be actually developed by the university evidences the change in the latter's social role in the twenty-first century (which we have previously made reference to).

CONCLUSIONS

In the last decades, cities and their suburban areas in Latin America have shown an accelerated growth which is entirely out of proportion, thus giving way to new neighborhoods, since the existing communities are overpopulated (Tonon, 2012, p. 3). Simultaneously, migrant populations from rural areas and/or neighboring countries have arrived in search of employment, and improvement of their living conditions. These reconfigured spaces include a great number of persons, characterized by their cultural heterogeneity, and it is in the course of this long process that the concept of community has undergone changes.

Table 25.1 Activities

Activity	Type
Courses and workshops in the community	Current themes which aim at preventing social problems
Joint programs to work with the institutions of the community	with schools (career assessment tests, activities to encourage young people to persevere in their university studies, thus avoiding doubts, or the idea of failure, and promoting the importance and actual possibility to continue with their studies – especially considering that university is free of charge[a] both for young people and senior citizens) with hospitals with homes for children and for the elderly with centers for prevention and control of substance addictions with the organizations that provide food for the poor
Community organization activities	sports and recreation activities in deprived areas festivals in squares open to the community community solidarity (different types of charity donations: food, clothing, toys and so on) legal services and counseling community health programs assistance for access to social services programs[b]

Notes:
[a] In Argentina, national state universities are free of charge. They are to be found in all the provinces of the country.
[b] In some cases, the social services programs office is centralized, which derives in the neighbors finding it difficult and costly to have access to these programs, for it requires them to go to the aforementioned offices on several occasions – moreover, the students make reference to the community members' need of assistance in this respect.

BOX 25.1 POLICIES FOR ACTION

- Assist the community whenever it is required.
- Work jointly in order to improve the persons' quality of life.
- Promote community participation.
- Succeed in adapting the university to the community.
- Ensure sustainability of the organized activities.

The university is a social actor that plays a leading role in these new processes, and today's challenge is centered in working on the construction of university-community relationships which may effectively place the knowledge generated by the university at the service of the social needs of the community.

In other words, the situation calls for concrete decisions which may effectively relate the university with the community through joint action aimed at community development.

REFERENCES

Barton, H. (ed.) (2000), *Sustainable Communities: The Potential for Eco-Neighbourhoods*. London: Earthscan.
Barton, H., Davis, G. and Guise, R. (1995), *Sustainable Settlements: A Guide for Planners, Designers and Developers*. Bristol: University of the West of England and The Local Government Management Board.
Bhattacharyya, J. (2004), Theorizing community development, *Journal of the Community Development Society*, *34* (2), 6–34. Retrieved 25 June 2018 from file:///C:/Users/Graciela/Downloads/Resident_Attitudes_Toward_a_Proposed_Limestone_Qua.pdf.
Braun, V. and Clarke, V. (2012), Thematic analysis, in H. Cooper (editor-in-chief), *APA Handbook of Research Methods in Psychology: Vol. 2 – Research Designs*. Washington: American Psychological Association, pp. 57–71. DOI: 10.1037/13620-004.
Briggs, X. de Souza (1997), Moving up versus moving out: neighborhood effects in housing mobility programmes, *Housing Policy Debate*, *8*(1), 195–234. Retrieved 24 June 2018 from file:///C:/Users/Graciela/Downloads/XBriggs_Movingupversusout_HPD_1997.pdf.
Brunner, J. (2007), *Universidad y sociedad en América Latina*, México, Universidad Veracruzana. Retrieved 26 June 2018 from file:///C:/Users/Graciela/Downloads/universidad_sociedad-def-libre.pdf.
Crocker, D. (2014), Participación deliberativa en el desarrollo local, in M. Nebel, P. Flores-Crespo and M.T. Herrera (coordinadores), *Desarrollo como libertad en América Latina: Fundamentos y aplicaciones*. México DF: Universidad Iberoamericana.
Galster, G. (2001), On the nature of neighbourhood, *Urban Studies*, *38*(12), 2111–24. Retrieved 28 June 2018 from https://www.researchgate.net/profile/George_Galster2/publication/247043318_On_the_Nature_of_Neighbourhood/links/5694235c08ae3ad8e33b63ee/On-the-Nature-of-Neighbourhood.pdf.
Hallman, H. (1984), *Neighborhoods: Their Place in Urban Life*. Beverly Hills, CA: Sage.
Jenks, M. and Dempsey, N. (2007), Defining the neighborhood: challenges for empirical research, *The Town Planning Review*, *78* (2), 153–78. DOI: 10.3828/tpr.78.2.4. Retrieved 22 June 2018 from https://www.researchgate.net/profile/Nicola_Dempsey/publication/250276900_Defining_the_neighbourhood_Challenges_for_empirical_research/links/568c56de08ae71d5cd04d634/Defining-the-neighbourhood-Challenges-for-empirical-research.pdf.
Kallus, R. and Law-Yone, H. (2000), What is a neighbourhood? The structure and function of an idea, *Environment and Planning B Planning and Design – November*, *27*, 815–26. DOI: 10.1068/b2636. Retrieved 23 June 2018 from https://www.researchgate.net/profile/Hubert_Law-Yone/publication/23541227_What_is_a_neighbourhood_The_structure_and_function_of_an_idea/links/5522b55a0cf2f9c130544550/What-is-a-neighbourhood-The-structure-and-function-of-an-idea.pdf.
Max-Neef, M., Elizalde, A. and Hopenhayn, M. (1993), Desarrollo y necesidades humanas, en M. Max-Neef, *Desarrollo a escala humana: Conceptos, aplicaciones y algunas reflexiones*. Montevideo-Barcelona: Editorial Nordan-Icaria Editorial, SA, pp. 37–82. Retrieved 1 July 2018 from https://repositories.lib.utexas.edu/bitstream/handle/2152/21625/Max_Neef-Desarrollo_a_escala_humana.pdf?sequence=2.
Mendoza Álvarez, A. (2006), El naufragio la universidad, en G. Plascencia Castellanos (coord.), *Palabra libre condición de la universidad*. México DF: Universidad Iberoamericana, pp. 99–119.
Mollis, M. (2006), Geopolítica del saber: biografías recientes de las universidades latinoamericanas, en H. Vessuri (compiladora), *Universidad e investigación científica*. Buenos Aires: CLACSO.
Murcia Peña, P. (2012), *Universidad y vida cotidiana: Imaginarios de profesores y estudiantes*. España: EAE.
Rothman, J. (1968). Three models of community organization practice, in *Social Work Practice 1968*. National Conference on Social Welfare. New York: Columbia University Press.
Sen, A. (2000), *Desarrollo y Libertad*. Bogotá: Editorial Planeta.
Sirvent, M.T. (1999), *Cultura popular y participación social: una investigación en el barrio de Mataderos (Buenos Aires)*; Buenos Aires-Madrid: Facultad de Filosofía y Letras, Universidad de Buenos Aires-Miño y Dávila.
Suttles, G. (1972). *The Social Construction of Communities*. Chicago, IL: University of Chicago Press.
Tonon, G. (2009), *Comunidad, participación y socialización política*. Buenos Aires: Espacio Editorial.
Tonon, G. (2012), Las relaciones universidad-comunidad: un espacio de reconfiguración de lo público, DOI: 10.4000/polis.6691. *Polis, Revista de la Universidad Bolivariana*, *11* (32), 511–20. Retrieved 18 August 2017 from http://www.scielo.cl/scielo.php?pid=S0718-65682012000200024&script=sci_arttext.
Tonon, G. (ed.) (2017), *Quality of Life in Communities of Latin Countries*, Community Quality of Life and Well-Being Series. Cham, Switzerland: Springer.

26. Community innovation and small liberal arts colleges: lessons learned from local partnerships and sustainable community development
Craig A. Talmage, Robin Lewis, Kathleen Flowers and Lisa Cleckner

For many communities, the local college and community are uniquely and dynamically intertwined and deeply entrenched with regard to their economic, environmental and social fates (CCAP, 2016; Gumprecht, 2003). In some cases, town-gown conflicts or divisions persist, while in other instances colleges and municipalities engage in ongoing collaborations and experimentation (Bombyk, 2003). Some researchers and practitioners claim small urban areas owe their resilience to their anchor educational institutions (Davis, 2016; Fischer, 2008); however, the relationship may be more mutually beneficial, if not reciprocal, as an increasing number of small colleges face enrollment challenges. Might these small urban areas serve as attractive differentiators for these institutions of higher education (Docking and Curton, 2015)?

This chapter analyses the complex relationships between small municipalities and small liberal arts institutions. It looks for community development and community innovation approaches found across and within colleges regarding working with their local communities. A macro-level literature review and internet search is undertaken to identify innovation and development themes found among small liberal arts schools and their small communities. A micro-level discussion of a college and its host community is presented to provide detailed examples of community development and community innovation. Finally, lessons learned for research and practice in higher education and community development are shared.

COMMUNITY DEVELOPMENT AND COMMUNITY INNOVATION

This chapter purports that all community innovations fall under community development, but not vice versa. Figure 26.1 outlines key themes highlighted in this chapter, which will later frame the unique, dynamic relationship between small liberal arts colleges and communities.

Community Development

Most definitions of community development view it as both a process and an outcome (Phillips and Pittman, 2009; Robinson and Green, 2011). Robinson and Green (2011) add programs to their definition of community development, which could be seen as

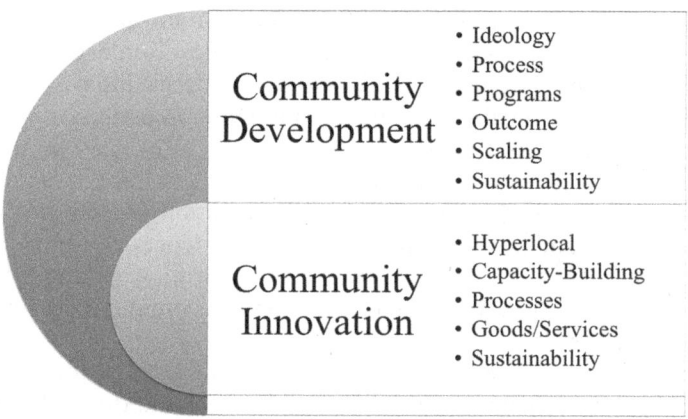

Figure 26.1   Community development and community innovation

part of processes, and they add ideologies, which also guide processes as part of their conceptualization of community development. Matarrita-Cascante and Brennan (2012) provide a more comprehensive definition of community development, arguing that:

> Community development is a process that entails organization, facilitation, and action, which allows people to establish ways to create the community they want to live in. It is desired goals associated with the promotion of efforts aimed at improving the conditions in which local resources operate. As a result, community developers harness local economic, human, and physical resources to secure daily requirements and respond to changing needs and conditions. (p. 297)

This definition of community development infuses Robinson and Green's (2011) additions of ideology and programs into process and outcome definitions of community development.

Bhattacharyya's (1995; 2004) work provides another definitional route. First, Bhattacharyya (2004) posits that community development must have distinct purposes and methods. He writes of solidarity and agency as crucial concepts for understanding purpose. To him, solidarity is an essential community characteristic (i.e., it is community), while the promotion of agency (i.e., development) is the purpose of community development. Second, community development's scope must be universal, extending its applications (i.e., purposes and methods) to other populations and settings (Bhattacharyya, 1995; 2004). While arguing much of community development to be context-bound, Bhattacharyya (1995; 2004) notes that a solidarity and agency approach allows community development to be universal. Furthermore, agency has been connected to sustainability in local development efforts (Newman and Dale, 2005).

Community Innovation

Community innovation has received far less attention regarding definition than community development. Early references to community innovation, as a concept, define community in terms of geography or municipality, while innovation is the sum of all innovative activity in those areas (Aiken and Alford, 1970). Meanwhile, more recent research suggests that the task of defining community innovation is often relegated to those doing the innovating and the context in which innovation occurs (Wachtendorf et al., 2018). While allowing those doing the innovating to define innovation may appear to be democratizing, it also contributes to greater ambiguity in definition.

Macro-level innovation, as a performance output, is influenced by political culture, concentration/diffusion of power, centralization of formal political structure, community differentiation and continuity, and community integration (Aiken and Alford, 1970). Political environments and government policies and practices continue to demonstrate their enabling and stifling effects on community innovation (Clammer, 2015). While macro-definitions are important, this chapter focuses on community innovation found at more local (or micro) levels.

On the micro-level, general definitions of community innovation consist of themes such as growth and change that help communities both survive and thrive (Klein and Knight, 2005; Westoby, 2017). Community innovation also requires creativity and discovery, generating a wide range of applications even beyond community development (Semali et al., 2015). Community innovation utilizes distinct processes, and these processes must be understood by community members and organizations (Adams and Hess, 2010). Furthermore, community innovation goes beyond helping individuals; it can be defined in terms of strengthening local organizations and building community capacity (Jackson, 2004). Thus, community innovation like community development is a type of process innovation that can lead to collective impact.

Community innovation differs from for-profit innovation. Roesler (2018) notes that social need(s) and ideology/ies drive community innovation, whereas profit drives for-profit innovations. Conversely, Westoby (2017) notes that social change can instigate collective action and community innovation. Community innovations are also different than social innovations in that they are small and do not require scaling up in large fashions (Baskaran and Mehta, 2016; Lord and Hutchison, 2007). Thus, more than community development itself, community innovation can take the form of a good (i.e., product) or service innovation with collective impact.

For the purposes of this chapter, community innovations are hyperlocal innovations in community services, goods, partnerships or processes that collectively impact community well-being (Dechief et al., 2008). These innovations need not be entirely new or purely grassroots, bottom-up ideas (Hobson et al., 2016). Similarly, community innovations may leverage existing resources and reappropriate them or use them better (Kiwanuka et al., 2015); however, knowledge gathering and knowledge sharing efforts are needed for successful execution (Dechief et al., 2008; Varuk et al., 2018). Finally, while community innovation may not require scalability, both processes and practices that are adaptable, resilient and long-lasting are essential to success (Ajmal et al., 2018).

COMMUNITY DEVELOPMENT, COMMUNITY INNOVATION AND LIBERAL ARTS COLLEGES

Much research in higher education and community development focuses on how best to enhance college-community partnerships in scope and outcomes (Sandy and Holland, 2006). Within such partnerships, experiential learning and service learning in practice span the spectrums of pure service and community engagement (Weerts and Sandmann, 2008). As this chapter will shed light upon, these forms of learning and other forms of community engagement often fall under the umbrella of community development; however, not all of these programs are particularly innovative in terms of processes and deliverables.

Liberal arts colleges have a long history of experiential learning and service learning as reflected in their curriculums (Barber and Battistoni, 1993; Kolb and Kolb, 2005). Over time, these institutions have sought to recognize and improve how experiential and service learning activities are impacting host communities and how local communities are impacting students and their experiences (Bromley, 2006; Lang, 1999). Additionally, entrepreneurship education at liberal arts institutions has drawn attention to how these institutions can foster community innovation (Walshok and Shapiro, 2014). Furthermore, colleges and communities have brought together their respective stakeholders under partnerships aimed at positive collective impact (Moore et al., 2015).

Innovation and its forms (e.g., goods, services and processes) have received less attention in the higher education literature in the context of community development and collective impact. Moreover, community innovation, as a practice and a process, is often overlooked by universities (Fortunato et al., 2015). The next section of this chapter showcases development and innovation work done by small liberal arts institutions in their local communities. The next section begins with an across-institution macro-approach and then transitions into a case study on one particular school in order to demonstrate how community innovation happens at the local (or micro) level.

MACRO-PERSPECTIVE ON LIBERAL ARTS AND COMMUNITY INNOVATION

Many colleges institutionalize experiential and service learning within campus centers for community engagement and service learning (Bringle and Hatcher, 2000). Other schools open entrepreneurial incubator spaces to engage local community members and issues (Mitchell and Levine, 2001) or host curricular programs focused on developing students' entrepreneurship and enterprise skills (Hines, 2005).[1] Some colleges engage municipalities through sustainable community development initiatives and other community-centered approaches (Helfrich, 2015). The next section covers various strategies for community development utilized by small liberal arts colleges and their communities as derived from the literature and mainstream media, while also searching for exemplars of community innovation.

MACRO-METHODS

The authors examined 20 private, non-Ivy League, non-religious and small liberal arts institutions (< 5,000 undergraduate students) in small towns and cities (< 40,000 residents). These institutions were conveniently selected from online lists of small liberal arts colleges (BestColleges.com, 2016; CVO, 2018; Wikipedia, n.d.), colleague recommendations and the New York Six Liberal Arts Consortium for comparison colleges to Hobart

420 *Research handbook on community development*

BOX 26.1 TWENTY LIBERAL ARTS INSTITUTIONS FOR EXPLORATION AND COMPARISON

Allegheny College	Flagler College	Scripps College
Bowdoin College	Franklin Pierce University	Skidmore College
Bucknell University	Hamilton College	St Lawrence University
Carleton College	Middlebury College	Washington College
Colgate University	Oberlin College	Wells College
College of Wooster	Pitzer College	Williams College
DePauw University	Pomona College	

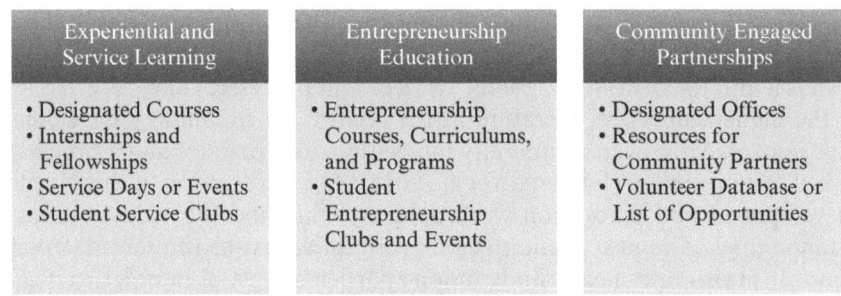

Figure 26.2 Overarching community development themes across institutions

and William Smith Colleges. One author explored the websites and publications of these institutions (see Box 26.1) to elucidate common practice themes found across institutions regarding experiential and service learning, entrepreneurship education and community engagement partnerships.

Figure 26.2 summarizes elucidated themes in the three aforementioned areas from a cursory look across the 20 institutions. The themes were peer debriefed among the authors (see Lietz and Zayas, 2010). Future research should take more in-depth rigorous approaches such as interviews with college and university leaders, faculty and students to confirm and expound on these themes. The authors also acknowledge that this method does not rigorously look at these areas from the local communities' perspectives, which should be investigated in the future.

EXPERIENTIAL LEARNING AND SERVICE LEARNING

Designated Courses

Designated courses are a popular option for schools yet terms for such courses differed across institutions: academic civic engagement courses, academic service learning courses, applied civic engagement courses, community-based courses, community-based learning courses, community engaged courses, experiential courses and service learning courses. Offerings of designated courses at these institutions ranged from 6 to 40 per semester. Some institutions focus on community-based research, while others rely more on community-based learning strategies. Some schools allow students to take independent study courses with these respective designations or complete community-based or -focused capstone (i.e., ending culminating experience) courses for juniors and seniors, while others do not. Often, such courses seek to inspire students to be more civically engaged citizens, but not all courses are particularly local-focused.

Internships and Fellowships

Some schools offer internships and fellowships to students interested in working with local community organizations. The language used to describe these programs ranges from *service* to *community engagement*. Schools also vary in the number of internships and fellowships available. None of the schools name substantial resources for these programs; however, some schools appear to have highly focused or rigorous models for these programs. For example, DePauw University has a Civic Fellows Program, which takes a cohort of 20 students each year who must apply for the program and demonstrate a high commitment to civic engagement and service learning (Depauw.edu, n.d.).

Service Days or Events

Many of the schools analysed offer regular days of service or scheduled small- and large-scale service events. Example event names include *Day of Service* or *Make a Difference Day*. These events recruit students to volunteer for one day in the local community through a single service project or service projects agreed upon between designated college/university offices and local groups/nonprofits. Many projects occur annually or biannually. Other events depicted on their websites include community yard sales and salvage programs, where proceeds go to local organizations.

Student Service Clubs

Most institutions support student clubs devoted to service with clubs ranging in their local foci and ties to national service organizations (e.g., Big Brothers Big Sisters, Kiwanis, Lions Club, Habitat for Humanity, Special Olympics, American Cancer Society, One Campaign, Rotary Club affiliates, fraternal/sororal service-focused organizations, among others). These organizations also range in their wordings, such as naming their clubs' dedication to service or (civic) engagement. These student clubs often sponsor and organize single or multiple service or community-based events throughout the academic year. Their foci range from: raising awareness of community issues (e.g., refugees, environmental stewardship, LGBTQ rights/support, among others), mentoring or tutoring local school children, fundraising for local organizations or initiatives, promoting volunteerism, among others.

ENTREPRENEURSHIP EDUCATION

Entrepreneurship Courses, Curriculums and Programs

Many of the schools have curriculums (i.e., minors and majors) and courses that focus on entrepreneurship knowledge and skills. Some of these programs house their entrepreneurship courses in business, management, economics or engineering departments, while others host courses outside of these common areas (e.g., environmental studies or political science). Some programs have capstone experiences with applied or service learning components to them. These programs also connect students to internships, fellowships, workshops, mentors and career development programs that help them develop and apply their entrepreneurial knowledge and skills. Some students receive funding to implement their entrepreneurial ideas and projects. Other programs offer idea competitions, pitch contests, startup showcases, startup bootcamps, startup bus programs, and speaker series. Often these types of programs are co- or extra-curricular, separated from curriculums and curricular offerings (i.e., formal courses). Some institutions only host entrepreneurship events and have no formal entrepreneurial curriculums or programs.

Student Entrepreneurship Clubs and Events

Much like student service clubs, student entrepreneurship clubs are found across institutions. These clubs do not often have an outward focus on the local community, but they aim to promulgate entrepreneurship skills and applications among their students. They host gatherings of students (e.g., events) regarding developing skills and knowledge, but the publicness of these programs is not well-noted.

COMMUNITY ENGAGEMENT PARTNERSHIPS

Designated Offices

Most schools have designed offices (sometimes called centers) devoted to community partnerships, and also link students, staff and faculty to experiential, service or community-based learning experiences. These offices range in names focused on citizenship, community and civic engagement, common good, opportunities, outreach, partnerships, service, learning in action and volunteerism. Often, offices are located in student affairs rather than academic affairs. Some are connected to career services.

These offices have multiple thematic aims or missions, such as addressing community and social problems, advocacy, benefiting others, citizenship, collaboration, community well-being, experiential learning, hands-on learning, lifelong learning, meaningful service, positive community impact and public engagement. These thematic areas often have national and global foci rather than local foci. Some offices support town-gown task forces to improve relationships between communities and institutions. Some note local foci in their missions and act as liaisons between the community and the institution. Many offices have designated staff, and one institution notes that their office is mostly staffed by student managers; however, most offices do not rely strongly on student staff.[2]

Resources for Community Partners

Some schools provide resources on their webpages for community partners. These pages offer community partners ways to put up offerings for students to volunteer. They feature pathways, guides and forms for agencies and organizations to formally connect with the higher education institution. Pages range in their ease of accessibility and immediate access. Some pages are devoted solely to community partners, while others have community partners simply featured or listed for public view.

Volunteer Database or List of Opportunities

Many schools have websites devoted to promoting volunteer opportunities or local organizations looking for student volunteers. These databases or lists serve to help organizations address volunteer needs and help students give of their time. In some cases, students have volunteer hour requirements; these lists or databases allow students to know how many hours are needed by the local organizations or for particular projects in the local community. These lists or databases feature ongoing or one-time projects in the local community. Notably, these lists and listings vary in formality and robustness across the institutions, and some institutions do not have pages for lists/databases.

IDEAS AND OPPORTUNITIES FOR COMMUNITY INNOVATIONS

The themes and examples elucidated above fit well in the frame of community development mostly from the perspectives of service and partnerships. Among these institutions, a few innovative programs and models were elucidated for discussion. These ideas and opportunities are offered as examples and opportunities only and not generalizable frameworks. These innovative ideas, opportunities and practices are presented in Table 26.1 for later discussion.

MICRO-PERSPECTIVE ON LIBERAL ARTS AND COMMUNITY DEVELOPMENT AND INNOVATION

This chapter utilizes a single case (one community, one college) to demonstrate how community development and innovation can occur in the context of a liberal arts college and its host community. Candid recollections of pivotal moments in this college-community relationship provide insight into how small colleges and communities can together turn challenges into opportunities through unique programs, processes, partnerships and collaborations that focus on reciprocity and respect. For this micro-case, the city is Geneva, New York, and the school is Hobart and William Smith Colleges (HWS).

Table 26.1 Highlighted ideas, opportunities, and practices for community innovation

Ideas, Opportunities, and Practices	Liberal Arts Institutions
Community engaged research and scholarship including faculty connections, grants and incentives	DePauw University, Oberlin College, Pitzer College, St Lawrence University
Community partnerships focused on environmental sustainability	Allegheny College, Washington College
Connections to college run farms and local food systems	Allegheny College, Carleton College, Pomona College, St Lawrence College
Dedicated teams and consulting groups	Carleton College, Colgate University, College of Wooster, DePauw University
Engagement with local entrepreneurs	Allegheny College, Oberlin College, Wells College
Entrepreneurship incubators (especially downtown)	Allegheny College, Pomona College, Scripps College
Focus on exploring questions instead of answers	DePauw University
Focus on the arts and engagement	Flagler College, Oberlin College
Focus on women entrepreneurs	College of Wooster, DePauw University
Intentional focus on understanding the whole region	Colgate College, Williams College
Intentional engagement pedagogies and principles	Carleton College
Long-term commitments and evaluation and measurement	Pitzer College, Skidmore College
Social entrepreneurship, social innovation and green entrepreneurship (courses, labs and projects)	Carleton College, Hamilton College, Oberlin College, Washington College
Summer programs	Colgate College, Middlebury College, St Lawrence College
Supplying innovative ideas from students to community partners	Pitzer College

HWS'S SUSTAINABLE COMMUNITY DEVELOPMENT (SCD) PROGRAM

Story

Spurred forward by conversations with recent alumni/ae, the SCD Program is an interdisciplinary minor that brings together students, faculty and staff from the Architectural Studies and Environmental Studies Programs at HWS to complete community-based research projects focused on local sustainability issues. This program reflects the institution's liberal arts focus, as well as growing interest in sustainability across campus and in the host community. The name, "Sustainable Community Development," highlights the program's central objective – providing students with the opportunity to develop the knowledge and skill sets necessary to work alongside local communities to forge a more sustainable future.

In addition to the minor, the SCD Program initially anchored its presence in the community in two main ways: (1) hosting a public lecture series and (2) establishing the Finger Lakes Community Development Center (FLCDC). A series of four to five lectures were hosted on the HWS campus each spring, providing the broader Geneva, New York community an opportunity to learn about important advances in the field. Meanwhile, the FLCDC served as a hub for the SCD Program's experiential approach to teaching/ learning. SCD faculty, staff and students alike work in the same space and cultivate a dynamic environment in which creative problem-solving and collaboration are the norm rather than the exception.

The SCD Program at HWS "is not only a differentiator that allows [the institution] to set itself apart from other colleges but also allows [us] to build on [our] unique and existing strengths" (Helfrich, 2015, para. 8) in demonstrating "the benefits of interdisciplinarity" (Helfrich, 2015, para. 17). The SCD Program extends the institution's history of service learning into a curriculum rich in experiential education and community-based learning opportunities. Students develop knowledge and skills "to help create a more livable present and future" (Helfrich, 2015, para. 17). The program uniquely combines curricular and co-curricular spaces to explore how to achieve SCD in the "real world" (Helfrich, 2015).

Curriculum

The first innovative aspect of the SCD Program concerns its academic curriculum. Students pursuing an academic minor in SCD complete a series of six courses designed to provide them with the background and skills necessary to pursue careers in the stewardship of the natural and built environments and the communities dependent on these resources (Figure 26.3). The gateway "Sustainable Communities" course "introduces students to the concept of sustainable development as applied to real world communities" (Helfrich, 2015, para. 5; see also Maurer, 2011, para. 3). "Sustainable Communities" students tackle a series of case studies to explore the opportunities and challenges communities encounter while attempting to address the complex and seemingly unsolvable day-to-day problems.

Armed with a strong foundation in sustainable community development and its underlying theories and practices, SCD students also sharpen their academic knowledge and skills by completing a series of two tools courses. They take a technical writing class focused on honing students' written communication skills (e.g., "Digital Journalism").

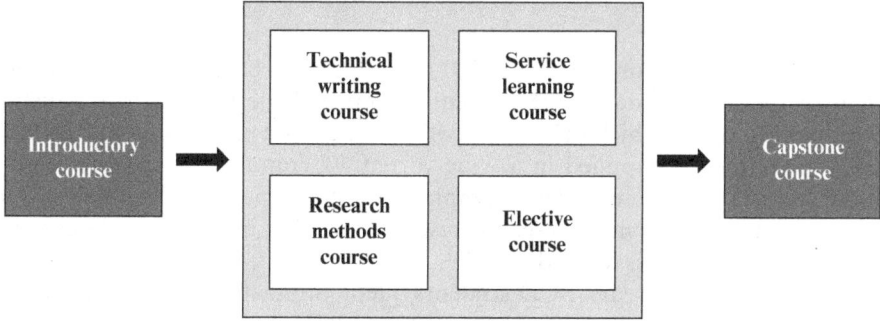

Figure 26.3 SCD minor at HWS curricular roadmap

The second tools course, which students typically complete in another disciplinary or interdisciplinary program, bolsters their research skills (e.g., "Qualitative Methods and the Community"). Rounding out SCD students' curricular experience is a designated service learning course in a related field of study (e.g., "Sociology of Community"), as well as one elective selected from our list of cognate courses in the natural sciences, social sciences and humanities (e.g., "Environmental Science," "Cultural Anthropology," "Environmental Ethics," respectively).

With the foundational knowledge of and requisite skills needed for sustainable community development work, junior and senior SCD minors from across campus come together to complete their capstone experience (Helfrich, 2015). The "Sustainable Community Development Capstone," typically co-taught by faculty members from the Architectural Studies and Environmental Studies Program and carrying a service learning designation, provides students with the opportunity to mobilize the knowledge and skills they gained in their previous SCD coursework by tackling a semester-long community-based research project in the Finger Lakes region, where Geneva, New York is located. While project topics vary from semester-to-semester, SCD capstone classes are tasked with developing a series of evidence-based recommendations for addressing the community issue at hand. These recommendations are initially summarized for partner(s) in a public presentation at the end of the semester. Once SCD faculty and staff edit the students' written work, community collaborators receive final recommendations in the form of a written report similar in quality to that produced by environmental consulting firms in the region.

Internships

Another innovative aspect of the SCD Program is its paid internships, which occur during both the academic year and summer months. Projects and interns alike are selected by the SCD Program Manager with input from the SCD Advisory Committee, a group of faculty and staff tasked with guiding the program and its work. Supervised by the SCD Program Manager and/or other SCD faculty and staff, interns complete one to two community-based research projects in the region over a period of 10 to 15 weeks. Similar to the capstone courses, SCD internships typically culminate in the production of a final report providing a series of evidence-based recommendations to help the community partner(s) move forward in their efforts to address the issues they face.

Outcomes

Since launching the program at HWS in 2011, 435 students have completed one or more SCD course. Of these students, 358 students completed the introductory "Sustainable Communities" course in which they are exposed to the basic principles of SCD and learn how these principles are applied in a wide variety of contexts around the world. The remaining 20 percent of these students completed the capstone course, gaining substantial experience conducting community-based research projects in the Finger Lakes region, where Geneva, New York is located.

Over the course of 6+ years, SCD students, faculty and staff have completed 13 different community-based research projects with ten different local and regional partners throughout the region (Table 26.2). Of these 13 projects, 7 were completed during the

Table 26.2 Summary of the SCD community-based research projects completed since 2011

Project Topic	Community Partner(s) Engaged
Class Projects	
Design proposals for vacant downtown properties	Municipal neighborhood resource center
Brownfield redevelopment and place-making	Municipal neighborhood resource center; local neighborhood association
Multimodal transportation planning and place-making	Local municipality
Stormwater management	Local municipality
Solid waste management	Local municipality
Enhancement of downtown amenities and streetscapes	Local municipality
Internship Projects	
Green infrastructure for historic districts	Regional planning council
Streetscape improvements for major arterial	Local municipalities
Sensory garden design	Local nonprofit organization
Urban green spaces; park master plans	Local municipality
Bicycle tour design	State government agency
Ice climbing feasibility study	Regional university
Scenario planning for repurposing a former military facility	County government

academic year or summer internships, providing 15 different undergraduate students with an opportunity to gain additional community-based research experience outside the confines of a traditional college classroom (Table 26.2). The remaining six projects, by comparison, served as an integral part of the student experience in the "Sustainable Community Development Capstone" class (Table 26.2).

Feedback from SCD students has been overwhelmingly positive to date. As one recent SCD capstone student reflects: "One of the most important things about the [SCD] minor is that we do this hands-on research that actually makes a difference" (Hibbard, 2016, para. 16) in local communities. Meanwhile, past interns point to their experiences as "not only allow[ing them] to hone [their] writing and research skills, but also [their] people skills" (Li et al., 2013, para. 14). Moreover, SCD students typically report gaining an "enhanced . . . ability to work in a cross-disciplinary, collaborative environment" (Varner, 2014, para. 11), an attribute that many graduates report as "opening up so many doors for [them]" and "put[ting them] ahead of the game" (HWS, 2015, para. 2; see also Li et al., 2013) when pursuing employment opportunities and/or applying to graduate school in related fields of study (Lewis, 2015).

With regard to community partners, representatives from the SCD Program have collaborated with a wide variety of organizations to date, ranging from municipal, county and state governments to neighborhood associations, nonprofit organizations and regional planning associations (Table 26.2). Similar to SCD students, community partners report positive collaboration outcomes. Reflecting on her work with a previous capstone class, one community member remarks how much she appreciated the fact that SCD students, faculty and staff "listened carefully to the members of the [project] steering committee and effectively integrated those ideas" (Anderson, 2014, para. 5) into their final site design and proposals. Likewise, a recent SCD project that focused on enhancing downtown amenities and streetscapes led another community partner to note that "the students put their hearts and souls into providing the community with thoughtful and creative designs that celebrate the social fabric and physical assets of [our community]" (HWS, 2018b, para. 5). Fortunately, impacts of SCD work also extend to the changes community partners make after collaborations with the SCD Program. Past partners report utilizing final reports in formulating subsequent municipal applications for grant funding to advance the ideas offered (Community partner A, personal communication, 2 August 2018; see also Murphy, 2015). Moreover, other community partners report that they are actively implementing recommendations (Community partner, B, personal communication, 14 November 2018).

ENTREPRENEURSHIP AT HWS

Story

The entrepreneurship programs at HWS also emerged from conversations in the mid-2010s with alumni regarding the need for technical writing, oral presentation and data analysis skills. During that time, HWS's Centennial Center for Leadership had already begun offering co-curricular programs focused on entrepreneurship. In the 2011–12 academic year, this leadership center began its entrepreneurial pitch competition where students could win a $10,000 prize for their winning ideas and plans. Leveraging student interest and enthusiasm, an entrepreneurial leadership course was soon developed and quickly filled to capacity each semester.

Curriculum

A new interdisciplinary minor in Entrepreneurial Studies (ENTR) officially launched in Spring 2016. A full-time faculty member was hired to anchor the program, with other faculty and staff from across HWS's departments and local adjuncts assisting with instructing the courses.[3] Unique course offerings expanded from entrepreneurial leadership to courses focused on economic principles, quantitative tools related to financial reporting and statistics, social innovation, and ethics. Today, students take core courses hosted within the Entrepreneurial Studies Program and also take courses offered by other departments as electives (Figure 26.4). The seniors also take courses out of a new downtown entrepreneurship center (HWS, 2018c; Wickenden, 2017).

Throughout their course of study, students receive instruction and experience in entrepreneurship theory and practice. They gain technical writing and oral presentation skills, which are built into all the core courses of the curriculum. They also learn software packages such as Microsoft Excel while simultaneously refining their data analysis skills. In both lower and upper level classes, Entrepreneurial Studies students collect data from potential customers and clients to substantiate the viability of their entrepreneurial ideas, forged from their work with others in the program.

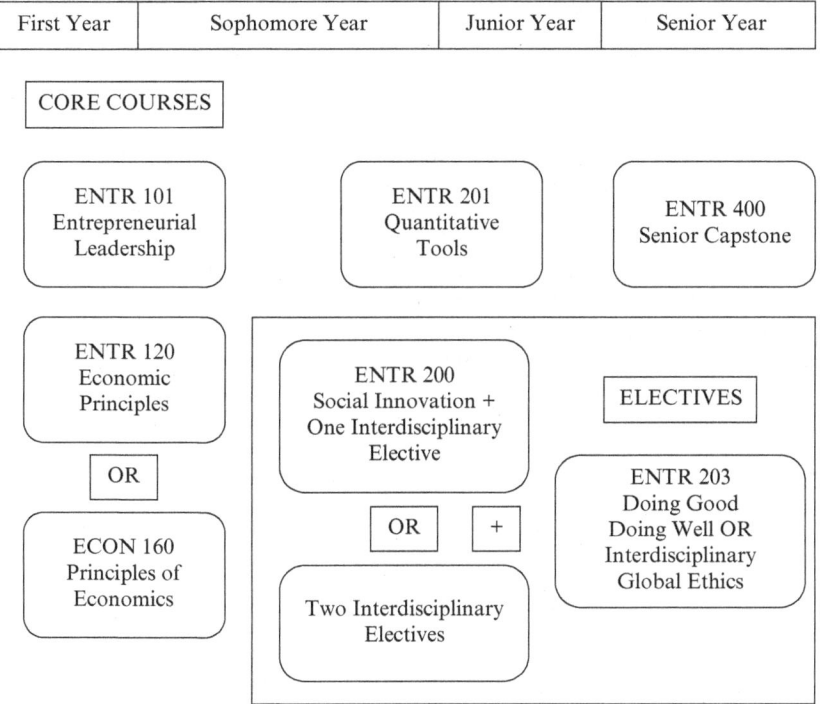

Figure 26.4 Pathways for the Entrepreneurial Studies minor

In the social innovation course and the senior capstone course, students take on a community-based research project that often focuses on market development, working and serving in a consulting role for local community partners. In order to be more accessible to community partners, the capstone class is held at the downtown entrepreneurship center location. Table 26.3 showcases example projects since the program's inception in 2016.

Outcomes

While the program is still new, student and community partner feedback has been positive. Students have generally enjoyed the consulting projects with local partners as it helps them gain skills to apply their entrepreneurial ideas and plans in the senior capstone. Anecdotally, alumni/ae have also noted the utility of the technical writing, oral presentation and Microsoft Excel skills that they have gained through the program. Community partners have shown gratitude for the knowledge, skills and abilities students bring to their enterprises, which have assisted them as they seek growth. Articles in local and campus news sources have highlighted the enthusiasm and impact of these projects (HWS, 2018d; Tulus, 2018).

Table 26.3 Example ENTR community-based research projects completed

Project Topic	Community Partner(s) Engaged
Class Projects	
Identify and demarcate accessible restrooms for individuals with physical challenges	Hobart and William Smith Colleges
Create marketing materials to explain vermiculture processes to encourage sustainable waste practices	Local vermiculture entrepreneur
Develop a marketing plan for pickled flower buds (e.g., substitute for capers on charcuterie boards)	Local chef
Develop a marketing plan for an apple-infused sausage	Local food retailer
Assess the financial and market viability of an affordable and healthy food market in a food desert	Geneva, New York City Manager's Office
Assess the financial and market viability of a grape-flavored sparkling/seltzer water and kefir product	Local vineyard owner

GENEVA 2020

Story

The City of Geneva, New York defines itself as "uniquely urban," but many of its problems are not unique. In 2010, only seven in ten high school students from Geneva's local high school graduated. The New York State Board of Education reported its concerns in early 2011, noting that the current high school graduation rates were not meeting expectations (NYSED/Geneva City School District, 2011). Community leaders decided to take action to address this well-documented problem more proactively and holistically, spurring the formation of an intentional partnership between HWS and the Geneva City School District. Inspired by the work of StriveTogether (n.d.), a "national, nonprofit network of nearly 70 community partnerships" (para. 2), the Geneva 2020 *cradle to career* collective impact initiative emerged.

After serving on the Obama Administration's "White House Council for Community Solutions" (HWS, 2010), HWS's President Mark Gearan brought StriveTogether's (n.d.) collective impact model to the local community leaders (e.g., education advocates, faith community leaders, business leaders, among others) and college leaders (e.g., students, faculty and staff). Community leaders in Geneva, New York wanted a broad and scalable approach to collective impact. They were informed by and leveraged Kania and Kramer's (2011) collective impact model and work, Turner, Merchant, Kania and Martin's (2012) commentary on collective impact and the StriveTogether (n.d.) network's resources.

Together, community leaders and college administration discussed the difference between powerful results from individual programs, the progress possible when individual programs coordinated efforts focused on a common goal and the potential that would be unleashed by the community and local college if resources, ideas and talent coalesced to fully embrace the tenets of collective impact. School district leadership set goals to amplify literacy programs, expand college and career readiness opportunities and improve the graduation rate. The school district leadership's guidance shaped the common agenda, the first of five conditions of collective impact (Kania and Kramer, 2011), with subsequent conditions including: the development of a shared measurement system(s) ensuring continuous communication to promote transparency and accountability; the coordination of mutually reinforcing activities; and the establishment of a backbone support organization (or anchor entity). HWS's president committed HWS to serve as the initiative's anchor entity, and staff members and leaders from the Center for Community Engagement and Service Learning, the communications office and the college advancement office aligned skill sets and capacity.

The Geneva 2020 steering committee was established. Committee members made a commitment that aligned with the scope of set "cradle to career" goals and with a near duplicate number of college students to the number of children enrolled in K-12, President Gearan encouraged staff from HWS to think beyond current youth tutoring programs to identify a tangible way that the institution could encourage college and career readiness. Every 2nd, 6th and 9th grader in Geneva, New York was invited to campus for programs that focus on literacy, science, technology, engineering, arts and mathematics (STEAM) and the college application process, respectively. College exposure opportunities, in particular, are a tangible way for HWS students to participate in supporting the higher education aspirations of children in the local community.

Action teams convened regarding priority areas, and these teams mobilized around data that shined a light on literacy, school attendance, college readiness and STEAM. They had to remember that change takes time and recognize that trusting partnerships developed through collaboration, challenging situations and successful resolutions lead to the indicators necessary to attain the "proof point" gateway (StriveTogether, 2017). They further explored (and continue to explore) data relevant to Geneva 2020's mission. Action team members developed and organized around a vision and strategy and a governance structure. They also refined communication and engagement strategies. A dedicated staff person was deemed necessary and the college advancement office secured funding to match HWS's financial commitment. Fortuitously, a current school board member applied and was hired to serve as the inaugural Geneva 2020 program coordinator. Through many inclusive efforts of dedicated individuals and organizations across four years, the graduation rate increased with progress made in key areas.

"The Geneva City School District has ranked in the top three percent of the most diverse school districts in New York State, with more than half of all students registered in demographic subgroups that have historically identified as minorities" (Geneva 2020 Community Report, 2018, p. 7). The median household income is $17,000 less than the county average, and 94 students qualify as homeless by McKinney-Vento Homeless Assistance Act of 1987 standards (Geneva 2020 Community Report, 2018). The Parental Appraisal of Children's Experience survey for each of the past three fall kindergarten enrollments gathered information about Adverse Childhood Experiences (ACE) Scores. For example, 10.2 percent of children in the school district have or had an incarcerated parent (Geneva 2020 Community Report, 2018). This data helps with intentionally crafting classrooms and connecting children and families with available resources.

Outcomes

Geneva 2020 has been hailed as a distinct model in NY state for school-community partnerships (District Administration, 2015; HWS, 2014). Action teams have emphasized and showcased accountability and transparency in community reports in 2017–18. From 2010 to 2018, high school graduation rates have increased from 71 percent to 83 percent, respectively, with hopes of further increasing rates to 90 percent by 2020 (Geneva 2020 Community Report, 2018). The College Bound Program has been particularly successful to date, with close to 500 students visiting the campus annually during each of the past six years. College students have contributed to the success of these visits in their roles as tour guides, program facilitators, lunch buddies and panelists highlighting study abroad experiences, the financial aid process and the importance of internships. Furthermore, HWS tripled the federal requirement of 7 percent of work study spent on community work study by hiring and supporting over 100 America Reads tutors at four schools, with additional funding supporting college students to serve as homework helpers at the Boys and Girls Club of Geneva, paid through a 21st Century Community Learning Center grant.

The Geneva 2020 stakeholders and leaders recognize more progress is needed, despite current successes. Stakeholders and leaders continue to monitor the Algebra Regent exam passing rates for students at the end of the 9th grade year, since that is an indicator of high school graduation (Rickles et al., 2017). Additionally, HWS remains deeply committed to advancing English language arts (ELA) scores. Additionally, staff from the Center for Community Engagement and Service Learning provide continued support for Geneva 2020 (Flowers et al., 2019).

LESSONS LEARNED

Universities are often criticized for focusing on outputs rather than processes regarding their community engagement (Fortunato et al., 2015). These critiques align with this chapter's notions that community innovation processes are understudied or not well understood in research and practice. In this chapter's cursory examination across institutions and communities and look within one institution and community, outcomes are easier to identify than processes, especially looking across institutions. In this final section, lessons learned are shared in two forms: (1) questions for future research and practice; and (2) gateways to innovation.

QUESTIONS FOR FUTURE RESEARCH AND PRACTICE

Returning to Figure 26.2, three sets of questions appear relevant to furthering research and practice. These questions are presented to instill greater innovation in community development efforts facilitated by higher education institutions with their community counterparts. These questions are based on the elucidated themes found across the 20 institutions cursorily investigated and the in-depth reflection on HWS. An overarching question is presented and a location for innovation is proposed. They are:

1. How can community-based and service learning be innovated, specifically in:
   a. Courses and curriculums;
   b. Capstone courses and projects;
   c. Internships and fellowships;
   d. Service days and events; and
   e. Student service clubs?
2. How can community-based and service learning be infused in entrepreneurship education, specifically in:
   a. Courses and curriculums;
   b. Capstone courses and projects;
   c. Competitions and co-/extra-curricular programs;
   d. Public events; and
   e. Student entrepreneurship clubs?
3. How can community engaged partnerships be innovated, specifically in:
   a. Designated offices;
   b. Resources for community partners; and
   c. Volunteer service opportunities?

Reflecting back on Figure 26.1, sustainable, hyperlocal capacity-building processes will be useful in community innovation.

GATEWAYS TO INNOVATION

Returning to Table 26.1 and reflections on HWS, a number of institutions have found different gateways to innovation. Established networks and incentives (intrinsic and extrinsic) remain needed to inspire community innovation from within the institution. Dedicated persons, centers and locations appear useful if not necessary to facilitate community innovation and long-term partnerships. The student consultant, evaluator and/or researcher model also can encourage innovative ideas and implement innovative community solutions. Moreover, students may be able to discover questions to be asked that local entrepreneurs and community members may not have previously generated or asked. Community innovation also appears to not be relegated to traditional for-profit-focused entrepreneurship programs, but can come from across departments and interdisciplinary programs. Community innovations need not come in for-profit business forms either; they can be artistic, social, environmental and nonprofit forms. Finally, a local or regional focus may help with developing long-term commitments and getting students out of the classroom to experience learning and service.

Entrepreneurship education and sustainable community development education programs may be two possible gateways for innovation at small liberal arts colleges. Community innovation can be pursued in courses, curriculums and other educational opportunities; however, community innovation should not be seen as only outputs (i.e., goods or services) as processes should be emphasized as well. Notably, interdisciplinarity appears to be a key attribute for successful innovation.

Within both forms of education programs showcased from HWS, students conduct data collection and analysis in collaboration with community partners. Data collection provides feedback to students and community members regarding the progress and potential future outcomes of their collaborations. Future research may focus more on innovations in data collection in student-community projects. Such work will not be easy; faculty, staff and students may also have to set aside their roles as experts in order to allow local community expertise to drive the data collection and subsequent community innovation (Fortunato et al., 2015).

Regarding community engaged partnerships, the outcomes are not what makes Geneva, New York remarkable. The local higher education institution commitment as the backbone or anchor institution was a unique model that few collective impact efforts have (District Administration, 2015; HWS, 2014), especially when taking into the account the current size of the StriveTogether network. Of the nearly 70 StriveTogether community partnerships, only five have higher education institutions serving as a backbone support to their community's collective impact effort, such as SUNY Plattsburgh, which co-anchors the North Country Thrive (n.d.). Virginia Commonwealth University serves as the backbone for Bridging Richmond (n.d.). Berea College (n.d.) anchors Partners for Education, and the University of Albany anchors The Albany Promise (n.d.). The backbone and anchor approach allows for students, faculty and staff to continue to partner in community development and innovation efforts with local community members and organizations and vice versa. For Geneva 2020, continued partnership regarding collective impact efforts is visible through the efforts of Action Teams, data collection and sharing and creatively establishing funding sources that endure.

Schools and universities in communities act as central hubs of engagement (Fortunato et al., 2015; Pstross et al., 2014; Talmage et al., 2015); they catalyse social and community innovation (Fortunato et al., 2015). Fortunato et al. (2015) note that universities first must inspire innovation from within before they can really impact their local communities. They go on to emphasize that community innovation must be co-created between universities and communities. These notions deserve further attention in research and practice.

CONCLUSION

Community development must be universal (Bhattacharyya, 1995; 2004) and build communities' capacity (Matarrita-Cascante and Brennan, 2012; Phillips and Pittman, 2009; Robinson and Green, 2011), agency (Bhattacharyya, 1995; 2004) and solidarity (Bhattacharyya, 1995; 2004). However, community innovation as part of community development in communities does not have to be necessarily scalable across communities, but should be sustainable (Ajmal et al., 2018). Much more research and focus is needed on community innovation, especially to inform community development education.

While community foci have been found across higher education institutions, community innovation has not. For liberal arts institutions, community innovation may be a new liberal art worthy of inclusion in curriculums and institutional programs. Community innovation can invigorate the experiential education ideologies and practices of liberal arts and other higher education institutions alike. This chapter serves as a call for further looks into these areas.

NOTES

1. Note that some liberal arts institutions view the establishment of a formal business or business-like curriculum as a potential threat to their missions (Neely, 1999), despite arguable opportunities for links between service learning and business education, which may benefit both (Zlotkowski, 1996).
2. Readers interested in learning more about the infrastructures and practices of designated offices like these might turn to the work of Welch and Saltmarsh (2013).
3. The program has two full-time, dedicated faculty members along with multiple adjuncts as of this writing.

REFERENCES

Adams, D., and Hess, M. (2010). Social innovation and why it has policy significance. *The Economic and Labour Relations Review*, *21*(2), 139–55.

Aiken, M., and Alford, R.R. (1970). Community structure and innovation: the case of public housing. *American Political Science Review*, *64*(3), 843–64.

Ajmal, M.M., Khan, M., Hussain, M., and Helo, P. (2018). Conceptualizing and incorporating social sustainability in the business world. *International Journal of Sustainable Development & World Ecology*, *25*(4), 327–39.

Anderson, J. (2014). SCD concepts: the East Lakeview transformation vision – transforming brownfield to park. Retrieved 9 Nov 2018 from https://www.hws.edu/ fli/pdf/proposal_sp14.pdf/.

Barber, B.R., and Battistoni, R. (1993). A season of service: introducing service learning into the liberal arts curriculum. *PS: Political Science & Politics*, *26*(2), 235–240.

Baskaran, S., and Mehta, K. (2016). What is innovation anyway? Youth perspectives from resource-constrained environments. *Technovation*, *52*(17), 4–17.

Berea College (n.d.). *Partners for Education*. Retrieved 7 Jan 2019 from https://www.berea.edu/pfe/.

BestColleges.com (2016). *America's Best Small Town Colleges*. Retrieved 7 Jan 2019 from https://www.bestcolleges.com/features/best-small-town-colleges/.

Bhattacharyya, J. (1995). Solidarity and agency: rethinking community development. *Human Organization*, *54*(1), 60–69.

Bhattacharyya, J. (2004). Theorizing community development. *Community Development*, *34*(2), 5–34.

Bombyk, M. (2003). University employees who live locally: bridging the town-gown divide. *Metropolitan Universities*, *14*(4), 22–8.

Bridging Richmond (n.d.). *What is Bridging Richmond?* Retrieved 7 Jan 2019 from http://bridgingrichmond.com/about/.

Bringle, R.G., and Hatcher, J.A. (2000). Institutionalization of service learning in higher education. *The Journal of Higher Education*, *71*(3), 273–90.

Bromley, R. (2006). On and off campus: colleges and universities as local stakeholders. *Planning, Practice & Research*, *21*(1), 1–24.

CCAP (2016). Are small town liberal arts colleges endangered? *Forbes*. 23 Dec. Retrieved 7 Jan 2019 from https://www.forbes.com/sites/ccap/2016/12/23/small-town-liberal-arts-college-r-i-p/#49663fdc4dff.

Clammer, J. (2015). Social economics and economic anthropology: challenging conventional economic thinking and practice. In N. Pun, Hok-Bun Ku, B., Yan, H., and Koo, A. (eds), *Social Economy in China and the World* (pp. 3–16). London and New York: Routledge.

College Values Online (CVO) (2018). *50 Best Small College Towns in America*. Retrieved 7 Jan 2019 from https://www.collegevaluesonline.com/features/best-small-college-towns-in-america/.

Davis, B. (2016). There's an antidote to America's long economic malaise: college towns. *Wall Street Journal*, 12 December. Retrieved 10 Jun 2016 from https://www.wsj.com/articles/theres-an-antidote-to-americas-long-economic-malaise-college-towns-1481558522/.

Dechief, D., Longford, G., Powell, A., and Werbin, K. (2008). Enabling communities in the networked city: ICTs and civic participation among immigrants and youth in urban Canada. In A. Aurigi and De Cindio, F. (eds), *Augmented Urban Spaces: Articulating the Physical and Electronic City* (pp. 155–70). London and New York: Routledge.

Depauw.edu (n.d.). *Civic Fellows*. Retrieved 7 Jan 2019 from https://www.depauw.edu/studentacademiclife/hartman/community-service/civic-fellows/.

District Administration (2015). Helping hand from higher ed: student achievement improves with local college and community collaboration. *District Administration*. March. Retrieved 12 Mar 2019 from http://www.nxtbook.com/nxtbooks/pmg/da201503/index.php?startid=48&qs=geneva+2020#/50.

Docking, J.R., and Curton, C.C. (2015). *Crisis in Higher Education: A Plan to Save Small Liberal Arts Colleges in America*. East Lansing, MI: MSU Press.
Fischer, K. (2008). Struggling communities turn to colleges. *The Chronicle of Higher Education*, *54*(36), A1.
Flowers, K., Wattles, J., Feinberg, S., and Sellers, A.J. (2019). Center for Community Engagement and Service Learning (CCESL): annual report 2017–18. Retrieved 11 Mar 2019 from https://www.hws.edu/academics/service/pdf/annual_report1718.pdf.
Fortunato, M.W.P., Alter, T.R., Frumento, P.Z., and Klos, J.M. (2015). Cultivating a culture of innovative university engagement for local entrepreneurship development in rural and distressed regions. *International Journal of Social Science Studies*, *3*(1), 122–38.
Geneva 2020 Community Report (2018). Retrieved 7 Jan 2018 from https://www.hws.edu/about/pdfs/Geneva2020_community_report2018.pdf.
Gumprecht, B. (2003). The American college town. *Geographical Review*, *93*(1), 51–80.
Helfrich, J. (2015). Sustainable community development education in the Finger Lakes. *Journal of Sustainability Education*, 10. Retrieved 10 Jun 2018 from http://www.jsedimensions.org/wordpress/content/sustainable-community-development-education-in-the-finger-lakes_2015_11/.
Hibbard, M. (2016). Solid ideas: HWS students work with town of Geneva on ways to reduce waste stream. *Finger Lakes Times*. 8 May. Retrieved 9 Nov 2018 from https://www.fltimes.com/news/solid-ideas-hws-students-work-with-town-of-geneva-on/article_baba8d24-14be-11e6-9f16-c3b86cb29f24.html.
Hines Jr, S.M. (2005). The practical side of liberal education: an overview of liberal education and entrepreneurship. *Peer Review*, *7*(3), 4–7.
Hobart and William Smith Colleges (HWS) (2010). Gearan serves on White House council. *The HWS Update*. 15 Dec. Retrieved 17 Dec 2018 from https://www2.hws.edu/article-id-13721/.
Hobart and William Smith Colleges (HWS) (2014). Gearan joins White House summit on higher ed. *The HWS Update*. 16 Jan. Retrieved 11 Mar 2019 from https://www2.hws.edu/article-id-17421/.
Hobart and William Smith Colleges (HWS) (2015). Interning for community development center. *The HWS Update*. 23 Jul. Retrieved 9 Nov 2018 from https://www2.hws.edu/article-id-18916/.
Hobart and William Smith Colleges (HWS) (2018a). Sustainable community development: local change in action. *The HWS Update*. 4 Dec. Retrieved 8 Dec 2018 from https://www2.hws.edu/change-in-action-with-community-based-design/.
Hobart and William Smith Colleges (HWS) (2018b). Sustainable community development rethinks Castle Street. *The HWS Update*. 1 Aug. Retrieved 11 Mar 2019 from https://www2.hws.edu/sustainable-community-development-rethinks-castle-street/.
Hobart and William Smith Colleges (HWS) (2018c). HWS dedicates Bozzuto Center for Entrepreneurship. *The HWS Update*. 17 Oct. Retrieved 7 Jan 2019 from https://www2.hws.edu/hws-dedicates-bozzuto-center-for-entrepreneurship/.
Hobart and William Smith Colleges (HWS) (2018d). Entrepreneurial capstone develops strategies for local change. *The HWS Update*. 18 Jul. Retrieved 7 Jan 2019 from https://www2.hws.edu/entrepreneurial-capstone-aids-local-business/.
Hobson, K., Hamilton, J., and Mayne, R. (2016). Monitoring and evaluation in UK low-carbon community groups: benefits, barriers and the politics of the local. *Local Environment*, *21*(1), 124–36.
Jackson, E.T. (2004). Community innovation through entrepreneurship: grantmaking in Canadian community economic development. *Community Development*, *35*(1), 65–81.
Kania, J., and Kramer, M. (2011). Collective impact. *Stanford Social Innovation Review*, *9*(1), 36–41.
Kiwanuka, S.N., Tetui, M., George, A., Kisakye, A.N., Walugembe, D.R., and Kiracho, E.E. (2015). What lessons for sustainability of maternal health interventions can be drawn from rural water and sanitation projects: perspectives from Eastern Uganda. *Journal of Management & Sustainability*, *5*(2), 97–107.
Klein, K.J., and Knight, A.P. (2005). Innovation implementation: overcoming the challenge. *Current Directions in Psychological Science*, *14*(5), 243–6.
Kolb, A.Y., and Kolb, D.A. (2005). Learning styles and learning spaces: enhancing experiential learning in higher education. *Academy of Management Learning & Education*, *4*(2), 193–212.
Lang, E.M. (1999). Distinctively American: the liberal arts college. *Daedalus*, *128*(1), 133–50.
Lewis, R.A. (2015). Sustainable communities: the Sustainable Community Development Program – reflections from within. *FLI Happenings*. 6 Oct. Retrieved 9 Oct 2018 from https://flihappenings.com/2015/10/06/sustainable-communities-the-sustainable-community-development-program-reflections-from-within/.
Li, A.Y., Markham, M., Mauch, J., Encababian, C., Combs, J., and Varner, D. (2013). Community Design Center: a summer of sustainable design. *Finger Lakes Institute's Happenings*. 30 Aug. Retrieved 9 Nov 2018 from https://flihappenings.com/2013/08/30/community-design-center/.
Lietz, C.A., and Zayas, L.E. (2010). Evaluating qualitative research for social work practitioners. *Advances in Social Work*, *11*(2), 188–202.
Lord, J., and Hutchison, P. (2007). *Pathways to Inclusion: Building a New Story with People and Communities*. Concord, ON: Captus Press.

Matarrita-Cascante, D., and Brennan, M.A. (2012). Conceptualizing community development in the twenty-first century. *Community Development*, *43*(3), 293–305.
Maurer, A. (2011). First course offering in the HWS Sustainable Community Development Program. Retrieved 9 Nov 2018 from https://flihappenings.com/2011/12/01/first-course-offering-in-the-hws-sustainable-community-development-program/.
McKinney-Vento Homeless Assistance Act 1987. (42 USC) § 11301 *et seq.* (USA).
Mitchell, P.T., and Levine, M.A. (2001). Leadership, engagement, and the small liberal arts college: Albion College and the smart community. *Metropolitan Universities*, *12*(3), 76–88.
Moore, C., Venezia, A., Lewis, J., and Lefkovitz, B. (2015). Organizing for success: California's regional education partnerships. *Education Insights Center*. California State University, Sacramento, Sacramento, CA. Technical Report. Retrieved 7 Jan 2019 from https://files.eric.ed.gov/fulltext/ED574459.pdf.
Murphy, M. (2015). Students helping chart town of Canandaigua's future course. *Daily Messenger*. 8 Mar. Retrieved 9 Nov 2018 from https://www.mpnnow.com/article/20150318/NEWS/ 150319619/1994/NEWS.
Neely, P. (1999). The threats to liberal arts colleges. *Daedalus*, *128*(1), 27–45.
Newman, L., and Dale, A. (2005). The role of agency in sustainable local community development. *Local Environment*, *10*(5), 477–86.
North Country Thrive (n.d.). *Partners Nurturing Growth Together*. Retrieved 17 Dec 2018 from http://www.northcountrythrive.org/partners/.
NYSED/Geneva City School District (2011). *Joint Intervention Team Report and Recommendations*. Technical Report. Retrieved 17 Dec 2018 from http://www.p12.nysed.gov/accountability/School_Improvement/Reports/GenevaHS.pdf.
Phillips, R., and Pittman, R.H. (2009). *An Introduction to Community Development*. New York: Routledge.
Pstross, M., Talmage, C.A., and Knopf, R.C. (2014). A story about storytelling: enhancement of community participation through catalytic storytelling. *Community Development*, *45*(5), 525–38.
Rickles, J., Heppen, J., Taylor, S., Sorenson, N., Walters, K., and Clements, P. (2017). Getting back on track, course progression for students who fail Algebra I in ninth grade. Retrieved 7 Jan 2019 from https://www.air.org/system/files/downloads/report/Course-Progression-for-Students-Who-Fail-Algebra-I-in-Ninth-Grade-June-2017.pdf/.
Robinson, Jr, J.W., and Green, G.P. (2011). *Introduction to Community Development: Theory, Practice, and Service Learning*. Thousand Oaks, CA: Sage Publications.
Roesler, T. (2018). Community resources for energy transition: implementing bioenergy villages in Germany. *Area*, *51*(2), 268–76.
Sandy, M., and Holland, B.A. (2006). Different worlds and common ground: community partner perspectives on campus-community partnerships. *Michigan Journal of Community Service Learning*, *13*(1), 30–43.
Semali, L.M., Hristova, A., and Owiny, S.A. (2015). Integrating Ubunifu, informal science, and community innovations in science classrooms in East Africa. *Cultural Studies of Science Education*, *10*(4), 865–89.
StriveTogether (n.d.). *About Us*. Retrieved 11 Mar 2019 from https://www.strivetogether.org/about/.
StriveTogether (2017). *Theory of Action: Creating Cradle to Career Proof Points*. Retrieved 17 Dec 2018 from https://www.strivetogether.org/wp-content/uploads/2017/03/StriveTogether-Theory-of-Action-2017.pdf.
Talmage, C.A., Dombrowski, R., Pstross, M., Peterson, C.B., and Knopf, R.C. (2015). Discovering diversity downtown: questioning Phoenix. *Metropolitan Universities*, *26*(1), 113–46.
The Albany Promise (n.d.). *What Does It Mean when a Community Comes together to Change its Future?* Retrieved 7 Jan 2019 from http://albanypromise.org/.
Tulus, S. (2018). BIGGER Picture: Sausage links… and more. *Finger Lakes Times*. 6 Mar. Retrieved 7 Jan 2019 from https://www.fltimes.com/opinion/bigger-picture-sausage-links-and-more/article_d525068c-1fc7-59cc-a275-5089b69ef148.html.
Turner, S., Merchant, K., Kania, J., and Martin, E. (2012). Understanding the value of backbone organizations in collective impact: Part 2. *Stanford Social Innovation Review*. 17 July (blog). Retrieved 24 Dec 2019 from http://www.leveragingourstrengths.ca/reading/Health_BackboneOrgsCollectiveImpact.pdf.
Varner, C. (2014). SCD concepts: Sampson State Park interpretive cell phone tours and the Geneva Parks Master Plan. Retrieved 9 Nov 2018 from https://flihappenings.com/2014/09/01/scd-concepts-sampson-state-park-interpretative-cell-phone-tours-and-the-geneva-parks-master-plan/.
Varuk, V.V., Kramarenko, A.O., Lunova, V.A., and Parkhomenko, O.S. (2018). Innovation as a factor of the development of territorial communities. *International Journal of Engineering & Technology*, *7*(4.3), 545–9.
Wachtendorf, T., Kendra, J.M., and DeYoung, S.E. (2018). Community innovation and disasters. In H. Rodriguez, Donner, W., and Traino, J.E. (eds), *Handbook of Disaster Research* (pp. 387–410). Cham, Switzerland: Springer.
Walshok, M.L., and Shapiro, J.D. (2014). Beyond tech transfer: a more comprehensive approach to measuring the entrepreneurial university. In A. Corbett, Siegel, D.S., and Katz, J. (eds), *Academic Entrepreneurship: Creating an Entrepreneurial Ecosystem* (pp. 1–36). Bingley, UK: Emerald Group Publishing Limited.
Weerts, D.J., and Sandmann, L.R. (2008). Building a two-way street: challenges and opportunities for community engagement at research universities. *The Review of Higher Education*, *32*(1), 73–106.

Welch, M., and Saltmarsh, J. (2013). Current practice and infrastructures for campus centers of community engagement. *Journal of Higher Education Outreach and Engagement, 17*(4), 25–56.

Westoby, P. (2017). *Soul, Community and Social Change: Theorising a Soul Perspective on Community Practice.* London and New York: Routledge.

Wickenden, A. (2017). HWS Entrepreneurial Studies launches downtown presence. *The HWS Update.* 30 Jun. Retrieved 7 Jan 2019 from https://www2.hws.edu/hws-entrepreneurial-studies-moves-downtown/.

Wikipedia (n.d.). *List of Liberal Arts Colleges in the United States.* Retrieved 7 Jan 2019 from https://en.wikipedia.org/wiki/List_of_liberal_ arts_colleges_in_the_United_States/.

Zlotkowski, E. (1996). Opportunity for all: linking service learning and business education. *Journal of Business Ethics, 15*(1), 5–19.

27. Sustaining an urban education pipeline: a case study of university and community development partnership
Gloria Bonilla-Santiago

INTRODUCTION

The Rutgers-Camden Community Leadership Center (CLC) has been a catalyst in redefining the role Rutgers University-Camden plays in collaborating with the local community in Camden, New Jersey (Community Leadership Center, 2018). The campus has become an anchor to launch successful community development projects that contribute to neighborhood transformation. The Rutgers CLC's work has enhanced these institutional efforts to improve the economic, social and cultural well-being of a now-vibrant Cooper Street Education Corridor. The CLC strengthened the collaborative ties with residents and institutions and has contributed to Rutgers-Camden's strategic vision of education, discovery, engagement and the development of networks of intellectual capital that reach into the community and around the globe. This case study captures and highlights the CLC's community landmark project during its 25-year trajectory of building and sustaining community development efforts that has led to improved educational outcomes for children from birth through adulthood and strengthened social and community capitals sustaining families and neighborhoods. The central research question being addressed is how a community development initiative creates a school that changes the educational conditions and sustains and produces high achieving students in one educational corridor for 25 years.

The successes and gains of the Rutgers CLC work in the last 25 years have helped solidify the neighborhood and university campus as anchors in transforming and galvanizing community development efforts. The impact of the work is evident in residents' improved quality of life and the promotion of solidarity and agency, which have allowed people in Camden to live and work according to their own meaning and arrangements. The Camden community with whom we work has a clear sense of purpose and participation where residents own their personal situation and are constantly working to rebuild and inspire others to pool their knowledge, talents, aspirations and political will toward the sustainable well-being of their community. Collectively, the Rutgers CLC has accomplished this by drawing on university-community assets, resources and pathways. The guiding theory of action centers on the belief that improving outcomes for young people and families with limited resources and opportunities necessitates the transformation of schools into entire communities where all children can experience an excellent educational system with caring adults in all areas of their lives, safe places, a healthy start and healthy development, and opportunities to learn, grow and prosper.

The Rutgers CLC's focus has been on communities that are in need or struggling to sustain themselves, whether they are located regionally or globally. We selected a city

(Camden) and place (Cooper Street) where the CLC and community could work closely together collectively to develop and implement meaningful programs that change people's lives. The CLC approach to transformational change is built upon six pillars: *(1) An infant to college educational continuum, (2) School as the nexus for community development and transformation, (3) Community wellness and leadership development, (4) Capacity building training and experiential learning programs for our college student population, (5) Research opportunities for our faculty and graduate students, and (6) Scaling up of best practices local and globally* (Community Leadership Center, 2018). A continuum of education running from early infancy through college and career readiness addresses the multiple and intersecting educational challenges in the Cooper neighborhood and city. Community wellness programs improve the health and leadership of the neighborhood's residents, which have positive ripple effects in areas ranging from school attendance to adult employability and productivity. Training and capacity programs for college students support employment and opportunities for the neighborhood, where the research supports new creation of knowledge to inform policy makers at all levels.

The work and evaluation of the CLC's impact has been framed using the Community Capitals Framework (CCF) developed by Flora et al. (2004), which provides a validated tool to plan strategically and to measure community change. The biggest impact has been on the children and families in our community. The CLC, in partnership with the university, parents, businesses, nonprofit organizations, legislators and local community members, founded the LEAP Academy University School in 1997 and Early Learning Research Academy in 2008 to build a comprehensive Cradle to College educational pipeline. LEAP, an official charter school district from Kindergarten to 12th grade, has achieved a 100 percent high school graduation and college acceptance rate for all 14 LEAP Academy graduating classes since 2005, and in this journey, thousands of families have benefited from our collaboration and provision of health care, parent development training, college preparation support, job placement assistance, and training that build business and entrepreneurial skills to enhance residents' long-term self-sufficiency. Moreover, the success in strengthening the university community's intellectual capital is evident in the use of the CLC's living labs to advance student teaching and high-quality research. Not only have graduate and undergraduate students gained invaluable, real-life teaching experiences in our zone of practice through internships, courses, fellowships, teacher training and research, but faculty and staff have also conducted and disseminated research based on our educational settings and best practices. Over the years, outstanding domestic and international scholars and practitioners, artists and educators have visited our lab to learn from our work.

BACKGROUND

In the global economy, education is one of the most important factors in the national production of a country (Welch, 1970). Those societies that have advanced the most in their social and economic aspects around the world have achieved incredible progress in their creation of knowledge, transformation of their K-12 education system, and contribution to new research, production, innovation and advancement in their competitive economies (Chmielewski and Reardon, 2016). Similarly, the social and cultural development of these countries and nations depends more today on the K-12 educational system for providing an education that will improve new economies, communication systems, innovation, merging technologies and STEM fields (Drew, 2011). The need for a new modern public instructional education system that is efficient, where children are the focus, where excellence is the destination and failure is not an option, is urgently needed so that we can develop and turn around our urban schools in the United States (Duncan and Murnane, 2011). A sustainable instructional educational system from infancy to college, where we can maximize available resources without compromising the future of our generation of children and families, must be a local and national imperative of our country (Ewell et al., 2003; Lawson, 2013).

Historically, the city of Camden, New Jersey, has gone through some of the worst fiscal and economic crises in the history of any city in the United States. These crises were caused partially by political corruption, ineffective local government, a lack of accountability and leadership on the part of school officials, high dropout rates and a laissez faire approach of state supervision, particularly as evidenced by a string of arrests of three mayors within a 20-year period and a lack of accountability and transparency (Gillette, 2006; Bonilla-Santiago, 2014; Seligsohn and Mazelis, 2014). Camden is one of the poorest cities in the United States, despite going through major economic changes such as business tax breaks to encourage the building of new facilities in the city. With a population of 77,000, it has a poverty level of 39.3 percent, which increased since the 2000 Census when it was 32.3 percent (US Census, 2016). The median household income is $26,200, which increased from $24,000 in 2000, and the percentage of people with a high school diploma is 68 percent, which increased from 51 percent in 2000 (US Census, 2016). Income and high school graduation have increased; poverty has not decreased.

Like many other cities that experienced deindustrialization and White relocation to the suburbs during the second half of the 20th century, Camden has had a troubling history, experiencing economic and social downward spirals between 1955 and the mid-1990s (McKinsey, 2001; Gillette, 2006; Seligsohn and Mazelis, 2014). Camden lost nearly 50,000 residents and tens of thousands of jobs (Seligsohn and Mazelis, 2014). As middle-class Whites, industries and jobs moved to surrounding suburbs, African Americans, Latinos and poor Whites remained behind in a jobless, decaying city without an economic base, turning to public assistance and perpetuating a cycle of poverty (CSUCL, 1995; Bonilla-Santiago, 2017). The city's tax base decreased and government capacity deteriorated (Gillette, 2006). The population rebounded following losses in the 1980s and 1990s, and again after the economic recession in the late 2000s. The Camden City School District population, on the other hand, has continuously dropped from 18,536 in 1998 to 7,941 in 2018, demonstrating detrimental effects of school age children population decline.

THE ROLE OF THE RUTGERS-CAMDEN COMMUNITY LEADERSHIP CENTER (CLC) IN ANCHORING LEAP ACADEMY UNIVERSITY SCHOOL

There have been community efforts dating back more than 100 years to transform distressed neighborhoods into places where residents lead healthy and thriving lives. Some of these efforts include the Settlement House movement of the early 1900s, the War on

Poverty in the 1960s and the rise of community development corporations (CDCs) in the 1980s. Throughout the 1980s and 1990s, interest and investment on the national, state, county and city levels led to a Comprehensive Community Initiative (CCI) movement among social change efforts, research and lessons learned from earlier segmented service sectors (Henig et al., 2015). This movement led to the earliest examples of large-scale, cross-sector collaborations to coordinate health, education, employment and housing resources for youth and communities. Concerned about the emergence of this movement, privately funded foundations, such as the Annie E. Casey Foundation, Ford Foundation, Hewlett Foundation and the John S. and James L. Knight Foundation, led similar initiatives aimed to address the complex, interrelated issues a community might face with complex and interrelated solutions. This influx of funding also included support for research on effective coalition functioning, decision-making and coalition grantee support (Za et al., 2016).

During the 1990s, as the Rutgers CLC was planning and designing the transformation of Cooper Street in Camden, similar efforts began to gain popularity in the country, particularly CCIs. CCIs arose as an ambitious strategy to address the needs of residents of poor communities and as an attempt to try to combat long-term poverty (Stagner and Duran, 1997). They intended to go beyond the achievements of existing community-based organizations, notably social service agencies and CDCs, by concentrating resources and combining the "best" of what had been learned from social, economic, physical and civic development in order to catalyse transformation of distressed neighborhoods. Unlike other community initiatives that focused on one intervention at a time, like the production of affordable housing units, CCIs adopted a comprehensive approach to neighborhood change and worked according to community building principles that value resident engagement and community capacity building.

By 2008, after the financial crisis and as the country was recovering, new community development funding streams were designated. Federal efforts included the creation of Promise Neighborhoods and Choice Neighborhoods in 2010 under the Obama Administration and the White House Council for Community Solutions, which led to the creation of the Aspen Forum for Community Solutions (White House Council for Community Solutions, 2012). The Promise Neighborhoods Initiatives, inspired by the Harlem Children's Zone, was created by the US Department of Education to encourage neighborhoods to support youth through their first two decades of life by creating a continuum of family, community and academic supports. Communities continued to pursue locally driven efforts consistent with the strategies adopted by Promise Neighborhood grantees despite not receiving federal resources to do so (Tough, 2009). Choice Neighborhood initiatives sought to improve the physical and economic infrastructure of neighborhoods while also providing direct services to youth and families (Schorr, 2009; Za, 2011; Smith, 2011).

These CCI initiatives inform our work with Rutgers CLC/LEAP Academy. The work of the CLC and LEAP collaboration adds value to the CCI research in that it creates new practices and approaches to community development. Our approach to the work is more comprehensive in scale, focusing on the relationships between different elements of the initiative, as we built social, human and built capital for the community, rather than just the component of the program or organization in isolation.

THE LEAP MODEL

The vision was not just to develop a school near a university, but also to form a community hub and comprehensive holistic model that provided a variety of services inside the school. Dr Gloria Bonilla-Santiago, the founder of the school, created the CLC to leverage Rutgers' intellectual assets (Middle States Commission on Higher Education, 2008) and channel university resources, both financial and academic, toward operations and structures of LEAP. She formed the university Centers of Excellence to complement the innovative STEM curriculum being implemented and built an environment conducive to supporting children and families holistically. The Centers are a college-access and culture center that begins at birth and ends in college, in which students and parents participate in college awareness seminars, dual courses and early college with local colleges, a family support center that provides workforce development, a co-op, microenterprise, and career training center for parents, a health and wellness center for families that provides legal and family support services and ESL (English as a second language) classes, and a STEAM (science, technology, engineering, arts, and mathematics) state-of-the-art fabrication lab for innovation and entrepreneurial education for children and families. All aspects connect to the university: the Rutgers School of Nursing assisted during development of the health and wellness center, the School of Law ran legal clinics, biology and chemistry students and faculty built curricula and projects in the fabrication lab and students and researchers from the Childhood Studies Department observe infants, toddlers and preschoolers in the Early Learning Research Academy (ELRA), the formative center in the birth-to-college pipeline. Figure 27.1 highlights the educational pipeline.

LEAP's current design and systems of operations are reflective of its collaborative path toward its formation. A LEAP working group established a guiding system by which all activities were accounted for based on three components: (1) a mission statement, (2) uniformly held beliefs by the group, and (3) guiding principles (Bonilla-Santiago, 2014). The group developed a focused and collaborative culture that allowed strategic planning to align with the vision and proceed with accomplishing the group's goals (Bonilla-

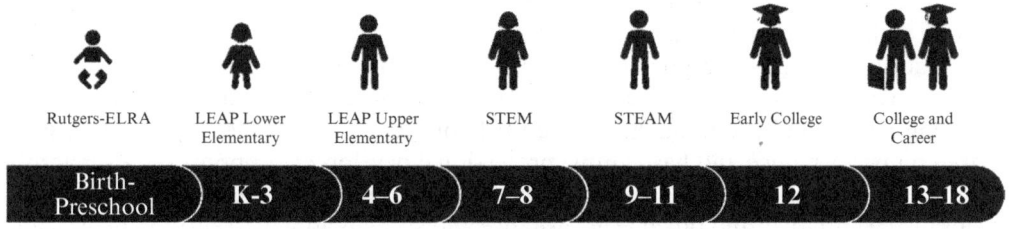

Source: Bonilla-Santiago (2014).

Figure 27.1 Rutgers/LEAP Cradle to College pipeline

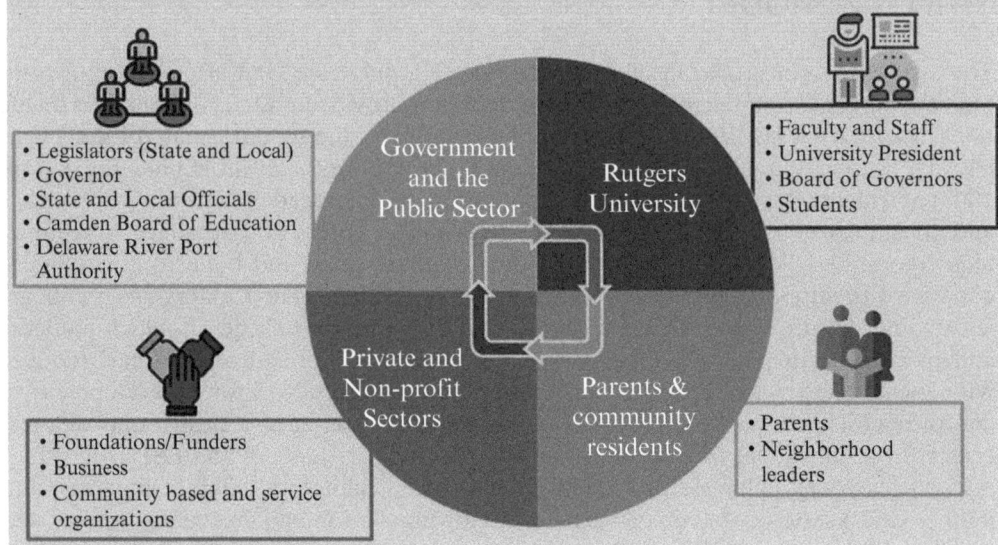

Source: Bonilla-Santiago (2014).

Figure 27.2 Rutgers/LEAP Cradle to College pipeline

Santiago, 2014). In a strategic plan titled *Camden Counts*, the mission was to "enhance opportunities for the children and families of Camden through the collaborative design, implementation, and integration of education, health, and human service programs and through community development" (CSUCL, 1995, pp. 2–7). Based on its collaborative nature, four categories of stakeholders, displayed in Figure 27.2, represent the indicative partnerships and alliances: (1) government and public sectors, (2) parents and community, (3) private entities, and (4) Rutgers University (Bonilla-Santiago, 1995).

Figure 27.3 is a model of four factors grounding a comprehensive approach to sustaining the pipeline: (1) organizational, (2) student, (3) teacher development, and (4) stakeholder/alliance factors. The four elements operate in tandem to ensure that synergies among children, families, teachers, organizations and curricula produce a focused, rigorous, data-driven and entrepreneurial environment conducive to teaching and learning in an urban context. Among the innovative factors are that students have an extended learning period and programs geared toward college access; teachers are compensated based on performance and have ample professional development opportunities; parents, institutions of higher learning and the greater community are actively involved in operations of the initiative; and pipeline governance and systems adhere to entrepreneurial and sustainable accountability measures.

LITERATURE REVIEW

During the planning stages in the early 1990s, the LEAP faculty planning team led focus groups with the aforenoted stakeholders and conducted asset mapping exercises that

Sustaining an urban education pipeline 445

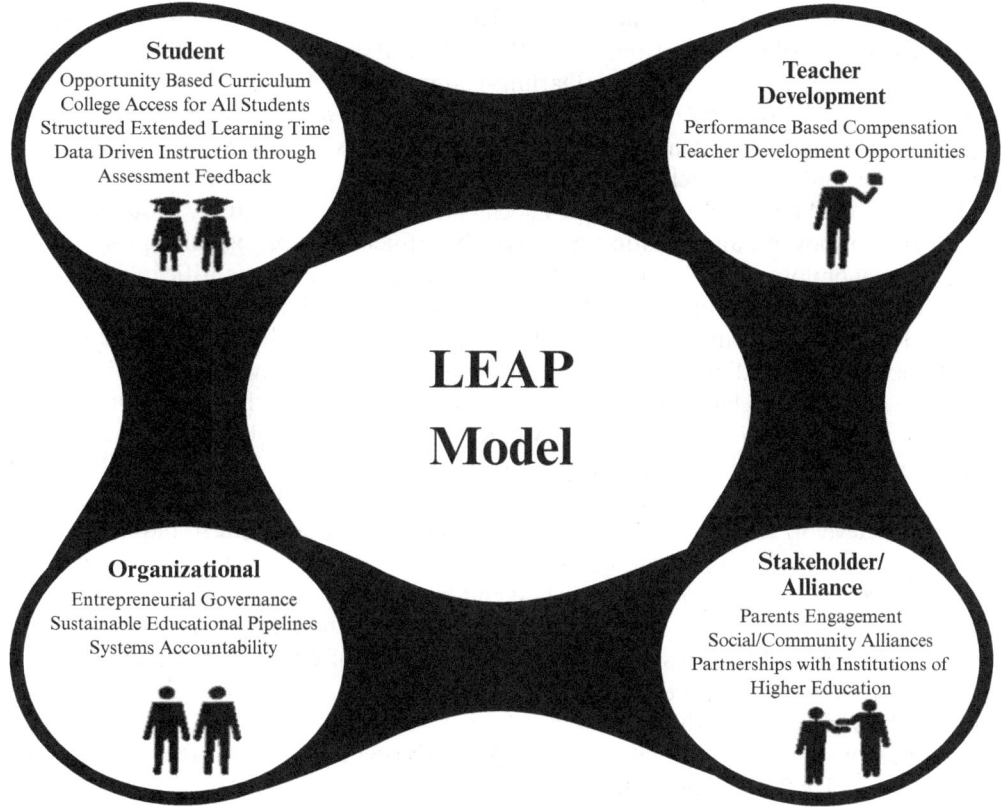

Source: Bonilla-Santiago (2014).

Figure 27.3 Rutgers/LEAP model

identified the strengths in the neighborhood, with the Appreciative Inquiry approach helping community members determine best strategies for investing in existing and future community assets (CSUCL, 1995).

When the initial community group planning process began, groups spoke about change in terms of what needs to change and how they were going to make those changes. Asset mapping and Appreciative Inquiry counteracted the negative conversations. The conversations began by asking participants to identify the positive aspects of the community instead of working from the negative ones. Assets are what we wanted to keep, build upon and sustain for future generations (Kretzmann and McKnight, 1993), either physical, like repurposing of a building, or social, like volunteers working to clean the streets.

The problems of urban public education that the Rutgers CLC addresses are emblematic of conditions nationwide, which require thoughtful critiques, reflections and engagement to overcome. Low test scores, high dropout rates, ineffective teachers and dilapidated buildings are common in urban schools (Noguera, 2003). Constraints to urban public schools are both external and internal to school systems (Noguera, 2003). Externally, they relate to the effects of poverty and social isolation on families in economically depressed inner-city neighborhoods (Wilson, 1987; Noguera, 2003). Internally, high turnover rates among school leaders and teachers (Darling-Hammond, 1997) and inadequate facilities create disorder (Payne, 1984) that hinders teaching and learning. Noguera (2003) calls for creative approaches and identification of effective schools that serve poor children and demonstrate students can achieve high success.

The CLC/LEAP work is influenced by social justice motivations and various sociological theories of poverty and isolation. Some of the influential pedagogical approaches and fundamental philosophical, sociological, psychological and neuroscientific reforms, theories and models throughout the educational movement at the time were: The *Connectivism* theory developed by Siemens (2005) and Downes (2010), where they suggest that the only way to learn in a post digital era is to consider the ecology of knowledge, the diversity of networks of human behaviors, personal relationships, space and the virtual global interconnectedness; *Cognitive/humanist* theory that Paulo Freire and John Dewey championed and who advocated for innovative methodologies like *Problem Based Learning* (PBL) and *Active Learning*; and *Constructivism* theory from Jean Piaget (1932), who advocated that we should develop students to trust their own ideas and allow them to develop so they can learn about themselves, make decisions, and accept their mistakes as part of creating new knowledge. There are not good or bad children in a school – children's capacities to adapt to a new learning style depend on the teachers and school methods of teaching and learning (Piaget, 1932); and social and cultural learning of everyone is affected by their social environment in which they develop (Vygotsky, 1978), all of which are foundational philosophies grounded in the CLC.

During a wave of educational reform attempts during the 1990s, at the time the CLC was forming LEAP Academy, other researchers recognized that new models of schooling had to emerge. Some have argued, "Salvation for our schools will not come from without but from within" (Lieberman, 1995). Lieberman (1995) recognizes that changing schools requires changing practices and structures around the whole school, rather than just individual projects and classrooms. Instead of reforming existing school systems, new school models in the form of community (Dryfoos, 1994) and charter schools (Budde, 1988) demonstrate the power and capacity to achieve these same goals by creating new structures and organizations. The goals of these new models are to decentralize administration to support democratic governance of schools, implement school-based decision-making and manage resources (Anyon, 1997).

In addition, capital theories of schools, poverty and community shaped how the CLC connected education and community development initiatives to use schools as channels for communities to overcome segregation and isolation. Social capital is the value that comes from connections within and between social networks of building trust, collective norms and reciprocating relationships (Walter, 2005). It represents the communal binding necessary for community capacity building and collective action and is a contributor to individual and community health and well-being (Kawachi et al., 2008; Walter, 2005). Glaeser (2001) combines social capital with human capital, defined as educational attainment, arguing that the "education-social connection relationship should probably be the most robust and most important fact about the formation of social capital" (p. 16); an educated person is an engaged person who drives transformative change (Helliwell and Putnam, 2007).

Wilson (1996) argues that schools should play a prominent role in designing policies that address concentrated poverty. Isolation of ethnic neighborhoods produces less social integration and increases disorder (Arum, 2000). Schools that do not produce adolescent attachment to conventional activities experience greater delinquency (Sampson and Laub, 1995). School is a forum for connecting youths to conventional adult norms and adapting them to mainstream societal and economic structures (Coleman, 1988; Arum, 2000). Arum (2000) extends this concept by suggesting "A school's relevant community is not just a neighborhood demographic environment, but equally an institutional environment" (p. 400). Expectations for success are institutionalized, and school challenges are addressed using intellectual, human and social capital available from researchers who share meaning and solidarity with a community (Giddens, 1984). Schools have innate abilities to foster relationships among various power structures of communities and elites, which symbolize a path to economic security because schools are integral during community development, influencing the shift in residents' perceptions of public institutions that are meant to serve them. Social capital regarding school development addresses broader issues of structural inequality, such as underfunding of school systems or school-to-prison pipelines. Warren (2005) defines this issue as relational power, in which community development should overcome external and internal isolation of urban public schools. An exodus of jobs destroys businesses, social institutions and youth socialization, leading to social isolation (Warren et al., 2001).

The Rutgers CLC anchors with LEAP to advance community development, empower residents and augment community control in urban revitalization. The CLC's mission resembles Stone et al.'s (2011) components of schools serving as community development vehicles: (1) they provide parents and others in poor communities with valuable experiences to interact with public agencies, (2) they increase skills and aptitudes of community residents for adults and children, (3) they strengthen social ties and the capacity for collective action in poor neighborhoods, and (4) they link neighborhoods with much-needed resources from communities.

The value of schools serving as vehicles for community development is virtuous, but implementation has challenged policy makers and practitioners. Instead of channeling efforts in bureaucracies of traditional public-school systems, university researchers and entrepreneurs recognize the possibility of building autonomous charter public schools to strengthen urban public education. Urban schools and the promise of educational attainment and success connected to a university attract poor families to neighborhoods and prevent families from leaving. New urban schools anchor community development to influence place-based decisions around economic revitalization (Taylor and Luter, 2013; Taylor et al., 2013; Bonilla-Santiago, 2017).

METHODOLOGY

To answer the research question of how a community development school initiative (LEAP Academy University School) in Camden changed the educational conditions and sustained high achieving students in one educational corridor, we used the CCF (Flora et. al, 2004) to guide a direct asset mapping and qualitative approach (Yin, 2017) to document and assess the emerging actions of the Rutgers CLC/LEAP pipeline and best practices that contribute to its sustainability. We conducted 25 semi-structured interviews with stakeholders involved in the formation, implementation and sustainability of the university-school partnerships, including former university and government officials, professors, former school leaders, parents, community leaders and alumni, all identified through a review of historical documents, strategic plans, state reports and newspaper clippings. Through the interviews, stakeholders narrated their own interpretations on the formation, implementation and sustainability that offered rich content (Borthwick et al., 2003). This stratified, purposeful and snowball sampling of stakeholders at various levels of the project (Miles and Huberman, 1994) was the most reliable form to assess the project from various perspectives. Each conversation was recorded and transcribed to capture the information for thematic analysis. Each participant completed and signed an interview consent form for audio/video recording. Upon completion of the interviews and review of documents, responses were coded manually to extrapolate themes that defined the development of the partnership (Miles and Huberman, 1994). Coding of interviews structured the themes and common issues throughout the content. Codes that represented the characteristics of the LEAP educational pipeline and the challenges of sustaining it allowed in-depth analyses. The codes were interpreted and structured to identify patterns, themes, surprises, challenges, and enabled dynamics to be explicated from the data.

FINDINGS

Due to LEAP's influence during the 1990s, Dr Santiago extended Rutgers' work to include Cooper Street as the strategic place of focus to transform the neighborhood. Cooper Street was characterized by dilapidated buildings, homeless people and drug use. Provost Roger Dennis and Dr Santiago decided to move their offices to Cooper Street in a historic building as the first step to be part of the community. The working group received grants from the Delaware River Port Authority, Tree Foundation, Prudential Foundation, Ford Foundation and others to rebuild and repurpose school buildings, plant vegetation and repave sidewalks to improve the physical conditions along the corridor, which attracted more capital. The growth of the LEAP schools along Cooper Street to encompass six major buildings, split between the ELRA, Lower Elementary, Upper Elementary, STEM building and 9–12 STEAM building, evidences the innovation and vision required to transform a neighborhood using schools as vehicles for physical development, combined with social and community development.

During the 1990s, the emotions among families toward the traditional school and classroom environment were of angst and major concern. Gang activities and drug dealing permeated the streets of Camden, leading to incessant crime and behavioral challenges, even inside schools. According to a founding staff member and parent of LEAP Academy:

> There was a lot of turmoil, a lot of things happening. And the people in the city felt helpless, hopeless. You could feel it in the educational system and the political system everywhere. It was a very terrible time in the city, for everyone. (Founding Staff Member Interview, 2017)

A parent of an inaugural year student recalls how parents were distanced from schools:

Camden school systems at that time . . . they're detached. They don't do much with parents. I guess you're not forbidden to go there but you're not welcome there, I guess, is the right . . . You don't feel welcomed. (Founding Parent Interview, 2017)

In the context of university-community relations, the relationship between Rutgers University-Camden and the community were uninviting. Signs posted on campus prohibited outsiders from walking through, and police patrolled the grounds looking for people who should not be there (Founding Staff Member Interview, 2017). The university environment was also unfriendly to minorities, since the number of African American and Latino students was low. Professors were unaccommodating, one founding staff member who lived in Camden had to take care of her children as a single mother, which involved constantly seeking alternative childcare options (Founding Staff Member Interview, 2017). According to Dr Deborah Bowles, a Rutgers-Camden associate chancellor and enrollment management leader recognized the imbalance between the university goals and results from the K–12 school district population:

A lot of people feeling that Rutgers was in the community but not of the community, and that Rutgers did not understand its obligation to try to bring along more students who could become eligible for admission to Rutgers or other universities. (Bowles, 2018)

In addition to creating new educational outcomes, including graduating high school and entering college, the motivation for LEAP was to provide a welcoming sense of community to parents and children in the city. Many policies precluded people from educational environments both in the Camden public-school system and at Rutgers. LEAP treated families with pride, dignity and respect. Families would be instrumental in shaping LEAP's structure and governance, and as an education program, to provide a sense of being and belonging that had never existed for minority families in Camden before. A founding parent and original board and planning member remarked how Dr Santiago gave parents a voice for stronger education:

[Dr Santiago] gave us opportunity to gain knowledge that we wouldn't have gotten in the Camden city school system . . . it was an opportunity to get out of poverty and get an education. Opportunity for me to want to be more educated, you know, and to have set higher goals for me and my children. And then while you're, you know, getting your education and you're improving your lifestyle, you also give back to the community where you came from and that's, you know, how I see her. (Founding Parent Interview, 2018)

As part of the LEAP planning process, the Parents as Partners for Educational Change Advocacy Council formed, as they became the first 500 parent group to legitimize the potential, knowledge and skill sets parents had to create change. It resembled a major shift in how parents were treated in the city, particularly regarding education. The school system tended to ignore parents and expected them to follow orders and not ask questions (Founding Staff Member Interview, 2017). Parents were conditioned to not participate in decision-making. Through the LEAP training program, parents were given learning assessments and the Parenting Stress Index to reflect on their own abilities and capacities as individuals. Seminars empower parents with new knowledge, approaches, techniques and competencies concerning children's ways of learning, their development and the education system (Bonilla-Santiago, 1995). Consequently, parents' roles as community change actors and drivers elevated. Dr Santiago capitalized on their strength as a constituent group to lobby state lawmakers as the charter school legislation blossomed. The presence of parents proved effective in moving the legislation to enactment. State associate Melanie Schulz reflects that parents "evolved like butterflies" through advocacy:

> [Gloria] gave them the experience of participating, made them proud of who they were, and gave them confidence. All of it changed the dynamics of shaping charter legislation, made a huge impression on legislators, and made the parents valuable participants in the process. That what she gave to them. (Bonilla-Santiago, 2014, p. 141)

The new school empowered parents to take hold of their own destinies, build new paths and open new doors for themselves. Dr Santiago intentionally organized parents in peaceful and civilly obedient ways to teach them to advocate and create structures to help themselves and their children. She explains:

> We went from advocacy in legislating change to how to conduct homework sessions at home, to how to present testimony and communicate with the legislature using peaceful means for getting people to respect the policy process. (Bonilla-Santiago, 2017)

The LEAP school changed the social and inter-relational dynamic of how parents were treated; they were now respected and treated with dignity in an environment where they have a voice. LEAP has an integrated holistic model that is unique to any school and throughout this process we collected best practices that are reflected in the school community.

RECOMMENDATIONS AND BEST PRACTICES

1. *Rigorous college preparation programs, access to quality education, and capacity building prepare students from cradle to college and develop soft skills necessary to achieve in higher education and into their careers.*
   The college-access focus throughout the entire Rutgers/LEAP pipeline makes the CLC's efforts strategically significant for Rutgers University. In anchoring our collective work, college and career readiness programs are purposely placed at the center of the school's organizational structure through a focus on high expectations for all students, educators and families. College readiness is not only embedded in the curriculum but is also the cornerstone of the development of programmatic structures that support the path and preparation of students through the pre-K-12 program to college. The Cradle to College and careers pathways on Cooper Street in Camden is the only intentional cohort effort in the country with a Latino and African American population achieving 100 percent high school graduation. Current LEAP Director of College Access and LEAP alum Marchelle Roberts comments:

   > My experience as a student at LEAP and alumni now, opened my eyes to how vital a college education is in this world and how necessary it is to show black and brown students that college is attainable, no matter where they come from, what they may lack, or who may doubt

them. It is important to me, that students of color are able to envision a world where anything is possible for them, and that starts with preparing for and succeeding in college. (Roberts, 2018)

As of 2018, 1,012 students, which is 100 percent of all senior classes, have graduated from LEAP Academy. Two hundred of LEAP high school graduates have attended one of Rutgers's three campuses in Camden, New Brunswick and Newark. Since 2013, data on our alumni have shown an 85 percent college retention rate and more than $30 million in scholarship funds to help students remove the financial barriers that often stand in the way of completing a bachelor's degree.

In 2017–18, the Rutgers CLC launched the Early College Program and engaged Rutgers and Rowan universities in articulating a program that provided the opportunity for the entire LEAP senior class to enroll in early college for the entire year, where 100 percent of the senior class was admitted to college with up to 30 college credits. The strategy of teaming up university faculty with LEAP teachers to co-teach as a recitation leader has proven to be an effective strategy for reinventing the senior year. More students will successfully complete college courses before graduating high school. Khary Golden, LEAP Director of Early College, expresses that the LEAP model informs that:

> There is one choice for LEAP students and families. That is to engage in higher education. We meet them where they are, and students are provided opportunities where they can learn about the avenues and pathways into higher education before they even become graduating seniors in high school or actual college standards. (Golden, 2018)

The students' biggest challenge with Early College is the rigor of college courses and adjusting to the early schedule, but the best is being part of a college culture and feeling welcomed by professors and staff who wanted so much to support them. A graduating student remarked, "This college program [gave] me the opportunity to show my best potential that there is a path to success no matter where you come from" (LEAP student, 2018).

2. *The Health and Community Wellness Clinic improves the quality of life for neighborhood residents and positively impacts school attendance, employability and productivity of adults.*

   Part of the Rutgers CLC mission is to deepen the public's knowledge of health and wellness issues by serving as a resource and care provider of "high quality ambulatory health care to the community" (Rivera, 2018). The Health and Community Wellness Clinic provides family medicine treatment to all LEAP children and families for asthma, diabetes, obesity, colds and other illnesses, and has a prevention wellness unit for children with traumas (both physical and psychological) including witnessing parents or family members being killed, traumatized from abandonment and neglect, domestic and sexual assault, and from being bullied. Many of these children and families experience these urban traumas based on squalid environmental and social conditions that negatively impact students' ability to focus and learn.

3. *Parent engagement builds school and community culture, creating accountability and transparency, and improving academic performance.*

The Family Support center serves as a community place for family education where parents are valued for their contributions to their child's education. The school community builds on individual strengths, assets and talents, and the combined capabilities of parents as stakeholders and partners. Using an asset-based community development approach, human and cultural capital of the schools coalesce to ensure students and parents reach their full potential. Jean Shepard, a founding parent and graduate of the first Parent Academy College program, remarks:

> LEAP was an opportunity to get out of poverty and get an education, for me to want to be more educated and to set higher goals for me and my children. While you're getting your education and you're improving your lifestyle, you also give back to the community where you came from. Even though you get a better education, better values, self-esteem and motivations, you'd come back, and you give back. (Shepard, 2018)

During the planning stages, parent focus group participants identified many important needs in their community. These discussions and a shared perspective that parental engagement is crucial to creating and sustaining the birth-to-college pipeline shaped the development of a citywide Parent Academy for School Reform that has engaged and trained thousands of parents. Since 1997, parents have led and enjoyed stronger family engagement facilitated by the school community as codified by a parent volunteer contract. Parents complete trainings and the majority report that their family's ability to read, write and perform better is due to increased educational engagement. The Parent Council is part of the school's governance and recommends policies for the school community and Board of Trustees.

4. *Academic pillars for learning should integrate entrepreneurship, data-driven instruction, problem solving, analytical research, applied experimentation, project-based learning, reflection and active learning.*

   Across grade levels, students and teachers use data-driven and applied learning modules that expose students to core content knowledge to raise awareness, interest and motivation of students in the STEM fields. The Fabrication Lab provides a zone of practice and the setting for multi-disciplinary projects for solving local problems. It offers an opportunity to develop replicable products that impact community and drive the curriculum. Its location within a school anchors it as a technical prototyping platform for innovation and invention that stimulates student engagement and creativity and promotes local entrepreneurship. Ricardo Miranda, Director of the Fabrication Lab, supports this:

   > the FabLab creates a synergistic effect that enriches and enhances the results of the projects . . . creat[ing] a new way to alternative, participatory, inclusive communities in the classroom, at school, locally and around the globe. (Miranda, 2018)

   This approach increases a sense of civic responsibility as students learn to become global citizens and contributors to solving local and global environmental problems such as water quality, food deserts, recycling and composting for organic urban gardens.

5. *The University-based CLC is an anchor for community transformation, giving access and parity for community to become partners to create new knowledge, innovations,*

a voice and seat at the table in the school governance and in community development efforts.
6. *Transformative school leadership is long, difficult and lonely work. Adults want to maintain the status quo and teachers are impatient for results. School leaders and teachers need to have resiliency and knowledge to know what to do, lead with courage and make difficult decisions, know who to listen to and how to manage critics along the way.*

It took five years to see signs of sustained improvement in the academic culture and climate of the school community. The journey has been challenging but the rewards for the community are transformative. There are now 14 years of having 100 percent of students graduate high school and attend college; 85 percent of parents engaged and 95 percent of staff have no absences.

DISCUSSION

This study places community development school practice in transforming a neighborhood at the forefront of a university partnership for high school and college completion to catalyse and champion policies, and systems and practices that ensure all students, particularly low income, first generation students, graduate from college and achieve their careers. The Rutgers CLC, through the LEAP pipeline, created a new theory of action using a school as a CCI integrating a holistic educational approach for children and families in a distressed urban setting. Asset-based community development informed the school model desired by the community and supported by the university, which led to policy outcomes of legislation creating charter schools in New Jersey. This case demonstrates that policies adapted from local grassroots innovation efforts with community ownership, rather than from government-mandated decisions, are transformational for communities and neighborhoods.

CONCLUSION

Implications for practice are that university-school partnerships need to be designed within an asset-based community development framework where community members are part of the planning stages and then are active and beneficiaries in its operations and governance. The Cradle to College pipeline is linked to the university and the relationship is reciprocal, collaborative and respectful of community life. Through CLC university oversight, management, and community ownership, the Rutgers/LEAP pipeline provides an integrated model and process for how it prepares students, trains families, builds community and remains sustainable.

Much can be learned from the LEAP educational pipeline. University faculty and senior staff need to embrace building tangible bridges through dual enrollment, early college and tuition-benefit opportunities with community members to instill a college-going culture within the community. Community, particularly parents, needs to be part of the planning process and governance structure to claim ownership and pride in the work, while holding everyone accountable for the work. The university needs to continue to provide leadership in the partnership by institutionalizing the project with ongoing community input, financial and academic support for long-term sustainability of the partnership. Having a university center and staff designated to provide oversight to the partnership was proven to be an asset in that it serves as mediator advocates and support for the partnership. The university and community partnership needs to continue to support innovative practices to provide multi-faceted services with the community and be flexible to experiment, listen, evolve, grow and adapt over time. Most importantly, with conviction, intention and guiding principles of social justice, equity and respect for community values, universities can be anchors of sustainable and meaningful change.

REFERENCES

Anyon, J. (1997). *Ghetto Schooling: A Political Economy of Urban Educational Reform*. New York, NY: Teachers College Press.
Arum, R. (2000). Schools and communities: Ecological and institutional dimensions. *Annual Review of Sociology*, *26*(1), 395–418.
Bonilla-Santiago, G. (1995). Testimony presented to the New Jersey Senate Education Committee in support of the Charter School Program Act of 1995 (S-1796).
Bonilla-Santiago, G. (2014). *The Miracle on Cooper Street*. Bloomington, IN: Archway Publishing.
Bonilla-Santiago, G. (2017). Personal interview. 6 October.
Borthwick, A.C., Stirling, T., Nauman, A.D., and Cook, D.L. (2003). Achieving successful school-university collaboration. *Urban Education*, *38*(3), 330–71.
Bowles, D. (2018). Personal interview. 9 May.
Budde, R. (1988). *Education by Charter: Restructuring School Districts. Key to Long-Term Continuing Improvement in American Education*. Andover, MA: The Regional Laboratory for Educational Improvement of the Northeast & Islands.
Chmielewski, A.K., and Reardon, S.F. (2016). Patterns of cross-national variation in the association between income and academic achievement. *AERA Open*, *2*(3), 2332858416649593.
Coleman, J.S. (1988). Social capital in the creation of human capital. *American Journal of Sociology*, *94*(1), S95–S120.
Community Leadership Center (2018). *Annual Report*. Camden, NJ. Retrieved 1 September 2019 from https://clc.rutgers.edu/annual-report-2018/.
CSUCL (Center for Strategic Urban Community Leadership) (1995). *Camden Counts: A Strategic Plan for the Project LEAP Academy*. Camden, NJ: CSUCL.
Darling-Hammond, L. (1997). *The Right to Learn: A Blueprint for Creating Schools that Work*. The Jossey-Bass Education Series. San Francisco, CA: Jossey-Bass.
Downes, S. (2010). New technology supporting informal learning. *Journal of Emerging Technologies in Web Intelligence*, *2*(1), 27–33.
Drew, D.E. (2011). *STEM the Tide: Reforming Science, Technology, Engineering, and Math Education in America*. Baltimore, MD: JHU Press.
Dryfoos, J.G. (1994). *Full-Service Schools: A Revolution in Health and Social Services for Children, Youth, and Families*. San Francisco, CA: Jossey-Bass.
Duncan, G.J., and Murnane, R.J. (eds) (2011). *Whither Opportunity? Rising Inequality, Schools, and Children's Life Chances*. New York, NY: Russell Sage Foundation.
Ewell, P.T., Jones, D.P., and Kelly, P.J. (2003). *Conceptualizing and Researching the College Student Pipeline*. Boulder, CO: National Center for Higher Education Management Systems.
Flora, C.B., and J.L. Flora, with S. Fey. (2004). *Rural Communities: Legacy and Change*, 2. Boulder, CO: Westview Press.
Giddens, A. (1984). *The Constitution of Society: Outline of the Theory of Structuration*. Cambridge, UK: Polity Press.
Gillette Jr, H. (2006). *Camden after the Fall: Decline and Renewal in a Post-Industrial City*. Philadelphia, PA: University of Pennsylvania Press.
Glaeser, E.L. (2001). The formation of social capital. *Canadian Journal of Policy Research*, *2*(1), 34–40.
Golden, K. (2018). Personal interview. 4 May.
Helliwell, J.F., and Putnam, R.D. (2007). Education and social capital. *Eastern Economic Journal*, *33*(1), 1–19.
Henig, J.R., Riehl, C.J., Rebell, M.A., and Wol, J.R. (2015). *Putting Collective Impact in Context: A Review of*

the Literature on Local Cross-Sector Collaboration to Improve Education. New York, NY: Teachers College, Columbia University, Department of Education Policy and Social Analysis.

Kawachi, I., Subramanian, S.V., and Kim, D. (2008). Social capital and health. In I. Kawachi, S. Subramanian and D. Kim (eds), *Social Capital and Health* (pp. 1–26). New York, NY: Springer.

Kretzmann, J.P., and McKnight, J.L. (1993). *Building Communities from the Inside Out: A Path Toward Finding and Mobilizing a Community's Assets*. Evanston, IL: Institute for Policy Research.

Lawson, H.A. (2013). Third-generation partnerships for P-16 pipelines and cradle-through-career education systems. *Peabody Journal of Education, 88*(5), 637–56.

Lieberman, A. (1995). Practices that support teacher development: Transforming conceptions of professional learning. *Innovating and Evaluating Science Education, 95*(64) 67–78.

Luter, D.G. (2016). Place-based school reform as method of creating shared urban spaces: What is it, and what does it mean for universities? *Metropolitan Universities, 27*(3), 156–77.

McKinsey (2001). *A Path forward for Camden*. Annie E. Casey Foundation. Retrieved 1 September 2019 from http://www.camconnect.org/datalogue/CamdenPath.pdf.

Middle States Commission on Higher Education (2008). *Community Engagement: Rutgers- New Brunswick and Rutgers-Camden*. (Section 8). New Brunswick, NJ: Rutgers, The State University of New Jersey.

Miles, M.B. and Huberman, A.M. (1994). *Qualitative Data Analysis: An Expanded Sourcebook*. Thousand Oaks, CA: Sage.

Noguera, P. (2003). *City Schools and the American Dream: Reclaiming the Promise of Public Education* (Vol. 17). New York, NY: Teachers College Press.

Payne, C.M. (1984). *Getting What We Ask For: The Ambiguity of Success and Failure in Urban Education*. Contributions to the Study of Education, Number 12. Westport, CT: Greenwood Press.

Piaget, J. (1932). *The Moral Judgment of Children*. Paris, France: Alcan.

Rivera, V. (2018). Personal interview with Chief Medical Officer, LEAP Academy. 13 May.

Roberts, M. (2018). E-mail communication. 17 July.

Saegert, S., Thompson, J.P., and Warren, M.R. (eds) (2002). *Social Capital and Poor Communities*. New York, NY: Russell Sage Foundation.

Sampson, R.J., and Laub, J.H. (1995). *Crime in the Making: Pathways and Turning Points through Life*. Cambridge, MA: Harvard University Press.

Sarason, S.B. (1996). *Revisiting "The Culture of the School and the Problem of Change"*. New York, NY: Teachers College Press.

Schorr, L. (2009). *Realizing President Obama's Promise to Scale up What Works to Fight Urban Poverty*. Washington, DC: Center for the Study of Social Policy. Retrieved 1 September 2019 from http://www.vacap.org/assets/content/Documents/ObamaPromiseMarch09.pdf.

Seligsohn, A., and Mazelis, J.M. (2014). The view from Camden. In S.N. Haymes, M.V. Haymes and R.J. Miller (eds), *The Routledge Handbook of Poverty in the United States* (pp. 19–26). New York, NY: Routledge

Shepard, J. (2018). Personal interview. 27 April.

Siemens, G. (2005). Connectivism: A learning theory for the digital age. *International Journal of Instructional Technology and Distance Learning, 2*(1), 3–10.

Smith, R.E. (2011). *How to Evaluate Choice and Promise Neighborhoods*. Washington DC: Urban Institute, Perspectives Brief 19. Retrieved 1 September 2019 from https://www.urban.org/sites/default/les/publication/32781/412317-How-to-Evaluate-Choice-and-Promise-Neighborhoods.PDF.

Stagner, M.W., and Duran, M.A. (1997). Comprehensive community initiatives: Principles, practice, and lessons learned. *The Future of Children, 7*(2), 132–40.

Stone, C., Doherty, K., Jones, C., and Ross, T. (2011). Schools and disadvantaged. In R.F. Ferguson and W.T. Dickens (eds), *Urban Problems and Community Development* (pp. 339–80). Washington DC: Brookings Institution Press.

Support for the Charter School Program Act of 1995 (S-1796): Hearing before the Committee on Education, New Jersey Senate (1995) (Testimony of Gloria Bonilla-Santiago).

Taylor Jr, H.L., and Luter, G. (2013). Anchor institutions: An interpretive review essay. *Anchor Institutions Task Force*, 14–18.

Taylor Jr, H.L., McGlynn, L., and Luter, D.G. (2013). Public schools as neighborhood anchor institutions: The choice neighborhood initiative in Buffalo, New York. In K. Patterson and R.M. Silverman (eds), *Schools and Urban Revitalization: Rethinking Institutions and Community Development* (pp. 109–35). New York, NY: Routledge.

Tough, P. (2009). *Whatever It Takes: Geoffrey Canada's Quest to Change Harlem and America*. New York, NY: Houghton Mifflin Harcourt.

US Census (2016). *American FactFinder*. Retrieved 1 September 2019, from factfinder2.census.gov.

Vygotsky, L.S. (1978). *Mind in Society*. Cambridge, MA: Harvard University Press.

Walter, C. (2005). Community building practice: A conceptual framework. In M. Minkler (ed.), *Community Organizing and Community Building for Health* (pp. 68–83). New Brunswick, NJ: Rutgers University Press.

Warren, M. (2005). Communities and schools: A new view of urban education reform. *Harvard Educational Review*, *75*(2), 133–73.
Warren, M.R., Thompson, J.P., and Saegert, S. (2001). The role of social capital in combating poverty. *Social Capital and Poor Communities*, *3*(1), 1–28.
Welch, F. (1970). Education in production. *Journal of Political Economy*, *78*(1), 35–59.
White House Council for Community Solutions (2012). *Final Report: Community Solutions for Opportunity Youth*. Washington DC: White House Council for Community Solutions.
Wilson, W.J. (1987). *The Truly Disadvantaged*. Chicago, IL: University of Chicago.
Wilson, W.J. (1996). When work disappears. *Political Science Quarterly*, *111*(4), 567–95.
Yin, R.K. (2017). *Case Study Research and Applications: Design and Methods*. Los Angeles, CA: Sage Publications.
Za, J.F. (2011). A cease and desist order for school reform: It's time for educational transformation. *Applied Developmental Science*, *15*(1), 1–7.
Za, J.F., Pufall Jones, E., Donlan, A.E., Lin, E.S., and Anderson, S. (2016). Comprehensive community initiatives creating supportive youth systems: A theoretical rationale for creating youth-focused CCIs. In J.F. Za, E. Pufall Jones, A.E. Donlan and S. Anderson (eds), *Comprehensive Community Initiatives for Positive Youth Development* (pp. 1–12). New York, NY: Routledge.

Index

Aberdeen University 128
Abu Dhabi 343
action perspective 4, 18, 19–20, 22, 245–6; *see also* reflection, action and knowledge creation: lived experience
active learning 446
actor network theory 84
ADESCOs (CD associations) 205, 207
Adler, P.S. 95
Aduwo, E.B. 303
advocacy 211–12, 320, 325–6, 333, 450
aesthetics and community development 8, 391–405
   Banff Centre for Arts and Creativity (Alberta, Canada) 398, 400–404
   case study limitations 404
   celebration 399
   co-creation 392
   collaborative creativity 394
   communication 394
   confusion 401
   creativity research and practice 391–4
      evaluative activities 393
      generative activities 393
      sociocultural activities 393–4
   curiosity 399, 401
   generosity 399
   grassroots creativity 391
   Hope Decoded 403–4
      cause champions: feedback 403–4
      open mic 403
   Japanese architecture and design: conceptions of space 392, 394–8, 402
      boundaries, ambiguous and contingent 395
      Buddhism 395–6
      interactions and uncertainty 395
      invisible structures 395
      knowledge mobilizing space (BA) 396–7
      location (TOKORO) 397
      negative space (MA) 397–8
      relational space (WA) 396
      Shinto 395
      Western ideas of space 396
   mechanics-dynamics-aesthetics (MDA) 398–400, 402
   newness, uniqueness and novelty 394
   paradox 401
   Peter Lougheed Leadership Initiative (University of Alberta) 400–403
      elder participation 401–2, 403
      fragmentation 402–3
   place-based interventions 391
   resource extraction and use 391
   social relations and communications 393
   structure building 391
AFINN dictionary 367
Africa 124, 226, 348–50
age factors 28, 32, 181, 283
   neighborhood environment satisfaction in deteriorated areas: impact of socioeconomic characteristics 303, 306, 307, 311, 312, 314
agency, *see* solidarity and agency
Aileniei, O. 94
Alberta University 400–403
Alinsky, S. 139, 140, 141
Álvarez, M. 407
Amabile, T.M. 393
Amin, A. 94, 321
anchor institutions 8, 53, 264, 282, 324
   community innovation: local partnerships and sustainable community development 416, 424, 428, 431, 434
   community-based organizations (CBOs) 111–12, 114, 116–17, 121
   public participation process and competing interests management: residential displacement (Buffalo, New York) 214, 216
   universities and community development partnerships 439, 441–2, 447, 450, 452, 454
Antonelli, F. 363
applied research 273
appreciative inquiry 3
approaches to community development 2–4
Argentina 408, 411
Armstrong Atlantic State University 244, 257, 258, 261
Arnstein, S.R. 76, 78–9, 82, 117, 212, 213
Arum, R. 447
Asia 226
assessment of community development 2

Asset-Based Community Development
    (ABCD) 2–3, 4, 13, 17, 19, 22, 58, 67–75
  associations 70–71, 72, 74
    Citizen Power Progression (Power Ladder)
       73
    co-producers 73
    connecting or mobilizing assets 72–3, 74
    consensus 70
    culture and stories 70, 71, 72, 74
    deindustrialization 67
    economic-based theory 72
    empowerment 73
    exchange 70, 71, 72, 74
    faith-based community development and
       evaluation capacity enhancement:
       ENLACE (El Salvador) 205
    food deserts: Utica, Mississippi 292
    funding 69, 74
    gifts of the head, hands and heart 70, 74
    half-empty/half-full glass 67
    identification of assets 72
    individuals 70, 71, 74
    institutions 70–71, 72, 73, 74
    land and physical environment 70, 71, 74
    leadership model 74
    multiplier effect 74
    needs-based approach 68–9, 74, 75
    paths or dilemmas 67
    physical space 71–2
    problem-oriented data 68
    reflection, action and knowledge creation:
       lived experience 13, 16–18
    religious, cultural and recreational activities
       70
    self-determination 74
    social service organizations 68
    triangle or hierarchy 71–2
    universities and community development
       partnerships 453
audio-visual data 235
Australia 343

Balduini, M. 363
Bangladesh, see group formation: Grameen
    Bank (Bangladesh); participatory action
    research: Bangladesh
Barrington-Leigh, C. 275
Barton, H. 409
baseline studies 207–8
Bates, L. 214
Beach, D. 110, 119
Beaulieu, L.J. 118
Beigi, G. 363
belonging, sense of 312, 409
Benjamin, G. 63

Bergdall, T. 15, 16, 17
Bergeron, S. 96
Berke, P. 51
Bertrand, K.Z. 364
Bhattacharyya, J. 12, 13–14, 15, 22, 410, 411,
    417
Bilodeau, A. 76, 79, 84–6
Bluestone, D. 364
Boden, M. 393–4
Bohl, C.C. 322
Bollen, J. 363
Bonilla-Santiago, G. 443, 448, 449–50
bottom-up approach 3
    community-based organizations (CBOs)
       116, 118
    faith-based community development and
       evaluation capacity enhancement:
       ENLACE (El Salvador) 208
    participatory action research: Bangladesh
       226
    photovoice and democratic research
       processes 383
    social economy, social capital and NGOs:
       gendered perspective 95
    social indicator projects for rural
       communities 276–7, 284
    well-being and technology: Mutaroni village
       (Kenya) case study 134
Bowles, D. 449
Bradbury, H. 226
Bradshaw, T. 25, 40–41
Bratt, R. 115
Braun, V. 412
Brazil 16
Brenman, M. 212
Brennan, M.A. 126, 417
Bridge, S. 94
Brint, S. 225
Broome, K. 296
Brown, D.L. 36
Brown, W. 77
Brundtland Commission 337
Brunick, N. 198
Buccus, I. 386
built capital 41, 205
built infrastructure 277, 278, 281, 343
Bunnell, G. 322
Burns, J.C. 228
Burris, M.A. 384
Burt, R.S. 95
Burundi 349, 350
Bush, G. 108

Cambridge University 128
Campbell, A. 303

Canada 157–60, 161–2, 205, 294; *see also* intentional community building: Reena Community Residence (Canada)
Candy, L. 2
capital theories of schools, poverty and community 446
case study design 233
Casper-Futterman, E. 114, 116
Castle, E.M. 29
Center for Rural Communities (CRC) 280, 283–4, 286, 287
Centers for Disease Control (CDC) (United States) 250, 265, 267
Chakraborty, A. 198
Chambers, R. 225
Chapple, K. 37
Chaskin, R.J. 212
Chaves, R. 94
Chetty, R. 25
China 343
Choice Neighborhoods 442
Christenson, J. 172
citizen participation ladder (Arnstein) 73, 76, 83, 117, 212, 213
civic engagement 4–5
Clarke, V. 412
Cloutier, S. 124, 132
cluster analysis 28
Coastal Georgia Indicators Coalition (CGIC): healthy community building 244–69
   500 Cities Projects 250, 265, 267
   Behavioral Risk Factor Surveillance Survey (2007) 250
   Centers for Disease Control (CDC) 250, 265, 267
   charitable contributions 251, 253–4
   Chatham County Blueprint 262–3, 264, 265, 266
   Chatham County Board of Commissioners 260, 269
   Chatham County Family Connections Collaboration 264–5
   coalition committee and project team members 266
   collective impacts 246, 255–7, 260–64, 268
   committee structure 263–4
   Community Advisory Council 267
   community development (definition) 245–6
   community needs 254
   community services 251, 252
   Community Summit (2014) 262
   Conduent Healthy Communities Institute 267
   content committee 263
   Data Center 267
   data-evaluation and survey committee 263
   'diabetes belt' 250
   disabled and elderly services cuts 251
   disparities dashboard 267
   Disparity data sets 265
   economic factors 254, 259, 263, 264
   education and youth development 251, 254, 259, 263, 264
   employment 251
   Executive Committee 266–7
   feasibility study 257–9, 261
   Form 990 (IRS) 253, 265, 269
   formation 244–5
   funding allocations 254
   Georgia Healthy Cities Project 250–51, 267
   Giving USA Reporting 253, 254
   Great Recession 244–5, 246, 253–4
   Health Tracker 265, 267
   health and wellness 250, 251, 252, 253, 254, 260, 262, 263, 264
   high/low power and high/low interest stakeholders 246
   housing 248–9
   income 251, 254
   indicator development 258–9
   indicator selection 259–60
   indicator system: planning and feasibility 257–8
   Jacksonville Community Council Inc. 262
   kidney disease, chronic 250–51
   Leadership/Executive Committee 263
   local government 251
   maturation from funder collaborative to collective impact initiative 260–64
      Chatham Community Blueprint priorities 263, 264
      governance initiatives 261–2
      partner expansion and recognized community brand increasing community participation 262–3
   Medicaid 249–50
   multi-stakeholder initiatives (MSIs) 255–7
   outreach committee 263
   partners and sponsors 265
   Patient Protection and Affordable Care Act (PPACA) (2010) 245, 249–51, 269
   poverty rates and poverty reduction strategies 251–2, 254
   priority area coalition teams 267
   process change formation 245–6
   ProPublica 269
   public–private partnerships 251, 256, 260, 264, 266, 268
   quality of life 244, 258, 263, 264
   regionalism 255, 260, 263

460 *Research handbook on community development*

replication best practices 268
resource allocation 267
Robert Wood Johnson Foundation 250, 265, 267
Savannah-Chatham background information 247
  economic factors and employment 247
Savannah-Chatham community 244, 246
Savannah-Chatham population demographics 247–51
  health and health inequities 249–51
  poverty rate, per capita income, education and wages 248–9
Savannah-Chatham public and philanthropic resources 251–4
  general fund expenditures 252
  Giving USA charitable contributions 254
  health and welfare expenditures 252
  IRS Form 990 data – United Way of the Coastal Empire 253
social services 251
social-needs index 267
steering/leadership committee 263
Step Up Savannah 252
summary reports 259
sustainability 265–7
top-down approach 258
tourism 247
United Way of the Coastal Empire 244–6, 251, 253, 254, 257–9, 260–61, 265
USA Giving Report 254, 269
Vision 20/20 Blueprint for Community Action report 256–7, 262
Vision 20/20 Commission 250, 255–6, 260, 263
vision statements 263
visioning to collective action 255–7
welfare expenditure 252
written reports 259
Cocoa360 NGO (Ghana) 7, 348–59
  background and history 351–2
  collaborative approach 349
  education 357–8
  farm-for-impact model 348, 351, 352–4, 356, 357, 358
    communal labor voucher card 353
    decision-making 354
    farm labor 352–3
    financial reporting and debriefing 354
  Free Compulsory Universal Basic Education (FCUBE) 351, 357
  future directions 358–9
  Ghana Cocoa Board 355–6
  Ghana Health Service (GHS) Community-Based Health Planning Services (CHPS) 352
  healthcare access 357–8
  impact of work to date 356–7
  Indego Africa (NGO) 350
  limitations and challenges 358
  Lwala Community Alliance (LCA) (Kenya) 350
  Ministry of Education (MOE) 355–6
  Ministry of Health (MOH) 352, 355–6
  One Acre Fund (NGO) 350
  scaling and replicability 358
  social innovation 349, 352
  stakeholders 355–6
    community leaders and community members 355
    funders 356
    Government of Ghana 355–6
    Parent-Teacher Association 355, 357
    Village Committee 352, 355
  Tarkwa Breman Center of Excellence 354–5, 356, 357, 358
  Tarkwa Breman Girls' School 348, 354, 356–7
  top-down approach 348, 349
  Village Health Works (VHW) (Burundi) 350
  Vodacom 356
  Warburg Pincus 356
cognitive/humanist theory 446
Cohen, A. 110, 138
collaborative decision-making 338
collaborative planning 175
collaborative research 228
collective action 255–7, 418
collective engagement 227
collective impacts 8, 430
  Coastal Georgia Indicators Coalition (CGIC): healthy community building 246, 255–7, 260–64, 268
Collins, J. 320
Colombia 343
community capacity building 41, 142
community capital 26, 342
Community Capitals Framework 2, 43
community (definition) 172–3
community design 142
  definition 140–41
  role players 142
Community Development Financial Institutions (CDFI) Fund 191
community development corporations (CDCs) 5, 104, 191, 442
community development defined 173–4, 245–6
community engagement 338
community innovation: local partnerships and sustainable community development 8, 416–35

collective impact 8
community development 416–17
community engagement partnerships 420, 422–3
　designated offices 422
　resources for community partners 423
　volunteer database of list of opportunities 423
community innovation 417–18
design 8
entrepreneurship education 8, 420, 422
　clubs and events 422
　courses, curriculums and programs 422
experiential learning and service learning 418–19, 420–21
　designated courses 420–21
　internships and fellowships 421
　service days or events 421
　student service clubs 421
future research and practice questions 432–3
Geneva 2020 430–32
　outcomes 432
　story 430–31
Hobart and William Smith Colleges (HWS) entrepreneurship 428–30
　curriculum 428–9
　outcomes 429, 430
　story 428
Hobart and William Smith Colleges (HWS) sustainable community development 423–8
　curriculum 425–6
　internships 426
　outcomes 426–8
　story 424–5
ideas, opportunities and practices 423, 424
innovation gateways 433–4
lessons learned 432
liberal arts colleges 418–19, 420, 423
macro-level innovation 416, 418, 419–20
micro-level innovation 416, 418, 423
service learning 418–19
Strive Together 430
community land trusts (CLTs) 196
community leaders 142
community learning 246
community mapping 235
community organizing 139–40
community productivity, *see* sustainable community development and community productivity
community, sense of 185, 292, 449
community social planning 142–3
community street festivals 158
community synonym terms 138

community-engaged mapping 228
community-based innovations 349
community-based organizations (CBOs) 5, 104–21
　Alphabet Scoop Creamery (ASC) (Manhattan) 108–9, 110, 112, 115, 119
　anchor institutions 111–12, 114, 116–17, 121
　anti-displacement policies 114
　art districts 106–7
　Aspen Forum for Community Solutions 442
　BCDI 116
　Benedict Allen (CDC) 107
　Black church CDCs 107–8, 110, 112, 113
　bottom-up approach 116, 118
　Bronx Community Development Initiative (BCDI) 114
　Bronx Zoo 114
　brownfield redevelopment 109
　Burton, Bell, Carr, Development Inc. 106
　capacity, factors affecting 104, 109–12
　　metropolitan networks 110–12
　　organizational factors 109–10
　　social capital 110
　Catholic Campaign for Human Development 112
　Catholic Charities 112
　Center for Neighborhood Development 111
　Central Baltimore Partnership 111
　citizen involvement 119
　citizen participation 104
　Citizens Advisory Committee (Boston) 118
　City of Cleveland Land Bank 105
　CityWide (Dayton, Ohio) 107
　Cleveland County Land Revitalization Corporation 105
　Cleveland Neighborhood Progress 111
　Collective Empowerment Group (CEG) 107–8
　community building 112, 113, 120
　community development corporations (CDS) 5, 104
　community development financial institution (CDFI) 118
　community enterprises (CEs) 104, 114–15
　community land trusts 115
　community organizing 112, 113–14
　community participation 114, 116–19
　comprehensiveness 104, 112–16
　cooperative housing 115
　Detroit Shoreway Community Development Organization (DSCDO) 106, 109–10, 111, 115, 118–19
　Diamond Neighborhood 108
　EcoCity (Cleveland) 106, 110, 111, 119
　economic development 104, 112, 115, 120

EcoVillage (Cleveland) 106, 110, 111, 116, 119
education 113
effectiveness in immigrant communities 110
entrepreneurial and community enterprise-like CDCs 108–9
environmental sustainability 112, 115, 120
equity 104, 112, 113, 114, 115, 120
Evergreen Laundry (Cleveland) 108, 114–15, 121
faith-based and minority-based CDCs 107–8
finance institutions 115
financial and economic feasibility 120
foreclosure crisis 105
Forest City Enterprises Inc. 111
gentrification 108–9, 113, 120
Gordon Square Art District (GSAD) (Cleveland) 106, 111, 119
government funding 115
Great Financial Crisis (GFC) 105
housing affordability 112–13, 116
housing production 105, 112, 113, 120
hybrid nature and institutional logics 104
Jackson Square CDC (Boston) 117–18
Jacobs Foundation 108
Jubilee Inc. 111
Lawrence Community Works (LCW) (Massachusetts) 116
lessons learned 120–21
Low-Income Housing Tax Credit (LIHTC) 115
Lowells' Coalition for a Better Acre (CBA) 105
Mary Queen of Vietnam Community Development Corporation (New Orleans) 106
Massachusetts Association of Community Development Corporations (MACD) 111
Metropolitan Community Health Services and Agape Community Clinic 110
Metropolitan Credit Union 110
moral embeddedness 115
nonprofits, workers cooperatives, community banks 104
Office for Community Development (OFCD) (Philadelphia) 107, 111–12, 113, 117
Partnership for New Homes (New York) 112
position and scope 104
production 113–14
Public Theatre and Near West Theatre (Cleveland) 111
regional CDCs 107
revitalization impacts 104, 119–20
RIK Enterprises LLC 111
Slavic Village Development (Cleveland) 105, 109, 111
SNAED (Baltimore) 109, 111, 113–14, 116, 120
social enterprises 104
social impact 120
social networks 110, 117
social services 112
strong leadership 110
The Neighborhood Developers (TND) 105
top-down approach 104, 114, 117, 118, 119
urban agriculture and environmental sustainability 106
Urban Innovation 21 112
viability 104, 112–16, 120
White House Office of Faith-Based and Neighborhood Partnerships 108
worker cooperatives 115, 116
competing interests management, *see* public participation process and competing interests management: residential displacement (Buffalo, New York)
component theory 393
Comprehensive Community Initiatives (CCIs) 43, 205, 442
concentric zone theory 37
connectivism theory 446
constructivism theory 446
content analysis 361, 363
Cooke, B. 79
Costa, P.M. 117–18
Craig, G. 226
Crandall, M.S. 28, 33
creativity 22; *see also* aesthetics and community development
critical consciousness 13, 16–18, 19, 21–2
critical social analysis 13, 17, 19
Crocker, D. 411
Crowley, N. 12
cultural capital 41, 205
cultural factors
  aesthetics and community development 392
  Asset-Based Community Development (ABCD) 70, 71, 72, 74
  neighborhood environment satisfaction in deteriorated areas: impact of socioeconomic characteristics 302, 303
  sustainable community development and community productivity 340, 341
  universities 407
  well-being and technology: Mutaroni village (Kenya) case study 124, 130, 132–3, 134
Cummins, R.A. 274–5
Cummins, S. 293

Curtis, K.J. 33
cycle of praxis 13, 18–22
  catalytic action 18–19
  walking the cycle 19–22
Czíkszentimihályi, M. 393

data collection and analysis framework 235, 238, 361
Davidoff, P. 211, 212
Davis, G. 409
Deal, N. 249, 269
DeFilippis, J. 114, 116
DeLugan, R.M. 244
democratic research processes, *see* photovoice and democratic research processes
Dempsey, N. 409
Denmark 343
Dennis, R. 448
Denzin, N.K. 388
depopulation hypothesis 38
descriptive methods 238, 304
descriptive results 306–8
Deutsch, F. 12, 15
Dewey, J. 446
diary and personal logs 228
Diaz, R. 325
Dickmann, P. 383
Diener, E. 362
Dong, H. 303
Downes, S. 446
Downey, L.H. 226–7
Dubb, S. 114–15
Dwyer, R.E. 26, 36, 38

East Africa 350
eco-efficiency theories 338
ecological factors 173, 175, 287
economic capital 143
economic factors
  aesthetics and community development 392
  Coastal Georgia Indicators Coalition (CGIC): healthy community building 247, 254, 259, 263, 264
  community innovation: local partnerships and sustainable community development 416
  community-based organizations (CBOs) 104, 112, 115, 120
  food deserts: Utica, Mississippi 294
  group formation: Grameen Bank (Bangladesh) 151
  neighborhood environment satisfaction in deteriorated areas: impact of socioeconomic characteristics 303

place attachment: inductive social media approach 361
  poverty in United States: scope, scale and place 25, 39
  social economy, social capital and NGOs: gendered perspective 5
  social indicator projects for rural communities 6, 277, 278, 280, 287
  sustainable community development and community productivity 338–9, 340
  well-being and technology: Mutaroni village (Kenya) case study 124, 128, 129–33
  see also socioeconomic factors
economic impact assessment 3
economic transformation or skills mismatch hypothesis 37
economic-based theory 72
Edmonds, E. 2
education factors 16, 36, 38, 40
  Coastal Georgia Indicators Coalition (CGIC): healthy community building 248–9, 251, 254, 259, 263, 264
  Cocoa360 NGO (Ghana) 357–8
  community-based organizations (CBOs) 113
  dialogical education 17–19
  entrepreneurship 8, 420, 422
  faith-based community development and evaluation capacity enhancement: ENLACE (El Salvador) 206
  food deserts: Utica, Mississippi 292, 299
  group formation: Grameen Bank (Bangladesh) 163
  inclusionary zoning and inclusionary housing 198
  neighborhood environment satisfaction in deteriorated areas: impact of socioeconomic characteristics 306, 307, 311, 312–13, 314
  participatory action research: Bangladesh 225, 226, 227
  photovoice and democratic research processes 382
  poverty in United States: scope, scale and place 25
  social indicator projects for rural communities 275, 276, 277, 280, 283
  universities and community development partnerships 441, 442
  well-being and technology: Mutaroni village (Kenya) case study 128, 131, 133
Eisenhower, D. 113
El Salvador, *see* faith-based community development and evaluation capacity enhancement: ENLACE (El Salvador)
Elizalde, A. 408

Ellen, I.G. 28
Ellickson, R.C. 198
employment factors 36, 40, 442
　Coastal Georgia Indicators Coalition (CGIC): healthy community building 247, 251
　neighborhood environment satisfaction in deteriorated areas: impact of socioeconomic characteristics 306, 307, 311, 312, 313, 314
empowerment 58
　Asset-Based Community Development (ABCD) 73
　nonprofit organizations and community participation 77, 78–9, 86
　participatory action research: Bangladesh 6, 225, 240
　photovoice and democratic research processes 383, 386–8
　place attachment: inductive social media approach 7
　public participation process and competing interests management: residential displacement (Buffalo, New York) 212
　reflection, action and knowledge creation: lived experience 4
　well-being and technology: Mutaroni village (Kenya) case study 124, 128, 133
environmental factors
　aesthetics and community development 392
　community innovation: local partnerships and sustainable community development 416
　food deserts: Utica, Mississippi 294
　group formation: Grameen Bank (Bangladesh) 151
　intentional community: Reena Community Residence (Canada) 173, 182–3
　neighborhood environment satisfaction in deteriorated areas: impact of socioeconomic characteristics 302, 303, 304, 309
　poverty in United States: scope, scale and place 25
　social indicator projects for rural communities 275, 276, 277, 278, 281
　sustainable community development and community productivity 338–9, 341
environmental health 304, 310, 310, 311, 312, 314
environmental justice 50, 64, 226
equity 6, 100, 321, 341, 454
　community-based organizations (CBOs) 104, 112, 113, 114, 115, 120
　public participation process and competing interests management: residential displacement (Buffalo, New York) 211–12, 213, 221
　see also under just communities: marginalized communities and regional sustainability planning
Escande, A. 275
ethics 96, 232, 235, 239, 388
ethnicity, *see* race/ethnicity
Etienne, H. 105
Europe 124
evaluation capacity enhancement, *see* faith-based community development and evaluation capacity enhancement: ENLACE (El Salvador)
evaluation and indicator models 16
Evans-Cowley, J.S. 364
evidence-based approaches 3
experience, shared 19–20
experiential knowledge 385–8
experiential learning 226; *see also under* community innovation: local partnerships and sustainable community development

face-to-face fieldwork 134
Facebook 132–3, 160
faith-based community development and evaluation capacity enhancement: ENLACE (El Salvador) 6, 204–9
　ADESCOs (CD associations) 205, 207
　asset-based approach 205
　baseline studies 207–8
　bottom-up approach 208
　catalysts for change 208
　challenges 206–7
　Church and Community Program 205, 208, 209
　community capacity building 207
　community *informes* (reports) 208
　Comprehensive Community Initiatives (CCIs) 205
　core programming 205, 208
　Department of Evaluation and Research 207
　evaluation and monitoring tools 208
　external focus 209
　information sharing and feedback 209
　internal focus 209
　internal ministries 205
　lessons learned 207–9
　　methodological rigor, maintaining 209
　　prioritization of strategic time-savers 208–9

Index 465

reciprocal relationships building 209
redirection of outcome assessment energies 208
repurposing of instruments for learning 207–8
local development initiatives 205
mobile data collection technologies (Magpi) 208
Multidimensional Poverty Index 206
organizational ties 205
outcome mapping approach 208
reflection, feedback and adaptation 207
strengthening community capacity 204–6
Favreau, L. 94
Fellin, P. 138
felt needs 15, 19–21, 22, 410
Fendley, K. 172
Fichter, R. 142
field notes 228, 231
field-based examples 240
filed selection 233
financial capital 41, 205
Fink, M. 332
Fisher, M.G. 29
Flickr 7, 365, 368, 371, 372
Flora, C.B. 41, 281, 440
Flora, J.L. 281
Florido, A. 108
Floro, M.S. 96
focus groups 257, 452
 participatory action research: Bangladesh 228, 230–31, 236–7
 public participation process and competing interests management: residential displacement (Buffalo, New York) 6, 215, 217–18, 220
Fontan, J.-M. 94
food deserts: Utica, Mississippi 6–7, 290–99
 background and history 291–3
 businesses, lack of 292
 Central Mississippi Civic Improvement Association 298
 community resources 298–9
 community, sense of 292
 Department of Agriculture (USDA) 290, 293, 294, 295
  Food and Nutrition Service (FNS) 295
 diversity, lack of 292
 educational attainment 292, 299
 environmental considerations 294
 Environmental Education Grants Program 295
 Environmental Protection Agency (EPA) – Environmental Education Division (EED) 295
 Federal Health Centers program 298
 food enterprises 295–6
 geographic location 299
 grassroots movement 298
 Green Thumb Challenge 295
 health center movement 298
 health disparities 293–4
 Hinds County Human Resource Agency (HCHRA) 296
 implementation grants 295, 296
 infrastructure limitations 299
 land use patterns 299
 Local Food Promotion Program (LFPP) 295–6
 nonprofit organizations 299
 Office of Children's Health Protection and Environmental Education 295
 overview of food deserts 293–4
 People's Garden School Pilot Program 295
 planning grants 295–6
 poverty rate 294
 school and community garden programs 295, 299
 senior citizens 296
 small size of Utica 292
 socioeconomic status 6–7, 299
 sustainability 292
 Title XX Transportation Program 296
 transportation 296–8
 undeveloped land 292
 young population 292
Fordham University 114
Fortunato, M.W.P. 434
Foster-Fishman, P. 387–8
Foucault, M. 79, 393–4
Fox, R. 48
Frascati, M. 110, 115
Freire, P. 15, 16–17, 226, 384, 446
Friedmann, J. 321
Fujii, Y. 105
funding 115, 198
 Asset-Based Community Development (ABCD) 69, 74
 Coastal Georgia Indicators Coalition (CGIC): healthy community building 254
 inclusionary zoning and inclusionary housing 191, 200
 nonprofit organizations and community participation 87, 89

Gallardo, R. 42
Galser, G. 409
Gearant, M. 430–31
gender factors 14, 25, 95–6

neighborhood environment satisfaction in deteriorated areas: impact of socioeconomic characteristics 303, 306, 307, 311, 312, 313, 314
  see also social economy, social capital and NGOs: gendered perspective
gentrification 50
  community-based organizations (CBOs) 108–9, 113, 120
  just communities: marginalized communities and regional sustainability planning 58, 60, 64
  public participation process and competing interests management: residential displacement (Buffalo, New York) 213, 214–15, 216, 217, 218, 219–21
German, L.A. 232, 240
Germany 343
Ghana 349, 350; *see also* Cocoa360 NGO (Ghana)
Gillis, A. 231
Glaeser, E.L. 446
Glover Blackwell, A. 48
Godschalk, D. 51
Goetz, S.J. 42
Golden, K. 451
Gomez-Novy, J. 325
Goodling, E. 50
Google 356
Gordon, T. 251
Grameen Bank, *see* group formation: Grameen Bank (Bangladesh)
Grant, U.S. 291
grassroots movement 4, 7, 298, 387, 388, 391
Green, G.P. 417
Greene, R.P. 38
Griffin, G. 364
group formation: Grameen Bank (Bangladesh) 5, 137–70
  Bidimalla (bylaws) 143–7, 152, 154, 156, 170
  Branch Manager 155
  Center Chief 144, 147, 153, 155, 157
  Chairman 143–5, 150
  Co-Center Chief 144, 155
  community building 142
  community design: definition 140–41
  community design: role players 142
  community organizing 139–40
  community social planning 142–3
  community synonym terms 138
  disruptive defaulters and expulsion or resignation 144–6
  duties and responsibilities 144–6
  education promotion 163
  Emergency Disaster Funds 146, 155–6, 163, 166–7
  field manager 155
  flexible loan (*chukti rin*) 166
  Form-1 150
  formation design 143–4
  functions of members 156–7
  general assembly (*shadaran parishhad*) 143
  grading system 169
  Grameen America 158
  Grameen Pension Scheme (GPS) 146, 165, 167
  Group Chairman 147, 150, 155
  Group Funds 145, 146, 150, 154–5, 163, 166
  Group Secretary 150
  group training work 152
  housing loans 160, 162–3
  insider information 142
  instalments 156
  insurance policies 166–7
  loans products and implementation strategies 156, 160–69
  Luhuria Center 143, 152–3
  Luhuria Center School 149, 153
  minimum membership 145
  moneylenders, elites and community agencies 150–54, 170
  new initiatives 156
  Nitimallas (operation instructions) 146
  objectives 137
  participatory decision-making process 138–9
  Phase-I 137, 154–6, 165, 166
  Phase-II 137, 154
  poverty rates 164
  project champion 142
  Purdah 151, 154
  repayment procedures 156, 166
  responsibility, sense of 156
  rules and procedures 166–7, 169
  Samassha (Problematic) Cell 165
  savings products 146, 154–5, 165, 167, 168–9
  Secretary 143–5, 155
  Sixteen decisions (social development charter) 163
  staff incentives and promotions 167, 169
  staff morale improvement 166
  structural faults 165
  subversive activities 145
  taxation 155
  timeline development 141
  training 147, 156
  Village Landless Association 143, 144
  women borrowers 152, 162
  working experience 146–54
    elites 153–4

faith-based issues and Purdah 151, 153–4
group and Center School organization 148–54
moneylender, Imam and Mattabarr (village leader) 150–54
survey and organization of members 147–8
written contract (*chukti*) 166
Guatemala 204
Guise, R. 409
Gunder, M. 50

Habraken, J. 142
Hall, E.T. 396
Hall, P. 97
Hallman, H. 409
Hannscott, L. 306
Hart, R.A. 78
Harvey, M.H. 118
Haveman, R. 41
health factors 27
   Coastal Georgia Indicators Coalition (CGIC): healthy community building 249–51, 252, 253, 254, 260, 262, 263, 264
   Cocoa360 NGO (Ghana) 357–8
   faith-based community development and evaluation capacity enhancement: ENLACE (El Salvador) 206
   food deserts: Utica, Mississippi 293–4, 298
   intentional community: Reena Community Residence (Canada) 172, 178, 179, 181, 185
   just communities: marginalized communities and regional sustainability planning (United States) 58
   nonprofit organizations and community participation 76, 77, 79
   photovoice and democratic research processes 382
   social economy, social capital and NGOs: gendered perspective 98
   social indicator projects for rural communities 277, 278
   universities and community development partnerships 442
   well-being and technology: Mutaroni village (Kenya) case study 128, 129
healthy community building, *see* Coastal Georgia Indicators Coalition (CGIC): healthy community building
Healy, S. 96
Heart, Father 119
Hendren, N. 25

Hendrickson, D. 293
Hernandez, M.D. 244
Hexter, K. 120
Hickey, R. 201
Hills, P. 320
Hirschl, T.A. 36
Hobart and William Smith Colleges (HWS), *see under* community innovation: local partnerships and sustainable community development
holistic approach 322, 337, 338
Holliday, A.L. 26, 36, 38
Holmqvist, A. 198
Hopenhayn, M. 408
Hossein, C.S. 96
housing, *see* inclusionary zoning and inclusionary housing
Howaldt, J. 349
Hu, L. 37
Hughen, W.K. 198
human capital 26, 27
   Asset Based Community Development (ABCD) 74
   faith-based community development and evaluation capacity enhancement: ENLACE (El Salvador) 205
   food deserts: Utica, Mississippi 292
   poverty in United States: scope, scale and place 42
   sustainable community development and community productivity 343
   universities and community development partnerships 446
Humphreys, J. 247
Hunting, D. 327–8, 332

Ibem, E.O. 303
ideologies 417–18
inclusionary zoning and inclusionary housing 5–6, 189–201
   affirmative measures 189
   Affirmatively Furthering Fair Housing (AFFH) rule (HUD) 198
   affordability 5, 189
   air pollution 189
   AMI 197
   anti-snob land use laws 190
   Assessments of Fair Housing (AFHs) 198
   below-market rates 195
   Budget Act (2011) 194
   business returns 6
   cash subsidies 194
   CDBG funding 198
   Community Development Block Grants (CDBG) program 191

Community Development Corporations (CDCs) 191
Community Economic Development (CED) initiatives 339
commuting 189
cost offsets 197
deed covenants and deeds of trust 195
density bonuses 193, 197, 199
Department of Housing and Urban Development (HUD) 191, 192
Department of Transportation (DOT) 192
Department of the Treasury 191
design and implementation 197
development fees, waived 194
economic integration 198
elasticity of supply and demand 197
Environmental Protection Agency (EPA) 192
essential workers 194
exclusionary zoning 189
expenditure-to-income ratio 201
fair share allocation systems 190
Federal Housing Administration (FHA) 195
Federal Rehabilitation Tax Credit (Historic Tax Credit) 191
first refusal rights 195
foreclosure 195
funding 191, 200
Historic Preservation Office (USA) 191
HOME Investment Partnership program 191
house prices 198
Housing Choice Voucher Program 192
housing market forces 189
Housing Opportunities Commission (HOC) (Maryland) 199
housing planning requirements 190
housing production 198
in-lieu fees 193
income averaging/tiering 197
inflexibility 200
inputs measurement 192–6
　affordability period 195–6
　developer requirements and incentives 192–5
Internal Revenue Service (IRS) 191
jobs imbalance 189
land dedications to land trusts 193
limited equity cooperatives (LECs) 196
local building code 193
local or regional economy 197
local worker shortages 189
low base zoning 199
Low Income Housing Tax Credit (LIHTC) program 191, 196, 201
mandatory policies 196–7
market-rate developments 193

market-rate units 191
mixed outcomes 198–9
mixed-income communities 190
National Low Income Housing Coalition (NLIHC) 196
National Park Service (USA) 191
negative outcomes 198–9
New Market Tax Credit (NMTC) 191
nonprofit housing developers 196
off-site development through linkages 193
opt out or cash out of a program 197
outcomes measurement 196–9
　beneficiary outcomes 197–8
　community outcomes 199
　housing stock outcomes 198–9
Palmeri/Sixth Street Properties, L.P. vs. the City of Los Angeles decision (2009) 194
Partnership for Sustainable Communities 192
permanent affordability 196
political will 197
positive outcomes 198–9
poverty rates 198, 199
pre-purchase and post-purchase stewardship practices 196
price controls 195, 200
public engagement 197
public policy 190–92
race/ethnicity issues 189, 198
Redevelopment Agencies (RDAs) (California) 194
relaxed planning or design standards 194
rent control laws, violating 194
resale formulas/restrictions 195
residual income approach 201
school attendance and test performance 198
set-asides 193, 194
socioeconomic integration 5, 190, 198
special accommodations 194
strategic partnerships 196
structural engineering 193
subsidies 191, 200
Sustainable Communities Regional Planning Grant program 192
Takings Clause (Fifth Amendment) 194
taxpayer expenditures 189
thresholds to exempt small developments 194
TIF 201
traffic congestion 189
transportation costs 189
urban sprawl 189
voluntary policies 196
income factors 25, 27, 28, 29–30

Coastal Georgia Indicators Coalition (CGIC): healthy community building 248–9, 251, 254
   inclusionary zoning and inclusionary housing 190, 197, 201
   neighborhood environment satisfaction in deteriorated areas: impact of socioeconomic characteristics 306, 307, 311, 312, 314
   social indicator projects for rural communities 280, 283
   universities and community development partnerships 441
inductive social media approach, *see* place attachment: inductive social media approach
Inglehart, R. 12, 15
inside-out community building 17
institutional ethnography 80
intentional community: Reena Community Residence (Canada) 5, 172–87
   Activities of Daily Living services 178
   administrative integration 185
   age range 181
   background 174–5
   the building, views on 178, 185
   Circle of Care 175, 185
   collaborative planning 175
   communication, internal and external 185
   community development defined 173–4
   community groups 173
   community, sense of 185
   confidence 182
   defining community 172–3
   ecological and sociological perspective 173, 175
   environmental characteristics 173
   findings 181–4
      limitations 184
      person-environment fit 182–3
      residential space 182
      strengths and weaknesses 183–4
   freedom 180
   genetic, lifestyle and social determinants 174
   health and well-being, self-perceived 172, 178, 179, 181, 185
   housing stability 172
   inclusion 174, 180
   independence 180
   integrated support services 174, 177
   intellectual/developmental disabilities 174–6
   lived-experience 176
   March of Dimes Canada 175, 185
   neighbourly reciprocity 184
   newsletter 185, 186
   nonprofit agencies 172
   open concept and fortress mentality 182
   performance 173
   person-environment fit 172–7, 181, 182–3, 184–5
   person-staff fit 173
   pets 178, 181, 185
   quality of life 172
   recommendations 186
   research project 175–81
      expected project outcomes 177
      interview results 178–9
      methodology 176–7
      observations 178–81
      scope and objectives 176
      tenant interviews 177–8
      tenants' views 179–80
   safety and security, smoking and fire drills 180
   St Elizabeth Health care 175, 185
   satisfaction 173
   screening for new tenants 186
   self-directed care 177, 180
   social action process 173
   social capital 172, 173, 175, 181, 182, 184, 185
   social interaction 184
   social participation 173, 175, 181, 184–6
   social vision 173
   spatial behavior and housing 172
   sports and recreation 178, 181, 185
   staff training and support 185, 186
   statement of values – CLEAR 174
   successful aging 174, 176, 177, 184, 186
   support services 178, 181, 185
   Tenant Council 180–81, 182, 186
      Principles 180
      roles 181
   tenant experiences 178, 185
   volunteerism 184
   well-being 172–3
interactional community development 3
interactional theory 391
interviews 228, 231–2, 235–6, 237–8, 448
Iran, *see* neighborhood environment satisfaction in deteriorated areas: impact of socioeconomic characteristics
Isserman, A.M. 35
Iwarsson, S. 173

Jackson, W. 231
Jenks, M. 409
Johns Hopkins University 111, 113, 116–17
Johnson, S.P. 115
Jokela, M. 303

Jurkowski, J.M. 387
just communities: marginalized communities and regional sustainability planning 4, 48–65
　APA criteria 58
　approach and methods 50–55
　　credibility 50–51
　　methods triangulation 50–51
　　mixed-methods approach 50
　　participant observation 51
　　plan evaluation methods 51
　　qualitative research 50–51
　　thematic analysis of reports: perspectives of HUD, grantees and capacity builders 51, 55
　assessments 55
　capacity building 62
　Capital Region grantee 61
　challenges and progress 64–5
　Chicago Metropolitan Agency for Planning 61
　Community Engagement Team (CET) 61
　data, deliberation and decision-making (three Ds) 49
　discrimination/structural racism 60, 62
　diverse regions and diverse community development needs 64
　engagement 51, 59, 61, 64
　　inclusive 51, 56, 58–9
　environmental justice 64
　equity 50, 51, 56, 57, 58, 61, 63, 64, 65
　Fair Housing Act 49
　Fair Housing and Equity Assessment (FHEA) 49, 58
　Final GTR (Grant Technical Representative) Performance Assessments 51, 55
　financial investments in engagement processes 61
　findings 55–61
　　plan evaluation: inclusion of equity 55–9
　　thematic analysis of equity concerns: grantee and HUD perspectives 59–61
　gentrification 58, 60, 64
　Grantee Fair Housing Equity Assessments 51
　Grantee Final Regional Plans 51
　Grantee Final Reports 51
　health 58
　high capacity/high performance 61
　'hot market' metropolitan regions 64
　housing affordability 48, 60
　housing, fair 64
　Housing and Urban Development (HUD) 48, 59, 61, 62, 63, 65
　Sustainable Communities Regional Planning and Community Challenge grant program 49
　immediate action 58
　livability principles 49
　localism 62
　member checking 51
　Metropolitan Council (Twin Cities) 59–60, 61
　metropolitan planning organizations (MPOs) 61
　narrative report 55
　Notice of Funds Available (NOFA) 59
　observation, persistent 51
　plan evaluation criteria and scoring rubric 53–5
　political challenges 60
　professional, fiscal, legislative and political constraints 50
　progressive regionalism and regional community development needs 4, 48–50
　racial equity 48
　racial or ethnic concentrated areas of poverty (RCAP/ECAP) 49
　regional equity 48, 62–3
　regional linkages 48
　resource sharing 48
　rural communities 64
　scoring guidelines 51
　segregation 48
　self-empowerment 58
　social equity 48, 59, 60, 62, 64
　Southeast Michigan Council of Governments (SEMCOG) 60, 63
　Southern Bankcorp (Arkansas) 58
　southern strategy race 62
　Sustainable Communities Initiative (SCI) 48–51, 52, 55, 58, 59, 61–2, 63, 64, 65
　sustainable development challenges 50
　Tea Party resistance 62
　thematic analysis of reports: perspectives of HUD, grantees and capacity builders 51, 55
　Thunder Valley plan for Pine Ridge Reservation 58–9
　transit-oriented development (TOD) 58, 60
　triangulation 51
　tribal areas 64
　tribal grantees 58
　'weak market' metropolitan regions 64
　'White flight' 62–3
just society 275
justice 4
　environmental 50, 64, 226
　health 76
　social 50, 82, 226, 446, 454

Kallus, R. 410
Kania, J. 244, 246, 255–6, 260–61, 263, 268, 430
Keating, D. 111, 115–16
Kellogg, W.A. 111, 115–16
Kenny, S. 1
Kenya 349, 350, 384; *see also* well-being and technology: Mutaroni village (Kenya) case study
Kesby, M. 79
key informants' interviews (KIIs) guideline 237–8
Keyes, L. 106
Khadduri, J. 196
Kirkpatrick, O. 115
Kirubi, C. 126
Kitzinger, J. 230
Knaap, G.-J. 198
knowledge 4, 21–2
knowledge creation 12, 18, 19, 20, 21, 224; *see also* reflection, action and knowledge creation: lived experience
Koch, T. 227
Kodama, M. 395
Kopp, R. 349
Kothari, U. 79
Kramer, A.D. 362
Kramer, M. 244, 246, 255–6, 260–61, 263, 268, 430
Kretzmann, J.P. 4, 67, 70, 72, 74
Krumholz, N. 118–19, 120, 211
Kwon, S.-W. 95

Landin, M.C. 291
Landolt, P. 97
Lao Tzu 74
Laos grass-roots cooperative case study 93, 97–101
Latin America 226, 407, 408, 411, 413
Laverack, G. 79
Law-Yone, H. 410
Lawler, E.E. 1–2
Lawton, M.P. 172–3
leadership model 74
learning 3, 4, 20–21
    active 446
    problem-based 446
    transformative 17
    see also experiential learning
Levesque, B. 93
Lewin, K. 224
Lichter, D.T. 33, 36
Lieberman, A. 446
Lincoln, Y.S. 388
LinkedIn 160

livability; *see* revitalization, livability and quality of life: Tuscon (Arizona)
lived experience 16, 17, 22; *see also* reflection, action and knowledge creation: lived experience
local government 142, 213–14, 216–17, 219–21, 251, 409
López-Ornelas, E. 364
Lovelace, R. 363
Lowndes, V. 97
Lowrie, K. 109
loyalty 225, 409
Lu, B. 313
Lynch, K. 26, 322

McAslan, D. 322
Macaulay, A.C. 227
McCallum, A. 79
MacDonald, C. 227
MacEachren, A.M. 363
McGovern, S.J. 212
McGrath, B. 1
Macintyre, S. 293
McKnight, J. 4, 70, 72, 74
McKnight, J.L. 67
McRoberts, O. 110
McTaggart, R. 238–9
Madar, J. 199
Maier, P.O'B. 198
Mali 384
Mandell, J. 116
Manpower Development Training Act 27
marginalized communities, *see* just communities: marginalized communities and regional sustainability planning
Markson, J. 182
Markusen, A. 320
Marshall, C. 231
Marston, S. 325–6
Martin, E. 430
Matarrita-Cascante, D. 126, 134, 417
Mattingly, M.J. 29
Max-Neef, M. 408
Mayo, E. 94
Merchant, K. 430
Michigan University 279
Miller, D. 125
Miller, K.K. 29, 35, 43
Miranda, R. 452
Mitchell, L. 364
mixed methods approach 3
mobile data collection technologies (Magpi) 208
Mohrman, S.A. 1–2
Molyneux, M. 97

Monzón, J.L. 94
Mook, L. 93
Morales Zaragoza, N. 364
Morgan, D.L. 231
Moulaert, F. 94
Mouratidis, K. 301
Mukhija, V. 199
Mulhall, A. 231
Mullolly, J.J. 244
multi-stakeholder initiatives (MSIs) 255–7
multivariate statistics 361

Narayan, D. 95
narrative report 55
Needleman, C.E. 211
Needleman, M.L. 211
neighborhood environment satisfaction in deteriorated areas: impact of socioeconomic characteristics 7, 301–15
　access to public facilities 304, 310, 311, 312, 314
　age 306, 307, 311, 312, 314
　Astan Quds 308, 315
　belonging, sense of 312
　character, culture and lifestyle 302
　community status 304, 310, 311, 312, 313, 314
　concentrations 7
　density 7
　descriptive results 306–8
　　physical–spatial characteristics of housing 308
　　socioeconomic (individual household) characteristics 306–7
　design-physical approaches 302, 315
　education 306, 307, 311, 312–13, 314
　employment status 306, 307, 311, 312, 313, 314
　Endowments Organization 308, 315
　environmental health 304, 310, 311, 312, 314
　environmental quality 302
　expectations, low 312
　family income 306, 307, 311, 312, 314
　gender 306, 307, 311, 312, 313, 314
　household size 311, 312, 314
　housing quality 304, 310, 311, 312, 313, 314
　housing satisfaction index 314
　identification and classification of satisfaction indicators 309
　indicators and physical variables 302
　individual characteristics 303, 314
　inferential results 308–14
　length of residence in neighborhood 311, 312, 313, 314
　living conditions 303
　marital status 306, 307, 311, 312, 313, 314
　Mashad city and Ab-Kooh neighborhood (Iran) study area 302, 305, 305, 308, 309, 311, 313
　meaningfulness of built environment 302
　mental and behavioral factors 302
　methods 304–5
　Neighborhood Environmental Satisfaction (NES) Indicators Measurement 308–11, 310, 311–14
　neighborhood security 304, 310, 311, 312–13, 314
　objectivity 303, 308
　perceived neighborhood environment and life satisfaction 303–4
　personality behaviors 304
　physical and environmental aspects 302
　product-oriented approach 315
　psychological–geographical approach 302–3
　qualitative characteristics 306
　relationships 7
　residential satisfaction and quality of housing 301
　risk-taking and security 301
　social environment 301–2
　socio-spatial approaches 302
　space behavior in social environment 303
　Statistical Center (Iran) 302
　subjective approach 302–3, 309
　transportation and mobility 301, 304, 310, 311, 312, 314
　users' needs and aesthetic values 315
neoliberalism 96, 104, 349
Nepal 97, 204
Netherlands 343
network theory of change 205
networked approach 3
Neuman, M. 339
Neumeier, S. 349, 352
Newman, O. 182
Ng, M.K. 320
Nielsen, F.A. 367
Nigeria 349
Nijstad, B.A. 394
Nitschke, G. 396
Nixon Administration 62
Noguera, P. 446
nonprofit organizations and community participation 4–5, 76–91
　accommodation 82–3, 84, 88
　accountability 78, 87, 89–90
　action committees 81
　analytical framework 80
　barriers to participation 84
　capacity building strategies 90

change 90
citizen control 78
citizen participation ladder 76, 83
co-design 82, 83, 84, 88–9
co-production 82, 83, 84, 88–9
consultation 83, 84, 88, 89
decoration and manipulation 83, 88, 89
delegated power 78, 82, 83
depth and breadth of participation 79
efficiency and professionalism 78
empowerment 77, 78–9, 86
enabling participation 80, 84
enclosure and dispossession 77
extractivism 77
funding 87, 89
governmentality framework 76, 79, 80
health care service-user participation 79
health justice 76
health promotion 77
informing 78–9, 83, 84, 88, 89
maintaining participation 87–9
managing participation 87, 88, 89
manipulation 83
meaningful participation 82, 83–4, 88, 89
Neighbourhood Table case study – residents versus agencies 80–84, 86, 86, 87, 89, 91
neoliberal restructuring 77–80, 90
non-participation 83, 88
objectives and indicators 84
participation 5, 78–9, 82, 86
participation fatigue 90
partnership 78, 82
partnership assessment wheel 85
performance measurement 89–90
placation 82, 83
policy decisions 87
politics of participation 78
power 76–9, 86–7
quality and levels of participation 83, 87, 88
self-evaluation tool for action in partnership 76, 84–5
social determinants of health 77
social justice 82
Steering Committee 81
sustaining participation 86, 87–9, 90
symbolic participation 83–4, 88
technologies of participation 89
understanding participation experiences for different types of partners 84–7
units of service 90
well-being 77
Northwoods Quality of Life (NWQoL) Database, *see under* social indicator projects for rural communities
Not in My Back Yard (NIMBY) 199

Obama Administration 4, 48–9, 108, 249, 430, 442
objective approach 9, 274, 303, 308, 342
observational methods 31, 235
    participant 51, 228, 231
O'Neill, T. 199
Oosterling, H. 396
O'Regan, K. 28
Orfield, M. 198
Organization for Economic Cooperation and Development (OECD) 124, 293
Orr, D. 341
Osborne, D. 261
outcome mapping approach 208
outcomes perspectives 141, 245, 246
outreach techniques 141

Pan, E. 139
participant observation 51, 228, 231
participatory action research: Bangladesh 6, 224–41
    action 226, 227
    active citizenship 225
    Advancement Project – Healthy City (2011) 228
    benefits 238–9
    bottom-up approach 226
    challenges 240
    collaborative research 228
    collective engagement 227
    common values 225
    commune and procedural level challenges 240
    community-based participatory research (CBPAR) 224, 228, 229, 230
    conceptual and theoretical context 225–7
    consent 235, 239
    cooperation 226
    data analysis and evaluation 232, 233, 234, 235, 238
    education 225, 226, 227
    empowerment 6, 225, 240
    equalities 226
    ethics 235, 239
    framework and methodology 228–32
        focus group discussions (FGD) 228, 230–31, 236–7
        interviews and in-depth interviews 228, 231–2, 235–6
        key informants' interviews (KIIs) guideline 237–8
        participant observation 228, 231
    guiding principles 227
    interviews 228, 235
    moderator 230

note taker 230–31
participation continuum 228, 230
planning step 232–3, 234
popular theater 226
preparation step 232, 234, 235
problem-solving capacity 227
public service provision 226
qualitative research 224, 227, 231, 233, 235, 240
research ethics 232
subjective experience and circumstance 224
team and partnership building 232
thematic approach 238
theory and practice 226
top-down approach 226
validation 235
Partridge, M.D. 28, 34, 39, 40
Pastalan, L. 172
Patel, L. 349
Paterson, R. 50
Patterson, K.L. 115, 117
Patton, M.Q. 246
Paulus, P.B. 394
Paumier, C. 322
pen-pictures 238
Peña, M. 407
performance measurement 89–90
Peter Lougheed Leadership Initiative (University of Alberta) 400–403
Peters, D.J. 28, 33–4, 39
Peterson, G. 319, 329
Pfeiffer, D. 124, 132
Phillips, R. 1, 41, 245
Phillips, S.D. 172
photovoice and democratic research processes 7–8, 382–8
  accountability 388
  best practices 383
  bottom-up approach 383
  centralized, hierarchical model 382
  community building 386
  consent 388
  empowerment 383, 386–8
  ethics 388
  experiential knowledge 385–8
    participant-directed agenda 386
    political advocacy 386–8
  feminism 384, 385
  formative and outcome evaluations 382
  grassroots activism 387, 388
  insider perspective 388
  international development 382
  needs assessment 382
  participatory models 383, 386
  poverty alleviation 382–3, 386–7
  problem-based inquiry 387
  public forum 387
  qualitative research 382, 387, 388
  quality of life 382, 386
  self-efficacy 386
  top-down approach 382–3
  United States 384
  Western Institutional Review boards 388
physical capital 26
Piaget, J. 446
Pittman, R. 41, 245
place attachment: inductive social media approach 7, 361–75, 378–81
  community feedback 371, 375
  content analysis 361, 363
  curatorial approach 364
  data collection and analysis framework 361
  'difficult places' 365
  Flickr 7, 365, 368, 371, 372
  geographic areas 363
  happiness 364
  Hedonometer website 362
  Holyoke (Massachusetts): case study 365–8, 366, 368, 369, 372, 373
  Instagram 367
  life satisfaction and affect 362
  microblogging 361, 363
  National Historic Preservation Act (1966) 361, 364
  National Register of Historic Places 361, 364, 367, 368, 369, 370, 371, 372
  Place Matters (New York) 365
  'places that matter' 365
  Potential Places that Matter 367–8, 371, 373, 374
  qualitative methods 361, 363, 371
  quality of life 362
  quantitative methods 361, 364, 371
  research design, data and methods 365–8
    framework development for significant properties identification 367–8
    sentiment monitoring through Twitter and Flickr posts 366–7
  results 368–71
    community engagement meetings 371
  sentiment analysis 367
  survey or questionnaire 362
  Twitter 7, 363–5, 368, 370, 371, 372, 374, 378, 379, 380, 381
  United States 361, 363, 364
  Urban Attitudes program 366
  well-being assessment 362–5
    historically significant properties, identification of 364–5
    microblog sentiment analysis 362–3

planning applications 363
  sentiment analysis, uses of 363–4
place, sense of 225, 341
place stratification model 37–8
Ploch, L. 245
political capital 41
political factors 151, 275, 294, 361, 407
Polyzoides, S. 325
population factors 25, 247–51, 280, 283
Portes, A. 97
poverty: scope, scale and place 4, 24–43
  block groups 36
  Census of Population and Housing 33–5
  central city poverty 38
  child poverty 24, 29, 36
  community capacity building 41
  community capitals framework 43
  comprehensive community initiatives (CCIS) 43
  Core Based Statistical Area (CBSA) classification 34–6
  counties 33
  culture of poverty 25
  cyclical and cumulative interdependencies model 25
  decisions and behaviors of individuals 25
  deep poverty 43
  demand-side policies 39–40
  dissimilarity index 29
  economic characteristics 25, 39, 42
  educational attainment 25
  employment growth 40
  environmental characteristics 25
  explanatory variables 25–6
  family structure 25
  gender 25
  geographic unit of analysis 25
  GeoLytics 34
  Global Workplace Analytics and FlexJobs report (2017) 42
  historical poverty rates 24
  household characteristics 25, 36
  implications 39–43
    capacity building 42
    economic opportunities 42
    engagement over long-haul 43
    external resources 43
    human capital 42
    inclusive community coordinating team 41–2
    role of community development practitioners 40–43
  income 25
  Index of Relative Rurality (IRR) 35
  individual deficiencies 25

job creation 41
lifestyle choices 25
location characteristics 25
Manufacturing Extension Partnership programs 42
mental maps 26
minor civil divisions (MCDs) 28, 33–4
national samples 26
neighborhoods 33
Office of Management and Budgets 34
people-based policy perspective 25–6
persistent poverty 29–30, 39
person-placed policies 31, 39–40
place-based research 25–6, 27–32, 33–4, 36, 39–40
  discipline approaches and methodologies 30–32
  people-focused approach 31
  place-focused approach 31
  society-focused approach 31
  types of poverty 28–30
population characteristics 25
poverty line 28, 29
poverty rate 30
poverty reduction policies 27, 41, 43
regional variations 39
rural areas 33, 36
Rural Urban Continuum Codes (RUCC) 35, 43
Rural Urban Density Code (RUDC) 35
Rural-Metropolitan Interface Levels classification 35
segregation indices 29–30
social, economic and political systems 25
Social Security and Unemployment Insurance system 27
space and poverty 32–9
  aggregation, unit of 32–5
  block groups and tracts 33–6
  geographic areal units: from nation to neighbourhood 32–4
  macro units 32–3
  meso units 32
  metropolitan patterns of poverty: spatial location and theories 36–8
  micro units 32–4, 36
  regional patterns of poverty 38–9
  urban-rural dichotomy 34–5
spatial aggregation bias 33
spatial approach 27
spatial and aspatial research 26
spatial concentration or incidence of poverty 25, 32
spatial structure 37–8
structural barriers 25

structural and social forces 25
sub-county poverty 33
supplemental poverty measure 43
supply-side policies 39–40
Tennessee Valley Authority 27
urban areas 33, 35, 36
Urban Influence Codes (UIC) 35, 43
Urban-Rural Typology (UA) classification 35
values, beliefs and norms 25
War on Poverty 27, 36
poverty: scope, scale and space
   Consumer Price Index 28
   Current Population Survey 28
   Department of Agriculture (USDA) – Economic Research Service (ERS) 29
   Earned Income Tax Credit 27
   Food Stamps/SNAP 27
   Head Start 27
   Model Cities program 27
   public housing and vouchers 27
   regional price parity data 28
   Social Security Administration 28
   Social Security payments 27
   spatial distribution 31
   spatial econometric analysis 31
   spatial mismatch hypothesis 37
poverty and poverty rates
   Coastal Georgia Indicators Coalition (CGIC): healthy community building 248–9, 251–2, 254
   food deserts: Utica, Mississippi 294
   group formation: Grameen Bank (Bangladesh) 164
   inclusionary zoning and inclusionary housing 198, 199
   photovoice and democratic research processes 382–3, 386–7
   universities and community development partnerships 441, 446
   see also poverty: scope, scale and place
Powell, B. 198
predictive modeling 3
Pretty, J.N. 78
primary data 6, 228, 281, 284
problem based learning (PBL) 446
public participation process and competing interests management: residential displacement (Buffalo, New York) 6, 211–22
   advocacy groups 211–12
   American Community Survey (ACS) 217
   analysis and recommendations 219
   anchor institutions 214, 216
   black power and civil rights 212
   business community 214, 216
   citizen participation ladder 212, 213
   citizen power 212–13
   city planners 217, 220
   coalitions 213–14
   community workshop 217–18
   competing interests 215
   data 216–17
   definitions 216–17
   displacement 214
   disputes 215
   draft report revisions 220
   empowerment 212
   equity goals 211–12, 213, 221
   feedback session 219
   final report 220–21, 222
   first tier of public participation 214–15, 216
   focus groups 6, 215, 217–18, 220
   gentrification 213, 214–15, 216, 217, 218, 219–21
   grassroots advocacy 212–13, 215–21
   inner-city revitalization 212–13
   institutional stakeholders 211, 213, 217, 219, 221
   lessons learned 221–2
   local government 213, 214, 216, 217, 219–21
   minority communities 212
   negative externalities 214
   neighborhood indicators 214–15, 216
   neighborhood-based organizations 215
   nonparticipation 212–13
   outcome orientation 213, 221
   participation, degree of 213
   peer review of draft project report 215, 219
   philanthropic organizations 214, 216
   planning department 218
   planning staff 219
   pluralism in planning 211
   policy implementation strategies 219
   public visibility 221
   qualitative data 220
   quantitative data 220
   recruitment strategy 217
   redistributive policies 214
   revitalization 216, 218, 221
   second tier of public participation 215, 218
   stakeholder engagement process 6
   stakeholder input, analysis and recommendations 220
   status quo policies 214
   third tier of public participation 215, 217–18, 219
   tokenism 212–13
   top-down approach 212
   trust-building 215, 218

Turning the Corner (TtC) 214, 219–20
typology 211–14
United States Census 219–20
Urban Institute 214, 220, 221
urban renewal 212–13
wildcard role 221
public–private partnerships 251, 256, 260, 264, 266, 268
Putnam, R.D. 95, 172, 184

Qin, B. 303
qualitative approach 3
just communities: marginalized communities and regional sustainability planning 50–51
neighborhood environment satisfaction in deteriorated areas: impact of socioeconomic characteristics 306
participatory action research: Bangladesh 224, 227, 231, 233, 235, 240
photovoice and democratic research processes 382, 387, 388
place attachment: inductive social media approach 361, 363, 371
public participation process and competing interests management: residential displacement (Buffalo, New York) 220
universities and community development 412
universities and community development partnerships 447
quality of life 3, 9
Coastal Georgia Indicators Coalition (CGIC): healthy community building 244, 258, 263, 264
intentional community: Reena Community Residence (Canada) 172
photovoice and democratic research processes 382, 386
place attachment: inductive social media approach 362
social indicator projects for rural communities 6, 273–5
universities and community development partnerships 439, 451
see also revitalization, livability and quality of life: Tuscon (Arizona)
quantitative approach 3, 220, 361, 364, 371
Quercia, D. 362, 363
questionnaires and surveys 228

race/ethnicity 14, 28, 32, 36, 38–9
community-based organizations (CBOs) 110
inclusionary zoning and inclusionary housing 189, 198
neighborhood environment satisfaction in deteriorated areas: impact of socioeconomic characteristics 303
public participation process and competing interests management: residential displacement (Buffalo, New York) 212
social indicator projects for rural communities 283
see also just communities: marginalized communities and regional sustainability planning
Raffel, J.A. 113
Rankin, K.N. 97
Razavi, S. 96
Read, D.C. 198
Reason, P. 226
Reena Community Residence, *see* intentional community building: Reena Community Residence (Canada)
reflection, action and knowledge creation: lived experience 4, 12–23
critical consciousness 13, 16–18, 19, 21–2
cycle of praxis 13, 18–22
catalytic action 18–19
walking the cycle 19–22
initiator 18–20, 22
insertion point 19, 22
shared experience 20
shared identity and norms 14
solidarity and agency 13–16, 18–20, 22
teleological practice and vision 12–13
theoretical foundations 13–18
World Values Survey (2019) 15
regional sustainability planning, *see* just communities: marginalized communities and regional sustainability planning
regionalism 255, 260, 263
relational community development 3
Renski, H. 364
residential displacement, *see* public participation process and competing interests management: residential displacement (Buffalo, New York)
revitalization, livability and quality of life: Tuscon (Arizona) 7, 319–33
analytical mechanism 320–22
Arizona Sun Corridor 319, 322, 327
charisma 321, 328, 330, 332
commercial urbanism 322
concentration 320, 328, 329–30, 332
cosmopolitanism 320, 328, 329, 332
currency 320, 327–9, 332
Good/Great City and Quality of Life theories 320, 321
greenhouse gas emissions 323

historic preservation 324
holistic and integral planning 322
home-grown advocacy 320, 325–6, 333
household income 329
I-10 freeway 323
ideological, rights and equity perspectives 321
leadership in governance 330, 332
Model Cities Program 325
morphological and socio-economic characteristics 321
Old Pueblo Redevelopment Plan 325
procedural decision-making 322
proximity to border 319–20, 332–3
public markets 322
railroad network 323
regional level 322
Rio Nuevo 325–7
suburbanization 324
synthesis of good/great city and quality of life theories and their applications 321
Tax Increment Financing (TIF) program 326
urban planning 322
urban renewal 320, 325–6, 333
urbanization 319, 332
Rich, M.A. 113–14, 117
Rickman, D.S. 28, 34, 39, 40
Robert Wood Johnson Foundation 250, 265, 267, 278
Roberts, M. 450–51
Roberts, N. 212
Robinson, J. 172
Robinson, Jr, J.W. 417
Roesler, T. 418
Roller, J. 118–19
Rosen, D. 198
Rossman, G. 231
Rowland, D. 291–2
Rupasingha, A. 42
Rwanda 343, 350

Saha, D. 50
Saint-Germain, M. 325–6
Sanchez, T.W. 212
Sander, P. 322–3
Sarkozy, N. 275
Sarmiento, C.S. 112–13
Sarrett, H.J. 291
Sattar, D. 94
Savitch, H.V. 320, 327, 329
scenario planning 3
Schuetz, J. 198–9
Schulz, M. 450
Schwartz, H.L. 199
Schwartz, K. 115

Schwarz, M. 349
Schweitzer, L. 364
secondary data 6, 281, 284, 287
Selenger, D. 238
self-determination 74
self-efficacy 386
self-expression values 15
self-help 15, 22, 142, 226, 227, 410
self-image 12, 14, 17
self-perception 12, 14
self-reflection 17
self-story 12, 14
Seligman, M.E. 362
semi-structured interviews 448
Sen, A. 225, 411
Senegal 349
sense-making 22
sentiment analysis 367
Shanks, T.R. 108
Shepard, J. 452
Shragge, E. 94
Siegmann, K.A. 97
Siemens, G. 446
Silverman, R.M. 115, 117
Sims, J.R. 112–13
Sirgy, M.J. 244–5, 257–8, 274, 275
Sirvent, M.T. 411
Smith, L. 388
Snapchat 125
snowball sampling 448
Sobering, K. 100
social action 173, 228
social capital 93, 95–7
    Coastal Georgia Indicators Coalition (CGIC): healthy community building 256
    community-based organizations (CBOs) 110, 112, 114
    emancipative 15
    faith-based community development and evaluation capacity enhancement: ENLACE (El Salvador) 205
    food deserts: Utica, Mississippi 293
    and gender 95–6
    group formation: Grameen Bank (Bangladesh) 5, 141, 142, 143
    intentional community: Reena Community Residence (Canada) 172, 173, 175, 181, 182, 184, 185
    place attachment: inductive social media approach 7
    poverty in United States: scope, scale and place 41–2
    social economy, social capital and NGOs: gendered perspective 93, 95–101

social indicator projects for rural
    communities 283
sustainable community development and
    community productivity 341, 343
universities and community development
    partnerships 446–7
social change
  Coastal Georgia Indicators Coalition
    (CGIC): healthy community building
    246, 255
  Cocoa360 NGO (Ghana) 349
  community innovation: local partnerships
    and sustainable community
    development 418
  participatory action research: Bangladesh
    224, 226
  photovoice and democratic research
    processes 386–8
  social indicator projects for rural
    communities 273
social ecological models 37–8
social economy 339; *see also* social economy,
    social capital and NGOs: gendered
    perspective
social economy, social capital and NGOs:
    gendered perspective 5, 93–101
  agency 100
  bottom-up approach 95
  caring/giving 96
  childhood health and development 98–9
  cohesion, trust and reciprocity 95
  cooperatives, nonprofits and worker
    collectives 96, 98
  economic, political and social norms,
    processes and power relations 96
  egalitarian or equity values 100
  ethics and values 96
  identity 94
  informal economy of care 96
  institutions 94
  intention 94
  Laos grass-roots cooperative case study 93,
    97–101
  market-based approach 99
  social capital 93, 95–101
  social and community value 96
  social economy 93–101
  social networks 97
  social provisioning approach 96
  socioeconomic differences 96
  top-down approach 95, 100, 101
  trust and reciprocity 97
  voluntary participation 101
  well-being 96
  women's work of care (children) 98

social environment 294, 301–2
social factors
  aesthetics and community development 392
  community innovation: local partnerships
    and sustainable community
    development 416
  group formation: Grameen Bank
    (Bangladesh) 151
  neighborhood environment satisfaction
    in deteriorated areas: impact of
    socioeconomic characteristics 303
  place attachment: inductive social media
    approach 361
  sustainable community development and
    community productivity 338–9, 340,
    341
  universities 407
  see also social indicator projects for rural
    communities
social indicator projects for rural communities
    6, 273–88
  American Community Survey (ACS) 286
  bottom-up approach 276–7, 284
  Bureau of Economic Analysis 279
  Bureau of Labor Statistics 279
  business options diversity 280
  Center for Rural Communities (CRC) 280,
    283–4, 286–7
  Citizen Engagement PACT of Jacksonville
    276
  Commission on the Measurement of
    Economic Performance and Social
    Progress 281
  community amenities, services and livability
    274
  community indicator projects in United
    States 275–9, 283, 284
  Comprehensive Quality of Life Inventory
    (ComQol) 274–5
  County Health Rankings & Roadmaps
    (CHRR) 273, 278
  county level indicators 284
  Department of Agriculture (USDA)
    Agriculture Census 281
    Economic Research Service (ERS) 286
  downtown vibrancy 283
  economic domain 6, 277, 278, 280, 287
  Economic Profile System 279
  education 275, 276, 277, 280, 283
  environment 275, 276
  Federal Government 273
  Federal Information Processing Standards
    (FIPS) code 286
  food infrastructure 280, 283
  geographic identifiers (GEOID) 286

Geological Survey (USGS) 281
Headwaters Economics (Montana) 279, 280
health 277, 278
income 280, 283
input process 286
intermediaries 277
Jacksonville Community Council Inc (JCCI) 276
Jacksonville *Quality of Life Progress* report 273, 276–7
Michigan University 279
Minnesota Compass 273, 277–8
National Neighborhood Indicators Partnership (NNIP) 273, 277
National Oceanic and Atmospheric Administration (NOAA) 281
National Science Foundation 273
natural environment and built infrastructure 277, 278, 281
Northwoods Quality of Life (NWQoL) Database 273, 279–87
  background 280–81
  category and domain selection process 284
  community characteristics and services 281, 282–3
  conceptual and operational framework 281–7, 282–3
  database structure, input and output 284, 286
  natural and built environment 281, 282
  people 281, 283
  region and population communities 285
  unit of analysis 284
  use and purpose 286–7
objective approach 274
Otto Bremer Trust 281
population 280, 283
Populations at Risk profile system 279
public assistance 277
quality of life 6, 273–5
race/ethnicity 283
recreational infrastructure 283
Roadmaps to Health concept 278
Russell Sage Foundation 273
safety and security 276, 277
subjective approach 274–5
top-down approach 276, 284
United Health Foundation – America's Health Rankings 278
United States 279
Urban Institute 277
Wilder Research/Amherst H. Wilder Foundation 277
Wisconsin University Population Health Institute (UWPHI) 278

social justice 50, 82, 226, 446, 454
social needs 226, 267, 418
social networks 3, 110, 117, 141
social planning approaches 142
social sector networks 260
socioeconomic factors 32, 34
  faith-based community development and evaluation capacity enhancement: ENLACE (El Salvador) 206
  food deserts: Utica, Mississippi 6–7, 299
  inclusionary zoning and inclusionary housing 5, 190, 198
  revitalization, livability and quality of life: Tuscon (Arizona) 321
  social economy, social capital and NGOs: gendered perspective 96
  social indicator projects for rural communities 275
  see also neighborhood environment satisfaction in deteriorated areas: impact of socioeconomic characteristics
solidarity and agency 1, 13–16, 18–20, 22
  community innovation: local partnerships and sustainable community development 417, 434
  group formation: Grameen Bank (Bangladesh) 141
  social economy, social capital and NGOs: gendered perspective 100
  universities and community development partnerships 439
Solitaire, L. 109
South Africa 349
South, S.J. 39
Speck, J. 322
Sperling, B. 322–3
Spradley, J. 231
Staab, S. 96
standpoint theory 215
statistical techniques 31
Stauffer, T. 325
Stiglitz, J.E. 275, 281
Stoecker, R. 112
Stone, C. 447
stories/storytelling 12–13, 20, 70, 71, 72, 74
strategic planning 3
Stringer, E.T. 231, 238
Stringham, E. 198
subjective approach 8–9, 224, 274–5, 302–3, 342
survey or questionnaire 257, 362
sustainability 3
  Coastal Georgia Indicators Coalition (CGIC): healthy community building 256, 265–7

community innovation: local partnerships and sustainable community development 434
community-based organizations (CBOs) 106, 112, 115, 120
  food deserts: Utica, Mississippi 292
  social indicator projects for rural communities 275
  universities and community development partnerships 448
  see also community innovation: local partnerships and sustainable community development; just communities: marginalized communities and regional sustainability planning; sustainable community development and community productivity
sustainable community development and community productivity 7, 337–44
  collaborative decision-making 338
  community engagement 338
  cultural productivity 342
  development of sustainable community 337–9
    importance of communities and urban areas 338
    theories and concepts 338–9
    weaknesses in current approaches 339
  Eco-District initiative (Portland, Oregon) 343
  Eco-localism initiatives 339
  economic productivity 342
  economic and resource circularity 341
  energy and built environment 343
  environmental equality 341
  holistic approach 337, 338
  human productivity 342
  infrastructure/energy resilience 342
  labour productivity 340–42
  natural environment protection and restoration 344
    agriculture 344
  natural productivity 342
  net-zero and net-positive design 341, 342
  objective information 342
  place, sense of 341
  potential 344
  productive community development 339–44
    conceptual issues 340–42
    international examples 343–4
    operational issues 342–3
  productivity 341
  regenerative design 341
  resource productivity 342
  resource regeneration 342
  social productivity 342
  subjective information 342
sustainable livelihoods approach 3
Suttles, G. 409
Sylvester, D.E. 244
systematic method 304
systems theory 342

Taylor, M. 226
Taylor, R.B. 26
technology, *see* well-being and technology: Mutaroni village (Kenya) case study
Teti, M. 384
theory 1–2, 238
Thieme, S. 97
Thomson, D.E. 105
Tickamyer, A.R. 41
Todd, H. 164
Toker, U. 139, 141
Toma, M. 269
top-down approach 3, 104, 114, 117, 118, 119
  Coastal Georgia Indicators Coalition (CGIC): healthy community building 258
  Cocoa360 NGO (Ghana) 348, 349
  participatory action research: Bangladesh 226
  photovoice and democratic research processes 382–3
  public participation process and competing interests management: residential displacement (Buffalo, New York) 212
  social economy, social capital and NGOs: gendered perspective 95, 100, 101
  social indicator projects for rural communities 276, 284
transformative learning 17
Tretter, E.M. 50
triangulation 50–51, 228, 238
Tritter, J.Q. 79
Trump administration 198
trust 16, 18
  Coastal Georgia Indicators Coalition (CGIC): healthy community building 262
  group formation: Grameen Bank (Bangladesh) 143, 166, 167
  poverty in United States: scope, scale and place 42
  public participation process and competing interests management: residential displacement (Buffalo, New York) 215, 218
  social economy, social capital and NGOs: gendered perspective 95, 97

Tsitsos, W. 113–14, 117
Turcotte, D.A. 105
Turner, J.F.C. 142
Turner, S. 430
Twitter 7, 125, 363–5, 368, 370, 371, 372, 374, 378, 379, 380, 381

Uganda 349, 384
United Kingdom 124, 125, 133, 343, 363
   Department of Environment (DoE) 322
United Nations
   Agenda 21 392
   development agenda 337
   Educational, Scientific and Cultural Organization (UNESCO) 2030 Sustainable Development Agenda 392
   Global Agenda for 2030 338
   Habitat The City We Need 2.0 report 341–2
   New Urban Agenda 337
   Sustainable Development Goals 2030 392
   Universal Declaration of Human Rights 382
United States 124, 141, 157, 205, 356
United Way of the Coastal Empire 244–6, 251, 254, 257–9, 260–61, 265
universality of community development 417, 434
universities and community development 8, 407–14
   activities 414
   analysis of research results: opinions of social work students 412–13
   block or area 409
   community and neighborhood 408–10
   concept of community development 410
   deliberative participation 411
   entire sector of city 409
   externality space 409
   human agency 411
   humanistic approach to neighborhood 410
   individual text 412
   instrumental approach to neighborhood 410
   local governmental body districts 409
   methodology 412
   negotiation 411
   neighborhood scales 409
   nominal participation 411
   participation, importance of 411–12
   passive participation 411
   petitionary participation 411
   phenomenological approach to neighborhood 410
   policies for action 414
   qualitative perspective 412
   real participation 411
   smallest area possessing corporate identity 409
   social participation 411
   social role 407–8
   spatial aspect 410
   specific location 410
   symbolic participation 411
   UNICOM 408
   see also universities and community development partnerships
universities and community development partnerships 8, 439–54
   advocacy 450
   Appreciative Inquiry approach 445
   asset mapping 445, 447
   asset-based community development 453
   background 440–41
   Board of Trustees 452
   Camden Counts 444
   Centers of Excellence 443
   charter schools 446
   Choice Neighborhoods 442
   civic responsibility 452
   Community Capitals Framework (CCF) 8, 440, 447
   community schools 446
   community, sense of 449
   Cooper Street Education Corridor 439, 448, 450
   Cradle to College educational pipeline 440, 443, 444, 450
   Department of Education 442
   Early College Program 451
   Early Learning Research Academy 440
   Family Support center 452
   findings 448–50
   Harlem's Children's Zone 442
   Health and Community Wellness Clinic 451
   LEAP Academy University School 8, 440–44, 445, 446–51
   literature review 444–7
   methodology 447–8
   organizational factors 444, 445
   Parent Academy College program 452
   Parent Academy for School Reform 452
   Parent Council 452
   parent focus groups 452
   Parenting Stress Index 449
   Parents as Partners for Educational Change Advocacy Council 449
   Promise Neighborhoods 442
   qualitative approach 447
   quality of life 439, 451

recommendations and best practices 450–53
Rutgers-Camden Community Leadership Center (CLC) (New Jersey) 8, 439, 441–2, 443, 444, 445, 446–53
semi-structured interviews 448
Settlement House movement 441
snowball sampling 448
social service agencies 442
stakeholder/alliance factors 444, 445
student factors 444, 445
teacher development 444, 445
War on Poverty 441–2
White House Council for Community Solutions 442

Valente, T.W. 95
valued-led community development 12
Van Kamp, I. 362
Van Poll, R. 313
Varady, D.P. 113
Vedral, C. 110
Vidal, A. 106

Waldorf, B. 35
Walker, R.E. 294
Walls, D. 212
Walsh, C.A. 239
Wang, C. 384
Wang, Jiun-Hao 362
Wang, M. 35
Warner, K. 50
Warren, M.R. 447
Weaver, J. 107, 110
Weber, B.A. 28, 29, 33, 35, 41, 42, 43
Weffer, S.E. 244
Welch, B.J. 117
well-being 9, 14, 142
    Coastal Georgia Indicators Coalition (CGIC): healthy community building 6
    intentional community: Reena Community Residence (Canada) 172–3, 178, 179, 181, 185
    neighborhood environment satisfaction in deteriorated areas: impact of socioeconomic characteristics 313
    nonprofit organizations and community participation 77
    place attachment: inductive social media approach 362–5
    social economy, social capital and NGOs: gendered perspective 96
    social indicator projects for rural communities 273, 283
    sustainable community development and community productivity 7, 339
    universities and community development 409
    see also well-being and technology: Mutaroni village (Kenya) case study
well-being and technology: Mutaroni village (Kenya) case study 5, 124–34
    bottom-up approach 134
    capacity building 128
    collective entertainment 129, 130
    connectedness 130
    cultural factors 124, 130, 132–3, 134
    directed model 126
    economic development 124, 128, 133
    educational dimension 128, 131, 133
    electrification 125–6, 128, 129–30, 132–4
    emic perspectives of participants 134
    empowerment 124, 128, 133
    entertainment 130
    etic perspectives of providers 134
    face-to-face fieldwork 134
    findings 128
    health dimension 128, 129
    home entertainment 129
    imposed method 126
    internet access 126, 128, 131, 132, 133
    Kenya Light and Power Company 127–8, 130
    lighting 129–31, 134
    M-Pesa banking and money transfer (Safaricom) 125, 127, 130
    methodology 128
    mobile phones 125, 127, 128, 132, 134
    music 130, 133, 134
    phone charging 129
    radio 128, 129, 130, 134
    rafkikis 128, 131–2, 134
    self-esteem, collective 133
    self-help projects 126
    social cohesion and social inclusion 124–5, 127, 130, 132–3
    social development 124
    social dimension 128
    social inclusion 124
    social media 125
    Social Quality paradigm 124
    solar and renewable energy 126, 129–30, 132
    television 128, 129, 130, 131, 132, 133, 134
    Ubuntu (NGO) 126, 127–8, 129–30, 131–2
    use of technology by the villagers 129–31
    well-being 124–5
    working hours, extension of and economic effects 129–33
Welzel, C. 12, 15
West, M. 111

Westoby, P. 418
Whatsapp 132
Whitacre, B. 42
White, S.C. 78
Williams, J. 325
Willis, M. 199
Willoughby, J. 96
Wilson, W.J. 38, 40, 447

Wisconsin University Population Health Institute (UWPHI) 278
Woolcock, M. 95
Worthy, W. 212

Yunus, M. 142, 144, 146, 154, 165–7, 169–70

Zhang, C. 313